P9-EAJ-837

Handbook of Experimental Pharmacology

Volume 92

Insulin

Contributors

J. Avruch · J.J. Bahl · D. Brandenburg · R. Bressler
P. Cuatrecasas · U. Derewenda · Z.S. Derewenda · G.G. Dodson
W.C. Duckworth · I.D. Goldfine · J.R. Gunsalus · M.D. Hollenberg
R.E. Hubbard · S. Jacobs · C.R. Kahn · T. Kono · J.M. Kyriakis
J. Larner · J.R. Levy · R.A. Liddle · W.J. Malaisse · J.L. Messina
J.M. Olefsky · D.J. Price · R.A. Roth · P. Rothenberg
R.J. Rushakoff · A.R. Saltiel · D.F. Steiner · H.S. Tager
H.E. Tornqvist · M.D. Walker · M.F. White · J.A. Williams
E.J. Yurkow · K. Zierler

Editors

P. Cuatrecasas and S. Jacobs

Springer-Verlag Berlin Heidelberg New York
London Paris Tokyo Hong Kong

LP

Professor PEDRO CUATRECASAS, M. D.
President, Pharmaceutical Research Division
Warner Lambert Company, 2800 Plymouth Road
Ann Arbor, MI 48105, USA

STEVEN JACOBS, M. D.
Director, Division of Cell Biology
Burroughs Wellcome Co.
Research Triangle Park, NC 27709, USA

With 86 Figures

ISBN 3-540-50319-6 Springer-Verlag Berlin Heidelberg New York
ISBN 0-387-50319-6 Springer-Verlag New York Berlin Heidelberg

Library of Congress Cataloging-in-Publication Data. Insulin/editors, Pedro Cuatrecasas and Steven Jacobs;
contributors, J. Avruch ... [et al.]. p. cm. – (Handbook of experimental pharmacology; v. 92)
ISBN 0-387-50319-6 (U.S.) 1. Insulin–Physiological effect. I. Cuatrecasas, P. II. Jacobs, Steven, 1946– .
III. Avruch, J. IV. Series. QP905.H3 vol. 92 [QP572.I5] 615'1 s–dc20 [615'.365] 89-21940 CIP

Typesetting, printing and bookbinding: Brühlsche Universitätsdruckerei, Giessen
2127/3130-543210 – Printed on acid-free paper

8/14/90

Preface

It is fourteen years since insulin was last reviewed in The Handbook of Experimental Pharmacology, in volume 32. The present endeavor is more modest in scope. Volume 32 appeared in two separate parts, each having its own subeditors, and together the two parts covered nearly all areas of insulin pharmacology. Such comprehensiveness seemed impractical in a new volume. The amount of information related to insulin that is now available simply would not fit in a reasonable amount of space. Furthermore, for better or worse, scientists have become so specialized that a volume providing such broad coverage seemed likely in its totality to be of interest or value to very few individuals.

We therefore decided to limit the present volume to the following areas: insulin chemistry and structure, insulin biosynthesis and secretion, insulin receptor, and insulin action at the cellular level. We felt these areas formed a coherent unit. We also felt, perhaps as much because of our own interests and perspectives as any objective reality, that these were the areas in which recent progress has been most dramatic, and yet, paradoxically and tantalizingly, these were the areas in which most has yet to be learned.

Even with this limited scope, there are some major gaps in coverage. Regrettably, two important areas, the beta cell ATP-sensitive potassium channel and the glucose transporter, were among these. Nevertheless, the authors who contributed have done an excellent job, and we would like to thank them for their diligence.

PEDRO CUATRECASAS
STEVEN JACOBS

List of Contributors

J. Avruch, Diabetes Unit, Massachusetts General Hospital, Harvard Medical School, MGH East, Building 149, 8th Floor, 13th Street, Charlestown, MA 02129, USA

J. J. Bahl, Department of Internal Medicine, The University of Arizona, Health Sciences Center, Tucson, AZ 85724, USA

D. Brandenburg, Deutsches Wollforschungsinstitut an der Technischen Hochschule Aachen, Veltmanplatz 8, D-5100 Aachen

R. Bressler, Department of Internal Medicine, The University of Arizona, Health Sciences Center, Tucson, AZ 85724, USA

P. Cuatrecasas, Pharmaceutical Research Division, Warner Lambert Company, 2800 Plymouth Road, Ann Arbor, MI 48105, USA

U. Derewenda, Department of Chemistry, University of York, Heslington, York Y01 5DD, Great Britain

Z. S. Derewenda, Department of Chemistry, University of York, Heslington, York Y01 5DD, Great Britain

G. G. Dodson, Department of Chemistry, University of York, Heslington, York Y01 5DD, Great Britain

W. C. Duckworth, Department of Internal Medicine, Veterans Administration Medical Center and University of Nebraska Medical Center, 42nd and Dewey Avenue, Omaha, NB 68105-1065, USA

I. D. Goldfine, Diabetes and Endocrine Research and Department of Medicine, Mount Zion Hospital and Medical Center, P. O. Box 7921, 1600 Divisadero Street, San Francisco, CA 94120, USA

J. R. Gunsalus, Howard Hughes Medical Institute Laboratories, Harvard Medical School, The Medical Services and Diabetes Unit, Massachusetts General Hospital and the Department of Medicine, Boston, MA 02114, USA

M. D. Hollenberg, Endocrine Research Group, Department of Pharmacology and Therapeutics, and Medicine, University of Calgary, Faculty of Medicine, 3330 Hospital Drive, Calgary, Alberta, Canada T2N 4N1

R. E. Hubbard, Department of Chemistry, University of York, Heslington, York Y01 5DD, Great Britain

S. Jacobs, Wellcome Research Laboratories, Burroughs Wellcome Company, 3030 Cornwallis Road, Research Triangle Park, NC 27709, USA

C. R. Kahn, Elliott P. Joslin Research Laboratory, Joslin Diabetes Center, One Joslin Place, Boston, MA 02215, USA

T. Kono, Department of Molecular Physiology and Biophysics, School of Medicine, Vanderbilt University, 21st Avenue, South, Nashville, TN 37232, USA

J. M. Kyriakis, Howard Hughes Medical Institute Laboratories, Harvard Medical School, The Medical Services and Diabetes Unit, Massachusetts General Hospital and Department of Medicine, Boston, MA 02114, USA

J. Larner, Department of Pharmacology, University of Virginia School of Medicine, Jefferson Park Blvd., Jordon Building, Charlottesville, VA 22901, USA

J. R. Levy, (111-P), McGuire Veterans Administration Medical Center, 1201 Broad Rock Blvd., Richmond, VA 23249, USA

R. A. Liddle, Department of Gastroenterology, Duke University Medical School, Durham, NC 27710, USA

W. J. Malaisse, Laboratory of Experimental Medicine, Brussels Free University, 115, Boulevard de Waterloo, B-1000 Brussels, Belgium

J. L. Messina, Department of Physiology, State University of New York Health Science Center at Syracuse, 750 East Adams Street, Syracuse, NY 13210, USA

J. M. Olefsky, Veterans Administration Medical Center, V-111G, 3350 La Jolla Village Drive, San Diego, CA 92161, USA

D. J. Price, Howard Hughes Medical Institute Laboratories, Harvard Medical School, The Medical Services and Diabetes Unit, Massachusetts General Hospital and the Department of Medicine, Boston, MA 02114, USA

R. A. Roth, Department of Pharmacology, Stanford University, School of Medicine, Stanford University Medical Center, Stanford, CA 94305, USA

P. Rothenberg, Department of Pathology and Laboratory Medicine, University of Pennsylvania School of Medicine, Philadelphia, PA 19104, USA

R. J. Rushakoff, Diabetes and Endocrine Research Laboratory and Department of Medicine, Mt. Zion Hospital and Medical Center, P. O. Box 7921, 1600 Divisadero Street, San Francisco, CA 94120, USA

A. R. Saltiel, The Rockefeller University, 1230 York Avenue, New York, NY 10021, USA

D. F. Steiner, Howard Hughes Medical Institute, Department of Biochemistry and Molecular Biology, The University of Chicago Hospitals, 5841 S. Maryland Ave., N216, Box 391, Chicago, IL 60637, USA

H. S. Tager, Department of Biochemistry and Molecular Biology, University of Chicago, 920 East 58th Street, Chicago, IL 60637, USA

H. E. Tornqvist, Department of Pediatrics, University Hospital, University of Lund, S-22362 Lund

M. D. Walker, Department of Biochemistry, Weizmann Institute of Science P. O. Box 26, Rehovot 76100, ISRAEL

M. F. White, Elliott P. Joslin Research Laboratory, Joslin Diabetes Center, One Joslin Place, Boston, MA 02215, USA

J. A. Williams, Department of Physiology, University of Michigan, Ann Arbor, MI 48109, USA

E. J. Yurkow, Howard Hughes Medical Institute Laboratories, Harvard Medical School, The Medical Services and Diabetes Unit, Massachusetts General Hospital and the Department of Medicine, Boston, MA 02114, USA

K. Zierler, The Johns Hopkins University School of Medicine, Department of Endocrinology and Metabolism, Room 918, Traylor Building, 720 Rutland Avenue, Baltimore, MD 21205, USA

Contents

CHAPTER 3

**Mutant Human Insulins and Insulin Structure–Function
Relationships**

Part II. Biosynthesis, Secretion, and Degradation

CHAPTER 4

The Biosynthesis of Insulin

CHAPTER 7

The Role of Cholecystokinin and Other Gut Peptides on Regulation of Postprandial Glucose and Insulin Levels
R. J. RUSHAKOFF, R. A. LIDDLE, J. A. WILLIAMS, and I. D. GOLDFINE

CHAPTER 8

Insulin-Degrading Enzyme
W. C. DUCKWORTH. With 3 Figures .

Part III. Insulin Receptor

CHAPTER 9

Insulin Receptor Structure
R. A. ROTH. With 3 Figures .

CHAPTER 10

Insulin Receptor-Mediated Transmembrane Signalling

M. D. HOLLENBERG. With 2 Figures 183

CHAPTER 11

The Insulin Receptor Tyrosine Kinase

P. ROTHENBERG, M. F. WHITE, and C. R. KAHN. With 1 Figure 209

CHAPTER 12

Receptor-Mediated Internalization and Turnover
J. R. LEVY and J. M. OLEFSKY. With 21 Figures 237

CHAPTER 13

Insulin-like Growth Factor I Receptors
S. JACOBS. With 2 Figures 267

Part IV. Effect of Insulin on Cellular Metabolism

CHAPTER 14

Second Messengers of Insulin Action
A. R. SALTIEL and P. CUATRECASAS. With 4 Figures 289

CHAPTER 15

Insulin Regulation of Protein Phosphorylation
J. AVRUCH, H. E. TORNQVIST, J. R. GUNSALUS, E. J. YURKOW,
J. M. KYRIAKIS, and D. J. PRICE. With 1 Figure

CHAPTER 16

Effects of Insulin on Glycogen Metabolism
J. LARNER. With 6 Figures

CHAPTER 17

Insulin-Sensitive cAMP Phosphodiesterase

CHAPTER 20

Insulin Regulation of Metabolism Relevant to Gluconeogenesis

Part I   Insulin

CHAPTER 1

Insulin Chemistry

D. BRANDENBURG

A. Introduction

Major aims of insulin chemistry are the large-scale production of the hormone for the treatment of diabetes mellitus as well as laboratory-scale syntheses of analogs for structure–function studies, of radioactive tracers, and of "tailor-made" special derivatives. Further, one should include the detection and isolation of new native insulins and related compounds. The total synthesis, accomplished 25 years ago by the groups of ZAHN (MEIENHOFER et al. 1963), KATSOYANNIS (1964), and in China marked the advent of a new era in peptide and protein chemistry. Remarkable progress has since been achieved through refinement of synthetic and semisynthetic procedures, the advances in recombinant DNA techniques, and high pressure liquid chromatography (HPLC).

Until 1981, the literature on insulin chemistry and structure–function relationships was reviewed annually (BRANDENBURG et al. 1983, and previous reports in this series). Much further information can be found in several books and conference proceedings (BRANDENBURG and WOLLMER 1980; KECK and ERB 1981; OFFORD 1980; DODSON et al. 1981; PETERSEN et al. 1982). Since no major review has appeared since then, it is the aim of this chapter to cover the literature from 1982 to mid-1988 as well as possible within the allotted space.

B. Analysis, Purification

I. HPLC

As in many other fields, reverse-phase HPLC brought about a remarkable advance in insulin chemistry due to its high resolution and sensitivity. Remarkable separations are possible of molecules with minute structural differences.

Applications include: analysis of insulin biosynthesis (HALBAN et al. 1987; DE GASPARO and FAUPEL 1986); the separation of insulins from different species (LLOYD and CORRAN 1982; RIVIER and McCLINTOCK 1983; McLEOD and WOOD 1984); and of two bovine proinsulin isomers (McLEOD and WOOD 1984). One should also mention characterization and purity control of new synthetic

---

Abbreviations: Abz, 4-azidobenzoyl; Acm, acetamidomethyl; Ahx, ε-amino-hexanoyl; Boc, t-butyloxycarbonyl; Fmoc, fluorenylmethoxycarbonyl; Msc, methylsulfonylethoxycarbonyl; Nap, 2-nitro-4-azidophenyl; Napa, 2-nitro-4-azidophenylacetyl; OBut, tert.butyl

analogs, such as Cys variants of human insulin (LLOYD and CALAM 1982) or an insulin–IGF hybrid (JOSHI et al. 1985a); of semisynthetic analogs (SHOELSON et al. 1983a; NAKAGAWA and TAGER 1986, 1987; CASARETTO et al. 1987), as well as chemically modified insulins (PANG and SHAFER 1983; MCLEOD and WOOD 1984); and, most recently, biosynthetic insulin analogs (MARKUSSEN et al. 1987a, b, 1988); isolation of homogeneous radioactive tracers, such as monoiodoinsulins (WELINDER et al. 1984; LINDE et al. 1986), tritiated insulins (see also Sect. C.V) and ^{125}I-labeled photoinsulins (HEIDENREICH et al. 1985; WEDEKIND et al. 1989).

Further important applications have been: optimization of chain combination (FRANK and CHANCE 1983), time course studies, e.g., disulfide formation in single-chain insulin (MARKUSSEN 1985), and fingerprint analysis of enzymatic digests, e.g., to establish disulfide bridges (FRANK and CHANCE 1983; GRAU 1985a).

For a discussion of methodology (columns, recovery, buffers) see WELINDER et al. (1984), for semipreparative fractionations VIGH et al. (1987), and for correlations with prediction rules and the three-dimensional structure of insulin MCLEOD and WOOD (1984). HPLC has also been indispensable in the following studies.

II. Isolation

In view of the discussion in Chap. 4, reference is only made to some recent reports. For example, the insulins from cat (HALLDÉN et al. 1986), ratfish (CONLON et al. 1986), sculpin (CUTFIELD et al. 1986a), and ostrich (EVANS et al. 1988) have been isolated and sequenced by automated Edman degradation of the S-sulfonate or S-pyridyl chains. The sequence of owl monkey was determined by cloning and DNA analysis (SEINO et al. 1987). In the search for mutant human insulins in serum of patients, pre-enrichment by immunoadsorption or hydrophobic adsorption, as well as RIA detection in the HPLC eluates are essential features (see SHOELSON et al. 1983b).

Three mutant insulins have been identified: insulin Chicago contains a $Phe^{B25} \rightarrow Leu$, insulin Los Angeles a $Phe^{B24} \rightarrow Ser$ (SHOELSON et al. 1983a), and insulin Wakayama a $Val^{A3} \rightarrow Leu$ replacement (NANJO et al. 1986; SAKURA et al. 1986). The structures were established applying semisynthesis (SHOELSON et al. 1983a) (see below) and DNA analysis (KWOK et al. 1983; HANEDA et al. 1983). A point mutation causing a $His10 \rightarrow Asp$ replacement impairs the conversion of proinsulin to insulin (CHAN et al. 1987).

III. Degradation

For a detailed discussion, see Chap. 8. Viewed from the chemical structure, enzymatic degradation can start either at disulfide bonds (glutathione insulin transhydrogenase, see DAWSON and VARANDANI 1987) or peptide bonds (insulin proteinases, see Chap. 8). Extensive studies in vitro and in vivo, using tracers with the label in various positions, high resolution separation techniques (increasingly HPLC), functional tests (receptor and antibody binding), and model peptides

(DAVIES et al. 1986) for comparisons, have identified sequential degradation pathways and several cleavage sites in the A and B chain.

The following tracers have been applied:

Isomers of iodoinsulin (A14, A19, B16, B26) (DUCKWORTH et al. 1988; SONNE 1987)

[Iodo-Tyr^{B1}] insulin (ASSOIAN et al. 1982)

[^{3}H-Phe^{B1}]insulin (MUIR et al. 1986; DAVIES and OFFORD 1988)

[^{3}H-Gly^{A1}]insulin (MUIR et al. 1986)

[^{3}H-Phe^{B1,B24,B25}]insulin (biosynthetically labeled; MISBIN et al. 1983)

Iodinated analogs, e.g., [Leu^{B24}] insulin (ASSOIAN et al. 1982; KOBAYASHI et al. 1982a).

IV. Stability

Chemical and physicochemical stability of insulin solutions are requirements for application in delivery systems. Since the hormone tends to aggregate and precipitate, its behavior under conditions simulating pump delivery has been examined (LOUGHEED et al. 1983; SATO et al. 1983; BRANGE 1987; GRAU 1985b; GRAU and SAUDEK 1987; SELAM et al. 1987). Contacts with surfaces and air play an important role in triggering aggregation. Insulin is stabilized in solution by detergents (LOUGHEED et al. 1983), calcium ions (BRANGE 1987), urea (SATO et al. 1983). For practical application, a copolymer of polyethylene/propyleneglycol is used (GRAU 1985a). Chemical modification with sulfate groups effectively inhibits aggregation (PONGOR et al. 1983; HUMBURG et al. in preparation).

C. Chemical and Enzymatic Modification, Semisynthesis

Domains of systematic modification and semisynthesis are the NH_2 and COOH terminal sequences outside the bicyclic disulfide structure. Procedures for constructive alterations (i.e., addition of substituents), destructive measures (i.e., shortening of the chains), or for a combination of both have been developed in several laboratories (for a review including practical examples see SCHÜTTLER et al. 1984; ZHU 1985).

I. Modification of Amino Groups and NH_2 Terminal Semisynthesis

The three amino groups, particularly at A1-glycine and B1-phenylalanine, are the most important functional groups (Fig. 1). Homogeneous mono-, di-, or trisubstituted derivatives are accessible through direct substitution and fractionation of the resulting mixtures. A recent example is the isolation of all seven biotinylinsulins by HPLC (PANG and SHAFER 1983). However, the application of amino protecting groups is often advantageous, and mandatory in sequential reactions. A1,B29-bis-Boc-insulin has been used for the synthesis of B1-peptidylinsulins (LOSSE and RADDATZ 1987), and a series of biotinyl-, dethiobiotinyl-, and related insulins containing spacers of varying length (HOFMANN et al. 1984). Based on the strong affinity to avidin, such derivatives are

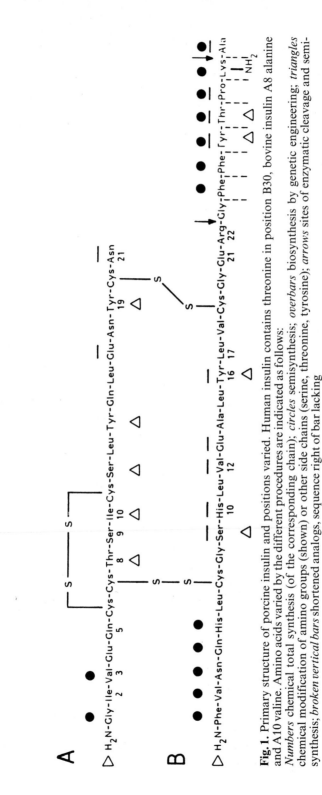

Fig. 1. Primary structure of porcine insulin and positions varied. Human insulin contains threonine in position B30, bovine insulin A8 alanine and A10 valine. Amino acids varied by the different procedures are indicated as follows: *Numbers* chemical total synthesis (of the corresponding chain); *circles* semisynthesis; *overbars* biosynthesis by genetic engineering; *triangles* chemical modification of amino groups (shown) or other side chains (serine, threonine, tyrosine); *arrows* sites of enzymatic cleavage and semisynthesis; *broken vertical bars* shortened analogs, sequence right of bar lacking

very useful for the isolation and characterization of insulin receptors (FINN et al. 1984a, b; HOFMANN et al. 1984, 1987).

The potential of the acid-labile citraconyl group (for citraconylinsulins, see NAITHANI and GATTNER 1982; ZAITSU et al. 1985) has not yet been fully explored. Single-chain intermediates (MARKUSSEN et al. 1987a) obviate protection at A1. α-Amino groups have also been protected in COOH terminal semisyntheses (see below). NH_2 terminal shortening by Edman degradation and semisynthesis requires temporary protection with acid-stable groups, such as the Msc and, most recently, the Fmoc moiety (INOUYE et al. 1985). Thus, CAO et al. (1986) have elongated bis-Msc-des(B1-B4)insulin with Boc-Gly-Pro-Glu(OBut) to give an insulin/IGF-I hybrid, and have replaced B5-histidine by alanine via sequential coupling of two peptides to bis-Msc-des(B1-B5)-insulin.

Semisynthesis at the A chain is more complicated and typically requires the following steps:

1. Temporary protection at A1 (Boc)
2. Protection at B1 and B29 (Msc)
3. Deblocking at A1
4. Edman degradation of glycine (and further cycles, if desired)
5. Substitution at the amino group
6. Deprotection

Following this protocol, TRINDLER and BRANDENBURG (1982) replaced Gly by several different residues, and DAVIES and OFFORD (1985) by [^{3}H]Gly. Somewhat different strategies were followed for the semisynthesis of L- or D-Trp A1-insulins (GEIGER et al. 1982), and [A1,B1 Ala] insulin (SAUNDERS and FREUDE 1982), as well as [Ala ^{A3}]insulin (INOUYE et al. 1985).

II. Hydroxy Amino Acids

Sulfated insulins exhibit a reduced aggregation tendency and are of interest for use in delivery systems. A low substitution degree is achieved by sulfonation with chlorosulfonic acid, or dicyclohexylcarbodiimide/sulfuric acid (PONGOR et al. 1983). Mono derivatives of porcine insulin with the sulfo esters of serine or threonine in position A8, A9, A12, B9, or B27 could be isolated by HPLC (HUMBURG et al. in preparation). Tyrosine residues could be substituted with the N-methylpyridinium group (DREWES et al. 1983), or cross-linked to yield a B16-B26 insulin dimer (CUTFIELD et al. 1986b) under the conditions of nitration.

III. COOH Terminal Shortening and Semisynthesis

In spite of much work, chemical coupling of amino acids or peptides to carboxyl groups of insulin or truncated insulins has been unsatisfactory. The use of proteolytic enzymes to form peptide bonds (review: JAKUBKE 1987) has opened the route to systematic alterations at the COOH terminus of the B chain.

Table 1. Analogs obtained through semisynthesis at B22

Parent compound	Position	Original amino acid replaced by
Insulin h	A3	Ala [1]
	B24	Ala [2] Leu [3] D-Phe [4] Ser [5]
	B25	Cys (Acm) [6] Cys dimer [6] hPhe [7] Leu [3, 7] n1-Ala [7] n2-Ala [7] D-Phe [4] Ser [5, 7]
	B29	Arg [8]
	B25 + B26	Ser, Ala [9]
	B25 + B27	Ser, Ala [9]
Des-B30-NH$_2$	B25	Ser [9]
Des (B29-B30)-NH$_2$	B25	Ser [9]
Des (B28-B30)-NH$_2$	B25 + B26	Ser, Ala [9]
	B25 + B27	Ser, Ala [9]
Des (B27-B30) [10]	B23 + B24	Ala, Phe (LL, LD, DL, DD) [11]
Des (B26-B30) [10, 12]	B25	Leu [13]
Des (B26-B30)-NH$_2$ [7, 14]	B24	D-Phe [15]
	B25	Ala [7] His [15] hPhe [7] Leu [7, 13] n2-Ala [7] D-Phe [15] Ser [7] Tyr [7, 15] Trp [15]
	B24 + B25	D-Phe, Tyr [15], D-Phe, D-Phe [15]
Des (B25-B30) [10, 16]	B23	D-Ala [17]
Des (B24-B30) [10]	B23	amide [7]

Abbreviations: h, homo; n1, 1-naphthyl; n2, 2-naphthyl
[1] INOUYE et al. 1985; [2] ASSOIAN et al. 1982; [3] KOBAYASHI et al. 1982a; [4] KOBAYASHI et al. 1982b; [5] SHOELSON et al. 1983a; [6] FISCHER et al. 1985b; [7] NAKAGAWA and TAGER 1986; [8] CAO et al. 1988; [9] NAKAGAWA and TAGER 1987; [10] RIEMEN et al. 1983; [11] RIEMEN et al. 1985; [12] KUBIAK and COWBURN 1986a; [13] FISCHER et al. 1986; [14] FISCHER et al. 1985a; [15] CASARETTO et al. 1987; [16] FAN et al. 1985; [17] ZHANG et al. 1983

1. Semisynthesis at B22 Arginine

Following the initial work of INOUYE (see SCHÜTTLER et al. 1984), a large number of analogs has been obtained (Table 1). The general methodology consists of coupling peptides (from solution or solid phase synthesis) in a 5- to 15-fold molar excess to des-(B23-B30)insulin in mixtures of aqueous buffer and organic cosolvents (dimethylformamide, glycerol, 1,4-butanediol) at pH 5.5–7.7 in the presence of trypsin. After 2–20 h at room temperature coupling yields are often as high as 90%. Amino protection by Boc groups, sometimes Msc groups (ZHANG et al. 1983; FISCHER et al. 1986), avoids side reactions such as intramolecular cyclization between B22 and A1, but can be omitted (KUBIAK and COWBURN 1986a, b). Alternatively, hydrazides or phenylhydrazides can be prepared, followed by chemical coupling of peptides (RIEMEN et al. 1983, 1985; GATTNER and SOMMER 1989).

2. Semisynthesis at B29 Lysine

Replacement of the COOH terminal alanine has allowed the large-scale conversion of porcine into human insulin and the preparation of many analogs. It can be accomplished either in a two-step reaction (GATTNER et al. 1981; SAKINA et al.

1986; DAVIES et al. 1987) by coupling amino acid derivatives (usually methyl ester, *t*-butyl esters, or amides) to des-Ala-insulin, or by a one-step transformation of insulin. Evidence both in favor (MARKUSSEN 1987) and against (ROSE et al. 1984a) a true transpeptidation has been produced. Thus, the mechanism may depend on the reaction conditions. These have been varied widely with respect to enzyme, solvent composition, pH, and temperature (see for instance ROSE et al. 1983; MARKUSSEN 1987; CAO et al. 1988; OBERMEIER and SEIPKE 1984). Besides trypsin, *Achromobacter* protease I (lysyl endopeptidase) is used which avoids undesired reactions at B22 arginine (MORIHARA et al. 1986; SAKINA et al. 1986; DAVIES et al. 1987), and carboxypeptidase Y has been explored (BREDDAM and JOHANSEN 1984). Although intramolecular A1-B29 cyclization can be achieved (MARKUSSEN 1985), amino protection is normally not necessary. As a large excess of amino component (60- to 100-fold) is generally used, yields are high, but have to be optimized. After COOH terminal deprotection and purification, laboratory yields of analogs are about 60%–80% (GATTNER et al. 1981, 1982; OBERMEIER and SEIPKE 1984; SAKINA et al. 1986), while the optimized industrial process for human insulin gives even higher yields. HPLC was an indispensable tool in most studies.

IV. Other Semisyntheses

An open-chain "mini-proinsulin" was obtained by coupling of Arg-Arg-X-Y-Lys-Arg A chain to B chain (BÜLLESBACH 1982). MARKUSSEN (1985) obtained single-chain des-B30-insulin via intramolecular joining of A1 glycine to B29 lysine with the aid of trypsin.

V. Labeling

Radioactive insulins are indispensable for studies of receptor binding, receptor structure, insulin metabolism, and for immunoassays.

1. Iodine

While partially purified iodoinsulins are acceptable for some studies, monoiodoinsulins are now state of the art. A14, A19, B16, and B26 ^{125}I-insulins are obtained after iodination (lactoperoxidase, chloramine-T) in carrier-free form (specific activity $\leq$ 2200 Ci/mmol) by reverse-phase HPLC in a variety of systems (WELINDER et al. 1984; LINDE et al. 1986). As biological properties may be impaired by material bleeding from columns, the purity of the final preparations should be carefully checked. Since A14 and B16 iodoinsulins exhibit full binding and biopotency, they are true tracers. The A19 isomer is about 50% active, the B26 iodoinsulin up to 200%, depending on the tissue (SONNE et al. 1983; Linde et al. 1986).

Position B1 can be labeled with iodine to high specific activity by semisynthetic replacement of phenylalanine (see Sect. C.I) for iododesaminotyrosine (Bolton-Hunter reagent) (BAHRAMI et al. 1980). ^{131}I is now rarely used. The short-lived isotope ^{123}I (half-life 12 min) is very useful for noninvasive in vivo imaging of insulin (SODOYEZ et al. 1983).

2. Tritium

Tracers were obtained by semisynthetic replacement of A1 glycine (specific activity 44 Ci/mmol; DAVIES and OFFORD 1985), and B1 phenylalanine in insulin (≤ 59 Ci/mmol; GRANT and VON HOLT 1987) and proinsulin (JONES et al. 1987). Enzyme-assisted semisynthesis gave [[^{3}H]Ala^{B30}] porcine insulin (specific activity ≤ 1.14 Ci/mmol; DAVIES et al. 1987). Reverse-phase HPLC was used for isolation. Reductive methylation was less satisfactory (MARSH et al. 1983).

3. Other Labeling

The stable isotope ^{18}O has been incorporated into B29 lysine by enzymatic semisynthesis in the presence of ^{18}O-labeled water (ROSE et al. 1984 b). For electron microscopic application, insulin has been coupled to colloidal gold (HELLFRITZSCH et al. 1986).

VI. Photoreactive Insulins

Several photoreactive derivatives have been prepared and these have allowed, after radiolabeling and covalent binding, analyses of receptor structure, function and fate (for reviews see SAUNDERS and BRANDENBURG 1984; BRANDENBURG and SAUNDERS 1985; YIP and YEUNG 1985). The most widely used derivatives have been B2-Napa-des-Phe^{B1}-insulin and B29-Abz-insulin. Several new photo-insulins have now been obtained via protected intermediates (listed below: 1, 3–5) or unprotected insulin and HPLC separation (2, 6):

1. B1-(4-azidosalicylamido-1,3-dithiopropionyl)insulin (KNUTSON 1987; subsequent iodination at ASA residue and tyrosine).
2. B29-(4-azido,3-iodophenylazobenzoyl)3-aminopropionylinsulin (NG and YIP 1985).
3. Des-(B26-B30)insulin-B25-(4-Nap-iminoethyleneamide) and
4. B29-B29′-Bis(Nap-Ahx)B1,B1-azelaoylinsulin dimer (AMBROSIUS et al. 1987).
5. B29-Nap-biocytinyl-[iodo-Tyr^{B26}]insulin. From A1,B29-di-Msc-insulin, subsequent iodination, and HPLC (WEDEKIND et al. 1989).
6. A14- and B26-iodo, B29-Abz-insulins (HEIDENREICH et al. 1985).

Derivatives 1 and 2 contain a cleavable disulfide or azo linkage; derivative 5 allowed complexing with avidin for the isolation of labeled receptor or its fragments; derivatives 1, 2, 5, and 6 have been used in receptor studies.

VII. Cross-linking

Insulin amino groups react with activated esters of dicarboxylic acids to give A1-B29-linked monomers (proinsulin models) and intermolecularly cross-linked derivatives (review: BRANDENBURG et al. 1977). Several new A1-B29 derivatives have been described: Diaminosuberoylinsulin with D-Ala in A1 and B1 (SAUNDERS and FREUDE 1982), succinoyl- and suberoyl-[Ser^{B25}]insulin (NAKAGAWA and TAGER 1987), and derivatives with cleavable bridges of the following type: oxaloyl-bis-methionine (SRINIVASA and CARPENTER 1983), and 1,2-bis-[o-(carbonyloxyethyl-(1))-phenoxy]ethane (LOSSE et al. 1982).

For the directed synthesis of symmetrical dimers, diprotected Msc-insulins were linked with *p*-nitrophenyl esters of oxalic, suberic, and sebacic acid. The three asymmetrical dimers (SCHÜTTLER and BRANDENBURG 1982) and insulin-des(B23-B30)insulin conjugates (TATNELL et al. 1983) became accessible in a two-step reaction. For the elucidation of cellular mechanisms of insulin action hybrid molecules have proved useful tools. Thus, insulin has been linked to ricin B chain (ROTH et al. 1983) or cholera toxin B chain (ROTH and MADDUX 1983). In situ cross-linking of ^{125}I-labeled insulin by means of di-*N*-hydroxysuccinylsuberate (DSS) has frequently been used to affinity label insulin receptors (see MASSAGUE and CZECH 1985 for review).

D. Chemical Synthesis of Insulin Analogs

Chemical total synthesis is the most versatile, but most laborious route to insulin analogs, because all positions, including those necessary to form the intra- and interchain bridges, can principally be varied by coded and noncoded amino acids. Since our last review (BÜLLESBACH and BRANDENBURG 1981) all analogs (Table 2) have been prepared from one synthetic chain and its natural counter-chain. With two exceptions (CASARETTO 1986) all chains have been synthesized in solution. The syntheses follow the patterns established earlier and differ with respect to strategy and protecting group tactics. Generally, the mode of SH protection determines the choice of the other blocking groups.

Table 2. Insulin analogs abtained by chemical synthesis

Position	Replacement
A2	Ala, Gly (KITAGAWA et al. 1984a), MeIle (OGAWA et al. 1987)
A3	MeVal (OGAWA et al. 1987)
A5	Leu (FERDERIGOS et al. 1983)
A8-10	His-Asn-Thr (WIENEKE et al. 1983a)
A19	Leu (KITAGAWA et al. 1984b), Phe (FERDERIGOS et al. 1983), Phe(F) (STOEV et al. 1988), Trp (OTAH et al. 1988), Tyr (I), Tyr (I$_2$) (WIENEKE et al. 1983b)
A20[a]	Amide, ethylamide, isopropylamide, 2-trifluorethylamide (CHU et al. 1987a, c)
A21	Diethylamide (CHU et al. 1987b)
B10	Asn (BURKE et al. 1984), Asp (SCHWARTZ et al. 1987) Lys (SCHWARTZ et al. 1982)
B12	Ala, Ile (CASARETTO 1986)
B16	Gln (SCHWARTZ et al. 1985)
B17	D-Leu, Nle[b] (KNORR et al. 1983)
B21	Pro (SCHWARTZ et al. 1983)
B22	D-Arg (KNORR et al. 1982)

[a] In des-A21-Insulin.
[b] Nle, norleucine.

I. Synthesis of A Chains

Benzyl (Ferderigos et al. 1983; Kitagawa et al. 1984a, b) or p-methoxybenzyl (Pmb) groups (Ogawa et al. 1987; Chu et al. 1987a, b; Othah et al. 1988) at cysteine were used in conjunction with (mainly) benzyl groups at COOH and OH functions. Boc groups served for temporary α-amino protection. The chains were assembled by joining tri- to pentapeptides to the COOH terminal peptide by azide coupling. In contrast, Wieneke et al. (1983a, b) and Stoev et al. (1988) used trityl groups for cysteines, t-butyl-type groups for side chains, and urethane-type (benzyloxycarbonyl, biphenylylpropyloxycarbonyl) for α-amino groups to prepare two large segments, 1-12 and 13-21, for the ultimate coupling. Final treatment with mixtures of trifluoroacetic acid and methane- or tri-fluoromethanesulfonic acid removed S-Pmb groups. S-Trityl and simultaneously all other protecting groups were cleaved with trifluoroacetic acid in the presence of thiols.

II. Synthesis of B Chains

While Knorr et al. (1982, 1983) used S-trityl protection and followed the same tactics as Wieneke et al. (1983a, b), Schwartz et al. (1982, 1983, 1985, 1987) and Burke et al. (1984) applied the diphenylmethyl group for cysteines, and benzyl as the other major protection. This necessitated final deblocking with liquid hydrogen fluoride, i.e., harsh conditions. The strategies for chain assembly were the synthesis of two large segments, 1-16 (from 1-8 and 9-16) and 17-30 (from 17-23 and 24-30) (Knorr et al. 1982, 1983), or the elongation of the decapeptide 21-30 by peptides 15-20, 9-14, and 1-8 (Schwartz et al. 1982, 1983, 1985, 1987; Burke et al. 1984).

Two B chain analogs with replacements of B12 valine have recently been obtained by solid phase synthesis using the "Fmoc strategy," i.e., base-labile amino protection, and acid-labile t-butyl-based side chain protection with Acm-S protection, on a polyacrylamide resin in high yields of 35% (Casaretto 1986).

III. Chain Combination

From all syntheses, and also biosyntheses (see Sect. E), chains have been isolated and purified as S-sulfonates. Correct joining of A and B chains has always been, and still is, the second problem of synthesis. Three procedures have been followed:

1. Reaction of reduced A chain (5 equiv.) with $B(SSO_3^-)_2$ according to Katsoyannis (see, for instance, Kitagawa et al. 1984b).
2. Coreduction and cooxidation of $A(SSO_3^-)_4$ and $B(SSO_3^-)_2$ in a molar ratio of 1:1 with intermediate gel filtration and buffer change after Gattner (see Stoev et al. 1988),
3. Coreduction of A and B chains (ratio 2.6:1) with dithiothreitol (1.2 equiv. per S-sulfonate group) and subsequent direct cooxidation (Frank and Chance 1983).

Relative yields (based on B chain) are generally low and rarely above 10%. In absolute terms, several hundred milligrams of a crude synthetic chain render, after combination, gel filtration, and ion exchange chromatography, and/or, most recently HPLC (OTAH et al. 1988), a few milligrams of analog. With the extensive use of HPLC, procedure 3 has been optimized for biosynthetic human insulin. In this special case yields of 60% could be achieved.

Positive correlations have been noted between yield, biopotency of analog, and potential of separated chains for interaction, but are not consistent (HUA et al. 1984; WANG and TSOU 1986; TIAN et al. 1987). A complete failure in chain combinations has sometimes been observed, and a "wrong," antiparallel analog been isolated in one case (KNORR et al. 1982). For direct separations, recycling studies (SCHARTMANN et al. 1983), and fingerprint analyses of enzymatic (particularly *Staphylococcus* protease) digests (FRANK and CHANCE 1983; GRAU 1985 b), HPLC has been invaluable.

In contrast to bimolecular SS bridging, intramolecular SH pairing proceeds with high yields, as confirmed with new A1 and B29 cross-linked derivatives (LOSSE et al. 1982; SRINIVASA and CARPENTER 1983) and single-chain molecules (MARKUSSEN et al. 1987 a, 1988; see Sect. E).

IV. Syntheses of Related Molecules

Applying similar procedures, the following two-chain hybrids have been synthesized (synthetic chains in italics):

1. *A(insulin)-D(IGF-I)*/B(insulin) KING et al. 1982
2. *A(insulin)-D(IGF-II)*/B(insulin) OGAWA et al. 1984
3. A(insulin)/*B(IGF-I)* JOSHI et al. 1985 a
4. *A(insulin)-D(IGF-II)*/*B(IGF-I)* JOSHI et al. 1985 b

For review, see KATSOYANNIS et al. 1987.

The extensive synthetic work directed towards the total synthesis of human proinsulin has been discussed by NAITHANI and ZAHN (1984). In spite of much effort, a homogeneous compound could not be obtained. In contrast, LI et al. (1983) accomplished the solid phase synthesis of IGF-I. A bicyclic minimal insulin proved to be biologically inactive (GALPIN et al. 1983).

E. Biosynthesis by Recombinant DNA Techniques

I. Human Insulin

Insulin was the first protein obtained by genetic engineering, and large-scale production began in 1982. In the initial process synthetic genes for A and B chains, preceded by a methionine codon and a promoter sequence such as tryptophan synthetase, were inserted into *Escherichia coli* plasmids. The chimeric proteins were obtained in separate *E. coli* K-12 fermentations. After cyanogen bromide cleavage the chains were isolated as S-sulfonates and combined according to procedure 3 of Sect. D.III. The second process, which is now used for production, is based on a semisynthetic proinsulin gene and yields the single-chain S-sulfonate. After reduction and SS bond formation (up to 70% yield), the connecting peptide is removed by sequential action of trypsin and carboxypeptidase B (FRANK and CHANCE 1983; JOHNSON 1983). A first analog, [Asp^{B16}]insulin (rat I sequence) has been obtained in *E. coli* K-12 via a mutated preproinsulin gene (VARLEY et al. 1988).

While the chimeric proteins are deposited within *E. coli*, systems have been developed in the meantime which secrete the desired compounds. THIM et al. (1986) have constructed recombinant plasmids from genes encoding fusion proteins with single-chain insulin precursors. These contained connecting peptides of 2–10 amino acids. Expression in yeast led to the processing of the fusion proteins and secretion of the insulin precursors with correct disulfide bonds. Conversion to insulin by treatment with trypsin and carboxypeptidase B required the presence of spacer between the basic amino acids in position B31 and A0.

II. Analogs

Single-chain des-B30-insulin and particularly single-chain insulins with a short spacer, such as B(1-29)-Ala-Ala-Lys-A(1-21), allow the preparation of human insulin and of analogs through a combination of bio- and semisynthetic procedures. In order to obtain insulins with higher isoelectric point, MARKUSSEN et al. (1987b, 1988) prepared the corresponding genes by site-directed mutagenesis of positions B13 and A17 (Glu→Gln), B27 (Thr→Lys or Arg). They were linked to promoter and terminator genes and incorporated into yeast plasmids. The secreted cyclic precursors were isolated by precipitation or crystallization, and then transformed into the two-chain analogs with simultaneous introduction of residue B30 by trypsin-assisted semisynthesis (see Sect. C.III) with threonine, its amide, or lysine amide.

Additional replacement of A21 asparagine by Asp, Gly, Ser, His, Thr, or Arg gave more stable analogs due to elimination of this acid-sensitive residue. A21 proline inhibited proper processing by the yeast cells (MARKUSSEN et al. 1988). Insulins with higher negative charge were obtained via the same route (BRANGE et al. 1988). For example, Asp was introduced into B9, B10, B28, and Glu into B12, B26, B27. Four analogs of IGF-I have recently been obtained from transformed yeast cells (CASCIERI et al. 1988).

F. Conclusions

Large-scale enzymatic semisynthesis and genetic engineering have made human insulin accessible for treatment, at the expense of porcine and, particularly, bovine insulin. However, the mode of administration remains unchanged, and the requirements of treatment have to be met through galenics (BRANGE 1987). Studies with many of the analogs described have contributed to our current picture of structure-function relationships, but it is yet incomplete (Chap. 3; GAMMELTOFT 1984). A rational drug design for insulins with improved therapeutic properties has to be built on a much broader basis. Leads with practical implications are insulins with increased solubility and faster onset of action (BRANGE et al. 1988), or basic insulins with decreased solubility and retarded action (MARKUSSEN et al. 1987a, b, 1988). With respect to unconventional administration, a model for release of a basic insulin (trilysylinsulin) under feedback control by glucose has been tested (FISCHEL-GHODSIAN et al. 1988).

Thanks to the preparative and analytical techniques now at hand, many new analogs will become available in a reasonably short time. Biosynthesis via DNA techniques is most promising, but is limited inter alia by the genetic code. Chemical synthesis is laborious, but universal, while chemical and enzymatic semisynthesis are valuable for restricted regions of macromolecules. Thus, the various methods all have their own merits and supplement each other well.

Acknowledgments. The author expresses his sincere thanks to C. Formen, U. Strack, and B. Pabsch for excellent assistance in preparing the manuscript and handling literature data. H.-G. Gattner is thanked for critically reading the manuscript. The chapter is dedicated to Professor Helmut Zahn on the occasion of the 25th anniversary of the total synthesis of insulin in Aachen (21 December 1963).

References

Ambrosius D, Bala-Mohan S, Behrendt C, Schäfer K, Schüttler A, Brandenburg D (1987) New photoreactive derivatives of insulin for affinity-labelling of the insulin receptor. In: Theodoropoulos D (ed) Peptides 1986. De Gruyter, New York, pp 521–523

Assoian RK, Thomas NE, Kaiser ET, Tager HS (1982) [LeuB24]Insulin and [AlaB24]insulin: altered structures and cellular processing of B24-substituted insulin analogs. Proc Natl Acad Sci USA 79:5147–5151

Bahrami S, Zahn H, Brandenburg D, Machulla H-J, Dutschka K (1980) [B1-^{125}I-Desaminotyrosine] insulin – a novel homogeneous insulin tracer. Radiochem Radioanal Lett 45:221–226

Brandenburg D, Saunders D (1985) Preparation and characterization of photoactivatable insulin analogues. In: De Pirro R, Lauro R (eds) Insulin receptors. Field Educational Italia, ACTA MEDICA, Rome, pp 20–32 (Handbook on receptor research)

Brandenburg D, Wollmer A (1980) (eds) Insulin: chemistry, structure and function of insulin and related hormones. De Gruyter, New York

Brandenburg D, Gattner H-G, Schermutzki W, Schüttler A, Uschkoreit J, Weimann J, Wollmer A (1977) Crosslinked insulins: preparation, properties, and applications. In: Friedman M (ed) Protein crosslinking: biochemical and molecular aspects. Plenum, New York, pp 261–282 (Advances in experimental medicine and biology, vol 86 A.)

Brandenburg D, Saunders D, Schüttler A (1983) Pancreatic hormones. In: Jones JH (senior reporter) Amino-acids, peptides and proteins, vol 14. Specialist periodical reports. Royal Society of Chemistry, London, pp 461–476

Brange J (1987) Galenics of insulin. The physico-chemical and pharmaceutical aspects of insulin and insulin preparations. Springer, Berlin Heidelberg New York Tokyo

Brange J, Ribel U, Hansen JF, Dodson G, Hansen MT, Havelund S, Melberg SG, Norris F, Norris K, Snel L, Sørensen AR, Voigt HO (1988) Monomeric insulins obtained by protein engineering and their medical implications. Nature 333:679–682

Breddam K, Johansen JT (1984) Semisynthesis of human insulin utilizing chemically modified carboxypeptidase Y. Carlsberg Res Commun 49:463–472

Büllesbach EE (1982) Semisynthesis of a shortened open-chain proinsulin. Tetrahedron Lett 23:1877–1880

Büllesbach EE, Brandenburg D (1981) Synthesis of insulin analogues. In: Keck K, Erb P (eds) Basic and clinical aspects of immunity to insulin. De Gruyter, New York, pp 395–419

Burke GT, Schwartz G, Katsoyannis PG (1984) Nature of the B10 amino acid residue. Requirements for high biological activity of insulin. Int J Pept Protein Res 23:394–401

Cao Q-P, Geiger R, Langner D, Geisen K (1986) Biological activity in vivo of insulin analogues modified in the N-terminal region of the B-chain. Biol Chem Hoppe Seyler 367:135–140

Cao Q-P, Zhang Y-S, Geiger R (1988) Comparison of reaction rates in trypsin-catalysed transamidation of porcine insulin and its B29-arginine analogue. Biol Chem Hoppe-Seyler 369:283–287

Casaretto M, Spoden M, Diaconescu C, Gattner H-G, Zahn H, Brandenburg D, Wollmer A (1987) Shortened insulin with enhanced in vitro potency. Biol Chem Hoppe-Seyler 368:709–716

Casaretto R (1986) Merrifieldsynthese von [Ala12]- und [Ile12]Insulin-B-Ketten und Kombination mit natürlicher A-Kette. Dissertation, RWTH Aachen

Cascieri MA, Chicchi GG, Applebaum J, Hayes NS, Green BG, Bayne ML (1988) Mutants of human insulin-like growth factor I with reduced affinity for the type 1 insulin-like growth factor receptor. Biochemistry 27:3229–3233

Chan SJ, Seino S, Gruppuso PA, Schwartz R, Steiner DF (1987) A mutation in the B chain coding region is associated with impaired proinsulin conversion in a family with hyperproinsulinemia. Proc Natl Acad Sci USA 84:2194–2197

Chu Y-C, Burke GT, Chanley JD, Katsoyannis PG (1987a) Possible involvement of the A20-A21 peptide bond in the expression of the biological activity of insulin. 2. [21-Asparagine diethylamide-A] insulin. Biochemistry 26:6972–6975

Chu Y-C, Wang R-Y, Burke GT, Chanley JD, Katsoyannis PG (1987b) Possible involvement of the A20-A21 peptide bond in the expression of the biological activity of insulin. 1. [21-Desasparagine, 20-cysteinamide-A]insulin and [21-desasparagine, 20-cysteine isopropylamide-A]-insulin. Biochemistry 26:6966–6971

Chu Y-C, Wang R-Y, Burke GT, Chanley JD, Katsoyannis PG (1987c) Possible involvement of the A20-A21 peptide bond in the expression of the biological activity of insulin. 3. [21-Desasparagine, 20-cysteine-2,2,2-trifluorethylamide-A]insulin. Biochemistry 26:6975–6979

Conlon JM, Dafgård E, Falkmer S, Thim L (1986) The primary structure of ratfish insulin reveals an unusual mode of proinsulin processing. FEBS Lett 208:445–450

Cutfield JF, Cutfield SM, Carne A, Emdin SO, Falkmer S (1986a) The isolation, purification and amino-acid sequence of insulin from the teleost fish Cottus scorpius (daddy sculpin). Eur J Biochem 158:117–123

Cutfield SM, Dodson GG, Ronco N, Cutfield JF (1986b) Preparation and activity of nitrated insulin dimer. Int J Pept Protein Res 27:335–343

Davies JG, Offord RE (1985) The preparation of tritiated insulin specifically labelled by semisynthesis at glycine-A1. Biochem J 231:389–392

Davies JG, Offord RE (1988) The chemical characterization of the radioactive products derived from [[^{3}H]PheB1] insulin in the circulation of the rat. Biochem J 250:389–393

Davies JG, Muir AV, Offord RE (1986) Identification of some cleavage sites of insulin by insulin proteinase. Biochem J 240:609–612

Davies JG, Rose K, Bradshaw CG, Offord RE (1987) Enzymatic semisynthesis of insulin specifically labelled with tritium at position B-30. Protein Engineering 1:407–411

Dawson DB, Varandani PT (1987) Characterization and application of monoclonal antibodies directed to separate epitopes of glutathione-insulin transhydrogenase. Biochim Biophys Acta 923:389–400

De Gasparo M, Faupel M (1986) Chromatographic characterization of adult and foetal rat insulin. J Chromatogr 357:139–146

Dodson G, Glusker JP, Sayre D (eds) (1981) Structural studies on molecules of biological interest. A volume in honour of Prof. Dorothy Hodgkin. Clarendon, Oxford

Drewes SE, Magojo HEM, Gliemann J (1983) (N-Methylpyridinium) insulins. Modification at the A19- and B16-tyrosines. Hoppe-Seyler's Z Physiol Chem 364:461–468

Duckworth WC, Hamel FG, Peavy DE, Liepnieks JJ, Ryan MP, Hermodson MA, Frank BH (1988) Degradation products of insulin generated by hepatocytes and by insulin protease. J Biol Chem 263:1826–1833

Evans TK, Litthauer D, Oelofsen W (1988) Purification and primary structure of ostrich insulin. Int J Pept Protein Res 31:454–462

Fan L, Cui DF, Zhang YS (1985) Enzymatic synthesis of deshexapeptide insulin in the presence of high concentration of 1,4-butandiol. Chin Biochem J 1:33–36

Ferderigos N, Burke GT, Kitagawa K, Katsoyannis PG (1983) The effect of modifications of the A5 and A19 amino acid residues on the biological activity of insulin. [Leu5-A] and [Phe19-A] sheep insulins. J Protein Chem 2:147–170

Finn FM, Titus G, Hofmann K (1984a) Ligands for insulin receptor isolation. Biochemistry 23:2554–2558

Finn FM, Titus G, Horstman D, Hofmann K (1984b) Avidin-biotin affinity chromatography: application to the isolation of human placental insulin receptor. Proc Natl Acad Sci USA 81:7328–7332

Fischel-Ghodsian F, Brown L, Mathiowitz E, Brandenburg D, Langer R (1988) Enzymatically controlled drug delivery. Proc Natl Acad Sci USA 85:2403–2406

Fischer WH, Saunders D, Brandenburg D, Wollmer A, Zahn H (1985a) A shortened insulin with full in vitro potency. Biol Chem Hoppe-Seyler 366: 521–525

Fischer WH, Saunders DJ, Zahn H, Wollmer A (1985b) Synthesis and properties of a disulphide-bridged insulin dimer. In: Deber CM, Hruby VJ, Kopple KD (eds) Peptides, structure and function. Pierce Chem Comp, Rockford, pp 301–304

Fischer WH, Saunders D, Brandenburg D, Diaconescu C, Wollmer A, Dodson G, De Meyts P, Zahn H (1986) Structure-function relationships of shortened [LeuB25]insulins, semisynthetic analogues of a mutant human insulin. Biol Chem Hoppe-Seyler 367:999–1006

Frank BH, Chance RE (1983) Two routes for producing human insulin utilizing recombinant DNA technology. Münch Med Wochenschr 125 [Suppl 1]:14–20

Galpin IJ, Hancock G, Kenner GW, Morgan BA (1983) The synthesis of an insulin active site analogue. Tetrahedron 39:149–158

Gammeltoft S (1984) Insulin receptors: binding kinetics and structure-function relationship of insulin. Physiol Rev 64:1321–1378

Gattner H-G, Sommer M-T (1989) Enzymatic preparation of hydrazides of insulin derivatives as intermediates for chemical syntheses In: König WA, Voelter W (eds) Chemistry of peptides and proteins, vol 4. Attempto, Tübingen, pp 111–117

Gattner H-G, Danho W, Knorr R, Naithani VK, Zahn H (1981) Trypsin catalyzed peptide synthesis: modification of the B-chain C-terminal region of insulin. In: Brunfeldt K (ed) Peptides 1980. Scriptor, Copenhagen, pp 372–377

Gattner H-G, Danho W, Knorr R, Zahn H (1982) Trypsin catalyzed peptide synthesis: modification of the B-chain C-terminal region of insulin. In: Voelter W, Wünsch E, Ovchinnikov YU, Ivanov V (eds) Chemistry of peptides and proteins, vol 1. De Gruyter, New York, pp 319–325

Geiger R, Geisen K, Summ H-D (1982) Austausch von A1-Glycin in Rinderinsulin gegen L- und D-Tryptophan. Hoppe-Seyler's Z Physiol Chem 363:1231–1239

Grant KI, von Holt C (1987) Improved preparation of semisynthetic PheB1-tritiated insulin. Biol Chem Hoppe-Seyler 368:239–248

Grau U (1985a) Fingerprint analysis of insulin and proinsulins. Diabetes 34:1174–1180

Grau U (1985b) Chemical stability of insulin in a delivery system environment. Diabetologia 28:458–463

Grau U, Saudek CD (1987) Stable insulin preparation for implanted insulin pumps. Laboratory and animal trials. Diabetes 36:1453–1459

Halban PA, Rhodes CJ, Shoelson SE (1986) High-performance liquid chromatography (HPLC): a rapid, flexible and sensitive method for separating islet proinsulin and insulin. Diabetologia 29:893–896

Halldén G, Gafvelin G, Mutt V, Jörnvall H (1986) Characterization of cat insulin. Arch Biochem Biophys 247:20–27

Hamel FG, Peavy DE, Ryan MP, Duckworth WC (1987) HPLC analysis of insulin degradation products from isolated hepatocytes. Effects of inhibitors suggest intracellular and extracellular pathways. Diabetes 36:702–708

Haneda M, Chan SJ, Kwok SCM, Rubenstein AH, Steiner DF (1983) Studies on mutant human insulin genes: identification and sequence analysis of a gene encoding [SerB24]insulin. Proc Natl Acad Sci USA 80:6366–6370

Heidenreich KA, Yip CC, Frank BH, Olefsky JM (1985) The preparation and characterization of mono-iodinated photoreactive analogs of insulin. Biochem Biophys Res Commun 126:1138–1145

Hellfritzsch M, Christensen EI, Sonne O (1986) Luminal uptake and intracellular transport of insulin in renal proximal tubules. Kidney Int 29:983–988

Hofmann K, Zhang WJ, Romovacek H, Finn FM, Bothner-By AA, Mishra PK (1984) Syntheses of biotinylated and dethiobiotinylated insulins. Biochemistry 23:2547–2553

Hofmann K, Romovacek H, Titus G, Ridge K, Raffensperger JA, Finn FM (1987) The rat liver insulin receptor. Biochemistry 26:7384–7390

Hua Q-X, Qian Y-Q, Tsou C-L (1984) The interaction of the S-thiomethyl insulin A and B chains in solution. Biochim Biophys Acta 789:234–240

Humburg E, Gattner H-G, Zahn H, Brandenburg D Synthesis and characterization of sulfated insulins (in preparation)

Inouye K, Watanabe K, Kanaya T, Tochino Y, Kobayashi M, Haneda M, Shigeta Y (1985) Insulin semisynthesis with N'B1-FMOC-DOI as intermediate. In: Izumiya HN (ed) Peptide chemistry 1984. Protein Research Foundation, Osaka, pp 193–198

Jakubke H-D (1987) Enzymatic peptide synthesis. In: Udenfried S, Meienhofer J (eds) The peptides: analysis, synthesis, biology, vol 9. Special methods in peptide synthesis, part C. Academic Press, New York, pp 103–165

Johnson IS (1983) Human insulin from recombinant DNA technology. Science 219:632–637

Jones RML, Rose K, Offord RE (1987) Semisynthetic human [[^{3}H2]Phe1]proinsulin. Biochem J 247:785–788

Joshi S, Burke GT, Katsoyannis PG (1985a) Synthesis of an insulin-like compound consisting of the A chain of insulin and a B chain corresponding to the B domain of human insulin-like growth factor I. Biochemistry 24:4208–4214

Joshi S, Ogawa H, Burke GT, Tseng L Y-H, Rechler MM, Katsoyannis PG (1985b) Structural features involved in the biological activity of insulin and the insulin-like growth factors: A27 insulin/B IGF-I. Biochem Biophys Res Commun 133:423–429

Katsoyannis PG (1964) The synthesis of the insulin chains and their combination to biologically active material. Diabetes 13:339–348

Katsoyannis PG, Schwartz GP, Burke GT, Joshi S (1987) Insulin-like growth factors: structure-activity relationships. In: Theodoropoulos D (ed) Peptides 1986. De Gruyter, New York, pp 43–60

Keck K, Erb P (eds) (1981) Basic and clinical aspects of immunity to insulin. De Gruyter, New York

King GL, Kahn CR, Samuels B, Danho W, Büllesbach EE, Gattner HG (1982) Synthesis and characterization of molecular hybrids of insulin and insulin-like growth factor I. J Biol Chem 257:10869–10873

Kitagawa K, Ogawa H, Burke GT, Chanley JD, Katsoyannis PG (1984a) Critical role of the A2 amino acid residue in the biological activity of insulin: [2-Glycine-A]- and [2-Alanine-A]insulins. Biochemistry 23:1405–1413

Kitagawa K, Ogawa H, Burke GT, Chanley JD, Katsoyannis PG (1984b) Interaction between the A2 and A19 amino acid residues is of critical importance for high biological activity in insulin: [19-Leucine-A]insulin. Biochemistry 23:4444–4448

Knorr R, Danho W, Büllesbach EE, Gattner H-G, Zahn H, King GL, Kahn CR (1982) [B22-D-Arginine]Insulin: synthesis and biological properties. Hoppe Seyler's Z Physiol Chem 363:1449–1460

Knorr R, Danho W, Büllesbach EE, Gattner H-G, Zahn H, King GL, Kahn CR (1983) [B17-D-Leucine]Insulin and [B17-Norleucine]Insulin: synthesis and biological properties. Hoppe-Seyler's Z Physiol Chem 364:1615–1626

Knutson VP (1987) The covalent tagging of the cell surface insulin receptor in intact cells with the generation of an insulin-free, functional receptor. J Biol Chem 262:2374–2383

Kobayashi M, Ohgaku S, Iwasaki M, Maegawa H, Shigeta Y, Inouye K (1982a) Characterization of [LeuB24]- and [LeuB25]-insulin analogues. Biochem J 206:597–603

Kobayashi M, Ohgaku S, Iwasaki M, Maegawa H, Shigeta Y, Inouye K (1982b) Supernormal insulin: [D-PheB24]-insulin with increased affinity for insulin receptors. Biochem Biophys Res Commun 107:329–336

Kubiak T, Cowburn D (1986a) Enzymatic semisynthesis of porcine despentapeptide (B26-30)insulin using unprotected desoctapeptide (B26–B30)insulin as a substrate. Int J Pept Protein Res 27:514–521

Kubiak T, Cowburn D (1986b) Trypsin-catalysed formation of pig des-(23-63)-proinsulin from desoctapeptide-(B23-30)-insulin. Biochem J 234:665–670

Kwok SCM, Steiner DF, Rubenstein AH, Tager HS (1983) Identification of a point mutation in the human insulin gene giving rise to a structurally abnormal insulin (insulin Chicago). Diabetes 32:872–875

Li CH, Yamashiro D, Gospodarowicz D, Kaplan SL, van Vliet G (1983) Total synthesis of insulin-like growth factor I (somatomedin C). Proc Natl Acad Sci USA 80:2216–2220

Linde S, Welinder BS, Hansen B, Sonne O (1986) Preparative reversed-phase high-performance liquid chromatography of iodinated insulin retaining full biological activity. J Chromatogr 369:327–339

Lloyd LF, Calam DH (1982) Separation of human insulin and some structural isomers by high-performance liquid chromatography. J Chromatogr 237:511–514

Lloyd LF, Corran PH (1982) Analysis of insulin preparations by reversed-phase high-performance liquid chromatography. J Chromatogr 240:445–454

Losse G, Raddatz H (1987) Synthese von B1-substituierten Peptidyl-Insulinen. J Prakt Chem 329:1–9

Losse G, Richter B, Naumann W, Mätzler G (1982) Neue bifunktionelle Brückenfunktionen zur A1-B29-Verklammerung der beiden Insulinketten. J Prakt Chem 324:993–1004

Lougheed WD, Albisser AM, Martindale HM, Chow JC, Clement JR (1983) Physical stability of insulin formulations. Diabetes 32:424–432

Markussen J (1985) Comparative reduction/oxidation studies with single chain des-(B30) insulin and porcine proinsulin. Int J Pept Protein Res 25:431–434

Markussen J (1987) Human insulin by tryptic transpeptidation of porcine insulin and biosynthetic precursors. MTP, Lancaster

Markussen J, Jørgensen KH, Sørensen AR, Thim L (1985) Single chain des-(B30)insulin. Int J Pept Protein Res 26:70–77

Markussen J, Diers I, Engesgaard A, Hansen MT, Hougaard P, Langkjaer L, Norris K, Ribel U, Sørensen AR, Søerensen E, Voigt HO (1987a) Soluble, prolonged-acting insulin derivatives. II. Degree of protraction and crystallizability of insulins substituted in positions A17, B8, B13, B27, and B30. Protein Engineering 1:215–223

Markussen J, Hougaard P, Ribel U, Sørensen AR, Sørensen E (1987b) Soluble, prolonged-acting insulin derivatives. I. Degree of protraction and crystallizability of insulins substituted in the termini of the B-chain. Protein Engineering 1:205–213

Markussen J, Diers I, Hougaard P, Langkjaer L, Norris K, Snel L, Sørensen AR, Sørensen E, Voigt HO (1988) Soluble, prolonged-acting insulin derivatives. III. Degree of protation, crystallizability and chemical stability of insulins substituted in positions A21, B13, B23, B27, and B30. Protein Engineering 2:157–166

Marsh JW, Nahum A, Steiner DF (1983) Reductive methylation of insulin. Int J Pept Protein Res 22:39–49

Massague J, Czech MP (1985) Affinity cross-linking of receptors for insulin and the insulin-like growth factors I and II. Methods Enzymol 109:179–187

McLeod A, Wood SP (1984) High-performance liquid chromatography of insulin. J Chromatogr 285:319–331

Meienhofer J, Schnabel E, Bremer H, Brinkhoff O, Zabel R, Sroka W, Klostermeyer H, Brandenburg D, Okuda T, Zahn H (1963) Synthese der Insulinketten und ihre Kombination zu insulinaktiven Präparaten. Z Naturforsch 18B:1120–1121

Misbin RI, Almira EC, Buynitzky SJ (1983) Insulin metabolism in rat hepatocytes. J Biol Chem 258:2157–2162

Morihara K, Ueno Y, Sakina K (1986) Influence of temperature in the enzymic semi-synthesis of human insulin by coupling and transpeptidation methods. Biochem J 240:803–810

Muir A, Offord RE, Davies JG (1986) The identification of a major product of the degradation of insulin by "insulin proteinase" (EC 3.4.22.11). Biochem J 237:631–637

Naithani VK, Gattner H-G (1982) Preparation and properties of citraconylinsulins. Hoppe-Seyler's Z Physiol Chem 363:1443–1448

Naithani VK, Zahn H (1984) Synthesis of proinsulin. In: Hearn MTW (ed) Peptide and protein reviews, vol 3. Dekker, New York, pp 81–146

Nakagawa SH, Tager HS (1986) Role of the phenylalanine B25 side chain in directing insulin interaction with its receptor. J Biol Chem 261:7332–7341

Nakagawa SH, Tager HS (1987) Role of the COOH-terminal B-chain domain in insulin-receptor interactions. J Biol Chem 262:12054–12058

Nanjo K, Sanke T, Miyano M, Okai K, Sowa R, Kondo M, Nishimura S, Iwo K, Miyamura K, Given BD, Chan SJ, Tager HS, Steiner DF, Rubenstein AH (1986) Diabetes due to secretion of a structurally abnormal insulin (insulin Wakayama). Clinical and functional characteristics of [LeuA3]insulin. J Clin Invest 77:514–519

Ng DS, Yip CC (1985) Peptide mapping of the insulin-binding site of the 130-kDa subunit of the insulin receptor by means of a novel cleavable radioactive photoprobe. Biochem Biophys Res Commun 133:154–160

Obermeier R, Seipke G (1984) Enzyme-catalyzed semisyntheses with porcine insulin. Process Biochem 19:29–32

Offord RE (1980) Semisynthetic proteins. Wiley, Chichester New York Brisbane

Ogawa H, Burke GT, Katsoyannis PG (1984) Synthesis and biological evaluation of a modified insulin incorporating the COOH-terminal hexapeptide ("D-region") of insulin-like growth factor II. J Protein Chem 3:327–348

Ogawa H, Burke GT, Chanley JD, Katsoyannis PG (1987) Effect of N-methylation of selected peptide bonds on the biological activity of insulin. Int J Pept Protein Res 30:460–473

Otha N, Burke GT, Katsoyannis PG (1988) Synthesis of an insulin analogue embodying a strongly fluorescent moiety, [19-Tryptophan-A]insulin. J Protein Chem 7:55–65

Pang DT, Shafer JA (1983) Stoichiometry for the binding of insulin to insulin receptors in adipocyte membranes. J Biol Chem 258:2514–2518

Petersen K-G, Schlüter KJ, Kerp L (eds) (1982) Neue Insuline. 1. Internationales Symposium, 4.–5. Dezember 1981, Freiburg

Pongor S, Brownlee M, Cerami A (1983) Preparation of high-potency, non-aggregating insulins using a novel sulfation procedure. Diabetes 32:1087–1091

Riemen MW, Pon LA, Carpenter FH (1983) Preparation of semisynthetic insulin analogues from bis(tert-butyl-oxycarbonyl)-desoctapeptide-insulin phenylhydrazide: importance of the aromatic region B24-B26. Biochemistry 22:1507–1515

Riemen MW, Hosoume JT, Hillyard NA, Allen MP, Carpenter FH (1985) Semisynthetic insulin analogs: substitution for Gly-B23 significantly alters the activity of destetrapeptide-insulin. In: Deber CM, Hruby VJ, Kopple KD (eds) Peptides 1985. Pierce Chem Comp, Rockford, pp 699–702

Rivier J, McClintock R (1983) Reversed-phase high-performance liquid chromatography of insulins from different species. J Chromatogr 268:112–119

Rose K, De Pury H, Offord RE (1983) Rapid preparation of human insulin and insulin analogues in high yield by enzyme-assisted semi-synthesis. Biochem J 211:671–676

Rose K, Gladstone J, Offord RE (1984a) A mass-spectrometric investigation of the mechanism of the semisynthetic transformation of pig insulin into an ester of insulin of human sequence. Biochem J 220:189–196

Rose K, Pochon S, Offord R (1984b) Oxygen-18 labeled human insulin: semisynthesis and mass-spectrometric analysis. In: Ragnarsson U (ed) Peptides 1984. Almqvist and Wiksell, Stockholm, pp 235–238

Roth RA, Maddux B (1983) Insulin-cholera toxin binding unit conjugate: a hybrid molecule with insulin biological activity and cholera toxin binding specificity. J Cell Physiol 115:151–158

Roth RA, Iwamoto Y, Maddux B, Golfine ID (1983) Insulin-ricin B chain conjugate has enhanced biological activity in insulin-insensitive cells. Endocrinology 112:2193–2199

Sakina K, Ueno Y, Oka T, Morihara K (1986) Enzymatic semisynthesis of [Leu B30]insulin. Int J Pept Protein Res 28:411–419

Sakura H, Iwamoto Y, Sakamoto Y, Kuzuya T, Hirata H (1986) Structurally abnormal insulin in a diabetic patient – characterization of the mutant insulin A3 (Val→Leu) isolated from the pancreas. J Clin Invest 78:1666–1672

Sato S, Ebert CD, Kim SW (1983) Prevention of insulin self-association and surface adsorption. J Pharm Sci 72:228–232

Saunders D, Freude K (1982) Cross-linked [D-Ala^{A1}]insulins. Evidence for a change in the conformation of the insulin monomer at its receptor. Hoppe Seyler's Z Physiol Chem 363:655–659

Saunders DJ, Brandenburg D (1984) Photoactivatable insulins and receptor photoaffinity labelling. In: Larner J, Pohl S (eds) Methods in diabetes research, vol I: laboratory methods, pt A. Wiley, New York, pp 3–22

Schartmann B, Gattner H-G, Danho W, Zahn H (1983) Erhöhte Insulinausbeuten durch "Recycling" der nichtkombinierten Ketten. Hoppe-Seyler's Z Physiol Chem 364:179–186

Schüttler A, Brandenburg D (1982) Preparation and properties of covalently linked insulin dimers. Hoppe Seyler's Z Physiol Chem 363:317–330

Schüttler A, Gattner H-G, Brandenburg D (1984) Preparation of selected insulin derivatives and analogues. In: Larner J, Pohl S (eds) Methods in diabetes research, vol I: Laboratory methods, pt A. Wiley, New York, pp 355–376

Schwartz G, Burke GT, Katsoyannis PG (1982) The importance of the B10 amino acid residue to the biological activity of insulin. [Lys10-B] Human insulin. J Protein Chem 1:177–189

Schwartz GP, Burke GT, Chanley JD, Katsoyannis PG (1983) An insulin analogue possessing higher in vitro biological activity than receptor binding affinity. [21-Proline-B]insulin. Biochemistry 22:4561–4567

Schwartz GP, Wong D, Burke GT, De Vroede MA, Rechler MM, Katsoyannis PG (1985) Glutamine B16 insulin: reduced insulin-like metabolic activity with moderately preserved mitogenic activity. J Protein Chem 4:185–197

Schwartz GP, Burke GT, Katsoyannis PG (1987) A superactive insulin: [B10-Aspartic acid]insulin (human). Proc Natl Acad Sci USA 84:6408–6411

Seino S, Steiner DF, Bell GI (1987) Sequence of a New World primate insulin having low biological potency and immunoreactivity. Proc Natl Acad Sci USA 84:7423–7427

Selam J-L, Zirinis P, Mellet M, Mirouze J (1987) Stable insulin for implantable delivery systems: in vitro studies with different containers and solvents. Diabetes Care 10:343–347

Shoelson S, Fickova M, Haneda M, Nahum A, Musso G, Kaiser ET, Rubenstein AH, Tager H (1983a) Identification of a mutant human insulin predicted to contain a serine-for-phenylalanine substitution. Proc Natl Acad Sci USA 80:7390–7394

Shoelson S, Haneda M, Blix P, Nanjo A, Sanke T, Inouye K, Steiner D, Rubenstein A, Tager H (1983b) Three mutant insulins in man. Nature 302:540–543

Sodoyez JC, Sodoyez-Goffaux F, Guillaume M, Merchie G (1983) [^{123}I]Insulin metabolism in normal rats and humans: external detection by a scintillation camera. Science 219:865–867

Sonne O (1987) Receptor-mediated degradation of insulin in isolated rat adipocytes. Formation of a degradation product slightly smaller than insulin. Biochim Biophys Acta 927:106–111

Sonne O, Linde S, Larsen TR, Gliemann J, Larso L (1983) Monoiodoinsulin labelled in tyrosine residue 16 or 26 of the B-chain or 19 of the A-chain. Hoppe Seyler's Z Physiol Chem 364:101–110

Srinivasa BR, Carpenter FH (1983) Intramolecular cross-linking of insulin. Int J Pept Protein Res 22:214–222

Stoev S, Zakhariev S, Golovinsky E, Gattner H-G, Naithani VK, Wollmer A, Brandenburg D (1988) Synthesis and properties of [A19-(p-fluorophenylalanine)] insulin. Biol Chem Hoppe-Seyler 369:1307–1315

Tatnell MA, Jones RH, Willey KP, Schüttler A, Brandenburg D (1983) Evidence concerning the mechanism of insulin-receptor interaction and the structure of the insulin receptor from biological properties of covalently linked insulin dimers. Biochem J 216:687–694

Thim L, Hansen MT, Norris K, Hoegh I, Boel E, Forstrom J, Ammerer G, Fiil NP (1986) Secretion and processing of insulin precursors in yeast. Proc Natl Acad Sci USA 83:6766–6770

Tian Y, Wang C-C, Tsou C-L (1987) Interaction and combination of separate A and B chains of insulin. Effects of D- and L-tryptophan in position A1. Biol Chem Hoppe-Seyler's 368:397–403

Trindler P, Brandenburg D (1982) Semisynthetic modification of the N-terminus of the insulin A-chain. In: Voelter W, Wünsch E, Ovchinnikov Yu, Ivanov V (eds) Chemistry of peptides and proteins, vol 1. De Gruyter, New York, pp 307–314

Varley JM, Davies JG, Shire D, Offord RE, Timmis KN (1988) Engineered rat I insulin analogue having a B16Tyr/Asp replacement exhibits unchanged susceptibility to cleavage by insulin proteinase. Eur J Biochem 171:351–354

Vigh Gy, Varga-Puchony Z, Szepesi G, Gazdag M (1987) Semi-preparative high-performance reversed-phase displacement chromatography of insulins. J Chromatogr 386:353–362

Wang C-C, Tsou C-L (1986) Interaction and reconstitution of carboxyl-terminal-shortened B chains with the intact A chain of insulin. Biochemistry 25:5336–5340

Wedekind F, Baer-Pontzen K, Bala-Mohan S, Choli D, Zahn H, Brandenburg D (1989) Hormone binding site of the insulin receptor: analysis using photoaffinity-mediated avidin complexing. Biol Chem Hoppe-Seyler 370:251–258

Welinder BS, Linde S, Hansen B (1984) Isolation of specific labeled insulin tracers. Comparison between RP-HPLC and disc electrophoresis – ion exchange chromatography. In: Larner J, Pohl S (eds) Methods in diabetes res, vol 1: laboratory methods, pt B. Wiley, New York, pp 341–354

Wieneke H-J, Wolf G, Wolff W, Büllesbach EE, Gattner H-G, Brandenburg D (1983a) Synthesis of a hybrid chicken/human insulin. In: Bláha K, Maloň P (eds) Peptides 1982. De Gruyter, New York, pp 367–370

Wieneke H-J, Danho W, Büllesbach EE, Gattner H-G, Zahn H (1983b) The synthesis of [A19-3-iodotyrosine] and [A19-3,5-diiodotyrosine]-insulin (porcine). Hoppe-Seyler's Z Physiol Chem 364:537–550

Yip CC, Yeung CWT (1985) Photoaffinity labelling of the insulin receptor. Methods Enzymol 109:170–179

Zahn H (1983) Insulin: Von der Strukturaufklärung zur chemischen Synthese. Münch Med Wochenschr 125 [Suppl 1]:3–13

Zaitsu K, Hosoya H, Hayashi Y, Yamada H, Ohkura Y (1985) High-performance liquid chromatographic separation of citraconylinsulins and preparation of GlyA1PheB1-dicitraconylinsulin. Chem Pharm Bull 33:1159–1163

Zhang Y-S, Cao Q-P, Li Z-G, Cui D-F (1983) Preparation of [B23-D-alanine]des-(B25-B30)-hexapeptide-insulin by a combination of enzymic and non-enzymic synthesis. Biochem J 215:697–699

Zhu S-Q (1985) Studies on structure and biological activity of insulin. In: Molecular architecture of proteins and enzymes. Academic Press, New York, pp 185–197

CHAPTER 2

Insulin Structure

U. Derewenda, Z. S. Derewenda, G. G. Dodson, and R. E. Hubbard

A. Introduction

The three-dimensional structure of insulin obtained by X-ray crystallographic analysis has made it possible to explain much of the molecule's chemistry, its behaviour in solution and a good deal of its biology (Blundell et al. 1972; Baker et al. 1988). For example, the antigenic surfaces, the exposed chemical groups and the surfaces buried by the hormone's assembly to dimer and hexamer have all been identified. Review of the effects that chemical modification has on assembly, solution structure and potency has helped delineate the hormone's biologically active surfaces. Chemical techniques for modifying proteins are becoming very much more powerful and are producing new evidence about the nature of insulin's binding surfaces. And the variation in the sequence of natural insulins and human mutant insulins have further focused our picture of the surfaces involved in assembly and in binding at the receptor. The structure has also allowed us to understand the biosynthesis of the hormone and the processing of its precursors to the finally active insulin molecule. It has even been possible to connect the biosynthetic processes in insulin to some of its homologues such as relaxin and the insulin-like growth factors whose structures are similar to insulin, but whose functions are profoundly different (Bradshaw et al. 1980; Steiner et al. 1986).

These deductions are now being given much more penetration through recent experiments in which the hormone's sequence is deliberately altered by the technique of site-directed mutagenesis (Brange et al. 1988). These manufactured mutants can be given new properties, their assembly can be modified, the stability of the monomer increased and the potency of the hormone enhanced or reduced. In all of these experiments the three-dimensional structure of the molecule is an essential starting point for the design and interpretation and modifications in properties.

Since the determination of insulin's structure in the 2Zn insulin crystal in 1969 there have been a series of crystallographic studies on other crystal forms. Some of these have been insulins from different animals which have allowed us to investigate how evolutionary change in sequence has affected the hormone's structure. In other analyses the hormone has been studied at different levels of assembly and in different conformations. These studies have revealed the molecule has an intrinsic mobility which undoubtedly plays an important part in governing its chemical, solution and biological behaviour (Dodson et al. 1983). In this chapter our current knowledge of the structures of the insulin molecule and their pattern of assembly will be described. This study of these structures gives some insight into the extent and structural behaviour of the active surface that binds to the receptor. From the flexibility of this surface can be drawn some inferences about the structural events associated with binding.

B. The Three-Dimensional Structure of the Insulin Molecule

Insulin exists as a monomer at low concentrations ($<10^{-7}$ mol) in neutral conditions (FRANK et al. 1972). At higher concentrations it dimerises and in the presence of zinc assembles further to a hexamer (BLUNDELL et al. 1972; MILTHORPE et al. 1976; GOLDMAN and CARPENTER 1974). The crystal structures of monomeric, dimeric and hexameric insulins have all been determined (Table 1). Although there are variations in the molecule's conformation in the different species and in different crystals, an essential common structure can be described.

Table 1. Crystal structures discussed in this chapter

Name	Aggrega-tion level	Comments	Resolu-tion (Å)	Crystallo-graphic R^{a}
2 Zn insulin	Hexamer	2 zinc stoichiometry	1.5	0.154
4 Zn insulin	Hexamer	4 zinc stoichiometry (Cl binds zinc)	1.5	0.18
Phenol Zn insulin	Hexamer	2 zinc stoichiometry (binds 6 phenol/hexamer)	1.6	0.22
Cubic insulin	Dimer	Neutral preparation	1.7	0.21
Orthorhombic insulin	Dimer	Acid preparation	1.9	0.19
Beef despenta-peptide insulin	Monomer	Modified (B26–B30 removed)	1.2	0.17

[a] $R = \dfrac{\sum ||F_{o}| - |F_{c}||}{\sum |F_{o}|},$

where $|F_{o}|$ is the observed structure amplitude; where $|F_{c}|$ is the calculated structure amplitude.

I. The Insulin Monomer

The molecule consists of two chains, the A chain of 21 amino acids and the B chain of usually 30 amino acids (BROWN et al. 1955). There are disulphide bonds linking the chains at A7-B7 and A20-B19; within the A chain there is a disulphide loop between A6 and A11. The full amino acid sequence for porcine insulin is shown in Fig. 1.

The X-ray analysis of 2Zn insulin reveals that there is helix in the B chain at B9-B20 and in the A chain at A1-A8 and A13-A20. The disulphide bonds A7-B7 and A20-B19 are at each end of the B chain helix and form a framework which then holds the A and B chains together. The A6-A11 disulphide bond is also associated with helical structure where it stabilises the folding of the A chain and,

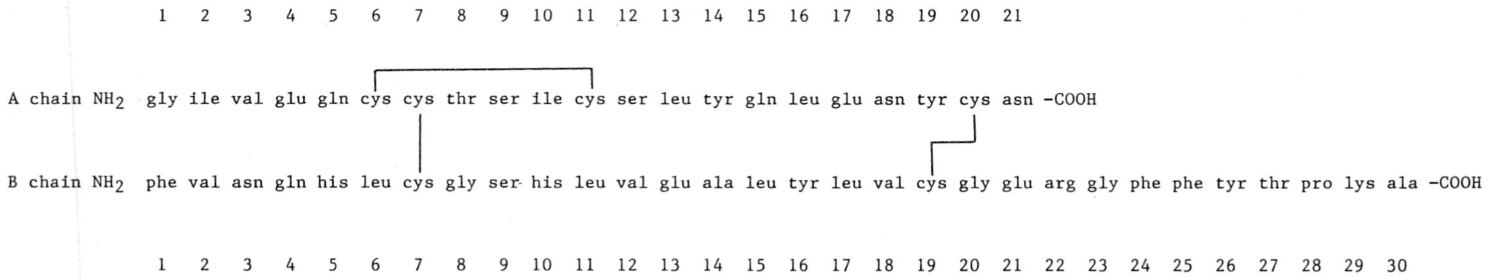

A chain NH$_2$ gly ile val glu gln cys cys thr ser ile cys ser leu tyr gln leu glu asn tyr cys asn –COOH
           1 2 3 4 5 6 7 8 9 10 11 12 13 14 15 16 17 18 19 20 21

B chain NH$_2$ phe val asn gln his leu cys gly ser his leu val glu ala leu tyr leu val cys gly glu arg gly phe phe tyr thr pro lys ala –COOH
           1 2 3 4 5 6 7 8 9 10 11 12 13 14 15 16 17 18 19 20 21 22 23 24 25 26 27 28 29 30

Fig. 1. The chemical sequence of porcine insulin

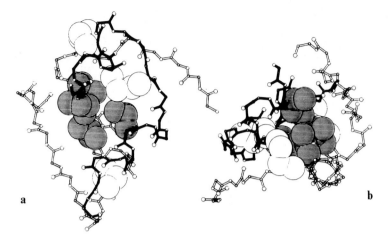

a   b

Fig. 2 a, b. Two views of molecule 1 of 2Zn insulin showing the packing of the helices around completely buried nonpolar residues. In both figures are shown the peptide backbone of the A chain (*thick solid bonds*) and that of the B chain (*open bonds*). Also shown are the side chain atoms of the cysteine residues (*open space-filling*) and the completely buried nonpolar residues (*shaded space-filling*). In **a** the A1-A8 and A12-A20 helices can be seen in the front of the molecule, **b** is a view down the axis of the B9-B19 helix revealing the relationship between the two chains and the nonpolar core. Note how the NH_2 and COOH termini partially enfold the A chain

together with a cluster of aliphatic side chains, forms a buried nonpolar core. The packing of the helices around the completely buried residues in the monomer is illustrated in Fig. 2.

The surface of the insulin molecule is covered by both polar and nonpolar residues. As can be seen in Fig. 3 there are mostly polar residues from the A chain covering a continuous surface on the molecule. This polar surface extends in one direction to a surface formed by the B chain NH_2 terminal residues (B5-B1) and continues to another polar surface formed on the B chain helix. This latter surface includes residues at the junction of the B chain helix and the extended sheet of the B chain COOH terminal residues. The more clearly defined region is that at the B chain COOH terminus also involving side chains on the B chain helix. This surface is mostly aromatic and is relatively flat with dimensions of about 400 $Å^2$. It is this surface that is buried when dimers are formed. The second nonpolar surface is more complex in structure and contains side chains from both the A and B chains. This surface is buried when dimers assemble to form hexamers. It is composed of both aromatic and aliphatic side groups and has a variable structure depending upon the molecule's level of assembly and solution conditions (BAKER et al. 1988).

In the monomeric insulin despentapeptide insulin (DPI) residues B26-B30 have been removed enzymatically. This abolishes the molecule's capacity to form dimers, but leaves the molecule still substantially active. The removal of these residues evidently has very little effect on the A chain structure and the B chain from B5 to B24, as seen in the dimeric and hexameric insulins (BI et al. 1984). This is remarkable since the absence of B26-B30 exposes much of the molecule's

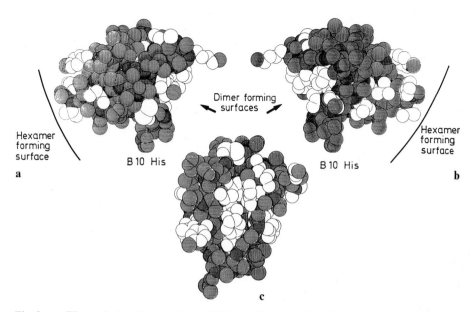

Fig. 3 a–c. Three views of molecule 1 of 2Zn insulin in which all atoms are drawn space-filling. The main chain atoms and all side chain atoms of polar residues are shaded; *horizontal shading* for B chain atoms, *vertical* for A chain atoms. Nonpolar side chains atoms are represented as *open van der Waals spheres*. **a** is a view up the threefold axis of the insulin molecule, **b** is a view down the threefold axis showing the other side of the molecule. In both figures the surfaces involved in forming dimers and which are buried on forming hexamers are labelled, **c** is a view perpendicular to the dimer-forming surface and to the threefold axis of the rhombohedral crystal

internal nonpolar core. The main changes are at B25 (now of course the B chain COOH terminal residue) and at the B chain NH_2 terminus where there are no restraints on the conformation (see Fig. 4). From these observations we can deduce that the incorporation of the monomer into the dimer and hexamer has largely peripheral effects on the structure.

II. The Insulin Dimer

The dimer, which is the predominant species in solution in the absence of metal ions, is organised with the two insulin molecules disposed about a twofold axis. In some crystals this axis is approximate; in others it is exact. The dimer within the hexamer is arranged in the same way, again sometimes with its twofold axis of symmetry exact and sometimes approximate.

The existence of two molecules in the dimer each of which can be different makes it necessary to define a convention for reference. It turns out that this is surprisingly simple; there is a persistent conformation in the insulin molecule which appears in nearly all crystal forms and which is very probably that of the circulating monomeric molecule. This is referred to as molecule 1; where there is a partner of significantly different conformation the molecule is referred to as molecule 2 (CUTFIELD et al. 1981).

Hexamer

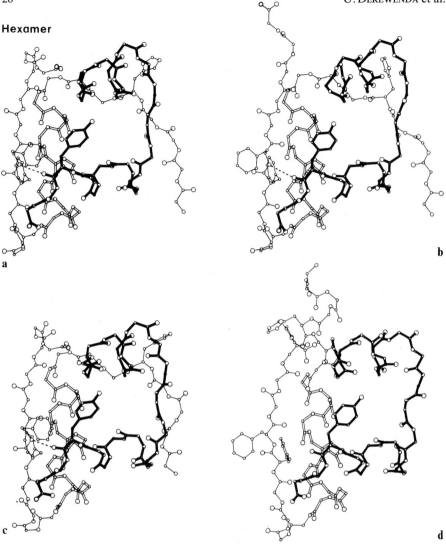

Fig. 4 a–l. Panel of 11 insulin molecules (optimised on B9-B19 backbone atoms) all viewed in the same direction. The molecules are arranged to show insulin found as a hexamer, as a dimer and as a monomer in the crystal. The peptide backbone of the A chain is drawn with *thick bonds*; that of the B chain helix with *open bonds*. The remainder of the B chain is drawn with *thin bonds*. The side chains B5 histidine, B24 and B25 phenylalanine and A19 tyrosine are also shown with the same bond type as the corresponding main chain. There is a structurally important H bond between A19 carbonyl and B25 peptide NH; it is shown as a *dashed line* where it exists. Note that B30 and B29 residues are not illustrated in the monoclinic insulin molecule 2 and in the cubic insulin structure – these are disordered in the crystal and it is not possible to determine accurate atomic coordinates for either the main chain or the side chains. **a** R3 2Zn insulin molecule 1, **b** R3 2Zn insulin molecule 2, **c** R3 4Zn insulin molecule 1, **d** R3 4Zn insulin molecule 2, **e** R3 4Zn insulin molecule 1 cross-linked A1-B29, **f** R3 4Zn insulin molecule 2 cross-linked A1-B29, **g** monoclinic phenol Zn insulin molecule 1, **h** monoclinic phenol Zn insulin molecule 2, **i** orthorhombic insulin molecule 1, **j** orthorhombic insulin molecule 2, **k** cubic insulin (two molecules of the dimer are identical), **l** beef despentapeptide insulin (monoclinic)

Hexamer

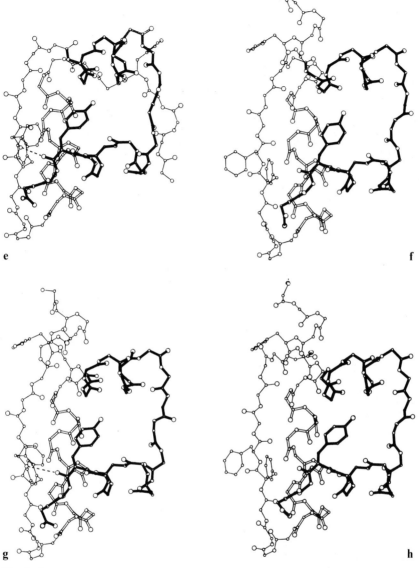

e

f

g

h

Fig. 4e–h

The two insulin molecules in the dimer are held together by nonpolar forces and four H bonds arranged as a β sheet structure between the two antiparallel COOH terminal strands of the B chain. The dimer is illustrated in Fig. 5 which shows the molecules viewed in the direction of the twofold axis. The hydrogen bonds, formed between residues B24 and B26 and equivalently between B26 and

Dimer

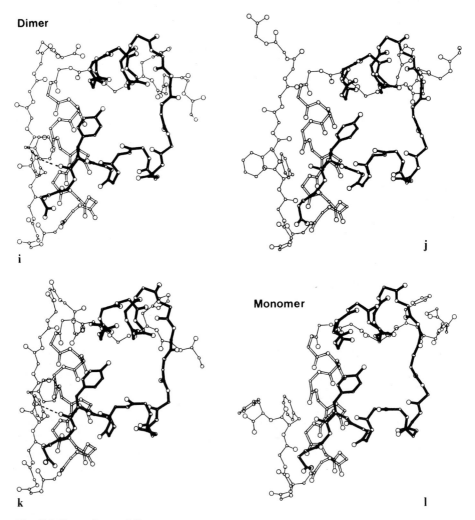

Monomer

Fig. 4i–l (Legend see p. 28)

B24, range between 2.8 and 3.2 Å. This pulls the two insulin molecules together compactly and enhances the van der Waals interactions. A large proportion of the side chains are aromatic; the two B24 Phe and the two B26 Tyr form a well-defined cluster at the centre of the interacting surfaces. The B12 Val fills in the gap between these and the two adjacent B16 Tyr. At the periphery of the dimer-forming surfaces is B28 Pro which in the dimer packs closely against B20 Gly. On the surface the two B25 Phe can fold back against the neighbouring A19 Tyr side chain. As a result of the burial of these hydrophobic residues in the dimer the surface is now more polar. Notice (see Fig. 4) the B chain NH_2 terminal residues (B1-B4) are free to take up varying conformations largely determined in the crystal by lattice contacts. Dimer formation brings the two B13 glu side chains together, they are however both negatively charged and repel each other.

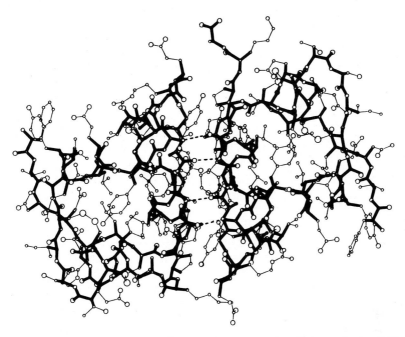

Fig. 5. The dimer from the 2Zn insulin crystal structure viewed down the twofold axis. Notice the loss of symmetry between the two molecules, mainly due to crystal packing forces. Main chain atoms are drawn with *thick bonds*; side chain atoms drawn with *thin bonds*. The *dashed lines* show the hydrogen bonds between the B24 and B26 peptides in one molecule and the corresponding B26 and B24 in the other

Two mammalian dimeric insulin structures have been determined (BENTLEY et al. 1976; DEREWENDA et al. 1988), one in crystals grown at neutral pH and the other in crystals grown at acid pH. The neutral pH crystals are cubic, and have one molecule in the asymmetric unit. This means the axis relating the two molecules in the dimer is used by the crystal. In this arrangement the two B25 Phe are both folded back from the twofold axis and contact A19 Tyr. The separation of B25 amide nitrogen and A19 carbonyl oxygen is 3.18 Å, and the conformations at the B chain NH_2 terminus and COOH terminus are identical. This set of conformations is characteristic of molecule 1.

In the orthorhombic crystals grown at acid pH the asymmetric unit is a dimer and the axis relating the two molecules in this case is not exact. In one of the molecules (molecule 1) B25 Phe is folded back against A19 Tyr; in the other molecule (molecule 2) B25 turns across the line of the twofold axis and contacts its partner. Associated with this difference is the expansion of the B25 amide nitrogen and A19 carbonyl oxygen distance from 3.18 to 3.78 Å, breaking the hydrogen bond. The possible significance of this structural change and the accompanying movement of B25 Phe side chain will be discussed again in Sect. C. There are also substantial differences in the conformations of the NH_2 terminal and COOH terminal B chain residues. In these two dimers the structure of the B chain residues B9-B24 and the A chain are closely preserved, reflecting their close packing and disulphide bond connections.

III. The Insulin Hexamer

In the presence of zinc ions three insulin dimers are assembled into a hexameric organisation in which the zinc ions are coordinated to imidazole groups of B10 or sometimes B5 histidine (CUTFIELD et al. 1981; BENTLEY et al. 1976). These zinc-containing hexamers are all similar in their symmetry and in the structure of the insulin molecules. The three dimers are related by a threefold axis, the local twofold axis within each dimer is perpendicular or approximately so to the threefold axis. The packing of the dimers around the zinc ions is associated with the burial of the second nonpolar surface which is made up of A13 Leu, A14 Tyr, B1 Phe, B2 Val, B14 Ala, B17 Leu, and B18 Val. This surface contains flexible structures which are seen to adapt in different hexamer organisations and the contacts between the dimers in the hexamer are considerably looser than those between the monomers in the dimer. There is only one rather weak hydrogen bond and the van der Waals separations tend to be large.

Three zinc insulin hexamer structures have been determined: that from the rhombohedral 2Zn crystal, that from the rhombohedral 4Zn insulin crystal and that from the monoclinic phenol insulin crystal (see Table 1). In Fig. 6 the three hexamers are seen represented in a simplified way in order to illustrate the progressive changes in their secondary structure and symmetry.

The 2Zn insulin hexamer contains threefold-related dimers. Their local axes are perpendicular to and intersect the threefold axis of the hexamer which means the dimers themselves are related by the same symmetry operation. There are two central zinc ions on the threefold axis each octahedrally coordinated to three B10 His imidazole nitrogens and three water molecules and related by the same twofold symmetry that relates the dimers. Between the dimers the B chain NH_2 terminal residues extend to the edge of the hexamer. Here the B1 Phe from each monomer is buried in a nonpolar environment (A13 Leu, A14 Tyr) in the adjacent dimer. This crossover lends extra stability to the dimer. Between these B chain NH_2 terminal extended segments are buried nonpolar residues whose contacts complete a hydrophobic circle which connects up the nonpolar cores of the monomers already in touch within the dimer (BAKER et al. 1988).

There is a polar channel at the centre of the hexamer made up of the coordinating B chain histidines, B9 Ser and B13 Glu. Associated with these groups is a well defined, but complex water structure which links the hydrogen bonding atoms in an elaborate chain of structure. These polar contacts serve to dissipate the charges at the glutamic acid and satisfy the hydrogen bonding contacts of the groups which are not in contact with protein. The structure of the six central glutamic acids is arranged as three hydrogen bonded pairs. At this pH glutamic acid is not normally protonated and able to make hydrogen bonds of this kind. The existence of this interaction suggests that the charge repulsion that would normally be operating is avoided by protonation; probably this potential repulsion is one factor that is important in driving the molecules apart rapidly, following release of the hexamers into the bloodstream.

In the 2Zn insulin crystal the hexamer constitutes the rhombohedral unit cell and the dimer the asymmetric unit. The two molecules in the dimer are mostly closely related by twofold symmetry. Along the axis within the dimer this

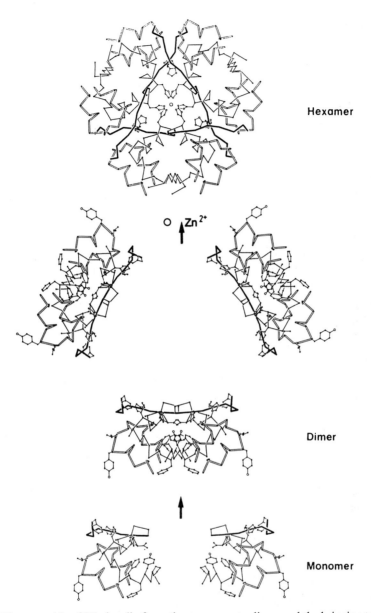

Hexamer

O ↑Zn^{2+}

Dimer

Monomer

Fig. 6a. The assembly of 2Zn insulin from the monomer to dimer and the bringing together of the dimers in the presence of zinc to form the hexamer. In these figures, only the C atoms of the main chain are drawn; the A chain is drawn with *open bonds*, the B chain with *thinner full bonds*, and B1–B8 by *thick bonds*. The atoms in the side chains of the dimer- and hexamer-forming residues are drawn bonded by *thin bonds*. These residues include B1, B5, B10, B12, B24, B25, B26, B28 and A13 and A14

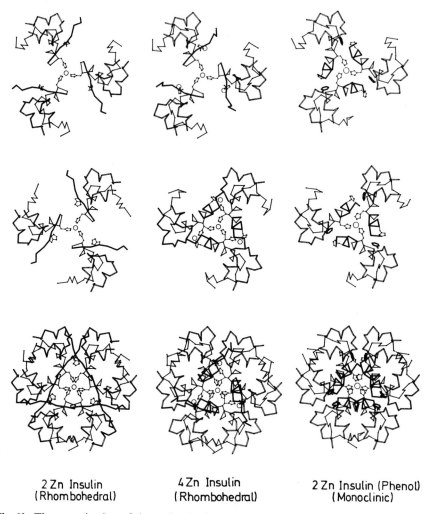

2 Zn Insulin   4 Zn Insulin   2 Zn Insulin (Phenol)
(Rhombohedral)   (Rhombohedral)   (Monoclinic)

Fig. 6 b. The organisation of the molecules in the three hexameric insulins discussed in the text. The most marked difference between the hexamers is the conformation of the B chain NH_2 terminus B1-B8 drawn with *thick bonds*. In 2Zn insulin residues B1-B8 are in extended form, in 4Zn insulin the B terminus of molecule 2 is folded in a helix and in monoclinic insulin both molecules have the NH_2 terminus of the B chain as an extension of the B9-B19 helix. The *circles* represent zinc ions; in 2Zn insulin and monoclinic insulin these bind only on the central axis of the hexamer

relationship is almost exact, at the surface however it breaks down. The B25 Phe side chain of one molecule is turned back over the A19 Tyr; this is molecule 1. The other B25 of the dimer is positioned away from A19 Tyr and contacts its B25 Phe partner; in doing this the phenyl ring lies directly on the line of the local twofold axis. (This molecule is referred to as molecule 2.) There are marked changes at the A chain NH_2 terminal residues caused by close lattice contacts by the B5 His side chains along the threefold crystal axis (Chothia et al. 1983;

BAKER et al. 1988). These adjustments at A1-A6 in molecule 2 are relayed to the B chain residues B28-B30 which are rotated around the B28-B27 peptide bond relieving otherwise close contacts between B28 carbonyl oxygen and the repositioned A4 glutamyl carboxylate group.

The insulin structure in the rhombohedral 4Zn insulin crystal is related to that in the 2Zn insulin hexamer. Each dimer contains one molecule essentially identical to an equivalent in the 2Zn insulin hexamer, this is molecule 1. The partner (molecule 2) in the 4Zn insulin dimer is however remarkably different at the B chain NH_2 and COOH termini. Thus, the 4Zn insulin hexamer – like the 2Zn insulin – is constructed from two types of insulin molecules (molecules 1 and 2). Each molecule is arranged as a trimer, that made up of molecule 1 is essentially the same in structure as its equivalent in 2Zn insulin (molecule 1) and three other molecules which are significantly different (molecule 2). This is illustrated in Fig. 6 which shows the structures of the two trimers and the whole hexamer. The X-ray analysis of 4Zn insulin shows that the residues B1-B8 which are extended in the 2Zn hexamer (molecule 2) switch into helix and form a continuation of the pre-existing helix at B9-B20. This rearrangement of structure in which the B chain is helical from B7 to B20 involves movements of more than 25 Å at B1. Associated with the change into helix at B1-B8 is a separation of the B chain COOH terminal residues B25-B30 from the A chain and the associated rotation of B25 Phe. In molecule 1 of 2Zn and 4Zn insulin the hydrogen bonds between A19 carbonyl oxygen and B25 amide nitrogen have a distance of 3.2 Å. In molecule 2 of 4Zn insulin the A19-B25 separation is increased to 4.8 Å and the hydrogen bond is again broken. This is parallel to the separation of the A and B chains seen in molecule 2 of the insulin dimer at acid pH. In contrast with the situation in 2Zn insulin crystal, the A chain NH_2 termini have the same conformation. This is a consequence of the extra helix structures extending from molecule 2 breaking the close crystal contacts that exist in 2Zn insulin.

The substantial rearrangements of the B chain NH_2 termini in molecule 2 in 4Zn insulin destroys the twofold symmetry between the dimers in the hexamer. By contrast, the symmetry immediately about the axis relating the monomers in the dimer is largely preserved. Away from the dimer axis the local symmetry obviously breaks down. In addition the twofold axis within the dimer no longer intersects the threefold axis, reducing the symmetry of the hexamer to simple threefold.

SCHLICHTKRULL (1958) showed that it was chloride ions that were responsible for the increased ability of the 4Zn insulin hexamer to bind zinc. The X-ray analysis revealed that the extra zinc sites were generated in molecule 2 by the B5 His, positioned on the helix, and the B10 His in the adjacent dimer which swivels away from the central threefold hexamer axis. A tetrahedral binding site is thus created, consisting of two histidyl N atoms, a chloride ion and a water molecule. These three tetrahedral sites and one octahedral axial site explain the stoichiometry of 4 zincs per hexamer observed by Schlichtkrull. The zinc binding site on the other insulin molecule (molecule 1) is unchanged. This arrangement of zinc sites further illustrates the breakdown of the twofold symmetry within the hexamer.

The third type of insulin hexamer structure has been discovered in monoclinic crystals grown in the presence of phenol (SCHLICHTKRULL 1958). In this crystal the hexamer constitutes the asymmetric unit and thus none of the symmetry elements present in the hexamer are used by the crystal. The X-ray analysis of these crystals has shown that all the molecules are helical at B1–B8 (DEREWENDA et al. 1988). They all show increased separation at B25-A19 which correlates with the variations in conformation of B25 Phe. Because the molecules now all have the same helical conformation, the contacts between the dimers are equivalent and this hexamer has reasonably good local twofold symmetry within the dimers and between the dimers. Thus, it has the same organisation of its symmetry elements as the rhombohedral 2Zn insulin hexamer.

The agent responsible for the helical structure turns out to be phenol molecules. These bind in a nonpolar cavity created by the packing of the B1-B8 helix aganist the A chain of the threefold-related dimer. There are two hydrogen bonds made by the hydroxyl group, one to A6 carbonyl oxygen, the other to the peptide nitrogen of A11. The aromatic ring of the phenol packs at van der Waals distance against the B5 His imidazole group. This interaction prevents the B5 His bonding to zinc. Consequently the zinc binding is restricted to the two axial sites where three symmetry-related B10 His imidazole nitrogens and an axial water molecule form a tetrahedral coordination structure. Thus, while the three hexamers form a progression in protein structure with increasing amounts of helix in the B chain, there is no corresponding increase in zinc binding.

C. Structural Variation in the Insulin Molecule

Comparison of the insulin molecules in the different insulin crystals reveals that there are certain structural elements in the molecule's structure which are stable and change little, and others which are variable and change a good deal (DODSON et al. 1983). The variation in the insulin molecules from different crystals is illustrated in Fig. 4 which compares the main chain structures of the monomer when crystallised as a monomer, dimer or hexamer. It can be seen that the B chain central helix B9-B20 is a very well preserved feature. The A chain helix A14-A20 is almost equally unaffected in its position relative to the helix B9-19. There are also few changes in the A chain NH_2 terminal helix, except in the 2Zn insulin hexamer molecule 2; these arise from tight crystal contacts and probably do not represent an intrinsic structural property. Thus, the B9-B20 helix and the A chain which are disulphide linked at B7-A7 and B19-A20 evidently form a stable structural core.

By contrast the B1-B8 and B25-B30 segments are very variable in conformation and there appears to be some correlation between their movements: the separation at A19 O and B25 NH is always increased when the B25 Phe is folded away. When in addition the B1-B8 is helical this effect appears to be further amplified.

The separation of B25 and A19 is most pronounced however in DPI (BI et al. 1984). Here the B25 Phe is a COOH terminal residue and is turned completely away from the body of the molecule with both its amide nitrogen and carboxylic group hydrated. Interestingly this movement has little effect on B24 Phe, the conformation of the B chain turn at B23-B20 or the conformation of the A chain

COOH terminus. In addition the loss of the 5 residues B26-B30 does not lead to any large adjustments. Since B26-B30 pack against the A chain buried residues and the NH_2 terminal helix, the absence of any significant adjustment following removal of B26-B30 is further evidence that the structural core is a stable arrangement.

The obvious notion that the active surface seen in the crystal structure binds directly to the receptor is challenged by two observations. The first is that several insulins with native-like structures, especially at the active surfaces, show reduced potency. One of these is the (des B30) insulin in which A1 α-amino is peptide bonded to B29, now the COOH terminal residue (MARKUSSON 1985). This molecule exhibits no detectable insulin-like activity although, as seen in Fig. 4, the residues in the identified active surfaces have an essentially identical structure to that seen in fully active insulins. An additional feature of the inactive cross-linked insulin is that it has crystallised in the 4Zn insulin structure. Thus, it can also be concluded that the capacity to make the conformational changes associated with the formation of helix at B1-B8 and the separation of A chain and B chain COOH terminal residues is not alone sufficient for the expression of biological activity.

The second observation relevant to insulin's active surfaces comes from the structure and potency of DPI. In this molecule, as has already been described, the removal of B30-B26 and the displacement of B25 from any contacts with the remainder of the molecule leaves the structural core essentially unaffected. The molecule's substantial potency of 40%, elevated to 100% upon amidation of the B25 carboxylate group, indicates that it contains all the active surfaces necessary for the expression of biological activity. One of these surfaces containing the A chain residues A2, A3, A19, and B15 is buried in native insulin by B30-B26. For this surface to be involved in binding in the native molecule the B chain COOH terminal residues must move away from the A chain. Indeed there is evidence from the crystal structures that this movement is intrinsic. Moreover it is perhaps significant that it is this surface whose exposure is prevented by the A1-B29 cross-link in the inactive single-chain insulin.

This allows us to speculate about the conformational changes associated with the binding of native insulin to its receptor. The proposal that best fits the structural and biological evidence is that the B chain COOH terminal residues are displaced from the A chain before or during binding. It is almost certain that this separation includes B25 Phe, but it may also extend to B24 Phe. Once these residues are displaced the newly exposed nonpolar surface would be available for interaction. In this connection the reduced potency of the mutant A3 Val→Leu is possibly significant (NANJO et al. 1986). This residue is much more accessible when B30-B28 is removed.

Equally suggestive is the observation that an insulin covalently cross-linked between B26 and B16 tyrosines has approximately 18% potency (CUTFIELD et al. 1986). This modification unequivocally demonstrates that the dimer-forming surfaces are not responsible for activity. It also suggests that the active surfaces include the A chain exposed on the dimer and that the surface revealed when the B chain separates from the A chain is also involved. It is also possible that the residues at B25-B20 interact with the receptor in some specific conformation.

D. Conclusions

The idea that insulin undergoes conformational changes in its interaction with the receptor considerably complicates the relationship between the hormone's structure and function. However, it may be that the active surfaces themselves are simply exposed at the receptor binding site and do not undergo any significant structural change once this has happened. Thus, it is possible to think in terms of modelling these surfaces in considering and even designing insulin analogues. More helpful must be the strategy of crystallising the insulin receptor and studying the complex it forms with the receptor. These experiments are formidably difficult, but the scientific consequences of their succeeding may be as important as Banting and Best's discovery of insulin itself.

Acknowledgments. This chapter contains the result of studies and discussions on insulin structure by many people. In particular we thank Professor D.C. Hodgkin and acknowledge the work of Mrs. E. Dodson, Dr. J. Brange (Novo), Dr. C.D. Reynolds (Liverpool Polytechnic), Dr. D. Smith (Medical Foundation, Buffalo, United States), Mrs. K. Sparks (Liverpool Polytechnic), Dr. S.P. Tolley, and Dr. D. Vallely.

References

Baker EN, Blundell TL, Cutfield JF, Cutfield SM, Dodson EJ, Dodson GG, Hodgkin DC et al. (1988) The structure of 2Zn pig insulin at 1.5 Å resolution. Philos Trans R Soc Lond [Biol] 319:369–450

Bentley GA, Dodson EJ, Dodson GG, Hodgkin DC, Mercola DA (1976) The structure of insulin in 4Zn insulin. Nature 261:166–168

Bi RC, Dauter A, Dodson EJ, Dodson GG, Giordano F, Reynolds CD (1984) Insulin's structure as a modified and monomeric monomer. Biopolymers 23:391–395

Blundell TL, Dodson GG, Hodgkin DC, Mercola DA (1972) Insulin: the structure in the crystal and its reflection in chemistry and biology. Adv Protein Chem 26:379–402

Bradshaw RA, Dodson GG, Isaacs NW, Niall HD (1980) Possible relationships in the processing, storage and secretion of insulin related peptides. In: Brandenburg D, Vollmer A (eds) Proceedings of the 2nd international insulin symposium. De Gruyter, Berlin, pp 695–701

Brange J, Ribel U, Hanson JF, Dodson GG, Hanson MT, Hawelind S, Melberg S et al. (1988) Monomeric insulins obtained by protein engineering. Nature 333:679–682

Brown H, Sanger F, Kitai R (1955) The structure of sheep and pig insulins. Biochem J 60:556–565

Chothia C, Lesk A, Dodson GG, Hodgkin DC (1983) Transmission of conformational change in insulin. Nature 302:500–505

Cutfield JF, Cutfield SM, Dodson EJ, Dodson GG, Reynolds CD, Vallely D (1981) Similarities and differences in insulin's crystal structures. In: Dodson GG, Glusker J, Sayre D (eds) Structural studies on molecules of biological interest. Oxford University Press, Oxford, pp 527–546

Cutfield JF, Cutfield SM, Dodson GG, Ronco N (1986) Preparation and activity of a nitrated insulin dimer. Int J Pept Protein Res 27:335–347

Derewenda U, Derewenda Z, Dodson EJ, Dodson GG, Reynolds CD, Sparks C, Swenson D (1989) Nature 338:594–596

Dodson GG, Hubbard RE, Reynolds CD (1983) Insulins's structural activity and its relation to activity. Biopolymers 22:281–292

Frank BH, Pekar AH, Veros AJ (1972) Insulin and proinsulin conformation in solution. Diabetes [Suppl 2] 21:486–491

Goldman J, Carpenter F (1974) Zinc binding insulin dichroism and equilibrium sedimentation studies on insulin (bovine) and several of its derivatives. Biochemistry 13:4566–4574

Markussen J (1985) Comparative reduction/oxidation studies with single chain des-(B30) insulin and porcine proinsulin. Int J Pept Protein Res 25:431–434

Milthorpe BK, Nicol LW, Jeffrey PD (1977) The polymerisation pattern of zinc (II) insulin at pH 7. Biochem Biophys Acta 495:195–202

Nanjo K, Sanke T, Miyano M, Okai K, Sowa R, Kondo M, Nishimura S, Iwo K, Miyamura K, Given BD, Chan SJ, Tager HS, Steiner DF, Rubenstein AH (1986) Diabetes due to secretion of a structurally abnormal insulin (Insulin Wakayama). J Clin Invest 77:514–519

Schlichtkrull J (1958) Insulin crystals. Munksgaard, Copenhagen

Steiner DF, Chan SJ, Welsh JM, Nielsen D, Michael J, Tager HS, Rubinstein AH (1986) Models of peptide biosynthesis: the molecular and cellular basis of insulin production. Chem Invest Med 9(4):328–336

CHAPTER 3

Mutant Human Insulins
and Insulin Structure–Function Relationships

H. S. TAGER

A. Introduction

Studies based on heritable defects in human function (sometimes called inborn errors of metabolism) have long played an important role to increase our understanding of mammalian biology in areas as diverse as biochemistry, cell and organismal physiology, and clinical medicine. In fact, studies over very many years of the hemoglobinopathies have formed a paradigm for the analysis of changes in molecular structure which both arise from genetic mutation and lead to important physiological consequences. Early work in the general area emphasized, for technical reasons alone, the results of mutations leading to: (a) structural change in the most abundant blood proteins; and (b) the replacement of amino acid residues that would change the properties of those proteins with respect to charge and electrophoretic mobility. Exceptions (including those involving important intracellular enzymes) are easily identified, but analysis of human tissues for genetic changes that result in concomitantly changed protein structure remained difficult for an extended period. Recent technical and methodological advances (including high performance liquid chromatography, instrumentation for ultramicro protein analysis, and the vast approaches of recombinant DNA analysis), however, have instilled renewed ability and interest to the field. The benefits that afford the study of genetic mutation in humans and of corresponding abnormal proteins are clear. They include determination of the genetic causes and implications of human disease and the importance of detailed protein structure in the attainment of correct protein, cellular, and organismal function. In many ways the analysis of human gene mutations helps to identify genes, proteins, and protein domains of special importance to normal and abnormal mammalian physiology.

Notwithstanding the complexity of metabolic regulation by the endocrine system, and the frequent existence of equally complex counterregulatory responses, the peptide hormones and their corresponding genes represent an especially fruitful area for study of the physiological implications of genetic mutation. In fact, it has been 20 years since ELLIOTT et al. (1966) and KIMMEL and POLLACK (1967) first considered that a structurally altered insulin might lead to abnormal glucose tolerance in at least some diabetic patients. Although insulin represents a favorable example for the study of the consequences of genetic mutation [since its actions in humans are critical to normal function, and since at least one consequence of its action (glucose homeostasis) is easily measured], it must be remembered that: (a) the normal fasting level of insulin in plasma is only about 1 ng/ml; (b) the hormone (like many others) is synthesized in and is secreted from a rather inaccessible endocrine organ; and (c) diabetes (simply considered as a sustained and inappropriate elevation of blood glucose level) can arise from many different causes.

We now know, of course, that insulin gene mutations do occur in humans, and that they can be said to be associated with (but not actually to cause) a relatively rare type of diabetes. Affected subjects show a variety of phenotypes, but the most severely affected generally exhibited mild to moderate degrees of glucose intolerance, higher than normal concentrations of immunoreactive insulin in the plasma, and normal or near normal responses to exogenously administered insulin. While the first two findings might be considered to arise from one sort or another of insulin resistance, the last identifies that affected subjects are apparently resistant only to their own insulin. It can thus be concluded in these cases that the structure of the subjects' insulin has somehow been modified, and that the hormones are not as potent as one would expect. The explanation, as it turns out, is that insulin gene mutations have resulted in the formation of abnormal insulins with altered structures. This chapter will summarize aspects of our knowledge of abnormal human insulins, the genetics and physiological consequences of abnormal insulin gene expression, and the importance of specific amino acid replacements to insulin structure–function relationships.

B. Structural Relationships Within the Human Insulin Gene and its Expressed Products

In order to understand fully the importance of insulin gene mutations and the formation of abnormal insulin gene products, it is important to consider the structures of both the relevant gene and the insulin precursor proinsulin. As has by now been amply documented (BELL et al. 1980; ULLRICH et al. 1980), the transcribed elements of the human insulin gene (illustrated diagrammatically in Fig. 1) consist of: (a) a 5′ untranslated region which is split by an intervening sequence; (b) coding sequences corresponding to the signal peptide (a sequence which is necessary for the cotranslational sequestration of the peptide within the B cell), the insulin B chain, the C peptide (a segment which links the two insulin chains, the gene sequence for which is again split by an intervening sequence), and the insulin A chain; and (c) a 3′ untranslated region. Transcription of the insulin gene yields initially an mRNA precursor which matures to the mRNA product through a process involving excision of the intervening sequences; further processing then yields a translationally effective mRNA (Fig. 1). Translation of the mRNA by ribosomes on the rough endoplasmic reticulum begins with the formation of the amino acid sequence corresponding to the signal peptide (a protein segment which is removed shortly thereafter by a signal peptidase; CHAN et al. 1976; STEINER 1977) and is followed by the formation of the amino acid sequence corresponding to the complete covalent structure of proinsulin. This single-chain precursor of the final hormone product contains structures corresponding to the insulin B chain, the C peptide and the insulin A chain (STEINER

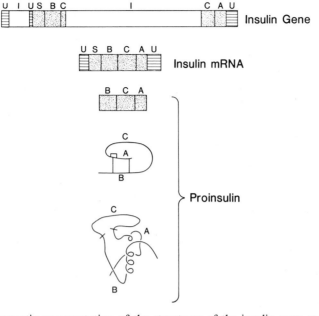

Fig. 1. Diagrammatic representation of the structures of the insulin gene and of its expressed products. The topmost entry shows the structure of the insulin gene; the second, the structure of the mature mRNA resulting from both transscription of the gene and excision of intervening sequences; and the third through fifth, various representations of the structure of proinsulin. Proinsulin is shown, from top to bottom, as the linear product of translation of insulin mRNA; as the folded structure, identifying both the relationships of the B chain, C peptide and A chain segments, and the correct formation of disulfide bonds; and as the crystal structure of insulin would predict. (The crystal structure of proinsulin has not been determined; the bottommost structure should therefore be considered to be very approximate.) Throughout: U untranslated sequence; I intervening sequence; S signal peptide; B insulin B chain; C C peptide; A insulin A chain

et al. 1967, 1969; STEINER 1977); once synthesized, the peptide folds to yield proinsulin in its final form, as illustrated in Fig. 1.

Conversion of proinsulin probably begins within the distal portion Golgi apparatus and proceeds to about 95% completion within secretion granules of the pancreatic B cell (STEINER 1977; ORCI et al. 1985). Whereas the enzymes actually participating in the conversion of proinsulin to insulin are not known with assurance, the chemistry of the process is known with clarity (Fig. 2). That is, two enzyme activities would be required (STEINER et al. 1969, 1980; STEINER 1977; TAGER et al. 1980a): one (an enzyme with trypsin-like specificity) would cleave the precursor at the paired dibasic amino acid junctions which connect the insulin chains to the C peptide (ArgArg or LysArg); the other (an enzyme with carboxypeptidase B-like activity) would remove COOH terminal basic residues from the B chain and the C peptide. Conversion intermediates corresponding to the structures shown in Fig. 2 are known to exist, and can actually be isolated in small amounts from whole pancreatic tissue (STEINER et al. 1969; STEINER 1977; GIVEN et al. 1985). The result of the overall processes described above is that the

Proinsulin

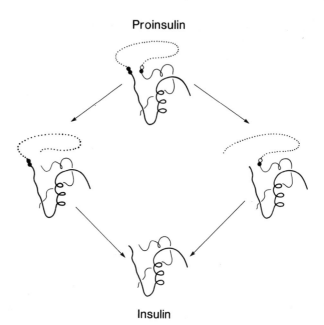

Insulin

Fig. 2. Diagrammatic scheme for the conversion of proinsulin to insulin. Proinsulin is shown at the top. The precursor is cleaved first at either the A chain–C peptide junction (left) or the B chain–C peptide junction (right) by a trypsin-like enzyme followed by the action of a carboxypeptidase B-like enzyme. Subsequent cleavage of the resulting processing intermediates (shown in the middle) on their contralateral sides by the same set of enzymes results in the formation of insulin (bottom) plus free C peptide (not shown). Throughout: *full circles* Arg present in the dibasic amino acid pair occurring at sites for precursor processing; *open circles* Lys present in the dibasic amino acid pair occurring at sites for precursor processing; *thin lines* insulin A chain; *thick lines* insulin B chain; *dotted lines* C peptide

single-chain precursor proinsulin is converted to the two-chain product insulin plus the C peptide.

Reference to Figs. 1 and 2 reveals that changes in structure of the human insulin gene (changes potentially resulting from single nucleotide replacements) could have multiple consequences for the synthesis of the normal gene product. First, changes in the promoter and other regulatory regions might alter the efficiency of overall gene transcription. Second, changes at any of the four intervening sequence–structural sequence junctions might result in improper excision of intervening segments; certain changes within an intervening sequence might be equally deleterious. Third, changes within the 5′ untranslated region might affect negatively mRNA ribosome binding sites and other interactions important for translation. Fourth, changes in the 3′ untranslated region affecting, for example, polyadenylation might decrease the stability and half-life of the translationally effective mRNA. Fifth, changes within the structural gene corresponding to the signal peptide might alter the intracellular movement of nascent proinsulin and the ability of the product eventually to become sequestered in secretion granules. Sixth, changes at sequences corresponding to the paired dibasic amino acids which connect the insulin chains to the C peptide might limit the ability of the precursor proinsulin to be converted to product. Seventh, changes within the structural regions corresponding to the A and B chains might result in amino acid replacements that alter the ability of insulin to combine with its receptor and to effect appropriate biological responses. Examples of insulin gene mutations leading both to improperly cleaved proinsulin conversion intermediates and to final products with decreased biological activity are now known.

C. Abnormal Insulin Gene Products

I. Mutant Human Proinsulins and Processing Intermediates

The syndrome called familial hyperproinsulinemia considers a heritable defect in which human subjects exhibit higher than normal serum levels of immunoactive insulin during fasting (about 100 μU/ml) and in which the vast majority of the immunoreactive insulin is shown by gel filtration to have a molecular weight nearer to that of proinsulin ($M_r = 9000$) than to that of insulin ($M_r = 6000$; GABBAY et al. 1976). The first example of familial hyperproinsulinemia could be traced through several generations, and, although the affected subjects were generally asymptomatic, the defect seemed to represent a situation where the insulin precursor proinsulin was incompletely processed to the expected product. Early studies demonstrated that the circulating material actually represented an intermediate of proinsulin processing, rather than the intact prohormone, and that the intermediate was resistant to conversion to insulin-sized products by the enzyme trypsin (under conditions where the normal precursor was readily processed; GABBAY et al. 1979). It thus seemed that: (a) the high molecular weight immunoreactive material in serum arose from the incomplete conversion of an abnormal proinsulin (proinsulin Boston); (b) the abnormal proinsulin contained an amino acid substitution which limited normal processing; and (c) the substituted product arose from the expression of a mutant human insulin gene. Characterization of the material was of course hindered by the limited amounts of the conversion intermediate which could be isolated from serum by immunoaffinity chromatography.

A second family exhibiting familial hyperproinsulinemia was identified several years later and was shown to have characteristics very similar to those occurring in the family expressing proinsulin Boston (ROBBINS et al. 1981). Application of ultramicro and immunometric methods to the high molecular weight immunoreactive serum insulin from subjects in the second family (subjects expressing proinsulin Tokyo) provided important clues as to the molecular defect in the proinsulin conversion intermediate that escaped complete processing to insulin. It was shown that the conversion intermediate contained a native insulin B chain and a C peptide domain which was connected at its COOH terminal region to the insulin A chain (ROBBINS et al. 1981). The conversion intermediate could thus be identified as an analog of the normal des-Arg[31],Arg[32] intermediate of proinsulin conversion (Fig. 2). Additional experiments, involving chemical modification of the intermediate and subsequent determination of its sensitivity to digestion by trypsin, suggested that the residue of lysine normally occurring in the Lys^{64}-Arg^{65} junction between the C peptide and the insulin A chain was indeed present in the abnormal form, whereas the adjoining residue of arginine was absent. We thus concluded that the intermediate had the general form C peptide-Lys-Xxx-A chain, with the B chain appearing in normal disulfide bond to the A chain (ROBBINS et al. 1981). Molecular cloning and subsequent analysis of the abnormal insulin allele from a subject expressing proinsulin Tokyo (SHIBASAKI et al. 1985) elegantly demonstrated that a single nucleotide change in the condon for Arg^{65} (CGC→CAC) had resulted in the replacement of Arg^{65} by His. As illustrated in Fig. 3, the processing intermediate arising from proinsulin Tokyo could thus be identified as human des-Arg^{31},Arg^{32}-[His^{65}]proinsulin.

Subsequent studies applied methods similar to those initially used for the characterization of the conversion intermediate arising from proinsulin Tokyo to the analysis of the intermediate arising from proinsulin Boston. By several criteria, including radioimmunometric assay for the insulin B chain, chemical modification and enzyme digestion, and reverse-phase, high performance liquid chromatography in two different solvent systems, it was found that the properties of the processing intermediates arising from proinsulin Boston and proinsulin Tokyo were indistinguishable (ROBBINS et al. 1984). It thus appeared that both intermediates contained an amino acid substitution at position 65 in the conversion intermediate des-Arg^{31},Arg^{32}-proinsulin. Since the abnormal allele corresponding to proinsulin Boston has not yet been subjected to molecular cloning and sequence analysis, we cannot be sure that the two abnormal proinsulins in fact contain the same amino acid substitution. Nevertheless, the similarity of the general structures of proinsulins Tokyo and Boston, as shown below, predicts that the two corresponding conversion intermediates will have arisen from related (if not identical) gene mutations and will exhibit similar (again, if not identical) biological properties.

C peptide-Lys-Xxx-insulin A chain
| |
insulin B chain

It is of course the case that dibasic amino acid pairs (Arg-Arg, Lys-Arg, Lys-Lys, or more rarely Arg-Lys) have long been known to occur at related processing sites in a great variety of peptide hormone precursors, and that such dibasic pairs have been held to direct the favorable and correct interaction of peptide hormone precursors with appropriate processing enzymes (STEINER et al. 1980; TAGER et al. 1980a; SCHWARTZ et al. 1983). In fact, the altered structures of proinsulins Tokyo and Boston, and the failure of both to be ultimately converted to insulin in vivo, are consistent with this proposal. Apparently the normal conversion site Arg^{31}-Arg^{32} (connecting the COOH terminus of the B chain to the NH_2 terminus of the C peptide) is correctly identified by the applicable processing enzyme in proinsulin Tokyo/Boston, whereas the modified conversion site Lys^{64}-His^{65} (connecting the COOH terminus of the C peptide to the NH_2 terminus of the A chain) is not. Thus, the identification and analysis of proinsulin

Mutant Proinsulins/Intermediates

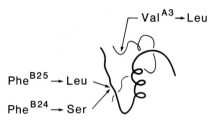

Arg^{65} → His

His^{B10} → Asp

Mutant Insulins

Val^{A3} → Leu

Phe^{B25} → Leu

Phe^{B24} → Ser

Fig. 3. Sites of amino acid replacements in mutant proinsulins, conversion intermediates, and insulins. The approximate locations of the specific replacements are shown on the diagrammatic structures. Note that the mutant products of the insulin gene contain single amino acid substitutions; several examples are shown on a single structure for convenience only. The sequence Arg-Arg in the C peptide–A chain junction (upper panel, left) is shown in parentheses to indicate that these residues have been excised in the conversion intermediate corresponding the proinsulin Tokyo/Boston (replacement of Arg^{65} by His); the exact form of the product resulting from the replacement of His^{B10} by Asp (proinsulin Providence) is not yet known. See text for details. Throughout: *full circles* Arg at proinsulin processing sites; *open circles* Lys at proinsulin processing sites; *thin lines* A chain; *thick lines* B chain; *dotted lines* C peptide

Tokyo/Boston provides an element of proof to the supposition that paired dibasic residues are indeed critical to the normal processing of proinsulin (and probably of other peptide hormone precursors as well) by endogenous converting enzymes. It must be noted in this regard that: (a) the dibasic pair appears to be necessary in the identification of appropriate sites for precursor processing, but appears in itself not to be sufficient (since related pairs in fact occur not infrequently within peptide hormones and other proteins as well); (b) proinsulin Tokyo/Boston, in contrast to proinsulin Providence (see later) contains all of the elements of normal insulin (but of course cannot realize the formation of the normal hormone due to the processing site defect); and (c) although proinsulin Tokyo/Boston is not converted to insulin in affected subjects, the proinsulin arising from the normal insulin gene allele in these heterozygous individuals apparently is processed as expected. The result is that both alleles are expressed, with one gene product eventually yielding des-Arg^{31},Arg^{32}-[His^{65}]proinsulin and the other yielding normal insulin (ROBBINS et al. 1984; SHIBASAKI et al. 1985). Given the relatively low biological activity of normal des-Arg^{31},Arg^{32}-proinsulin

(PEAVY et al. 1985), it is likely that the abnormal circulating proinsulin intermediate arising from proinsulin Tokyo/Boston could contribute only marginally to glucose homeostasis in affected subjects. This marginal contribution, together with the expression of normal insulin from the normal insulin allele (perhaps in conjunction with a greater rate of secretion of total insulin gene-derived products from the pancreatic B cell) no doubt gives rise to the frequently observed normal or near normal glucose tolerance of subjects with familial hyperproinsulinemia.

A third abnormality of insulin gene structure giving rise to elevated levels of 9000 molecular weight insulin-immmunoreactive material in serum concerns the case of proinsulin Providence (GRUPPUSO et al. 1984). The laboratory findings on subjects who express proinsulin Providence are slightly different from those who express proinsulin Tokyo/Boston in that the fraction of insulin-immunoreactive material having high molecular weight ($M_r = 9000$) is somewhat lower (60%–75% of the total, rather than 90%–95%), and that this fraction is readily digested by trypsin to yield insulin-sized immunoreactive peptides. Most important, and again in contrast to findings on subjects with classical familial hyperproinsulinemia, analysis of immunoaffinity-purified material from serum by high performance liquid chromatography suggested that the vast majority of the high molecular weight, insulin-like peptide had the properties of proinsulin rather than those of an intermediate of proinsulin processing. As is often the case in similar studies, insufficient amounts of the peptide were available for detailed chemical analysis. Although it seemed likely at the time that the genetic defect in the example of proinsulin Providence might be related to a structural change in the enzyme participating in proinsulin conversion (rather than to one in the expressed product of the insulin gene), subsequent findings proved quite the opposite. That is, molecular cloning and sequence analysis of both insulin gene alleles from two members of the affected kindred demonstrated that a single base change had occurred within the coding region of one allele in the heterozygous individuals. Surprisingly, the point mutation occurred at the codon normally representing His^{B10} (CAC→GAC), rather than one representing one or another of the paired dibasic residues which occur at sites for precursor processing (CHAN et al. 1987). The result predicts that His^{B10} in the product of the abnormal allele would be replaced by Asp, as illustrated diagrammatically in Fig. 3.

The biochemical and biological consequences of expression of the abnormal insulin gene coding for proinsulin Providence are more difficult to understand than those associated in expression of the abnormal gene coding for proinsulin Tokyo/Boston. As already noted, the majority of 9000 molecular weight insulin-related material in the serum of affected subjects possessing the allele for proinsulin Providence seems to represent intact proinsulin, rather than an intermediate of conversion; small amounts of conversion intermediates, however, are also present (GRUPPUSO et al. 1984). High performance liquid chromatographic analysis of insulin-immunoreactive serum material further indicated that the 9000 molecular weight, proinsulin-related component had the characteristics of normal human proinsulin, and that the 6000 molecular weight insulin present in the same serum had the characteristics of normal human insulin. It must be understood, however, that whereas differences in peptide mobility during high performance liquid chromatography indeed demonstrate important alterations in structure, identical peptide mobilities during equivalent analysis do not necessarily prove structural identity. Unexpected, the biological activity of synthetic human [Asp^{10}]insulin (a possible, but as yet unconfirmed component of the B cell secretions of individuals expressing proinsulin Providence) is actually several times greater than that of native human insulin (SCHWARTZ et al. 1987).

The set of data accumulated in several laboratories on the abnormal gene coding for proinsulin Providence and on the nature of relevant, potentially secreted products, evokes more questions than it does answers. Very important matters remaining to be investigated on the matter of proinsulin Providence include determination of whether: (a) the abnormal and normal insulin alleles in affected subjects are condominantly expressed; (b) the insulin-immunoreactive material present in serum actually represents normal human proinsulin and normal human insulin; (c) the proinsulin precursor corresponding to the abnormal allele is processed in vivo to yield the abnormal insulin [Asp^{B10}]insulin (a form with greater than normal potency); and (d) expression of the abnormal allele somehow interferes with the processing of normal proinsulin to yield normal insulin, as it might in fact interfere with the processing of the abnormal hormone precursor. An intriguing question overall concerns how an insulin gene mutation resulting in replacement of His^{B10} by Asp (a site relatively distant from the paired dibasic amino acid sites intimitely involved in precursor processing) actually modifies the efficiency of conversion of the hormone precursor (perhaps both normal and abnormal precursors) to the final product. Recent studies have demonstrated that D-Ala and Gly serve nearly as well as D-Phe and Phe at position B24 and have emphasized the importance of main chain flexibility in achieving a high affinity state of insulin-receptor interactions (MIRMIRA and TAGER 1989). The answers may take a long time in coming, but they will surely provide important information on insulin gene expression, on proinsulin and insulin structure, and on the biochemical and cellular mechanisms that serve during proinsulin processing.

II. Mutant Human Insulins

The mutant human insulins, like the mutant human proinsulins and related conversion intermediates, arise from single nucleotide changes within the human insulin gene. In these cases, however, the mutations and the corresponding amino acid replacements in the peptide products do not affect precursor processing. The result is a 6000 molecular weight abnormal insulin which, as far as can be told, is secreted from the pancreatic B cell along with normal human insulin in heterozygous individuals. The first detailed analysis of the expressed product of a mutant human insulin gene in fact involved the study of a 6000 molecular weight mutant insulin. The applicable subject exhibited fasting hyperglycemia, fasting insulin levels of about 90 μU/ml, and no signs of insulin resistance; most important, he responded normally to exogenously administered procine insulin (GIVEN et al. 1980). Although small amounts of insulin purified from the patient's serum by immunoaffinity chromatography and quantitated by radioimmunoassay showed normal electrophoretic mobility on nondenaturing polyacrylamide gels, the same insulin exhibited markedly decreased potency (relative to insulin purified from the sera of normal subjects) in binding to the insulin receptors on several different cell types and in stimulating both glucose transport and glucose oxidation in isolated adipocytes.

Evidence demonstrating that the subject identified above actually secreted an abnormal insulin resulting from gene mutation depended, however, on analysis of insulin purified from a 3-g biopsy of the patient's pancreas. The purified pancreatic insulin exhibited a biological potency lower than that measured for insulin purified from a specimen of pancreas obtained from a normal subject at the same time (TAGER et al. 1979). As important, analysis of the pancreatic insulin for its amino acid composition demonstrated that the insulin from the patient suspected of having an insulin gene mutation was relatively deficient in phenylalanine whereas it exhibited a relative excess of leucine. Since phenylalanine was readily identified in the pancreatic insulin at position B1 by NH_2 terminal analysis involving the use 2-dimethylaminonaphthylsulfonyl chloride, and since the COOH terminal tryptic peptide of the insulin B chain could be shown to contain leucine (a residue normally not present in this region of the molecule), we reasoned that the patient's insulin corresponded to a mutant hormone in which leucine replaced either Phe^{B24} or Phe^{B25}. Quantitative amino acid analysis actually suggested that the patient's pancreatic insulin most likely represented an equimolar mixture of normal insulin and the abnormal form which contained a leucine for phenylalaline substitution (insulin Chicago). It thus seemed probable that the isolated pancreatic insulin arose from the codominant expression of both normal and mutant insulin alleles in a heterozygous individual (TAGER et al. 1979).

Recognition that the nucleotide sequence TCTTC was contained within the paired codons for Phe^{B24}-Phe^{B25} (TTCTTC) in the human insulin gene, and that the pentanucleotide sequence represented a site for cleavage of the corresponding DNA by the restriction enzyme Mbo II, permitted further characterization of the insulin genes from the subject secreting insulin Chicago by Southern analysis. Indeed, it was found by Steiner and his colleagues that, whereas one insulin allele in the subject's leukocyte DNA was cleaved by the restriction enzyme (the normal allele), the other (the abnormal allele) was not (KWOK et al. 1981). The result confirmed the existence of an insulin gene mutation in the structural region corresponding to the codons for the sequence Phe^{B24}-Phe^{B25}.

Experiments designed to identify the exact site of replacement of Phe by Leu in insulin Chicago, and to identify the relevant specific genetic change, took two courses. In the first, the method developed by INOUYE et al. (1979) for the semisynthesis of insulin analogs containing amino acid replacements in the COOH terminal region of the insulin B chain was used to construct the two possible abnormal insulins [Leu^{B24}]insulin and [Leu^{B25}]insulin), either one of which could have corresponded to insulin Chicago (TAGER et al. 1980b; INOUYE et al. 1981; KOBAYASHI et al. 1982a). Simultaneous development of a method for the separation of insulins differing from one another by single amino acid replacements (by use of reverse-phase, high performance liquid chromatography and an isocratic solvent system) permitted the use of the semisynthetic insulins as standards for the identification of the natural mutant insulin (Fig. 4; SHOELSON et al. 1983a). Results showed that normal human insulin, human [Leu^{B24}]insulin and human

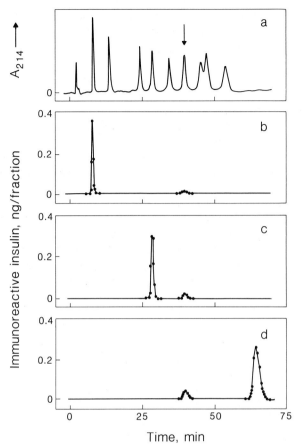

Fig. 4 a–d. Analysis of mutant insulins by reverse-phase high performance liquid chromatography. Details of the chromatographic system have already been described (SHOELSON et al. 1983 a). **a** series of standards comprising various insulins differing from one another by single amino acid replacements; the *downward arrow* indicates the elution position of normal human insulin, **b–d** analysis of serum insulin from patients expressing insulin Los Angeles, insulin Chicago, and insulin Wakayama, respectively. Insulins were identified by optical absorbance (**a**) or by radioimmunoassay (**b–d**); for ease of presentation, data points corresponding to the absence of detectable insulin have been omitted from **b–d**. The figure is similar to one published previously (SHOELSON et al. 1983 a)

[Leu^{B25}]insulin were clearly separated by the chromatographic method, and that the mutant insulin derived from the affected subject cochromatographed with [Leu^{B25}] insulin. Insulin Chicago could thus be identified as human [Leu^{B25}]insulin. The second experimental course used in identifying insulin Chicago involved the molecular cloning of the abnormal insulin allele from the affected subject, and subsequent nucleic acid sequence analysis. KWOK et al. (1983) indeed demonstrated that a single base change had occurred in the patient's abnormal insulin allele in the region coding for Phe^{B25} (TTC→TTG). This result thus agreed with the assignment made on the basis of peptide chemistry, and identified the specific nucleotide change that: (a) occurred as the result of genetic mutation in the affected subject; (b) yielded a modified insulin in which Phe^{B25} had been replaced by Leu; and (c) resulted in the secretion of insulin Chicago (Fig. 3).

Preliminary analysis of the second family associated with the expression of a mutant insulin gene yielding an abnormal insulin resulted in findings very similar to those identified for the first. Immunoaffinity-purified serum insulin showed greatly decreased biological activity (HANEDA et al. 1984) and, indeed, the abnormal serum insulin was eluted during reverse-phase, high performance liquid chromatography at a position very different from that taken by either normal human insulin or insulin Chicago (human $[Leu^{B25}]$insulin; SHOELSON et al. 1983 a). Whereas the elution position of the new abnormal insulin (insulin Los Angeles) indicated that it could be characterized as having an exceedingly hydrophilic nature relative to other forms (Fig. 4), in the absence of further information, the change in hormone hydrophobicity could have been attributed to any one of dozens of potential structural alterations. Fortuitously, however, analysis of the patient's leukocyte DNA, by digestion with the restriction enzyme Mbo II and by Southern analysis, showed that one of the two insulin alleles was cleaved whereas the other was not. The failure of Mbo II to cleave one of the two alleles revealed that the uncut allele was in fact abnormal, and that (for reasons addressed earlier) the abnormal allele had undergone a mutation in the gene sequence which normally codes for the insulin sequence Phe^{B24}-Phe^{B25} (SHOELSON et al. 1983 a). Assuming that only a single base change had occurred, we reasoned from knowledge of the genetic code that only a Ser for Phe substitution at position B24 or B25 could have altered the mobility of insulin Los Angeles on reverse-phase columns in the direction and to the extent observed. While the normal and abnormal insulin alleles from the subject expressing insulin Los Angeles were being studied by molecular cloning and sequence analysis, human $[Ser^{B24}]$insulin and human $[Ser^{B25}]$insulin were prepared by semisynthetic methods. Essentially at the same time, comparisons of the elution position of the natural abnormal insulin with the elution positions of the semisynthetic insulin standards by use of high performance liquid chromatography (SHOELSON et al. 1983 B), and gene sequence analysis (HANEDA et al. 1983) demonstrated that insulin Los Angeles was human $[Ser^{B24}]$insulin. A single nucleotide change in the codon for Phe^{B24} (TTC→TCC) in the abnormal insulin allele thus led to expression of a mutant insulin in which Ser replaced Phe at position B24 (see Fig. 3).

As described above for the first two examples of mutant human insulins, serum insulin from the third affected family also had decreased biological activity. In this case, however, the mutant insulin (insulin Wakayama) appeared to be much more hydrophobic than normal human insulin by reverse-phase high performance liquid chromatographic analysis (Fig. 4; SHOELSON et al. 1983 a), and, in this case, there was no alteration in the ability of the restriction enzyme Mbo II to cleave either insulin allele in the DNA extracted from the leukocytes of affected patients. No definitive structural information was available, and determination of the site and nature of amino acid replacement in insulin Wakayama had to await the elucidation of the insulin gene sequence corresponding to the abnormal allele. Indeed, such analysis demonstrated the existence of a single base change in the abnormal allele corresponding to the codon for Val^{A3}

(GTG→TTG), a change which would result in the replacement of Val at position A3 by Leu (NANJO et al. 1986a). It can thus be said that insulin Wakayama, the third abnormal human insulin, is human [Leu^{A3}]insulin (see Fig. 3). Interestingly, two additional examples of insulin Wakayama have been identified in Japanese subjects who are apparently unrelated to the propositus who formed the basis for these studies (NANJO et al. 1986b; IWAMOTO et al. 1986a, b). In one case, the molecular abnormality in the mutant insulin was identified by molecular cloning and DNA sequence analysis, whereas in the other, the abnormality was identified by protein sequence analysis of insulin isolated from a specimen of pancreas. While the characterization of three possibly separate occurrences of mutant insulin alleles expressing insulin Wakayama (human [Leu^{A3}]insulin) is itself of great interest, it is also noteworthy that analysis of the isolated pancreatic insulin from one patient secreting insulin Wakayama (like that from the patient secreting insulin Chicago) demonstrated the existence of equimolar amounts of normal and mutant insulins. It thus appears that heterozygosity and codominant expression of normal and abnormal alleles can be considered to be the rule, rather than the exception, in cases of the mutant insulin syndrome.

D. Insulin Structure–Function Relationships and Mutant Insulins

As noted earlier, the abnormal insulins arising from mutations in the human insulin gene, including insulin Chicago (human [Leu^{B25}]insulin), insulin Los Angeles (human [Ser^{B24}]insulin) and insulin Wakayama (human [Leu^{A3}]insulin), exhibit markedly decreased biological activity (usually only about 1% of that found for normal human insulin). Most important, all three of the mutant human insulins have been synthesized by chemical or combined chemical and enzymatic means, and all three show decreases in biological activity parallel to corresponding decreases in receptor binding affinity. That is: (a) the insulin receptors on a variety of cells exhibit decreased affinity for the abnormal insulins; (b) the abnormal insulins stimulate appropriate cellular responses fully (given high enough concentrations of the hormone analogs); and (c) the decreased biological potencies of the insulins arise specifically from their decreased affinities for the insulin receptor. Thus, the sites of amino acid replacement corresponding to insulins Chicago, Los Angeles and Wakayama identify what might be considered as "hot spots" for biologically important regions of the molecule. Indeed amino acid replacements in these mutant insulins occur within two essentially invariant tetrapeptide sequences: Gly^{B23}-Phe^{B24}-Phe^{B25}-Tyr^{B26} and Gly^{A1}-Ile^{A2}-Val^{A3}-Glu^{A4}. (The only exceptions to the invariance are the insulins from the coypu and dogfish, in which Phe^{B25} has essentially been deleted, and those from a few other species where Glu^{A4} is replaced by Asp.) Both of these insulin domains, as well as a domain corresponding to the extreme COOH terminal region of the insulin A chain, have long been known to play important roles in conferring biological activity on the hormone (Fig. 5a). All considered, the contributions of abnormal insulins to diabetes or to glucose intolerance could only arise from circumstances in which the mutant gene products have been altered at critical sites and in important ways.

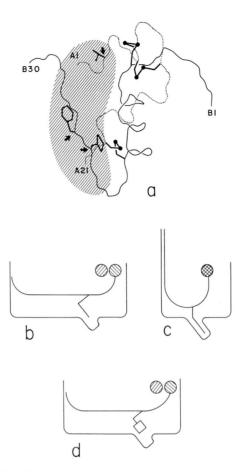

Fig. 5 a–d. Aspects of insulin–receptor interactions. **a** structure of insulin showing by *hatching* the domains thought to be most important in conferring receptor binding and biological potency; the side chains of residues replaced in three mutant insulins are shown explicitly, and are identified by *arrows*, **b–d** diagrammatic representation of potential insulin–receptor interactions and related conformational changes, **b** insulin in low affinity association with receptor, a state resulting from recognition of the ligand and from a potential negative effect of the COOH terminal region of the B chain on ligand–receptor interaction, **c** insulin in high affinity association with receptor, a state resulting from concerted conformational changes involving both receptor and ligand, and requiring the presence of Phe^{B25}, **d** an insulin analog containing an amino acid replacement at position B25 in low affinity association with receptor. (Concerted conformational change in the interaction shown in **d** is not possible due to the absence of a β-aromatic ring in the side chain at position B25.) The *hatched areas* in **b–d** are meant to indicate potential, but unidentified, binding sites on insulin and receptor. Aspects of these diagrams have been taken from published work (DeMeyts et al. 1978; Blundel et al. 1972; Dodson et al. 1983; Nakagawa and Tager 1986)

I. Substitutions at Position A3 (Insulin Wakayama)

The NH_2 terminal region of the insulin A chain possesses several structural attributes of importance to insulin-receptor interactions. These include: (a) a requirement for a free α-amino group; (b) a proscription against side chain mass in the L configuration at position A1; (c) a helical region spanning residues A2-A8; (d) a potential electrostatic interaction between the side chain carboxylate of residue A4 and the amino group of Gly^{A1} or Lys^{B29}; and (e) a potential van der Waals interaction between the side chains of residues Ile^{A2} and Tyr^{A19} (BLUNDELL et al. 1972). The NH_2 terminal region of the insulin A chain (a region which contains the replacement of Val^{A3} by Leu in insulin Wakayama) seems also to dictate the conformation of other aspects of the insulin molecule (since, for example, structural manipulations in the region can alter the ability of the hormone to aggregate into dimers; KITAGAWA et al. 1984a). Considerable attention has been placed, in fact, on the synthesis and study of insulin analogs containing modifications of the α-amino function at residue Gly^{A1} or containing substitutions of residue Gly^{A1} or residue Ile^{A2}. Combined results demonstrate that the potential for positive charge at the α position of residue A1 is important for activity and that, whereas replacement of Gly^{A1} by a variety of L-amino acids has deleterious effects on biological activity, similar replacements with D-amino acids are actually well tolerated (COSMATOS et al. 1978). Replacement of Ile^{A2} by Ala or even by the isoleucine isomer norleucine causes very significant decreases in receptor binding and biological potency in the corresponding analogs (KITAGAWA et al. 1984a, b). While we know, of course, that replacement of Val^{A3} by Leu (in insulin Wakayama) causes a very marked decrease in biological activity (to about 1% of that of normal insulin; KOBAYASHI et al. 1984a), few related analogs have been examined. The substitution of Val with Leu certainly would not alter the general hydrophobic character of the region, but it could (perhaps by the elimination of side chain mass at the β position) induce a significant change in the conformation of the A chain helix, modify the potential for interaction of residue A2 with Tyr^{A19}, or otherwise alter the conformational relationships between the main chains and side chains of the insulin molecule.

II. Substitutions at Position B24 (Insulin Los Angeles)

As noted earlier, Phe^{B24} lies within a tetrapeptide sequence of insulin (Gly^{B23}-Phe^{B24}-Phe^{B25}-Tyr^{B26}) that has remained essentially invariant throughout evolution. In fact, replacements of Phe^{B24} with Ser (in insulin Los Angeles), Ala, Leu, or other residues all cause marked decreases in receptor binding and biological activity (to about 1%–10% of the corresponding activity of insulin, depending on the nature of the replacement). Given the very different character of side chains replacing the benzyl group in these low activity analogs, it seemed likely that steric considerations (rather than those relating to specific interactions or to the hydrophobic effect) would eventually explain the importance of Phe^{B24} in conferring high affinity to insulin–receptor interactions. Most important in this regard, early studies on relevant semisynthetic analogs (studies involving the analysis of circular dichroic spectra and of rates of tyrosine radioiodination) demonstrated that insulins substituted at position B24 exhibit a conformation in solution significantly different from that of the native hormone (WOLLMER et al. 1981; ASSOIAN et al. 1982). The applicable conformational change actually interferes with the ability of the hormone to dimerize in the usual way, and most probably results from the propagation of structural change to regions of the molecule quite distant from the site of amino acid replacement.

Reference to the crystallographic structure of insulin reveals that the side chain of residue Phe^{B24} lies at the surface of a hydrophobic pocket formed by side chains of both the insulin A chain and the insulin B chain (BLUNDELL et al. 1972; DODSON et al. 1983; GAMMELTOFT 1984). Careful analysis of the potential structures of insulin analogs containing position B24 replacements by WOLLMER et al. (1981) in fact identified the probable structural cause for the low biological activities of B24-substituted insulins. It seems that the packing of side chains within this region would not allow replacement of Phe^{B24} by Leu unless there were concomitant changes in the conformations of the main chain or of other side chains, particularly the side chain of Leu^{B15}. That is: (a) simple replacement of Phe^{B24} by Leu would result in an unacceptable close contact between the side chains of Leu^{B24} and Leu^{B15}; (b) structural changes would be necessary to accommodate the two side chains; and (c) conformational changes could well be propagated to distant regions of the molecule by both side chain rotations and main chain movements. The γ tetrahedral carbon of Leu^{B24} (a carbon the equivalent of which in phenylalanine is planar) seems to play an especially important role in affecting the fit of side chains within the region. Since replacements of Phe^{B24} by residues with smaller side chains cause equivalent, or even more severe, decreases in the ability of insulin to combine with receptor (and induce related structural perturbations throughout the molecule), it seems that the overall fit of side chains within the region may be critical to insulin's structure and potential for receptor binding (SHOELSON et al. 1983b; KOBAYASHI et al. 1984a). Although Phe^{B24} appears to be uniquely suited to maintaining the normal structure of insulin (and to orienting, as necessary, the COOH terminal domain of the B chain in relation to the rest of the molecule), it has been shown that [D-Phe^{B24}]insulin exhibits an affinity for receptor interaction nearly twice that of the normal hormone (KOBAYASHI et al. 1982b). Given the apparent importance of the L-phenylalanine side chain at the same position in normal insulin, the high activity of insulin substituted with the D-amino acid is surprising. The explanation is not yet known, but will undoubtedly provide further insights into the important contributions of the position B24 side chain to normal insulin structure–function relationships. Recent studies have demonstrated that the amino acid substitution unexpectedly directs both the rapid secretion of the unprocessed precursor from the B cell via an unregulated pathway and the appearance of the mutant proinsulin in the extracellular space (CARROLL et al. 1988).

III. Substitutions at Position B25 (Insulin Chicago)

Although Phe^{B25} occurs within the same invariant insulin sequence as Phe^{B24}, the structural dispositions of the two relevant side chains differ considerably. That is, whereas the side chain of Phe^{B24} is located partly in association with insulin's hydrophobic pocket, that of Phe^{B25} lies on the monomer surface (BLUNDELL et al. 1972); in fact, the Phe^{B25} side chain is in a position suitable for direct contact with receptor. As important when considering the structural and functional impact of molecular replacements at position B25: (a) B25-substituted insulins (unlike their B24-substituted counterparts) seem to retain normal conformation (WOLLMER et al. 1981; ASSOIAN et al. 1982); and (b) insulins in which Phe^{B25} is replaced by Leu (a branched-chain, hydrophobic residue, as in insulin Chicago) or by Ser (a small hydrophilic residue) show nearly equivalent and severe decreases in receptor binding potency (to about 1% of that of normal insulin; SHOELSON et al. 1983b; KOBAYASHI et al. 1984a). The apparent absence of propagated structural changes in the case of insulin Chicago allowed us to evaluate in some detail the functional and structural importance of the Phe^{B25} side chain in insulin action.

Since the loss of the phenylalanine side chain in position B25-substituted insulins (rather than the nature of the replacement) seemed to be most important in determining related losses in receptor binding potency, we initially prepared and examined a series of analogs in which Phe^{B25} had been replaced by a variety of unnatural aromatic amino acid residues. Our results showed that replacement of Phe by α- or β-naphthylalanine or by p-methylphenylalanine resulted in analogs with nearly normal receptor binding potency, whereas replacement of the same residue by homophenylalanine (an amino acid containing an additional methylene group between the β-carbon and the aromatic ring of phenylalanine) yielded an analog exhibiting only 1% of normal receptor binding and biological potency (NAKAGAWA and TAGER 1986). These studies indicated that an aromatic residue at position B25 was indeed beneficial to normal insulin-receptor interactions, but that the relevant aromatic ring was effective only when it was located in the β position. Related experiments involved the study of insulin analogs in which amino acid residues were deleted from the COOH terminal domain of the insulin B chain. Results showed that the COOH terminal pentapeptide of the B chain could be deleted without loss of activity, but that deletion of Phe^{B25} (in addition to residues B26-B30) gave rise to an analog with as much as 6% of normal potency (NAKAGAWA and TAGER 1986). Although the decrease in activity was expected (since the importance of Phe^{B25} in insulin-receptor interactions had already been demonstrated), the limited magnitude of the decrease was surprising. That is, deletion of residues B25–B30 resulted in an analog with greater potency than one produced by simple substitution of residue B25 in a full-length insulin (for example, with Leu in insulin Chicago). It thus seemed that a functional relationship (actually a negative one) might exist between an amino acid substitution at position B25 and the seemingly otherwise unnecessary COOH terminal domain of the B chain.

To test further the hypothesis presented above, we examined additional full-length and truncated insulin analogs bearing amino acid substitutions at position B25. Our results showed, as now expected, that deletions involving the COOH terminal pentapeptide of the insulin B chain did not alter the receptor binding properties of insulin analogs when position B25 was filled with a residue bearing a β-aromatic ring. Nevertheless equivalent deletions in analogs containing position B25 replacements actually increased the potencies of the parent insulin analogs very significantly (NAKAGAWA and TAGER 1986). For example, deletion of residues B26-B30 in [Ser^{B25}]insulin (a full-length analog with only 1% of normal potency) resulted in a truncated analog (des-(B26-B30)-[Ser^{B25}]insulin) with 50% of the receptor binding and biological potency of normal insulin. It thus seems that a potentially negative relationship indeed exists between the structures of the position B25 side chain and the COOH terminal B chain domain, but that the relationship is expressed only when position B25 is filled by a residue other than Phe or one of its β-aromatic cognates (NAKAGAWA and TAGER 1986). Further experiments attempted to identify the molecular locus for the negative contribution of the COOH terminal B chain domain to insulin-receptor interactions (NAKAGAWA and TAGER 1987). Two sets of analogs were prepared: the first included insulins that were substituted at position B25 with Ser and were shortened from their carboxy termini by one, two, three, four, or five residues; the second included insulins that were substituted with Ser at position 25 and also with Ala at position B26 or B27. Results showed that: (a) up to three residues could be removed from the carboxy terminus of the B chain in [Ser^{B25}]insulin without major effect on insulin–receptor interactions; (b) removal of residues Thr^{B27} and Tyr^{B26} resulted in incremental increases in receptor binding potency; and (c) replacement of Thr^{B27} or Tyr^{B26} by Ala had no effect to reverse the negative contribution of the COOH terminal B chain domain in insulin-receptor interactions involving analogs containing Ser for Phe substitutions at position B25. It was thus concluded that the negative contribution of the insulin B chain derives from the region containing residues B26 and B27, and most likely involves the participation of the Tyr^{26}-Thr^{27} peptide bond, rather than either of the two relevant side chains (NAKAGAWA and TAGER 1987).

All considered, the detailed study of insulins substituted at position B25 has contributed in several ways to our understanding of mutant human insulins and of insulin structure–function relationships. First, it has provided a still preliminary, but plausible, analysis of the molecular causes of altered function in patients secreting insulin Chicago and in others who might be identified in the future to secrete different insulins in which Phe^{B25} is replaced by any of a variety of amino acids. Second, it explains, to some degree, the functional and structural attributes of insulin that limit the kinds of structural changes that can be accepted during evolution. On one hand, accumulated results suggest that some replacements at position B25 might be tolerated, but only under a condition where the very COOH terminal B chain domain has been deleted; the probability of these related changes occurring simultaneously (especially while still allowing the correct folding and processing of the resulting shortened proinsulin), however, would be very low. On the other hand, related results help to explain how the insulin from the coypu (one in which Phe^{B25} has been deleted rather than having been replaced) retains appropriate activity; it appears that Tyr^{B25} in coypu insulin can take the place of Phe^{B25} (since it possesses a β-aromatic ring), and that the remaining residues in the COOH terminal region of the B chain are actually of very little specific importance. Third, the detailed study of insulins substituted at position B25 has provided a model for insulin–receptor interactions that involves molecular movements among different domains of the insulin molecule, and concerted conformational changes that involve both hormone and receptor

(Fig. 5 b–d). While the model is still tentative, it does account for experimental findings, and it provides a functional framework for understanding both how insulin achieves a high affinity state of interaction with receptor and how it effects the transduction of its signal across the plasma membrane through ligand–receptor interactions. Importantly, the flexibility of the insulin molecule and its potential for propagated structural change have long been recognized (DODSON et al. 1979; CHOTHIA et al. 1983; SMITH et al. 1984). In fact, several crystallographic structures of insulins are known [including those corresponding to molecules I and II of both 2Zn and 4Zn insulin, and to des-(B26-B30)insulin], and each has been found to differ from the others in moderate to very substantial ways. Many of these conformers of insulin in fact exhibit pronounced side chain rotations at Phe^{B25}, in addition of course to other rotations and to main chain movements. Nevertheless, it is not yet clear whether the proposed conformational changes in insulin as analyzed by peptide chemistry actually mirror those that have been identified by crystallographic methods.

E. Physiology, Genetics, and Clinical Aspects

While a review of the genetics, physiology and clinical implications of mutant insulin genes and abnormal insulin or proinsulin secretion is not a major purpose of this chapter, a few related topics should be addressed (see also GIVEN et al. 1980; HANEDA et al. 1984). First, as noted above, all affected individuals studied to date have been heterozygous for normal and abnormal insulin alleles; given the very low biological activity of the mutant gene products (generally only a few percent or less of that of normal insulin, see earlier) it seems probable that any extremely rare case of homozygosity would result in very severe metabolic disturbance. Second, the pattern of inheritance of the abnormal allele in those families spanning two or more generations is that of an autosomal gene; the insulin gene is present on chromosome 11. Interestingly, in the family expressing the gene for insulin Los Angeles (one in which three generations have been studied): (a) the mutant allele was traced to the father of the propositus; (b) all three siblings of the propositus (a sister and two brothers) inherited the paternal mutant allele; and (c) the sole member of the third generation was also found to have inherited the abnormal allele (SHOELSON et al. 1984; HANEDA et al. 1984). Third, as noted above, in the two cases for which the opportunity for study has existed (one involving insulin Chicago and the other insulin Wakayama), evidence has been obtained for the codominant expression of normal and abnormal alleles. Fourth, in a single study involving unbiased screening for potential abnormal insulin alleles within 400 members of the diabetic population (a study accomplished by use of restriction enzymes, cleavage of leukocyte DNA, and Southern analysis), a single mutant allele was identified; nevertheless this allele represented a replacement of one phenylalanine codon for another (a silent mutation) and did not result in the production of a mutant insulin (SANZ et al. 1985). Fifth, given the apparently codominant expression of normal and abnormal alleles in affected subjects, it must be recognized that the hyperinsulinemia which is characteristic of the mutant human insulin syndrome arises exclusively from the presence of very high levels of mutant insulin in the circulation (SHOELSON et al. 1984). We now know that the low receptor binding potency of abnormal insulin gene products, coupled with the overwhelming importance of ligand-receptor interactions in insulin degradation (TERRIS and STEINER 1975, 1976), leads: (a) to decreased degradation of the mutant hormones at insulin target tissues; (b) to prolongation of their half-lives and rates of metabolic clearance; and (c) to their accumulation in the blood (KOBAYASHI et al. 1982a, 1984b; ASSOIAN et al. 1982; SHOELSON et al. 1984; HANEDA et al. 1985). Concomitantly secreted normal insulin, as the result of its high affinity for the insulin receptor and its high biological activity, is apparently cleared from the blood at normal rates, and (in the absence of complicating factors) appears in the blood in expected amounts.

Last, it is important to consider the association of abnormal insulin gene products with diabetes. While the products of abnormal insulin genes within each kindred have most often been identified initially in a diabetic or glucose-intolerant subject, the penetrance of diabetes in genetically affected individuals is incomplete. That is, the phenotype of diabetes does not itself follow a clear autosomal dominant pattern of inheritance (GIVEN et al. 1980; HANEDA et al. 1984). For example, in the case of insulin Los Angeles, only two of six individuals expressing the abnormal insulin gene can be considered to be overtly diabetic. One of these individuals (the propositus) was shown to have a markedly blunted insulin secretory response to oral glucose, whereas two of her asymptomatic siblings (individuals with at worst borderline glucose intolerance) were shown to respond to oral glucose with the robust and sustained secretion of both normal and abnormal forms of the hormone (SHOELSON et al. 1984). It thus appears that the latter individuals were able to compensate for the fact that half of their secreted insulin was effectively biologically inactive by secreting a greater amount of the total hormone. Whether or not these genetically affected subjects will retain this compensating ability throughout their lives remains to be seen. All considered, it must be said that abnormal insulin genes and their products can be associated with diabetes, but that they do not cause the disease. Affected individuals suffer essentially a 50% deficit in the biological activity of secreted insulin. Overt diabetes seems to occur, however, only in the presence of a compounding defect in the pancreas-insulin-target cell axis, whether the compounding abnormality is at the level of the pancreatic B cell (a secretion defect) or at that of the periphery (a defect in insulin sensitivity).

F. Concluding Remarks

While our knowledge of mutant human insulin genes and of mutant human insulins has accumulated over a period of less than 10 years, it already seems difficult to summarize their impact and implications. As noted in Sect. A, very substantial improvements in technique and experimental approach have made possible the determination of related gene and protein structures, the chemical synthesis of abnormal hormones, the determination of physiological responses to coequally secreted normal and mutant insulins, and the thorough investigation of the clinical correlates of abnormal insulin and proinsulin secretion. It must be said, in addition, that the speed at which our understanding of the area has advanced had depended greatly on scientific teamwork and on cooperation among clinicians, biochemists, and molecular biologists in the United States, Japan, Germany, England, and elsewhere. There seem to be relatively few examples as clear where a complex clinical syndrome has led so quickly to understanding at the genetic level and to the asking of questions fundamental to many aspects of biological function. While the story of mutant human insulins is not yet complete, one can hope that equivalent understanding will soon develop on the matter of relevant gene mutations and abnormal hormone production as applied to the many scores of peptide hormones and other bioactive peptides that are known to exist and to play critical roles in human physiology.

Acknowledgments. I thank the many colleagues and co-workers who have contributed to our understanding of abnormal human insulin genes and their products. Particular thanks are due to Drs. Kenneth Polonsky, Arthur Rubenstein and Donald Steiner. Work performed in this laboratory was supported by grants DK 18347 and DK 20595 from the National Institutes of Health.

References

Assoian RK, Thomas NE, Kaiser ET, Tager HS (1982) [Leu^{B25}]insulin and [Ala^{B24}]insulin: altered structures and cellular processing of B24-substituted insulin analogs. Proc Natl Acad Sci USA 79:5147–5151

Bell GI, Pictet RL, Rutter WJ, Cordell B, Tischer E, Goodman HM (1980) Sequence of the human insulin gene. Nature 284:26–32

Blundell T, Dodson G, Hodgkin D, Mercola D (1972) Insulin: the structure in the crystal and its reflection in chemistry and biology. Adv Protein Chem 26:279–402

Carroll RJ, Hammer RE, Chan SJ, Swift HH, Rubenstein AH, Steiner DF (1988) A mutant human proinsulin is secreted from islets of langerhans in increased amounts via an unregulated pathway. Proc Natl Acad Sci USA 85:8943–8947

Chan SJ, Keim P, Steiner DF (1976) Cell-free synthesis of rat preproinsulins: characterization and partial amino acid sequence determination. Proc Natl Acad Sci USA 73:1964–1968

Chan SJ, Seino SU, Gruppuso PA, Schwartz R, Steiner DF (1987) A mutation in the B chain coding region is associated with impaired proinsulin conversion in a family with hyperproinsulinemia. Proc Natl Acad Sci USA 84:2194–2197

Chothia C, Lesk AM, Dodson G, Hodgkin DC (1983) Transmission of conformational change in insulin. Nature 302:500–505

Cosmatos A, Cheng K, Okada Y, Katsoyannis PG (1978) The chemical synthesis and biological evaluation of [1-L-alanine-A]- and [1-D-alanine-A]insulins. J Biol Chem 253:6586–6590

DeMeyts P, van Obberghen E, Roth J, Wollmer A, Brandenburg D (1978) Mapping of the residues responsible for the negative cooperativity of the receptor-binding region of insulin. Nature 273:504–509

Dodson EJ, Dodson GG, Hodgkin DC, Reynolds CD (1979) Structural relationships in the two-zinc insulin hexamer. Can J Biochem 57:469–479

Dodson EJ, Dodson GG, Hubbard RE, Reynolds CD (1983) Insulin's structural behavior and its relationship to activity. Biopolymers 22:281–291

Elliott RB, O'Brien D, Roy CC (1966) An abnormal insulin in juvenile diabetes mellitus. Diabetes 14:780–787

Gabbay KH, DeLuca K, Fisher NJ Jr, Mako ME, Rubenstein AH (1976) Familial hyperproinsulinemia: an autosomal dominant defect. N Engl J Med 249:911–915

Gabbay KH, Bergenstal RM, Wolff J, Mako ME, Rubenstein AH (1979) Familial hyper-proinsulinemia: partial characterization of circulating proinsulin-like material. Proc Natl Acad Sci USA 76:2882–2885

Gammeltoft S (1984) Insulin structure and function. Physiol Rev 64:1321–1378

Given BD, Mako ME, Tager HS, Baldwin D, Markese J, Rubenstein AH, Olefsky J et al. (1980) Diabetes due to secretion of an abnormal insulin. N Engl J Med 302:129–135

Given BD, Cohen RM, Shoelson SE, Frank BH, Rubenstein AH, Tager HS (1985) Biochemical and clinical implications of proinsulin conversion intermediates. J Clin Invest 76:1398–1405

Gruppuso PA, Gorden P, Kahn RC, Cornblath M, Zeller WP, Schwartz R (1984) Familial hyperproinsulinemia due to a proposed defect in conversion of proinsulin to insulin. N Engl J Med 311:629–634

Haneda M, Chan SJ, Kwok SCM, Rubenstein AH, Steiner DF (1983) Studies on mutant insulin genes: identification and sequence analysis of a gene encoding [Ser^{B24}]insulin. Proc Natl Acad Sci USA 80:6366–6370

Haneda M, Polonsky KS, Bergenstal RM, Jaspan JB, Shoelson SE, Blix PM, Chan SJ et al. (1984) Familial hyperinsulinemia due to a structurally abnormal insulin: definition of an emerging new clinical syndrome. N Engl J Med 310:1288–1294

Haneda M, Kobayashi M, Maegawa H, Watanabe N, Takata Y, Ishibashi O, Shigeta Y, Inouye K (1985) Decreased biological activity and degradation of human [Ser^{B24}]insulin, a second mutant insulin. Diabetes 34:568–573

Inouye K, Watanabe K, Morihara K, Tochino Y, Kanaya T, Emura J, Sakakibara S (1979) Enzyme-assisted semisynthesis of human insulin. J Am Chem Soc 101:751–752

Inouye K, Watanabe K, Tochino Y, Kobayashi M, Shigeta Y (1981) Semosynthesis and properties of some insulin analogs. Biopolymers 20:1845–1858

Iwamoto Y, Sakura H, Yui R, Fujita T, Sakamoto Y, Matsuda A, Kuzuya T (1986a) Identification and characterization of a mutant insulin isolated from the pancreas of a patient with abnormal insulinemia. Diabetes [Suppl 1] 35:77A

Iwamoto Y, Sakura H, Ishii Y, Yamamoto R, Kumakura S, Sakamoto Y, Masuda A, Kuzuya T (1986b) Radioreceptor assay for serum insulin as a useful method for detection of abnormal insulin with a description of a new family of abnormal insulinemia. Diabetes 35:1237–1242

Kimmel JR, Pollack HG (1967) Studies of human insulin from nondiabetic and diabetic pancreas. Diabetes 16:687–694

Kitagawa K, Ogawa H, Burke GT, Chanley JD, Katsoyanis PG (1984a) Critical role of the A^2 amino acid residue in the biological activity of insulin. Biochemistry 23:1405–1413

Kitagawa K, Ogawa H, Burke GT, Chanley JD, Katsoyanis PG (1984b) Interaction between the A^2 and A^{19} amino acid residues is of critical importance for high biological activity in insulin. Biochemistry 23:4444–4448

Kobayashi M, Ohgaku S, Iwasaki M, Maegawa H, Shigeta Y, Inouye K (1982a) Characterization of [Leu^{B24}]- and [Leu^{B25}]insulin analogs. Biochem J 206:597–603

Kobayashi M, Ohgaku S, Iwasaki M, Maegawa H, Shigeta Y, Inouye K (1982b) Supernormal insulin: [D-Phe^{B24}]insulin with increased affinity for insulin receptors. Biochem Biophys Res Commun 107:329–336

Kobayashi M, Haneda M, Maegawa H, Watanabe N, Takato Y, Shigeta Y, Inouye K (1984a) Receptor binding and biological activity of [Ser^{B24}]insulin, an abnormal mutant insulin. Biochem Biophys Res Commun 119:49–57

Kobayashi M, Haneda M, Ishibashi O, Takata Y, Maegawa H, Watanabe N, Shigeta Y (1984b) Prolonged disappearance rate of a structurally abnormal mutant insulin from the blood. Diabetes [Suppl 1] 33:17A

Kobayashi M, Takata Y, Ishibashi O, Sasoka T, Iwasaki M, Shigeta Y, Inouye K (1986) Receptor binding and negative cooperativity of a mutant insulin [Leu^{A3}]insulin. Biochem Biophys Res Commun 137:250–257

Kwok SCM, Chan SJ, Rubenstein AH, Poucher R, Steiner DF (1981) Loss of restriction endonuclease cleavage site in the gene of a structurally abnormal insulin. Biochem Biophys Res Commun 98:844–849

Kwok SCM, Steiner DF, Rubenstein AH, Tager HS (1983) Identification of the mutation giving rise to insulin Chicago. Diabetes 32:872–875

Mirmira RG, Tager HS (1989) Role of the phenylalanine B24 side chain in directing insulin interaction with its receptor. J Biol Chem 264:6349–6354

Nakagawa SH, Tager HS (1986) Role of the phenylalanine B25 side chain in directing insulin interaction with its receptor. J Biol Chem 261:7332–7341

Nakagawa S, Tager HS (1987) Role of the COOH-terminal B-chain domain in insulin-receptor interactions. J Biol Chem 262:12054–12058

Nanjo K, Sanke T, Miyano M, Okai K, Sowa R, Kondo M, Nishimura S et al. (1986a) Diabetes due to secretion of a structurally abnormal insulin (insulin Wakayama). J Clin Invest 77:514–519

Nanjo K, Given B, Sanke T, Kondo M, Miyano M, Okai K, Miyama K et al. (1986b) Pancreatic function in the mutant insulin syndrome. Diabetes [Suppl 1] 35:77A

Orci L, Ravazzola M, Amherdt M, Madsen O, Vassalli J-D, Perrelet A (1985) Direct identification of prohormone conversion site in insulin-secreting cells. Cell 42:671–681

Peavy DE, Brunner MR, Duckworth WC, Hooker CS, Frank BH (1985) Receptor binding and biological potency of several split forms (conversion intermediates) of human proinsulin. J Biol Chem 26:13989–13994

Robbins DC, Blix PM, Rubenstein AH, Kanazawa Y, Kosaka K, Tager HS (1981) A human proinsulin variant at arginine 65. Nature 291:679–681

Robbins DC, Shoelson SE, Rubenstein AH, Tager HS (1984) Familial hyper-proinsulinemia: two cohorts secreting indistinguishable type II intermediates of proinsulin conversion. J Clin Invest 73:714–719

Sanz N, Karam JH, Horita S, Bell GI (1985) DNA screening for insulin gene mutations in non-insulin-dependent diabetes mellitus (NIDDM). Diabetes [Suppl 1] 34:85A

Schwartz GP, Burke GT, Katsoyanis PG (1987) A superactive insulin: [B10-aspartic acid]insulin (human). Proc Natl Acad Sci USA 84:6408–6411

Schwartz TW, Wittels B, Tager HS (1983) Hormone precursor processing in the pancreatic islet. In: Hruby VJ, Rich DH (eds) Peptides: structure and function. Pierce Chemical Company, Rockford, pp 229–238

Shibasaki Y, Kawakami T, Kanazawa Y, Akamura Y, Takaku T (1985) Posttranslational cleavage of proinsulin is blocked by a point mutation in familial hyperproinsulinemia. J Clin Invest 76:378–380

Shoelson S, Haneda M, Blix P, Nanjo K, Sanke T, Inouye K, Steiner D et al. (1983a) Three mutant insulins in man. Nature 302:540–543

Shoelson S, Fickova M, Haneda M, Nahum A, Musso G, Kaiser ET, Rubenstein AH, Tager HS (1983b) Identification of a mutant insulin predicted to contain a serine-for-phenylalanine substitution. Proc Natl Acad Sci USA 80:7390–7394

Shoelson SE, Polonsky KS, Zeidler A, Rubenstein AH, Tager HS (1984) Human insulin (Phe→Ser): secretion and metabolic clearance of the abnormal insulin in man and in a dog model. J Clin Invest 73:1351–1358

Smith GD, Swenson DC, Dodson EJ, Dodson GG, Reynolds CD (1984) Structural stability in the 4-zinc human insulin hexamer. Proc Natl Acad Sci USA 81:7093–7097

Steiner DF (1977) Insulin today. Diabetes 26:322–340

Steiner DF, Cunningham DD, Spigelman S, Aten B (1967) Insulin biosynthesis: evidence for a precursor. Science 157:697–700

Steiner DF, Clark JL, Nolan C, Rubenstein AH, Margoliash E, Aten B, Oyer PE (1969) Proinsulin and the biosynthesis of insulin. Recent Prog Horm Res 25:207–282

Steiner DF, Quinn PS, Chan SJ, Marsh J, Tager HS (1980) Processing mechanisms in the biosynthesis of proteins. Ann NY Acad Sci 343:1–16

Tager HS, Given B, Baldwin D, Mako M, Markese J, Rubenstein AH, Olefsky J et al. (1979) A structurally abnormal insulin causing human diabetes. Nature 281:122–125

Tager HS, Palzelt C, Assoian RK, Chan SJ, Duguid JR, Steiner DF (1980a) Biosynthesis of islet cell hormones. Ann NY Acad Sci 343:133–147

Tager HS, Thomas N, Assoian R, Rubenstein A, Saekow M, Olefsky J, Kaiser ET (1980b) Semisynthesis and biological activity of porcine [Leu^{B24}]insulin and [Leu^{B25}]insulin. Proc Natl Acad Sci USA 77:3181–3185

Terris S, Steiner DF (1975) Binding and degradation of ^{125}I-insulin by rat hepatocytes. J Biol Chem 250:8389–8398

Terris S, Steiner DF (1976) Retention and degradation of ^{125}I-insulin by perfused livers from diabetic rats. J Clin Invest 57:885–896

Ullrich A, Dull TJ, Gray A, Brosius J, Sives I (1980) Genetic variation in the human insulin gene. Science 209:612–615

Wollmer A, Strassburger W, Glatler V, Dodson GG, McCall M, Danho W, Brandenburg D et al. (1981) Two mutant forms of human insulin: structural consequences of the substitution of invariant B24 or B25 by leucine. Hoppe Seylers Z Physiol Chem 362:581–592

Part II Biosynthesis, Secretion, and Degradation

CHAPTER 4

The Biosynthesis of Insulin

D. F. STEINER

A. Introduction

Insulin is a compact globular protein of molecular weight 6000 that is made up of
two short polypeptide chains stabilized by two interchain disulfide bridges. Its
characteristic secondary and tertiary structures are essential for its biological activity and are highly conserved in evolution. Separated insulin chains can be
recombined in the laboratory and this approach has been succesfully refined for
the commercial production of human insulin. However, in the β cells of the pancreas the hormone is synthesized as a single-chain precursor or proinsulin, which
readily folds to form the correct disulfide bridges of the native hormone with high
efficiency (STEINER and OYER 1967; STEINER and CLARK 1968). Early studies of
the biosynthesis and maturation of proinsulin to insulin revealed important new
aspects of the subcellular organization and processing mechanisms of peptide-
secreting cells (STEINER et al. 1969, 1970; ROBBINS et al. 1984b). With the subsequent discovery that precursors similar to proinsulin exist for many other peptide hormones, as well as for neuropeptides and many larger proteins, it has become apparent that such forms fulfill many functions in the biosynthesis,
transport, and correct targeting of their products to appropriate locations within
their cells of origin (DOUGLASS et al. 1984).

In recent years several other secreted products of the β cell have been
identified. These are produced at levels ranging from one to three orders of magnitude lower than insulin. Chromogranin A is an 88 kdalton granule protein,
which is processed proteolytically before secretion into the peptides betagranin
and pancreastatin (O'CONNOR and DEFTOS 1986; HUTTON et al. 1988). The
recently identified islet amyloid peptide (IAP), a homolog of the neuropeptide
CGRP (calcitonin gene-related peptide) is also derived from a larger precursor by
processing at paired basic residues in a manner similar to that for the conversion
of proinsulin to insulin (SANKE et al. 1988). IAP is found in increased amounts in
amyloid deposits in the islets of patients with type II diabetes (WESTERMARK et al.
1987). Thyrotropin-releasing hormone (TRH) has also been tentatively identified
as a β cell product, but is more abundant in fetal than adult islets (MARTINO et al.
1978; DOLVA et al. 1983).

B. Insulin Precursors

I. Preproinsulin: Its Role in Insulin Biosynthesis

The initial precursor of insulin is preproinsulin, a 12 kdalton single-chain molecule (Fig. 1) that consists of 9 kdalton proinsulin extended by a 24 amino acid prepeptide, or signal peptide, at its amino terminus (CHAN et al. 1976). This NH_2 terminal sequence of amino acids, with its characteristic cluster of hydrophobic residues, specifies the entry of proinsulin into the secretory pathway, where it will undergo conversion and storage, and ultimately be released as mature insulin and C peptide in response to glucose and other signals. A complex series of molecular interactions involving the signal peptide result in the translocation of the nascent proinsulin chain from the cytosol across the membrane of the rough endoplasmic reticulum (RER) into its luminal compartments, or cisternae (STEINER et al. 1980; KREIL 1981). Recent work on this mechanism of segregation has led to the formulation of a model for the insertion of presecretory

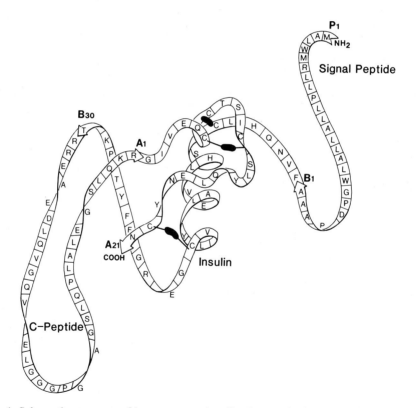

Fig. 1. Schematic structure of human preproinsulin. Removal of the first 24 amino acids (signal peptide) gives rise to proinsulin. Cleavage after B30 (*T*) and before A1 (*G*) gives rise to insulin and C peptide. (See text for details and DAYHOFF 1973 for explanation of single-letter amino acid code)

proteins into RER membranes which involves a cycle of interactions of the nascent signal peptide, first with the signal recognition particle (SRP) and then with its receptor, or "docking protein," in the RER (WALTER et al. 1984). The SRP–docking protein cycle may be a point of control in the regulation of proinsulin biosynthesis by glucose in islets (see Sect. G). During, or shortly after, translocation, the signal sequence is cleaved by signal peptidase, a specialized protease that is located on the luminal surface of the RER membrane (KREIL 1981). Following translocation and cleavage of the signal sequence, the proinsulin molecules fold, undergo rapid formation of disulfide bonds to gain their native structure and are transported to the *cis* Golgi for further processing and packaging, as discussed below. The signal peptide is rapidly degraded in the RER and is not a normal secretory product of β cells (PATZELT et al. 1978 b).

Proinsulin contains the B and A chains within a single 9 kdalton polypeptide chain (STEINER and OYER 1967; CHANCE et al. 1968; STEINER et al. 1969). After its intracellular transfer through the Golgi apparatus into condensing vacuoles, proinsulin is cleaved to yield insulin and a 26-31 residue peptide fragment designated the C peptide. Insulin and the C peptide are stored together in the secretion granules along with small amounts of residual proinsulin and intermediate cleavage forms (STEINER et al. 1972). Such intracellular cleavage of precursor peptides differs in several respects from the proteolytic activation of zymogen proteins, which is a predominantly extracellular process catalyzed by serine proteases related to trypsin. Precursor processing has proven to be an important feature of almost all peptide-producing endocrine or neural cells and also occurs in many other tissues not normally associated with the neuroendocrine system, e.g., liver (QUINN et al. 1975; RUSSELL and GELLER 1975; GORDON et al. 1984), heart (CANTIN and GENEST 1985), and lung (LAUWERYNS et al. 1987), as well as in the biosynthesis and maturation of many viruses (JACOBSON and BALTIMORE 1968; KIEHN and HOLLAND 1970; ARNOLD et al. 1987). The intracellular localization and mechanism of action of the proteolytic processing enzymes thus represents a problem of rather general significance and is discussed in the section on the conversion of proinsulin to insulin.

II. Structure and Properties of Proinsulin

Methods for the isolation of proinsulin and related peptides have been reviewed elsewhere by STEINER et al. (1977). Mammalian proinsulins range in size from 81 (cow) to 86 (human, horse, rat) amino acid residues (STEINER 1984). This difference in size is a consequence of variations in the length of the connecting polypeptide (C peptide) which links the COOH terminus of the insulin B chain to the NH$_2$ terminus of the insulin A chain (Fig. 1). All known proinsulins have pairs of basic residues linking the C peptide to the insulin chains. These residues are removed during the proteolytic conversion of proinsulin to insulin (STEINER et al. 1971) and the resulting products are native insulin and C peptide.

Despite its larger size, proinsulin is very similar to insulin in many properties, including solubility, isoelectric point (STEINER et al. 1972), self associative properties (FRANK and VEROS 1968), and reactivity with insulin antisera (STEINER and OYER 1967; RUBENSTEIN et al. 1969 b, 1970). These observations, and others, strongly suggest that the conformation of the insulin moiety in proinsulin is near-ly identical to that of insulin itself (Steiner et al. 1972). The length of the con-necting peptide is much greater than is required to bridge the short 8 Å gap between the ends of the B and A chains in the native insulin molecule (Fig. 1). Although the connecting peptide may be folded over a portion of the surface of the insulin monomer, it does not completely mask the "active site," since intact proinsulin possesses 3–5% biological activity and is a full agonist (Gliemann and Sørensen 1970; Freychet et al. 1974; Kitabchi 1977). It is unlikely that any sig-nificant cleavage or "activation" of proinsulin occurs in the circulation or tissues to account for this level of intrinsic activity (Lazarus et al. 1970). The surfaces of the insulin monomer that interact to form dimers and hexamers are not con-strained from interacting by the connecting peptide in proinsulin (Frank and Veros 1970; Steiner 1973). A hypothetical arrangement of the connecting pep-tide in a hexamer of proinsulin is shown in Fig. 2. This hexameric structure, with the C peptide arranged on the outside, may play a role in the efficient conversion of proinsulin to insulin in the β cells. The three-dimensional structure of proinsulin has not yet been determined, despite the successful crystallization of the prohormone in several laboratories (Blundell and Wood 1982; Fullerton et al. 1970; Rosen et al. 1972; Low et al. 1974). The availability of biosynthetic human proinsulin and insulin, as well as of human C peptide, has opened many new possibilities for studies of the role, metabolism, and antigenicity of these peptides (Glauber et al. 1987; Madsen et al. 1983, 1984).

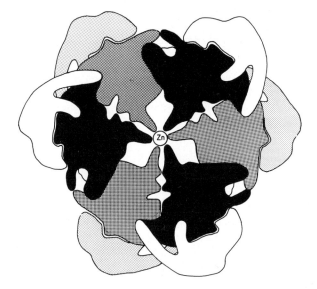

Fig. 2. Hypothetical 2Zn proinsulin hexamer as viewed along the threefold axis. The con-necting peptide is shown in *light gray* and *white* around the periphery of the *darker* outline of an insulin hexamer arranged according to the data of Blundell et al. (1972). The central density represents two zinc atoms on the threefold axis in coordination linkage to the six (three above and three below) histidine side chains at position 10 in the B chain

III. Proinsulin is the Immediate Precursor of Insulin

The precursor–product relationship between proinsulin and insulin has been demonstrated in a variety of studies (STEINER et al. 1967; STEINER 1967; LIN and HAIST 1969; TANESE et al. 1970; ORCI et al. 1985). With isolated rat islets it has been shown that the conversion of proinsulin to insulin begins only after a delay of about 20 min and continues over a period of several hours as a pseudo-first-order reaction with a half-life of 30–60 min (STEINER et al. 1967; STEINER 1967; SANDO et al. 1972; NAGAMATSU et al. 1987). Proteolytic conversion of proinsulin to insulin does not require continued protein synthesis (STEINER et al. 1967), but glucose enhancement of proinsulin conversion is inhibited by cyclohexamide during preincubation of islets with glucose (NAGAMATSU et al. 1987). As intracellular processing of proinsulin proceeds, intermediate cleavage products are generated, but the mature secretory granules normally contain only small amounts (1%–2%) of proinsulin or such intermediate materials. Consequently, secreted insulin normally contains only small amounts of these precursor-related peptides (SANDO et al. 1972). Newly synthesized insulin is preferentially released to a small extent, but most of the material secreted in response to glucose consists of stored hormone and C peptide (SANDO et al. 1972; SANDO and GRODSKY 1973; GOLD et al. 1982). Many studies on the biosynthesis, isolation, and characterization of intermediate forms of mammalian proinsulins or C peptides have been reported (CHANCE 1971; CLARK and STEINER 1969; TAGER et al. 1973; TUNG and YIP 1969; NOLAN et al. 1971; KITABCHI 1977; KUZUYA et al. 1978 b).

Comparative studies of insulin biosynthesis in the principal islets (Brockman bodies) of teleost fishes such as the cod (GRANT and COOMBS 1971) or the anglerfish (YAMAJI et al. 1972; HOBART et al. 1980), as well as in the islet organs of primitive vertebrates such as the hagfish, a cyclostome (STEINER et al. 1973; EMDIN and FALKMER 1977; CHAN et al. 1981), are all consistent with the synthesis and cleavage of proinsulins that are similar in size to the mammalian prohormones. Recent studies have also revealed the synthesis of insulin-like proteins in both molluscs (SMIT et al. 1988) and insects (NAGASAWA et al. 1986). These proteins are also derived from proinsulin-like precursors that have basic residues at cleavage sites.

C. Cell Biology of Insulin Biosynthesis

The β cells of the islets of Langerhans are organized similarly to many other cells that secrete proteins, as illustrated in Fig. 3. The participation of the Golgi apparatus in the formation of β cell storage granules was observed in early studies (HARD 1944). Later studies made possible by the advent of electron microscopy confirmed that secretory granule formation occurs within the Golgi apparatus (MUNGER 1958). Munger identified "progranules" with altered morphology near the Golgi body. Pulse-chase data from biosynthetic studies also pointed to a site distal from the RER, such as the Golgi apparatus and/or early secretory granules, as the possible site of intracellular proteolytic processing of proinsulin in the β cell (STEINER et al. 1969, 1970). Many recent studies have amply con-

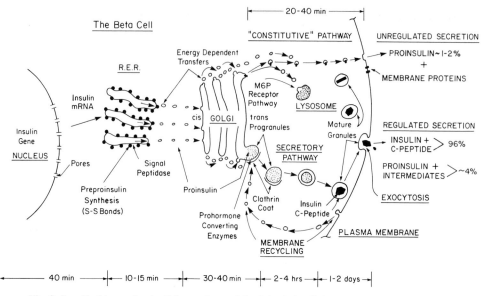

Fig. 3. Insulin biosynthesis. Schematic model of the subcellular transport of preproinsulin after its synthesis in the rough endoplasmic reticulum (RER) and rapid cleavage to proinsulin (within 1–2 min). Proinsulin is then released into the cisternal spaces of the RER where it folds to form the native disulfide bonds of insulin, and is then transported to the Golgi apparatus. The clathrin-coated early granules budding from the *trans* Golgi cisternae are rich in proinsulin and contain the converting proteases. Processing occurs mainly, if not exclusively, in the granules (Orci et al. 1985; Steiner et al. 1970), giving rise to the more condensed mature granules. Recent studies in our laboratory using fractionation techniques have confirmed that the mature granule-dense cores consist almost entirely of insulin, often in crystalline arrays (see Fig. 5), while the granule-soluble phase that surrounds the inclusion consists mainly of C peptide and small amounts of proinsulin (Michael et al. 1987). The release of newly synthesized proinsulin and insulin begins only about 1 h after synthesis in the RER, and hence granules must undergo a maturation process that renders them competent for secretion. Only very small amounts of proinsulin or insulin are released via constitutive or unregulated pathways (upper pathway). Exocytosis of mature secretory granules is regulated by glucose and many other factors, and in humans and other species results in the release of insulin and C peptide in equimolar proportions under both basal and stimulated conditions (Polonsky and Rubenstein 1985)

firmed that the Golgi body plays an essential role in the processing and sorting of proteins in all secretory cells (Farquhar and Palade 1981). The *trans* Golgi is probably the site where prohormones and their converting proteases are brought together as secretion granules are formed (Steiner et al. 1980, 1984; Orci et al. 1985). Immunocytochemical studies using monoclonal antibodies specific for intact proinsulin have shed new light on events occurring in the Golgi apparatus during the transfer of proinsulin from the RER into early secretory vesicles (Orci et al. 1985). These indicate that proinsulin may be transferred through the *cis* and *mid* Golgi in small membrane-enclosed vesicles to the *trans* Golgi cisternae where it is concentrated into budding prosecretory vesicles (Orci et al. 1986). Earlier studies indicated that energy is required for proinsulin transport to the site of conversion (Steiner et al. 1970), but not for its subsequent proteolytic conversion. More recent findings suggest that energy is required not only for proinsulin's transport from the RER, but also for the intercisternal transfers of proinsulin within the Golgi stack from *cis* to *trans* (ORCI et al. 1985). The chemical basis of the energy requirement for the intracellular translocation of secretory proteins (JAMIESON and PALADE 1968) has been studied by Rothman and co-workers. ATP hydrolysis appears to be required for the budding and/or fusion of small vesicles which transport the secretory product from the *cis* through the *trans* Golgi cisternae (WATTENBERG and ROTHMAN 1986; CHAPPELL et al. 1986). All of these findings taken together are consistent with the hypothesis that proinsulin conversion is initiated at the *trans* Golgi stage and proceeds for several hours within newly formed secretion granules as these mature in the cytosol and acquire the ability to be secreted (STEINER et al. 1970). The observation of ORCI et al. (1985) of a clathrin coat on the budding and early (pro)secretory granules in the β cell is of interest (Fig. 3). Although its significance is not known, it is tempting to speculate that it may be related to the concentration of the granule contents or their restructuring, to the budding process or to proteolytic processing.

Figure 3 summarizes schematically the intracellular pathway followed by newly synthesized proinsulin and indicates the times required for transit at various stages, as well as the sites where proteolytic processing occurs. As mentioned earlier, the conversion of proinsulin to insulin in intact rat islets begins about 30 min after biosynthesis of the peptide chain and resembles a first-order reaction having a half-time of about 30 min to 1 h (NAGAMATSU et al. 1987; STEINER 1967; D. F. STEINER 1985, unpublished work). Peak labeling of proteins in the Golgi apparatus is observed 30–40 min after biosynthetic labeling of islets with tritiated amino acids and relatively little radioactivity remains in this region after 1 h (HOWELL et al. 1969; ORCI et al. 1971). Electron microscopic immunocytochemistry using antisera specific for uncleaved proinsulin reveals a similar pattern (ORCI et al. 1985). Clearly, in view of the relatively slow rate of proinsulin conversion this process must continue for several hours as the newly formed secretion granules, or "progranules," mature biochemically in the cytosol.

D. Mechanism of Conversion of Proinsulin to Insulin

The major types of proteolytic cleavage required for processing of proinsulin to insulin are summarized in Fig. 4. This schema requires the conjoint action of a trypsin-like protease with another having specificity like that of carboxypeptidase B. The latter enzyme removes the COOH terminal basic residues left after tryptic cleavage, giving rise to the main naturally occurring products – C peptide and native insulin. It has been shown that suitable mixtures of pancreatic trypsin and carboxypeptidase B can quantitatively convert proinsulin to insulin in vitro (KEMMLER et al. 1971). This model system can explain how the known major intermediate forms and products that occur naturally in pancreatic extracts are generated (STEINER et al. 1971; NOLAN et al. 1971). Studies with isolated islet secretion granule fractions have confirmed that these are major sites of proinsulin conversion (STEINER et al. 1975; DAVIDSON et al. 1987). Recently,

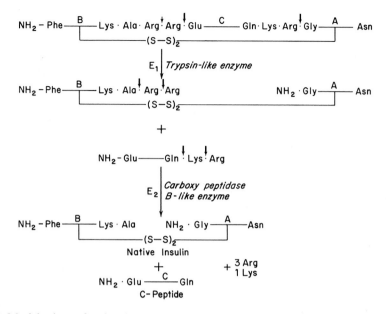

Fig. 4. Model scheme for the cleavage of proinsulin to insulin by the combined action of trypsin-like and carboxypeptidase B-like proteases. (See text for details)

Davidson et al. (1988) have reported the identification of two endopeptidase activities in insulinoma secretory granules that are both stimulated by calcium ions. Type I requires 1 mM Ca^{2+} and cleaves exclusively at Arg31-Arg32 in proinsulin while type II requires 0.1 mM Ca^{2+} and cleaves predominantly, but not exclusively at Lys64-Arg65. Both enzymes have pH optima near 6.0, but the type II enzyme is also active in the neutral range.

By labelling proinsulin in intact islets with [^{3}H]arginine, it can be shown that subsequent conversion in the isolated granule fraction in vitro leads to the release of free arginine rather than of basic dipeptides (Kemmler et al. 1973), indicating the presence of both trypsin-like endopeptidase and carboxypeptidase B-like exopeptidase activities in maturing secretory granules (Zühlke et al. 1976). The carboxypeptidase B-like "trimming" exopeptidase has been identified as a soluble secretory granule constituent in a variety of neuroendocrine tissues. This enzyme, now known as carboxypeptidase H (Skidgel 1988), is stimulated by cobalt and exhibits other unique features that distinguish it from other known carboxypeptidases (Docherty and Hutton 1983). Molecular cloning has shown it to have structural homology with pancreatic carboxypeptidases A and B (Fricker et al. 1986), suggesting a common evolutionary origin.

In yeast, recent studies have led to the identification of genes encoding an endopeptidase and a carboxypeptidase which process the α mating factor precursor (Fuller et al. 1988). Both are membrane-bound proteases and the endopeptidase, which is encoded by the *kex-2* gene, exhibits a striking preference for cleavage at pairs of basic residues, similar to the islet-converting peptidases described above. It appears to be a subtilisin-like serine protease (Mizuno et al.

1988) while the carboxypeptidase (*kex-1*) is related to the vacuolar carboxypeptidase Y (FULLER et al. 1988).

In some mammals, such as rats and probably also in pigs and humans, additional maturation cleavages occur in the C peptide region of proinsulin due to a chymotrypsin-like activity (CHANCE 1971; TAGER et al. 1973; KUZUYA et al. 1978 a). Their importance is unclear since they seem to occur only in species in which the C peptide contains sites of high chymotrypsin sensitivity. In the dog, the C peptide is cleaved at a single arginine residue to produce an NH_2-terminally truncated C peptide having only 23 residues (KWOK et al. 1983 a). All these findings taken together suggest that the β granules contain a mixture of proteases having specificity for double and/or single basic amino acids, as well as for residues such as leucine, glutamine or alanine, in rare instances.

E. Formation of Insulin Secretory Granules

As mentioned earlier, morphological studies of newly formed secretory granules suggest that these undergo a series of biochemical changes as they mature in the cytoplasm of the β cell. Early secretory granules characteristically are less dense than mature granules and have a uniform pale density throughout (MUNGER 1958; ORCI et al. 1985). Electron microscopic examination of mature insulin secretory granules indicate that these have a dense central core which often appears to be crystalline in appearance (Fig. 5) having repeat unit spacings that are closely similar to those observed in ordinary zinc insulin crystals (GREIDER et al. 1969; LANGE 1974; MICHAEL et al. 1987). These observations suggest that as insulin is liberated from proinsulin, it tends to crystallize with zinc that is taken up by the β granules. Biochemical fractionation of islet secretory granules confirms that the cores consist mainly of insulin while the soluble C peptide remains in the clear fluid space surrounding the crystalline core (MICHAEL et al. 1987). There is no evidence for cocrystallization of the C peptide with insulin either in vivo or in vitro. However, low levels of proinsulin can cocrystallize with insulin, probably by forming mixed hexamers with insulin (STEINER 1973). Perhaps because of this tendency, the granule cores usually contain 1–2% of intact or partially processed precursor forms (MICHAEL et al. 1987).

In addition to zinc, secretory granules are rich in calcium (HOWELL et al. 1978), which may play a role in regulating conversion, as discussed earlier, and also may contribute to the crystallization of insulin (PALMIERI et al. 1988). The mechanisms for uptake of zinc and calcium and their roles in secretory granule formation are not clear (HOWELL et al. 1978; EMDIN et al. 1980). Most of the zinc in islets is present in the β granules and is liberated proportionately with insulin during secretion (LOGOTHETOPOULOS et al. 1964; FALKMER 1971). Accumulation of zinc within the granules may be a passive process due to the ability of both proinsulin and insulin to bind zinc (FRANK and VEROS 1970; GRANT et al. 1972; HOWELL et al. 1978). However, the insulins of a few species, including the guinea pig, coypu, and other hystricomorph rodents (BLUNDELL and WOOD 1982; SMITH 1966) and the hagfish (PETERSON et al. 1975; CUTFIELD et al. 1979) lack the histidine residue at position 10 of the B chain required for zinc binding during the

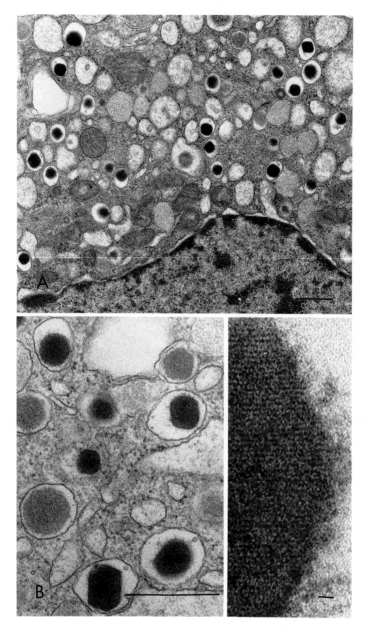

Fig. 5. A Photomicrograph of normal rat β cell. *Bar* = 1 μm, × 10 100. **B** Higher magnification view showing morphology of mature secretory granules; *Bar* = 1 μm, × 25 200. **C** High magnification view showing repeat unit structure of a crystalline granule core. *bar* = 0.01 μm, × 325 000. Samples were fixed with Karnovsky's solution and stained with osmium tetroxide. (Electron micrographs courtesy of Hewson H. Swift)

association of insulin dimers into hexamers (BLUNDELL et al. 1972) and hence do not hexamerize. As illustrated in Fig. 2, most mammalian proinsulins are able to form hexamers stabilized by two zinc atoms coordinated with the six B10 histidines, as in the case of insulin (FRANK and VEROS 1970). Proinsulin hexamers can also bind zinc at additional sites without precipitating from solution (GRANT et al. 1972) and this property may allow proinsulin to play a role in zinc accumulation in the islet cells. The uptake of calcium by granules, on the other hand, may be an active process that is carried out by an ion pump.

The pH of the interior of the mature secretory granule appears to be about 5.0–5.5 (KEMMLER et al. 1973; ORCI et al. 1985), an optimal pH for insulin crystallization in vitro. On the other hand disulfide exchange reactions are favored by alkaline conditions and it is thus likely that the pH in the cisternal spaces of the RER, where proinsulin folding and sulfhydryl oxidation occur, is somewhat above neutrality. As the secretory products move through the Golgi apparatus and into granules the pH probably becomes more acidic. Proton pumps in the granules may also aid in the displacement of the cationic arginine and lysine residues liberated during conversion, which are then replaced by hydrogen ions and other cations as they diffuse out of the granules, resulting in a further downward shift in the intragranular pH. These ideas are supported by immunocytochemical studies showing that progranules are only mildly acidic or neutral and undergo acidification as the granules mature (see Fig. 3; ORCI et al. 1985). These pH changes tend to create appropriate conditions for the formation of the crystalline zinc insulin inclusions (STEINER and RUBENSTEIN 1973) and may possibly serve to regulate proinsulin conversion. Thus, the processes related to the biosynthesis of insulin via preproinsulin and proinsulin, and their intracellular transport, proteolysis, and ultimate storage in secretory granules are remarkably well integrated, both topologically and biochemically, within the β cells. This delicately poised integration of processes leading to the formation and storage of insulin is disturbed in islet cell tumors which often show unregulated release of insulin together with large amounts of proinsulin; measurements of the latter can provide a useful diagnostic indicator (RUBENSTEIN et al. 1977; KITABCHI 1977; ROBBINS et al. 1984b; COHEN et al. 1986; STEINER et al. 1989).

F. The C-Peptide as a Product of Proinsulin Transformation

Due to the localization of proinsulin conversion within secretory granules, the C peptide accumulates with insulin in equimolar amounts (STEINER 1984) and is secreted along with the hormones during exocytosis of the granule contents (RUBENSTEIN et al. 1969a). Representative vertebrate C peptide amino acid sequences are compared in Fig. 6. These peptides exhibit a 15-fold higher rate of mutation acceptance than do the corresponding insulins, a finding consistent with the likelihood that this region in proinsulin does not have any hormonal function. Only the fibrinopeptides have a higher rate of mutation acceptance than the proinsulin C peptides. Nevertheless, certain regions of the relatively large connecting peptide probably serve specific functions, such as facilitating the folding of the proinsulin polypeptide chain and the formation of the correct di-

```
              1  2  3  4  5  6  7  8  9 10 11 12 13 14 15 16 17 18 19 20 21 22 23 24 25 26 27 28 29 30 31 32 33 34 35 36 37 38

HUMAN         Glu-Ala-Glu-Asp-Leu-Gln-Val-Gly-Gln-Val-Glu-Leu-Gly-Gly-Gly-Pro-Gly-Ala-Gly-Ser-Leu-Gln-Pro-Leu-Ala-Leu-Glu-Gly-Ser-Leu-Gln

MONKEY        Glu-Ala-Glu-Asp-Pro-Gln-Val-Gly-Gln-Val-Glu-Leu-Gly-Gly-Gly-Pro-Gly-Ala-Gly-Ser-Leu-Gln-Pro-Leu-Ala-Leu-Glu-Gly-Ser-Leu-Gln

HORSE         Glu-Ala-Glu-Asp-Pro-Gln-Val-Gly-Glu-Val-Glu-Leu-Gly-Gly-Gly-Pro-Gly-Leu-Gly-Gly-Leu-Gln-Pro-Leu-Ala-Leu-Ala-Gly-Pro-Gln-Gln

PIG           Glu-Ala-Glu-Asn-Pro-Gln-Ala-Gly-Ala-Val-Glu-Leu-Gly-Gly-Gly-Leu-Gly — Gly — Leu-Gln-Ala-Leu-Ala-Leu-Glu-Gly-Pro-Pro-Gln

COW, LAMB     Glu-Val-Glu-Gly-Pro-Gln-Val-Gly-Ala-Leu-Glu-Leu-Ala-Gly-Gly-Pro-Gly-Ala-Gly-Gly-Leu — — — — — Glu-Gly-Pro-Pro-Gln

RABBIT        Glu-Val-Glu-Glu-Leu-Gln-Val-Gly-Gln-Ala-Glu-Leu-Gly-Gly-Gly-Pro-Gly-Ala-Gly-Gly-Leu-Gln-Pro-Ser-Ala-Leu-Glu — Ala-Leu-Gln

DOG           Glu-Val-Glu-Asp-Pro-Gln-Val-Arg-Asp-Val-Glu-Leu-Ala-Gly-Ala-Pro-Gly-Glu-Gly-Gly-Gly-Leu-Gln-Pro-Leu-Ala-Leu-Glu-Gly-Ala-Leu-Gln

RAT I         Glu-Val-Glu-Asp-Pro-Gln-Val-Pro-Gln-Leu-Glu-Leu-Gly-Gly-Gly-Pro-Glu-Ala-Gly-Asp-Leu-Gln-Thr-Leu-Ala-Leu-Glu-Val-Ala-Arg-Gln

RAT II        Glu-Val-Glu-Asp-Pro-Gln-Val-Ala-Gln-Leu-Glu-Leu-Gly-Gly-Gly-Pro-Gly-Ala-Gly-Asp-Leu-Gln-Thr-Leu-Ala-Leu-Glu-Val-Ala-Arg-Gln

GUINEA PIG    Glu-Leu-Glu-Asp-Pro-Gln-Val-Glu-Gln-Thr-Glu-Leu-Gly-Met-Gly-Leu-Gly-Ala-Gly-Gly-Leu-Gln-Pro-Leu-Ala-Leu-Glu-Met-Ala-Leu-Gln

CHINCHILLA    Glu-Leu-Glu-Asp-Pro-Gln-Val-Gly-Gln-Ala-Asp-Pro-Gly-Val-Val-Pro-Glu-Ala-Gly-Arg-Glu-Leu-Pro-Leu-Ala-Leu-Glu-Met-Thr-Leu-Gln

DUCK          Asp-Val-Glu-Gln-Pro-Leu-Val-Asn-Gly-Pro — Leu-His-Glu-Glu-Val-Gly-Glu — — Leu-Pro-Phe-Gln-His-Glu-Glu — — Tyr-Gln

CHICKEN       Asp-Val-Glu-Gln-Pro-Leu-Val-Ser-Ser-Pro — Leu-Arg-Gly-Glu-Ala-Gly-Val — — Leu-Pro-Phe-Gln-Gln-Glu-Glu-Tyr-Glu-Lys-Val

ANGLERFISH    Asp-Val-Asp-Gln-Leu-Leu-Gly-Phe-Leu-Pro-Pro-Lys-Ser-Gly-Gly-Ala-Ala-Ala-Ala-Gly-Ala-Asp-Asn-Glu-Val-Ala-Glu-Phe-Ala-Phe-Lys-Asp-Gln-Met-Glu-Met-Met-Val

HAGFISH       Asp-Thr-Gly-Ala-Leu-Ala-Ala-Phe-Leu-Pro-Leu-Ala-Tyr-Ala-Glu-Asp-Asn-Glu-Ser-Gln-Asp-Asp-Glu-Ser-Ile-Gly-Ile-Asn-Glu-Val-Leu-Lys-Ser

MOLLUSC       Asn-Ala-Glu-Thr-Asp-Leu-Asp-Asp-Pro-Leu-Arg-Asn-Ile-Lys-Leu-Ser-Ser-Glu-Ser-Ala-Leu-Thr-Tyr-Leu-Thr
```

Fig. 6. Compilation of amino acid sequences of proinsulin C peptides in vertebrates and a mollusc (for sources see DAYHOFF 1973; STEINER 1984; MARKUSSEN and SUNDBY 1972; TAGER and STEINER 1972, SUNDBY and MARKUSSEN 1970; OYER et al. 1971; PETERSON et al. 1972; SMIT et al. 1988). The guinea pig C peptide sequence corresponds to that predicted from the nucleotide sequence of the guinea pig insulin gene. (CHAN et al. 1984)

sulfide bonds (STEINER and CLARK 1968), or guiding the enzymatic cleavage of proinsulin to insulin by orienting the basic residue pairs for cleavage (THIM et al. 1986; STEINER 1984). Several acidic residues are consistently present in mammalian connecting peptides (Fig. 6). These offset the positive charges due to the basic residues at the cleavage sites such that the isoelectric pH of proinsulin is nearly the same as that of insulin, i.e., in the range of pH 5.1–5.5 for most mammalian forms (STEINER et al. 1972; KOHNERT et al. 1972).

Recent studies have shown that relatively small bifunctional cross-linking reagents inserted between the amino group of A1 glycine and the ε-amino group of B29 lysine of insulin (see Fig. 1) can replace the C peptide in promoting the correct reoxidation of the sulfhydryls in high yields after complete reduction and denaturation (BRANDENBURG et al. 1977; BUSSE et al. 1974). Miniproinsulins with greatly shortened connecting peptide segments also readily oxidize to form the correct disulfide bridges and can be cleaved to yield insulin (THIM et al. 1986; MARKUSSEN 1979). Moore and her co-workers (POWELL et al. 1988) have shown that deletion of the entire C peptide region of proinsulin does not alter its biosynthesis, folding to an immunocompetent form, or targeting to secretory vesicles in AtT20 pituitary hormone-secreting cells. Thus, the C peptide region in proinsulin must serve functions in the biosynthetic process other than promoting sulfhydryl oxidation and/or correct intracellular targeting. These might include regulation of biosynthesis at the translational level, optimization of proteolytic processing, and the formation of stable storage vesicles.

G. The Regulation of Insulin Production

Although the rate of secretion of insulin is subject to elaborate control by glucose and other nutrients as well as hormones and probably neurotransmitters (HEDESKOV 1980; MEGLASSON and MATSCHINSKY 1986; PRENTKI and MATSCHINSKY 1987), the renewal and regulation of the granular stores of hormone in the β cells is an important aspect of normal homeostasis. It should be emphasized that little, if any, direct secretion of proinsulin from the RER to the plasma membrane or via other "unregulated" pathways occurs (KELLY 1985; RHODES and HALBAN 1987). Thus, calcium-dependent (GRODSKY 1970) exocytosis of preformed storage granules appears to be the main source of both basal and glucose-stimulated insulin release in vivo (STEINER 1967; TANESE et al. 1970; SANDO et al. 1972; SANDO and GRODSKY 1973; RHODES and HALBAN 1987). We might ask how this granular compartment is maintained and regulated by biosynthesis versus degradation. The chief positive effectors for biosynthesis that have been identified thus far are glucose, augmented by cyclic AMP, which may also be generated by a mechanism coupled to glucose metabolism in the β cell (VALVERDE et al. 1983). Secretion, however, is not a direct stimulus to insulin biosynthesis, as can be proven by blocking the secretory process by lowering external calcium levels or using inhibitors of exocytosis. These do not impair the biosynthetic response to glucose (STEINER et al. 1972). Moreover, in fetal and newborn rat islets glucose stimulates insulin biosynthesis although it has little effect on insulin secretion (ASPLUND 1973). During prolonged periods without stimulation, or if secretion is inhibited, intracellular degradation (autophagy) of granules occurs (ORCI et al. 1985; RHODES and HALBAN 1988). It is possible that both autophagy of granules as well as increased secretion of proinsulin or partially processed materials occur in islet cell tumors (CREUTZFELDT et al. 1973).

It is well known that glucose directly stimulates insulin biosynthesis via stimulation of mRNA translation (PERMUTT and KIPNIS 1971; STEINER et al. 1972; ITOH and OKAMOTO 1980). Recent work has shown that this response is complex, entailing not only effects of glucose on both the initiation and elongation of proinsulin chains, but also on the duration of SRP–signal peptide-mediated arrest of translation of nascent preproinsulin chains prior to docking protein interaction in the early phases of RER membrane insertion (WELSH et al. 1985 b). In addition to these very rapidly acting translational control mechanisms the rate of transcription of insulin mRNA is also closely regulated by glucose and cyclic AMP (NIELSEN et al. 1985; WELSH et al. 1985 a). Insulin mRNA is normally quite stable, turning over very slowly with a half-life of about 30 h at normal or below normal levels. However, elevated glucose levels increase its half-life dramatically (approximately threefold) and this action, in combination with increased rates of transcription, can effect significant increases in insulin mRNA levels over 48- to 72-h periods. Thus, prolonged glucose stimulation can greatly augment insulin production and may eventually lead to increased β cell mitosis and hyperplasia of islets (RABINOVITCH et al. 1980; SWENNE 1982).

The multifaceted control mechanism for insulin biosynthesis described above is appropriate considering the vital importance of insulin to the organism, especially the necessity for prompt responses to glucose challenges. Operation of

these various regulatory mechanisms in an integrated fashion can give rise to a
more than 20-fold increase in insulin biosynthesis in response to glucose in
minutes and even greater increases over periods of hours. Moreover, should any
of these mechanisms become defective through disease or mutations, the β cell
will still retain considerable capacity to synthesize and secrete insulin in sufficient
amounts to meet normal demands. However, under conditions of insulin
resistance, failure of one or more of these control mechanisms could contribute
to the development of diabetes.

It should be noted that the details of regulation of mRNA levels, translation
rates, and secretory release or degradation (BIENKOWSKI 1983) of hormones may
differ quite significantly in various endocrine or neural cells, depending on
unique secretory patterns that may occur in some systems. Thus, in the
parathyroid gland it is not the rate of synthesis of parathyroid hormone that is
regulated by the level of ionized calcium, but rather the rate of intracellular
degradation of newly formed hormone (POTTS et al. 1982). As noted earlier,
glucose also regulates the turnover of insulin in the β cells (RHODES and HALBAN
1988), but this effect seems to play a less significant regulatory role under
normoglycemic conditions.

H. The Insulin Gene Family

The fundamental genetic mechanisms underlying the expression of endocrine and
neural regulatory peptides in the organism have been extensively explored in
recent years. The gene for insulin was among the first to be isolated (ULLRICH et
al. 1977). Its structure in humans (BELL et al. 1980) and several other species
(STEINER et al. 1985) is summarized in Fig. 7. The single copy human gene is located on the short arm of chromosome 11 in the region *p15* (OWERBACH et al.
1981; HARPER et al. 1981). It is flanked on the 5' side by a unique polymorphic
region composed of tandem repeats (Fig. 8) that does not seem to influence its expression (WELSH et al. 1985a), but provides a useful marker for genetic linkage
analysis (BELL et al. 1982). Earlier correlations of the presence of larger (class 3)

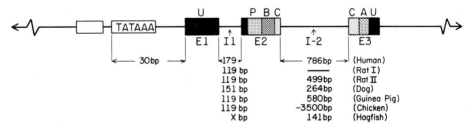

Fig. 7. Diagrammatic representation of the insulin gene in vertebrates. Regions (exons) appearing in mature preproinsulin mRNA are shown as *bars* (*E 1–3*) and the sizes of the two
introns or intervening sequences (*I*) in various species are tabulated below. *U* untranslated
region; *P* prepeptide coding region; *B* B chain coding region; *C* C peptide coding region;
and *A* A chain coding region. A typical TATA box signaling transcription initiation is
shown approximately 30 base pairs upstream from the messenger start site

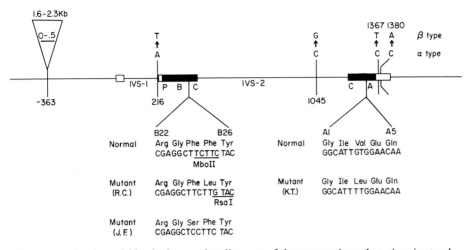

Fig. 8. Localization within the human insulin gene of three mutations that give rise to abnormal insulins with reduced receptor binding affinity. A highly polymorphic locus on the 5′ side, represented by the *triangle*, is made up of unique tandem 14 base pair repeats (BELL et al. 1982) and numbers indicate the approximate sizes of the "large" or "small" inserts found in most human populations. In addition, two allelic variants of the human gene have been described, designated α or β, and differing at the indicated positions (ULLRICH et al. 1980). Note that two of the three mutations modify a region that normally makes up a recognition site for the restriction enzyme Mbo II, giving rise to cleavage defects demonstrable on Southern blot analysis (KWOK et al. 1981), while the third exhibits no change in restriction sites. On the other hand, in mutant R.C., loss of the Mbo II cleavage site is offset by the acquisition of a new *Rsa I* site

versus smaller (class 1 or class 2) repeats in this region (ROTWEIN et al. 1983) with the incidence of type II diabetes were confounded by ethnic differences in the distribution of tandem repeats (BELL et al. 1987). Further analysis of larger populations failed to support the earlier conclusion, but have revealed that class I alleles and genotypes are significantly more frequent in white people with type I diabetes than type II diabetes or controls (BELL et al. 1987). This allele may thus be a marker for a nearby susceptibility gene for type I diabetes.

There has also been rapid progress in the detection and cloning of the genes for the insulin-related growth factors IGF-I and -II, as well as the ovarian hormone relaxin (for a review see STEINER et al. 1985). The genomic sequences encoding these various peptides substantiates the view that they are all related to insulin and are appropriately considered members of an insulin superfamily of hormones and growth factors. It is thus feasible to construct a more complete picture of the biosynthesis and processing of insulin-related peptides in the body, and this emerging schema indicates the participation of many extrapancreatic tissues as well as hypothalamically regulated hormones, e.g., pituitary growth hormone and perhaps others, in the regulation and integration of organismic growth and metabolism (STRAUS 1984). However, in contrast to IGF-I and -II, which are synthesized by most tissues throughout life, or at some period during development, insulin appears to be produced only in the β cells of the islets of Langerhans in the adult organism (GIDDINGS et al. 1985). Thus, disruption or disorganization of the β cells through a variety of pathophysiologic mechanisms and the resulting perturbation of insulin biosynthesis would be expected to affect glucose homeostasis and thereby lead to diabetes. Much needs to be learned, especially in the genetic area, as to the nature of abnormalities that are responsible for impairment of insulin production, secretion, or action in the organism.

J. Defects in the Insulin Gene: The Insulinopathies

The new biochemical and genetic tools have allowed us to begin to identify and study genetic variants in insulin and proinsulin structure in greater depth than was hitherto possible. Howard S. Tager and his associates were the first to successfully identify a structurally abnormal insulin in the circulation and pancreas of a patient with mild diabetes associated with elevated insulin (TAGER et al. 1979). The development of HPLC systems capable of resolving plasma insulin components then led to the further identification of abnormal insulins differing in hydrophobic character in two additional unrelated lineages (SHOELSON et al. 1983b). These and similar studies have led to the definition of a new clinical syndrome analogous to the hemoglobinopathies – the insulinopathies, i.e., molecular defects involving the insulin molecule. Five families have been identified thus far, all having the syndrome of mild hyperinsulinemic diabetes that is similar, in some respects, to type II or non-insulin-dependent diabetes (HANEDA et al. 1984; NANJO et al. 1986). The affected individuals have high circulating insulin levels with a distorted C peptide/insulin ratio resulting most likely from the delayed turnover, in vivo of circulating insulin variants due to their impaired receptor binding properties (SHOELSON et al. 1984). The disorder is inherited in an autosomal dominant fashion within families, consistent with the mendelian distribution of a defective allele.

The insulin genes (both alleles) have been cloned from affected individuals in these families, and in all five cases studied (Fig. 8) a single nucleotide substitution has occurred in only one of the alleles, leading to a single amino acid replacement within the receptor binding region of the insulin molecule (NANJO et al. 1986; KWOK et al. 1983b; HANEDA et al. 1983). The abnormal insulins generated by these missense mutations are all characterized by a very low binding potency, below 5% of normal, as demonstrated by direct assays (ASSOIAN et al. 1982; SHOELSON et al. 1983a). However, the replacements occur at different sites within the insulin molecule (at residues B24, B25, and A3) and the affected individuals thus far have all been heterozygous for the defective gene. These cases represent the first hormonal variants to be identified in humans that lead to a clinically identifiable syndrome. Although not all affected individuals have overt diabetes, it is evident from the high incidence of mild diabetes or glucose intolerance among the affected individuals in these families that the presence of a defective insulin allele can be a significant predisposing factor to the development of diabetes. Hence such mutations, or still others that might reduce the level of expression of the insulin gene (i.e., leading instead to hypoinsulinemia), could give rise to a picture indistinguishable from the fairly common type II, or non-insulin-dependent, form of diabetes.

In addition to molecular variants involving insulin, others have been identified that give rise to elevated circulating proinsulin, with or without clinically significant carbohydrate intolerance (GABBAY et al. 1979; KANAZAWA et al. 1979). In these families an autosomal dominant pattern of inheritance is again evident and in two cases the defect has been localized to the conversion site in the proinsulin molecule (at the C peptide–A chain junction) where the arginine of the Lys-Arg pair recognized by the converting enzyme has been replaced by another amino acid, rendering this site uncleavable (ROBBINS et al. 1981, 1984a). In one of these, molecular cloning has shown substitution of a histidine for arginine at position 65 (see Fig. 1; SHIBASAKI et al. 1985).

An additional family with hyperproinsulinemia in which a point mutation changes the histidine at position 10 of the B chain to aspartic acid has also been identified (GRUPPUSO et al. 1984; CHAN et al. 1987). This is a particularly interesting mutation since the resultant proinsulin molecule retains the paired basic residues. Studies of its biosynthesis have been possible through expression of the mutant gene in transgenic mice (CARROLL et al. 1988). These studies have revealed that Asp^{10} human proinsulin is processed normally within the β cells of the transgenic mice. However, a much larger fraction of the mutant prohormone (approximately 15%) is secreted from the cells rapidly after biosynthesis via an unregulated, or constitutive (BURGESS and KELLY 1987) pathway, suggesting that the efficiency of sorting of this molecule into newly forming secretory vesicles in the *trans* Golgi region is decreased. Further study of this mutation may shed new light on mechanisms of protein sorting in secretory cells.

The study of these insulin variants has not only confirmed previous theories regarding the location of the receptor binding region in the insulin molecule, but have also provided direct evidence that receptor-mediated uptake and degradation of insulin (TERRIS and STEINER 1975, 1976; TERRIS et al. 1979) is a major pathway of insulin metabolism in vivo (HANEDA et al. 1984; SHOELSON et al. 1984). If additional variants involving mutations in regions of the insulin gene outside the hormone and prohormone coding regions can be identified these might provide insights into the mechanisms regulating insulin gene expression.

K. Summary

Insulin is the product of a single copy gene (in most species) which encodes the 12 kdalton precursor, preproinsulin. Expression of the insulin gene occurs only in the β cells of the islets in the adult organism and is regulated by glucose and cAMP levels. The translation of insulin mRNA is also regulated selectively by glucose through several control mechanisms, such that very rapid increases in insulin biosynthesis can occur in response to glucose challenges. Preproinsulin is introduced into the secretory pathway via its NH_2 terminal signal sequence, which is then rapidly removed in the RER by signal peptidase. The 9 kdalton proinsulin rapidly folds to assume its disulfide-bridged native structure in the lumen of the RER and is then transferred to the Golgi apparatus where, beginning about 40 min after its synthesis, it is concentrated into clathrin-clad early secretory vesicles. As these vesicles mature in the cytosol proinsulin is cleaved, acidification occurs, zinc and calcium are taken up, and the released insulin forms small dense crystalline inclusions surrounded by the more soluble C peptide; this peptide links the B and A chains in proinsulin, as follows: B chain Arg-Arg C peptide Lys-Arg A chain. The processing proteases for proinsulin are associated with the secretory granules and include endoproteases that cleave on the carboxyl side of the pairs of basic residues and carboxypeptidase H, an exopeptidase, structurally related to pancreatic carboxypeptidase B, that removes the COOH terminal basic residues. Newly formed insulin secretory granules become competent for secretion after about 2 h and then appear to be released somewhat preferentially relative to "older" mature granules in response to glucose and other signals. Unused secreted granules are removed by autophagy, an intracellular degradative process that is stimulated by low glucose or inhibitors of insulin secretion. The release of insulin storage granules occurs by exocytosis and the secreted material consists of insulin and C peptide in equimolar amounts along with a few percent of uncleaved or partly cleaved intermediate forms and small amounts of other unrelated peptides. Several point mutations in the insulin gene have been identified in patients with elevated proinsulin or insulin levels. These cause single amino acid substitutions that affect either the conversion and/or secreted of proinsulin or the receptor binding affinity of the resultant insulin molecules.

Acknowledgments. It is a pleasure to acknowledge the many valuable contributions to various aspects of this work of a large number of students and colleagues over a period spanning more than two decades. I especially wish to thank Drs. Shu Jin Chan, Arthur H. Rubenstein, Graeme I. Bell, Howard Tager, Susan Terris, Cecilia Hofmann, Åke Lernmark, Sture Falkmer, Stefan Emdin, Kishio Nanjo, Tokio Sanke, Steven P. Smeekens, Michael Welsh, and David Nielsen for their enthusiastic support and inspiration. I also thank Florence Rozenfeld for her expert assistance in preparing this manuscript. Work of this laboratory has been supported by the Couston Memorial Fund, the Kroc Foundation, the Howard Hughes Medical Institute and by the USPHS (grants DK 13914 and 20595).

References

Arnold E, Luo M, Vriend G, Rossman MG, Palmenberg AC, Parks GD, Nicklin MJH, Wimmer E (1987) Implications of the picornavirus capsid structure for polyprotein processing. Proc Natl Acad Sci USA 84:21–25

Asplund K (1973) Effects of glucose on insulin biosynthesis in foetal and newborn rats. Horm Metab Res 5:410–415

Assoian RK, Thomas NE, Kaiser ET, Tager HS (1982) [Leu^{B24}] insulin and [Ala^{B24}] insulin: altered structures and cellular processing of B24-substituted insulin analogs. Proc Natl Acad Sci USA 79:5147–5151

Bell GI, Pictet RL, Rutter WJ, Cordell B, Tischer E, Goodman HM (1980) Sequence of the human insulin gene. Nature 284:26–32

Bell GI, Selby MJ, Rutter WJ (1982) The highly polymorphic region near the human insulin gene is composed of simple tandemly repeating sequences. Nature 295:31–35

Bell GI, Xiang K, Horita S, Sanz N, Karam JH (1987) The molecular genetics of diabetes mellitus. Ciba Found Symp 130:167

Bienkowski RS (1983) Intracellular degradation of newly synthesized secretory proteins. Biochem J 214:1–10

Blundell T, Wood S (1982) The conformation, flexibility, and dynamics of polypeptide hormones. Annu Rev Biochem 51:123–154

Blundell TL, Dodson GG, Hodgkin DC, Mercola D (1972) Insulin: the structure in the crystals and its reflection in chemistry and biology. Adv Protein Chem 26:279–402

Brandenburg D, Gattner H-G, Schermutzki W, Schüttler A, Uschkoreit J, Weimann J, Wollmer A (1977) Crosslinked insulins: preparation, properties, and applications. In: Friedman M (ed) Protein crosslinking, part A. Plenum, New York, p 261

Burgess TL, Kelly RB (1987) Constitutive and regulated secretion of proteins. Annu Rev Cell Biol 3:243–293

Busse WD, Hansen SR, Carpenter FH (1974) Carbonylbis (L-methionyl) insulin. A proinsulin analog which is convertible to insulin. J Am Chem Soc 96:5949–5950

Cantin M, Genest J (1985) The heart and the atrial natriuretic factor. Endocr Rev 6:107–127

Carroll RJ, Hammer RE, Chan SJ, Swift HH, Rubenstein AH, Steiner DF (1988) A mutant human proinsulin is secreted from islets of Langerhans in increased amounts via an unregulated pathway. Proc Natl Acad Sci USA 85:8943–8947

Chan SJ, Keim P, Steiner DF (1976) Cell-free synthesis of rat preproinsulins: characterization and partial amino acid sequence determination. Proc Natl Acad Sci USA 73:1964–1968

Chan SJ, Emdin SO, Kwok SCM, Kramer JM, Falkmer S, Steiner DF (1981) Messenger RNA sequence and primary structure of preproinsulin in a primitive vertebrate, the Atlantic hagfish. J Biol Chem 256:7595–7602

Chan SJ, Episkopou V, Zeitlin S, Karathanasis SK, MacKrell A, Steiner DF, Efstratiadis A (1984) Guinea pig preproinsulin gene: an evolutionary compromise? Proc Natl Acad Sci USA 81:5046–5050

Chan SJ, Seino S, Gruppuso PA, Gordon P, Steiner DF (1987) A mutation in the B chain coding region is associated with impaired proinsulin conversion in a family with hyperproinsulinemia. Proc Natl Acad Sci USA 84:2194–2197

Chance RE (1971) Chemical, physical, biological and immunological studies on porcine proinsulin and related polypeptides. In: Rodriques RR, Vallance-Owen JJ (eds) Proceedings of the 7th congress of the International Diabetes Federation. Excerpta Medica, Amsterdam, p 292

Chance RE, Ellis RM, Bromer WW (1968) Porcine proinsulin: characterization and amino acid sequence. Science 161:165–167

Chappell TG, Welch WF, Schlossman DM, Palter KB, Schlesinger MJ, Rothman JE (1986) Uncoating ATPase is a member of the 70 kdalton family of stress proteins. Cell 45:3–13

Clark JL, Steiner DF (1969) Insulin biosynthesis in the rat: demonstration of two proinsulins. Proc Natl Acad Sci USA 62:278–285

Cohen RM, Given BD, Licinio-Paixao J, Provow SA, Rue PA, Frank BH, Root MA et al. (1986) Proinsulin radioimmunoassay in the evaluation of insulinomas and familial hyperproinsulinemia. Metabolism 35:1137–1146

Creutzfeldt C, Track NS, Creutzfeldt W (1973) In vitro studies of the rate of proinsulin and insulin turnover in seven human insulinomas. Eur J Clin Invest 3:371–384

Cutfield JF, Cutfield SM, Dodson EJ, Dodson GG, Emdin SO, Reynolds CD (1979) Structure and biological activity of hagfish insulin. J Mol Biol 132:85–100

Davidson HW, Peshavaria M, Hutton JC (1987) Proteolytic conversion of proinsulin into insulin. Biochem J 246:279–286

Davidson HW, Rhodes CJ, Hutton JC (1988) Intraorganellar calcium and pH control proinsulin cleavage in the pancreatic β cell via two distinct site-specific endopeptidases. Nature 333:93–96

Dayhoff MO (ed) (1973) Atlas of protein sequence and structure, vol 5. Biomedical research Foundation, Bethesda

Docherty K, Hutton JC (1983) Carboxypeptidase activity in the insulin secretory granule. FEBS Lett 162:137–141

Dolva LO, Nielsen JH, Welinder BS, Hanssen KF (1983) Biosynthesis and release of thyrotropin-releasing hormone immunoreactivity in rat pancreatic islets in organ culture – effects of age, glucose, and streptozotocin. J Clin Invest 72:1867–1873

Douglass J, Civelli O, Herbert E (1984) Polyprotein gene expression: generation of diversity of neuroendocrine peptides. Annu Rev Biochem 53:665–715

Emdin SO, Falkmer S (1977) Phytogeny of insulin. Some evolutionary aspects of insulin production with particular regard to the biosynthesis of insulin in *Myxine glutinosa*. Acta Paediatr Scand [Suppl] 270:15–23

Emdin SO, Dodson GG, Cutfield JM, Cutfield SM (1980) Role of zinc in insulin biosynthesis. Diabetologia 19:174–182

Falkmer S (1971) Sulfhydryl compounds and heavy metals in islet morphology and metabolism. In: Rodrique RR, Vallance-Owen JJ (eds) Proceedings of the 7th congress of the International Diabetes Federation. Excerpta Medica, Amsterdam, p 219

Farquhar MG, Palade GE (1981) The Golgi apparatus (complex)-(1954–1981)- from artifact to center stage. J Cell Biol 91:77s–103s

Frank BH, Veros AJ (1968) Physical studies on proinsulin: association behavior and conformation in solution. Biochem Biophys Res Commun 32:155–160

Frank BH, Veros AJ (1970) Interaction of zinc with proinsulin. Biochem Biophys Res Commun 38:284–289

Freychet P, Brandenburg D, Wollmer A (1974) Receptor-binding assay of chemically modified insulins. Comparison with in vitro and in vivo bioassays. Diabetologia 10:1–5

Fricker LD, Evans CJ, Esch FS, Herbert E (1986) Cloning and sequence analysis of cDNA for bovine carboxypeptidase E. Nature 323:461–464

Fuller RS, Sterne RE, Thorner J (1988) Enzymes required for yeast prohormone processing. Annu Rev Physiol 50:345–362

Fullerton WW, Potter R, Low BW (1970) Proinsulin crystallization and preliminary x-ray diffraction studies. Proc Natl Acad Sci USA 66:1213–1219

Gabbay KH, Bergenstal RM, Wolff J, Mako ME, Rubenstein AH (1979) Familial hyperproinsulinemia: partial characterization of circulating proinsulin-like material. Proc Natl Acad Sci USA 76:2881–1885

Giddings SO, Chirgwin J, Permutt MA (1985) Evaluation of rat insulin messenger RNA in pancreatic and extrapancreatic tissues. Diabetologia 28:343–347

Glauber HS, Henry RR, Wallace R, Frank BH, Galloway JA, Cohen RM, Olefsky JM (1987) The effects of biosynthetic human proinsulin on carbohydrate metabolism in non-insulin-dependent diabetes mellitus. N Engl J Med 316:443–449

Gliemann J, Sørenson HH (1970) Assay of insulin-like activity by the isolated fat cell method. IV. The biological activity of proinsulin. Diabetologia 6:499–504

Gold G, Gishizky ML, Grodsky GM (1982) Evidence that glucose "marks" β cells resulting in preferential release of newly synthesized insulin. Science 218:56–58

Gordon JI, Sims HF, Edelstein C, Scanu AM, Strauss AW (1984) Human proapolipoprotein A-II is cleaved following secretion from Hep G2 cells by a thiol protease. J Biol Chem 259:15556–15563

Grant PT, Coombs TL (1971) Proinsulin, a biosynthetic precursor of insulin. In: Campbell PN, Greville GD (eds) Essays in biochemistry, vol 6. Academic, London, p 69–92

Grant PT, Coombs TL, Frank BH (1972) Differences in the nature of the interaction of insulin and proinsulin with zinc. Biochem J 126:433–440

Greider MH, Howell SL, Lacy PE (1969) Isolation and properties of secretory granules from rat islets of Langerhans. II. Ultrastructure of the beta granule. J Cell Biol 41:162–165

Grodsky GM (1970) Insulin and the pancreas. Vitam Horm 28:37

Gruppuso PA, Gorden P, Kahn CR, Cornblath M, Zeller WP, Schwartz R (1984) Familial hyperproinsulinemia due to a proposed defect in conversion of proinsulin to insulin. N Engl J Med 311:629–634

Haneda M, Chan SJ, Kwok SCM, Rubenstein AH, Steiner DF (1983) Studies on mutant human insulin genes: identification and sequence analysis of a gene encoding [Ser^{B24}] insulin. Proc Natl Acad Sci USA 80:6366–6370

Haneda M, Polonsky KS, Bergenstal RM, Jaspan JB, Shoelson SE, Blix PM, Wishner WB et al. (1984) Familial hyperinsulinemia due to a structurally abnormal insulin: definition of an emerging new clinical syndrome. N Engl J Med 310:1288–1289

Hard L (1944) The origin and differentiation of the alpha and beta cells in the pancreatic islets of the rat. Am J Anat 75:369–403

Harper ME, Ullrich A, Saunders GF (1981) Localization of the human insulin gene to the distal end of the short arm of chromosome 11. Proc Natl Acad Sci USA 78:4458–4460

Hedeskov CJ (1980) The mechanisms of glucose-induced insulin secretion. Physiol Rev 60:442–509

Hobart PM, Shen L-P, Crawford R, Pictet RL, Rutter WJ (1980) Comparison of the nucleic acid sequence of anglerfish and mammalian insulin mRNAs from cloned cDNAs. Science 210:1360–1363

Howell SL, Kostianovsky M, Lacy PE (1969) Beta granule formation in isolated islets of Langerhans: a study by electron microscopic radioautography. J Cell Biol 42:695–705

Howell SL, Tyhurst M, Duvefelt H, Andersson A, Hellerström C (1978) Role of zinc and calcium in the formation and storage of insulin in the pancreatic β-cell. Cell Tissue Res 188:107–118

Hutton JC, Nielsen E, Kastern W (1988) The molecular cloning of the chromogranin A-like precursor of β-granin and pancreastatin from the endocrine pancreas. FEBS Lett 236:269–274

Itoh N, Okamoto H (1980) Translational control of proinsulin synthesis by glucose. Nature 283:100–102

Jacobson MR, Baltimore D (1968) Morphogenesis of poliovirus. I. Association of the viral RNA with coat protein. J Mol Biol 33:369–378

Jamieson JD, Palade GE (1968) Intracellular transport of secretory proteins in the pancreatic exocrine cell. IV. Metabolic requirements. J Cell Biol 39:589–603

Julius D, Brake A, Blair L, Kunisawa R, Thorner J (1984) Isolation of the putative structural gene for the lysine-arginine-cleaving endopeptidase required for processing of yeast prepro-β factor. Cell 37:1075–1089

Kanazawa Y, Hayashi M, Ikeuchi M, Kasuga M, Oka Y, Sato H, Hiramatsu K, Kosaka K (1979) Familial proinsulinemia: a rare disorder of insulin biosynthesis. In: Baba S, Kaneko T, Yanaihara N (eds) Proinsulin, insulin, C-peptide. Excerpta Medica, Amsterdam, p 262

Kelly RB (1985) Pathways of protein secretion in eukaryotes. Science 230:25–32

Kemmler W, Peterson JD, Steiner DF (1971) Studies on the conversion of proinsulin to insulin. I. Conversion in vitro with trypsin and carboxypeptidase B. J Biol Chem 246:6786–6791

Kemmler W, Steiner DF, Borg J (1973) Studies on the conversion of proinsulin to insulin. III. Studies in vitro with a crude secretion granule fraction isolated from rat islets of Langerhans. J Biol Chem 248:4544–4551

Kiehn ED, Holland JJ (1970) Synthesis and cleavage of enterovirus polypeptides in mammalian cells. J Virol 5:358–367

Kitabchi AE (1977) Proinsulin and C-peptide: a review. Metabolism 26:547–587

Kohnert KD, Ziegler M, Zühlke H, Fiedler H (1972) Isoelectric focusing of proinsulin and intermediates in polyacrylamide gel. FEBS Lett 28:177–182

Kreil G (1981) Transfer of proteins across membranes. Annu Rev Biochem 50:317–348

Kuzuya OH, Blix PM, Horwitz DL, Rubenstein AH, Steiner DF, Binder C, Farber OK (1978a) Heterogeneity of circulating human C-peptide. Diabetes [Suppl 1] 27:184–191

Kuzuya J, Chance RE, Steiner DF, Rubenstein AH (1978b) On the preparation and characterization of standard materials for natural human proinsulin and C-peptide. Diabetes [Suppl 1] 27:161–169

Kwok SCM, Chan SJ, Rubenstein AH, Poucher R, Steiner DF (1981) Loss of a restriction endonuclease cleavage site in the gene of a structurally abnormal human insulin. Biochem Biophys Res Commun 98:844–849

Kwok SCM, Chan SJ, Steiner DF (1983a) Cloning and nucleotide sequence analysis of the dog insulin gene: coded amino acid sequence of canine preproinsulin predicts an additional C-peptide fragment. J Biol Chem 258:2357–2363

Kwok SCM, Steiner DF, Rubenstein AH, Tager HS (1983b) Identification of a point mutation in the human insulin gene giving rise to a structurally abnormal insulin (insulin Chicago). Diabetes 32:872–875

Lange RH (1974) Crystalline islet B-granules in the grass snake [*Natrix natrix* (L.)]: tilting experiments in the electron microscope. J Ultrastruct Res 46:301–307

Lauweryns JM, de Bock V, Decramer M (1987) Effects of unilateral vagal stimulation in intrapulmonary neuroepithelial bodies. J Appl Physiol 63:1781–1787

Lazarus NR, Penhos JE, Tanese T, Michaels L, Gutman R, Recant L (1970) Studies on the biological activity of porcine proinsulin. J Clin Invest 49:487–496

Lin BJ, Haist RE (1969) Insulin biosynthesis: effects of carbohydrates and related compounds. Can Physiol Pharmacol 47:791–801

Logothetopoulos J, Maneko M, Wrenshall GA, Best CH (1964) Zinc, granulation, and extractable insulin of islet cells following hyperglycemia or prolonged treatment with insulin. In: Falkmer S, Hellman B, Täljedal I (eds) The structure and metabolism of the pancreatic islets. Pergamon, Oxford, p 333 (Wenner-Gren Center international symposium series, vol 3)

Low BW, Fullerton WW, Rosen RS (1974) Insulin/proinsulin, a new crystalline complex. Nature 248:339–340

Madsen OD, Cohen RM, Fitch FW, Rubenstein AH, Steiner DF (1983) Production and characterization of monoclonal antibodies specific for human proinsulin using a sensitive micro-dot assay procedure. Encodrinology 113:2135–2144

Madsen OD, Frank BH, Steiner DF (1984) Human proinsulin specific antigenic determinants identified by monoclonal antibodies. Diabetes 33:1012–1016

Markussen J (1979) Proteolytic degradation of proinsulin and of the intermediate forms: application to synthesis and biosynthesis of insulin. In: Baba S, Kaneko T, Yanaihara N (eds) Proinsulin, insulin, C-peptide. Excerpta Medica, Amsterdam-Oxford, p 50

Markussen J, Sundby F (1972) Rat proinsulin C-peptides. Amino acid sequences. Eur J Biochem 25:153–163

Martino E, Lernmark Å, Seo H, Steiner DF, Refetoff S (1978) High concentration of thyrotropin-releasing hormone in pancreatic islets. Proc Natl Acad Sci USA 75:4265–4267

Meglasson MD, Matschinsky FM (1986) Pancreatic islet glucose metabolism and regulation of insulin secretion. Diabetes Metab Rev 2:163–164

Michael J, Carroll R, Swift H, Steiner DF (1987) Studies on the molecular organization of rat insulin secretory granules. J Biol Chem 262:16531–16535

Mizuno K, Nakamura T, Ohshima T, Tanaka S, Matsuo S, Matsuo H (1988) Yeast KEX2 gene encodes an endopeptidase homologous to subtilisin-like serine proteases. Biochem Biophys Res Commun 156:246–254

Munger BL (1958) A light and electron microscopic study of cellular differentiation in the pancreatic islets of the mouse. Am J Anat 103:275–312

Nagamatsu S, Bolaffi JL, Grodsky GM (1987) Direct effects of glucose on proinsulin synthesis and its processing during desensitization. Endocrinology 120:1225–1231

Nagasawa H, Kataoka H, Isogai A, Tamura S, Suzuki A, Mizoguchi A, Fujiwara Y et al. (1986) Amino acid sequence of a prothoracicotropic hormone of the silkworm Bombyx mori. Proc Natl Acad Sci USA 83:5840–5843

Nanjo K, Sanke T, Miyano M, Okai K, Sowa R, Kondo M, Nishimuar S et al. (1986) Diabetes due to secretion of a structurally abnormal insulin (insulin Wakayama): clinical and functional characteristics of [Leu A3] insulin. J Clin Invest 77:514–519

Nielsen DA, Welsh M, Casadaban MJ, Steiner DF (1985) Control of insulin gene expression in pancreatic β-cells and in an insulin-producing cell line, RIN-5F cells. I. Effects on the transcription of insulin mRNA. J Biol Chem 260:13585–13589

Nolan C, Margolish E, Peterson JD, Steiner DF (1971) The structure of bovine proinsulin. J Biol Chem 246:2780–2795

O'Connor D, Deftos L (1986) Secretion of chromogranin a by peptide-producing endocrine neoplasms. N Engl J Med 314:1145–1151

Orci L (1985) The insulin factory: a tour of the plant surroundings and a visit to the assembly line. Diabetologia 28:528–546

Orci L, Lambert AE, Kanazawa Y, Amherdt M, Rouiller C, Renold AE (1971) Morphological and biochemical studies of B cells in fetal rat endocrine pancreas in organ culture. Evidence for proinsulin biosynthesis. J Cell Biol 50:565–582

Orci L, Ravazzola M, Amherdt M, Madsen O, Vassalli J-D, Perrelet A (1985) Direct identification of prohormone conversion site in insulin-secreting cells. Cell 42:671–681

Orci L, Glick BS, Rothman JE (1986) A new type of coated vesicular carrier that appears not to contain clathrin: its possible role in protein transport within the Golgi stack. Cell 46:171–184

Owerbach D, Bell GI, Rutter WJ, Brown JA, Shows TB (1981) The insulin gene is located on the short arm of chromosome 11 in humans. Diabetes 30:267–270

Oyer PE, Cho E, Peterson JD, Steiner DF (1971) Studies on human proinsulin. Isolation and amino acid sequence of the human pancreatic C-peptide. J Biol Chem 246:1375–1386

Palmieri R, Lee RW-K, Dunn MF (1988) ^{1}H fourier transform NMR studies on insulin: coordination of Ca^{2+} to the Glu(B13) site drives hexamer assembly and induces a conformation change. Biochemistry 27:3387–3397

Patzelt C, Chan SJ, Duguid J, Hortin G, Keim P, Heinrikson RL, Steiner DF (1978a) Biosynthesis of polypeptide hormones in intact and cell-free systems. In: Magnusson S, Ottesen M, Foltmann B, Danø K, Neurath H (eds) Regulatory proteolytic enzymes and their inhibitors. Pergamon, New York, p 69

Patzelt C, Labrecque AD, Duguid JR, Carroll RJ, Keim P, Heinrikson RL, Steiner DF (1978b) Detection and kinetic behavior of preproinsulin in pancreatic islets. Proc Natl Acad Sci USA 75:1260–1264

Permutt MA, Kipnis DM (1971) Insulin biosynthesis. I. On the mechanism of glucose stimulation. J Biol Chem 193:265–275

Peterson JD, Nehrlich S, Oyer PE, Steiner D (1972) Determination of the amino acid sequence of the monkey, sheep and dog proinsulin C-peptides by a semi-micro Edman degradation procedure. J Biol Chem 247:4866–4871

Peterson JD, Steiner DF, Emdin SO, Falkmer S (1975) The amino acid sequence of the insulin from a primitive vertebrate, the Atlantic hagfish (*Myxine glutinosa*). J Biol Chem 250:5183–5191

Polonsky K, Rubenstein AH (1986) Current approaches to measurement of insulin secretion. Diabetes Metab Rev 2:315–329

Potts JT Jr, Kronenberg HM, Rosenblatt M (1982) Parathyroid hormone: chemistry, biosynthesis, and mode of action. Adv Protein Chem 35:323–396

Powell SK, Orci L, Craik CS, Moore H-PH (1988) Efficient targeting to storage granules of human proinsulins with altered propeptide domain. J Cell Biol 106:1843–1851

Prentki M, Matschinsky FM (1987) Ca^{2+}, cAMP, and phospholipid-derived messengers in coupling mechanisms of insulin secretion. Physiol Rev 67:1185–1248

Quinn PS, Gamble M, Judah JD (1975) Biosynthesis of serum albumin in rat liver – isolation and probable structure of proalbumin from rat liver. Biochem J 146:389–393

Rabinovitch A, Blondel B, Murray T, Mintz DH (1980) Cyclic adenosine-3′,5′-monophosphate stimulates islet B cell replication in neonatal rat pancreatic monolayer cultures. J Clin Invest 66:1065–1071

Rhodes CJ, Halban PA (1987) Newly synthesized proinsulin/insulin and stored insulin are released from pancreatic B cells predominantly via a regulated, rather than a constitutive, pathway. J Cell Biol 105:145–153

Rhodes CJ, Halban PA (1988) The intracellular handling of insulin-related peptides in isolated pancreatic islets. Biochem J 251:23–30

Robbins DC, Blix PM, Rubenstein AH, Kanazawa Y, Kosaka K, Tager HS (1981) A human proinsulin variant at arginine 65. Nature 291:679–681

Robbins DC, Shoelson SE, Rubenstein AH, Tager HS (1984a) Familial hyperproinsulinemia: two cohorts secreting indistinguishable type II intermediates of proinsulin conversion. J Clin Invest 73:714–719

Robbins DC, Tager HS, Rubenstein AH (1984b) Biologic and clinical importance of proinsulin. N Engl J Med 310:1165–1175

Rosen LS, Fullerton WW, Low BW (1972) Proinsulin: further crystallization and x-ray crystallographic studies of bovine and porcine prohormone. Arch Biochem Biophys 152:569–573

Rotwein PS, Chirgwin J, Province M, Knowler WC, Pettitt DJ, Cordell B, Goodman HM, Perutt MA (1983) Polymorphism in the 5′ flanking region of the human insulin gene: a genetic marker for non-insulin-dependent diabetes. N Engl J Med 308:65–71

Rubenstein AH, Clark JL, Melani F, Steiner DF (1969a) Secretion of proinsulin C-peptide by pancreatic B cells and its circulation in blood. Nature 224:697–699

Rubenstein AH, Melani F, Pilkis S, Steiner DF (1969b) Proinsulin: secretion, metabolism, immunological and biological properties. Postgrad Med J [Suppl] 45:476–481

Rubenstein AH, Welbourne WP, Mako M, Melani F, Steiner DF (1970) Comparative immunology of bovine, porcine, and human proinsulin and C-peptides. Diabetes 19:546–553

Rubenstein AH, Steiner DF, Horwitz DL, Mako ME, Block MB, Starr JI, Kuzuya H, Melani F (1977) Clinical significance of circulating proinsulin and C-peptide. Recent Prog Horm Res 33:435–475

Russell JH, Geller DM (1975) The structure of rat proalbumin. J Biol Chem 250:3409–3413

Sando H, Grodsky GM (1973) Dynamic synthesis and release of insulin and proinsulin from perifused islets. Diabetes 22:354–359

Sando H, Borg J, Steiner DF (1972) Studies on the secretion of newly synthesized proinsulin and insulin from isolated rat islets of Langerhans. J Clin Invest 51:1476–1485

Sanke T, Bell GI, Sample C, Rubenstein AH, Steiner DF (1988) An islet amyloid peptide is derived from an 89-amino acid precursor by proteolytic processing. J Biol Chem 263:17243–17246

Shibasaki Y, Kawakami T, Kanazawa Y, Akanuma Y, Takaku F (1985) Posttranslational cleavage of proinsulin is blocked by a point mutation in familial hyperproinsulinemia. J Clin Invest 76:378–380

Shoelson S, Fickova M, Haneda M, Nahum A, Musso G, Kaiser ET, Rubenstein AH, Tager H (1983a) Identification of a mutant human insulin predicted to contain a serine-for-phenylalanine substitution. Proc Natl Acad Sci USA 80:7390–7394

Shoelson S, Haneda M, Blix P, Nanjo A, Sanke T, Inouye K, Steiner D et al. (1983b) Three mutant insulins in man. Nature 302:540–543

Shoelson SE, Polonsky KS, Zeidler A, Rubenstein AH, Tager HS (1984) Human insulin B24 (Phe-Ser): secretion and metabolic clearance of the abnormal insulin in man and in a dog model. J Clin Invest 73:1351–1358

Skidgel R (1988) Basic carboxypeptidases: regulators of peptide hormone activity. Trends Pharmacol Sci 9:299–304

Smit AB, Breugdenhil E, Ebberink RHM, Geraerts WPM, Klootwijk J, Joosse J (1988) Growth-controlling molluscan neurons produce the precursor of an insulin-related peptide. Nature 331:535–538

Smith LF (1966) Species variation in the amino acid sequence of insulin. Am J Med 40:662–666

Steiner DF (1967) Evidence for a precursor in the biosynthesis of insulin. Trans NY Acad Sci 30:60–68

Steiner DF (1973) Cocrystallization of proinsulin and insulin. Nature 243:528–530

Steiner DF (1984) The biosynthesis of insulin: genetic, evolutionary and pathophysiologic aspects. In: The Harvey lectures, series 78. Academic, New York, pp 191–228

Steiner DF, Clark JL (1968) The spontaneous reoxidation of reduced beef and rat proinsulins. Proc Natl Acad Sci USA 60:622–629

Steiner DF, Oyer PC (1967) The biosynthesis of insulin and a probable precursor of insulin by a human islet cell adenoma. Proc Natl Acad Sci USA 57:473–480

Steiner DF, Rubenstein AH (1973) Recent studies on the biosynthesis, secretion and metabolism of proinsulin and C-peptide. In: Musacchia XJ, Breitenbach RP (eds) Proceedings of the 8th midwest conference on endocrinology and metabolism. University of Missouri-Columbia, pp 43–59

Steiner DF, Cunningham DD, Spigelman L, Aten B (1967) Insulin biosynthesis: evidence for a precursor. Science 157:697–700

Steiner DF, Clark JL, Nolan C, Rubenstein AH, Margoliash E, Aten B, Oyer PE (1969) Proinsulin and the biosynthesis of insulin. Recent Prog Horm Res 25:207–268

Steiner DF, Clark JL, Nolan C, Rubenstein AH, Margoliash E, Melani F, Oyer PE (1970) The biosynthesis of insulin and some speculations regarding the pathogenesis of human diabetes. In: Cerasi E, Luft R (eds) The pathogenesis of diabetes mellitus. Almqvist and Wiksell, Stockholm, p 57 (Novel symposium 13)

Steiner DF, Cho S, Oyer PE, Terris S, Peterson JD, Rubenstein AH (1971) Isolation and characterization of proinsulin C-peptide from bovine pancreas. J Biol Chem 246:1365–1374

Steiner DF, Kemmler W, Clark JL, Oyer PE, Rubenstein AH (1972) The biosynthesis of insulin. In: Steiner DF, Freinkel N (eds) Endocrinology I. Williams and Wilkins, Baltimore, p 175 (Handbook of physiology)

Steiner DF, Peterson JD, Tager H, Emdin S, Falkmer S (1973) Comparative aspects of proinsulin and insulin structure and biosynthesis. Am Zool 13:591–604

Steiner DF, Kemmler W, Tager HS, Rubenstein AH, Lernmark Å, Zühlke H (1975) Proteolytic mechanisms in the biosynthesis of polypeptide hormones. In: Reich E, Rifkin D, Shaw E (eds) Proteases and biological control. Cold Spring Harbor Laboratory, Cold Spring Harbor, p 531

Steiner DF, Terris S, Chan SJ, Rubenstein AH (1977) Chemical and biological aspects of insulin and proinsulin. Acta Med Scand [Suppl] 601:53–107

Steiner DF, Quinn PS, Chan SJ, Marsh J, Tager HS (1980) Processing mechanisms in the biosynthesis of proteins. Proc NY Acad Sci 343:1–16

Steiner DF, Docherty K, Carroll R (1984) Golgi/granule processing of peptide hormone and neuropeptide precursors (A minireview). J Cell Biochem 24:121–130

Steiner DF, Chan SJ, Welsh JM, Kwok SCM (1985) Structure and evolution of the insulin gene. Annu Rev Genet 19:463–484

Steiner DF, Chan SJ, Welsh JM, Nielsen D, Michael J, Tager HS, Rubenstein AH (1986) Models of peptide biosynthesis – the molecular and cellular basis of insulin production. Clin Invest Med 9:328–336

Steiner DF, Bell GI, Tager HS (1989) Chemistry and biosynthesis of pancreatic protein hormones. In: DeGroot L (ed) Endocrinology. Saunders, Philadelphia, p 1263

Stevens T, Esmon B, Schekman R (1982) Early stages in the yeast secretory pathway are required for transport of carboxypeptidase Y to the vascule. Cell 30:439–448

Straus DS (1984) Growth-stimulatory actions of insulin in vitro and in vivo. Endocr Rev 5:356–369

Sundby F, Markussen J (1970) Preparation method for the isolation of C-peptides from ox and pork pancreas. Horm Metab Res 2:17–20

Swenne I (1982) Effects of cyclic AMP on DNA replication and protein biosynthesis in fetal rat islets of Langerhans maintained in tissue culture. Biosci Rep 2:867–876

Tager HS, Steiner DF (1972) Primary structures of the proinsulin connecting peptides of the rat and horse. J Biol Chem 247:7936–7940

Tager HS, Emdin SO, Clark JL, Steiner DF (1973) Studies on the conversion of proinsulin to insulin. II. Evidence for a chymotrypsin-like cleavage in the connecting peptide region of insulin precursors in the rat. J Biol Chem 248:3476–3482

Tager HS, Given B, Baldwin D, Mako M, Markese J, Rubenstein AH, Olefsky J et al. (1979) A structurally abnormal insulin causing human diabetes. Nature 281:122–125

Tanese T, Lazarus NR, Devrim S, Recant L (1970) Synthesis and release of proinsulin and insulin by isolated rat islets of Langerhans. J Clin Invest 49:1394–1404

Terris S, Steiner DF (1975) Binding and degradation of ^{125}I-insulin by rat hepatocytes. J Biol Chem 250:8389–8398

Terris S, Steiner DF (1976) Retention and degradation of ^{125}I-insulin by perfused rat livers. J Clin Invest 57:885–896

Terris S, Hofmann C, Steiner DF (1979) Mode of uptake and degradation of ^{125}I-labelled insulin by isolated hepatocytes and H4 hepatoma cells. Can J Biochem 57:459–468

Thim L, Hansen MT, Norris K, Hoegh I, Boel E, Forstrom J, Ammerer G, Fiil NP (1986) secretion and processing of insulin precursors in yeast. Proc Natl Acad Sci USA 83:6766–6770

Tung AK, Yip CC (1969) Biosynthesis of insulin in bovine fetal pancreatic slices: the incorporation of tritiated leucine into a single-chain proinsulin, a double-chain intermediate, and insulin in subcellular fractions. Proc Natl Acad Sci USA 63:442–449

Ullrich A, Shine J, Chirgwin J, Pictet R, Tischer E, Rutter WJ, Goodman HM (1977) Rat insulin genes: construction of plasmids containing the coding sequences. Science 196:1313–1319

Ullrich A, Dull TJ, Gray A, Brosius J, Sures I (1980) Genetic variation in the human insulin gene. Science 209:612–615

Valverde I, Garcia-Morales P, Ghiglione M, Malaisse WJ (1983) The stimulus-secretion coupling of glucose-induced insulin release. LIII. Calcium dependency of the cyclic AMP response to nutrient secretagogues. Horm Metab Res 15:62–68

Walter P, Gilmore R, Blobel G (1984) Protein translocation across the endoplasmic reticulum. Cell 38:5–8

Wattenberg BW, Rothman JE (1986) Multiple cytosolic components promote intra-Golgi protein transport: resolution of a protein acting at a late state, prior to membrane fusion. J Biol Chem 261:2208–2213

Welsh M, Nielsen DA, MacKrell AJ, Steiner DF (1985a) Control of insulin gene expression in pancreatic β-cells and in an insulin-producing cell line, RIN-5F cells. II. Regulation of insulin mRNA stability. J Biol Chem 260:13590–13594

Welsh M, Scherberg N, Gilmore R, Steiner DF (1985b) Translational control of insulin biosynthesis: evidence for regulation of elongation, initiation and signal recognition particle-mediated translational arrest by glucose. Biochem J 235:459–467

Westermark P, Wilander E, Westermark GT, Johnson KH (1987) Islet amyloid polypeptide-like immunoreactivity in the islet B cells of type 2 (non-insulin-dependent) diabetic and non-diabetic individuals. Diabetologia 30:887–892

Yamaji K, Tada K, Trakatellis AC (1972) On the biosynthesis of insulin in anglerfish islets. J Biol Chem 247:4080–4088

Zühlke H, Steiner DF, Lernmark Å, Lipsey C (1976) Carboxypeptidase B-like and trypsin-like activities in isolated rat pancreatic islets. Ciba Found Symp 41:183

CHAPTER 5

Insulin Gene Regulation

M. D. WALKER

A. Overview

Insulin genes from several mammalian species have been isolated and characterized. This has permitted significant progress towards understanding the mechanisms involved in regulation of expression of the gene. In particular, it has been
established that specific DNA sequences located in the 5' flanking region of the
gene are responsible for restricting transcription exclusively to β cells of the
endocrine pancreas. Selective transcription appears to involve recognition of
these sequences by protein factors present in β cells. The characterization of such
sequence-specific DNA binding proteins and the genes which encode them will
contribute decisively to our understanding of differentiation and development of
the endocrine pancreas.

B. Introduction

The expression of genetic information is a complex multistage process. In principle, the rate of synthesis of a protein can be controlled by modulation of any
one of several steps, and indeed multiple modes of control have been documented
(reviewed in DARNELL 1982). However, a key control point in both prokaryotic
and eukaryotic cells is at the stage of transcription initiation. In mammals, insulin genes are present in all cell types examined, yet are transcribed exclusively in
β cells of the endocrine pancreas. The detailed mechanisms underlying transcriptional control of the insulin gene, and indeed of other genes which are expressed
in a restricted subset of cells, are unknown.

    In prokaryotic systems, where powerful genetic tools have permitted a
detailed analysis of regulatory mechanisms, transcription initiation is controlled
by the association of specific proteins with DNA sequences located immediately
upstream of the transcribed DNA (JACOB and MONOD 1961; PTASHNE 1986). This
interaction can lead to activation or repression of transcription, often as a consequence of direct contact between the specific DNA binding protein and RNA
polymerase.

    For complex multicellular organisms, comparable genetic tools are not avail-
able. However, in recent years it has become possible to isolate eukaryotic genes,
to mutate them in vitro in precise fashion, and to observe the expression of the
gene in cultured cells (PELLICER et al. 1980) or intact animals (PALMITER and
BRINSTER 1986). Such analyses of transcription control regions of protein-

encoding genes of mammalian cells, have led to the identification of two major classes of control element (reviewed in McKNIGHT and TJIAN 1986; MANIATIS et al. 1987). The promoter is a DNA element located proximal to the transcription start site that is capable of initiating transcription; its activity is dependent on position and orientation relative to the transcribed sequence. In many genes, the promoter contains a consensus TATA sequence located about 20–25 bp upstream from the transcription initiation site (BREATHNACH and CHAMBON 1981). This sequence is necessary for selection of the appropriate start site and maximal rates of initiation. Promoters often contain additional short sequence elements (upstream promoter elements) which are required for optimal rates of initiation. In contrast to promoters, enhancers are incapable of initiating transcription, but rather increase the activity of homologous or heterologous promoters relatively independently of orientation or precise positioning with respect to the transcription unit. Enhancers were initially identified in genes of animal viruses (KHOURY and GRUSS 1983) and subsequently found in several cellular genes (SERFLING et al. 1985). The enhancers associated with genes expressed in specific cell types are themselves active only in the appropriate cell; immunoglobulin gene enhancers are active only in lymphoid cells (GILLIES et al. 1983; BANERJI et al. 1983), whereas the insulin gene enhancer is active only in pancreatic β cells (EDLUND et al. 1985). Thus, the enhancer often represents an important component involved in directing the specificity of expression of the associated gene.

Both promoters and enhancers are believed to function by interacting directly with sequence-specific DNA binding proteins. For some mammalian transcription control elements, the cognate binding factors have been identified and the genes encoding them molecularly cloned (EVANS 1988; SANTORO et al. 1988). This has permitted analysis of structure–function relationships. In some cases these proteins have been shown to contain distinct functional domains, one capable of recognizing the appropriate DNA sequence and another capable of transcription activation (KEEGAN et al. 1986; EVANS 1988). This supports the idea that transcription activation occurs as a result of direct protein–protein interactions between DNA binding proteins and components of the transcription machinery, possibly RNA polymerase (SIGLER 1988).

C. Insulin Gene Structure

Insulin genes have been isolated and characterized from primates, carnivores, rodents, birds, and fish (reviewed in STEINER et al. 1985). In addition genes coding for insulin-related peptides have been isolated from insects (NAGASAWA et al. 1986) and molluscs (SMIT et al. 1988). Most mammalian species contain a single insulin gene consisting of three exons and two introns. Rats and mice contain an additional nonallelic gene which is closely related except for the loss of the second intron (LOMEDICO et al. 1979; SOARES et al. 1985). In mice the two nonallelic genes are located on separate chromosomes, whereas in rats both genes reside on chromosome 1, separated by about 100 000 kb (SOARES et al. 1985). The insulin gene in humans is located on chromosome 11 (OWERBACH et al. 1980) 2.7 kb downstream of the tyrosine hydroxylase gene (O'MALLEY and ROTWEIN 1988) and 1.6 kb upstream of the insulin-like growth factor II gene (Fig. 1; BELL et al.

1985; DE PAGTER-HOLTHUIZEN et al. 1987). The human insulin gene contains a highly polymorphic region located upstream of nucleotide −363 in the 5′ flanking region (BELL et al. 1981) and consisting of varying numbers of a 14–15 bp repeat (BELL et al. 1982). Reports linking classes of length variant to the incidence of forms of diabetes (ROTWEIN et al. 1983; BELL et al. 1984a) have yet to be confirmed.

Sequence comparison between the nonallelic genes suggest that the additional rodent gene was derived by a process of reverse transcription of a rare, partially spliced germ line RNA (SOARES et al. 1985). Initiation of this transcript apparently occurred several hundred base pairs upstream of the normal start site since the rat insulin I gene (the product of the reverse transcription event) shows clear similarity with the rat insulin II gene for at least 500 bp upstream of the start site of transcription. The rat insulin I and II genes are approximately equally expressed (CLARK and STEINER 1969) in normal islet cells (though not in some tumors; CORDELL et al. 1982), indicating that essentially all sequences involved in determining the appropriate expression pattern of the gene were present on the original partially spliced transcript.

Sequence comparisons among the various vertebrate genes (SOARES et al. 1985) have shown that mutations leading to amino acid substitutions are much less frequent in regions coding for the A and B peptides than regions coding for the C peptide and leader peptide. Predictably however, silent mutations are observed at similar rates in the various protein-coding regions, consistent with the notion that amino acid substitutions are strongly selected against in the A and B peptides, but not in the C and leader peptides. BELL et al. (1982) first noted that the 5′ flanking regions of the rat and human insulin genes can be aligned to reveal extensive sequence conservation from the transcription start site to about 300 bp upstream, which was initially interpreted as indicating an important function for these sequences. Later, as the sequences of additional insulin genes became available, the extent of substitutions among the 5′ flanking sequences (37%) was found to be similar to that observed among introns (38%) or 3′ flanking sequences (37%) (SOARES et al. 1985). If one assumes that introns and 3′ flanking sequences are not constrained by selection pressures, one might conclude that 5′ flanking regions are evolving neutrally, which is unexpected given the role of 5′ flanking sequence in transcriptional control (see Sect. E). However, the sequence comparisons are difficult to interpret unambiguously given the uncertainty concerning the possible role of specific sequences within introns and 3′ flanking DNA. Furthermore, although the overall rate of substitution in the 5′ flanking region of rat, mouse, and human genes is 37%, distinct portions of the flank show very much lower rates of substitution. Indeed, these sequences correspond to some degree with the sequences which were subsequently identified as transcriptional control elements using in vivo assay (see Sect. E.V).

D. Experimental Approaches

The unambiguous definition of transcription control elements requires a functional assay in which putative regulatory DNA sequences are introduced to live cells and their ability to direct expression measured. The use of hybrid genes (reviewed in KELLY and DARLINGTON 1985) containing putative regulatory sequences linked to a reporter gene encoding an easily measureable product provides an elegant and convenient method for establishing such a functional assay. Choice of an appropriate reporter gene permits sensitive detection of activity of the regulatory region independently of expression of the endogenous counter-

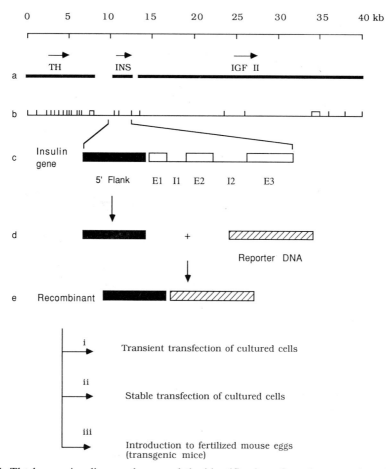

Fig. 1. The human insulin gene locus and the identification of regulatory regions. Row *a* shows the transcribed regions (exons and introns) and direction of transcription of the tyrosine hydroxylase gene (*TH*), the insulin gene (*INS*), and the insulin-like growth factor gene (*IGF-II*). Row *b* shows the position of exons in the locus. Rows *c–e* show the stages in construction of a hybrid gene comprising 5′ flanking regions of the insulin gene. E1–3 represent the exons and I1–2 the introns of the insulin gene. For transient assays the use of an easily assayable enzyme as reporter is common. For stable assays, it is necessary to use an enzyme whose expression can be selected for. However, the gene of interest can be one for which no selective criteria are available if transfection is performed in the presence of an additional marker for which selection conditions are available (GORMAN 1986). For transgenic mice, the reporter function can be an oncoprotein, enzyme, or any gene product whose expression can be readily assayed

part gene. Three major types of experimental approach are commonly utilized for evaluating the expression of such hybrid genes (Fig. 1).

I. Transient Assay

(Reviewed in GORMAN 1986)

Hybrid genes are introduced to cultured cells by exposure to DNA in the presence of calcium phosphate, DEAE dextran, or by direct electroporation. A small percentage of cells take up and express the DNA. However, since integration of the DNA to the host genome is a rare event, transcription is transient and after a few days expression levels become undetectable. Therefore, expression of the reporter is usually measured within about 48 h of exposure of the cells to DNA. This method permits rapid and convenient delineation of control elements which function in established cell lines.

II. Stable Assay

(Reviewed in GORMAN 1986)

Occasionally the exogenous DNA in a transfected cell becomes integrated into the chromosome of the host. If this DNA contains a gene encoding a selectable marker, for example an enzyme required for survival in the presence of antibiotic or antimetabolites (MULLIGAN and BERG 1981; SOUTHERN and BERG 1982), then application of this selection will permit isolation of those cells containing and expressing the exogenous gene. Although the selection procedure takes considerably longer to carry out than a typical transient transfection (2–4 weeks depending on the growth rate of the cells) an advantage is that a stable line is generated which can subsequently be characterized in detail.

III. Transgenic Animals

(Reviewed in PALMITER and BRINSTER 1986)

Established cell lines are valuable models for the corresponding cell type in intact animals. However, they differ significantly from the normal counterpart, perhaps most strikingly in their ability to grow indefinitely in culture, and usually also in their more limited expression of the differentiated phenotype. Are transcription control elements defined by transfection of established cell lines equally active in the natural, in vivo, setting? The production of transgenic animals (most commonly mice) permits this issue to be addressed directly. DNA is injected by glass micropipette to fertilized one-cell embryos. Injected embryos are implanted to the oviduct of a foster mother, and are permitted to develop to term. The expression of the injected DNA (transgene) is then measured in the cell types of interest. Since every cell in a given transgenic mouse contains the transgene in an identical chromosomal position, every possible cell type is theoretically available for comparison of expression, a situation which does not apply for cell culture experiments. Furthermore, the transgene is present from the one-cell stage onward and therefore is exposed to similar influences during development as the normal endogenous gene.

E. Identification of Control Sequences

I. Role of 5′ Flanking DNA

Cell-specific control sequences in insulin genes were initially identified (WALKER et al. 1983) using a transient transfection procedure based on the use of chloramphenicol acetyl transferase (CAT), a reporter function developed by GORMAN et al. (1982a). Hybrid genes were constructed containing 5′ flanking sequences of the rat insulin I gene, the human insulin gene, the chymotrypsin B gene (a gene expressed in the exocrine pancreas only; BELL et al. 1984b) and the Rous sarcoma virus LTR (a promoter–enhancer complex which is active in a wide range of cell types; GORMAN et al. 1982b). These sequences were fused to the DNA encoding CAT, and hybrid genes were introduced to cultured cell lines derived from endocrine pancreas (HIT cells; SANTERRE et al. 1981), exocrine pancreas (JESSOP and HAY 1980), or nonpancreas cells (CHO). Constructions containing 5′ flanking sequences of the rat or human insulin gene were 50- to 200-fold more active in a line of cells derived from the pancreas than from fibroblasts or pancreatic exocrine cells (Fig. 2; WALKER et al. 1983). This result demonstrated several significant points:

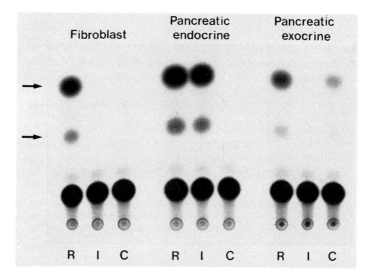

Fig. 2. Chloramphenicol acetyltransferase (CAT) activity directed by eukaryotic 5′ flanking sequences in different cells. Recombinant plasmids containing 5′ flanking sequences were applied to tissue culture cells. After 48 h, CAT activity from extract containing 25 µg protein was determined. The photograph shows an autoradiogram of the thin layer chromatography separation of CAT reaction products. The cell types used were fibroblasts (CHO), pancreatic endocrine (HIT) cells, and pancreatic exocrine (AR4-2J) cells (JESSOP and HAY 1980). The upper two spots, indicated by *arrows*, correspond to the two isomers of [¹⁴C]chloramphenicol monoacetate and the lower intense spot corresponds to unreacted [¹⁴C]chloramphenicol. The 5′ sequences are *R* Rous sarcoma virus LTR; *I* rat insulin I; *C* chymotrypsin

1. At least a portion of the information responsible for determining cell-specific expression is present in a relatively short length of DNA (in the case of the rat insulin I gene, 410 bp of 5′ flanking DNA sequence).
2. A rapid and convenient experimental protocol can be used to identify these sequences.
3. Since under the conditions of these experiments the transfected DNA is not located in its normal position in the genome, and in fact does not integrate into the chromosome to any significant extent, the positional requirements for reconstituting the cell-specific effect are relatively nonstringent.

The important role of 5′ flanking sequence in cell-specific expression has been confirmed using two independent approaches. EPISKOPOU et al. (1984) generated stable cell lines using retrovirus vectors in which the 5′ flanking region of the rat insulin II gene was linked to the DNA encoding the xanthine–guanine phosphoribosyl transferase (*gpt*) gene. Infection of cells with this marker, followed by application of appropriate selection conditions, led to the conclusion that efficient expression occurs in insulin-producing cell lines, but not in fibroblasts. HANAHAN (1985) prepared transgenic mice containing the promoter of the rat insulin II gene linked to the DNA encoding an oncoprotein, the SV40 T antigen. Such mice developed tumors of the endocrine β cells, and in no other cell, suggesting that expression of the transgene was limited to this cell type.

Deletion analysis of the flanking DNA regions (WALKER et al. 1983) revealed that for the rat and human insulin genes, sequences upstream of about −300, including the polymorphic region present on the insulin gene (BELL et al. 1981) had no effect on expression of the linked gene in cultured HIT cells. However, removal of sequences located downstream of −302 led to dramatic losses in activity.

II. Cell-Specific Enhancer and Promoter

The type of regulatory signals involved were further examined in a second series of experiments (EDLUND et al. 1985) designed to identify transcriptional enhancer sequences in the 5′ flanking DNA. The analysis revealed that the 5′ flanking DNA of the rat insulin I gene does contain a cell-specific enhancer located between about −103 and −333. This enhancer is capable of strongly activating (40-fold) the thymidine kinase (TK) promoter in β cells. No activation was seen in non-β cells, indicating that the enhancer represents an important component of the machinery determining cell-specific expression of the insulin gene.

Although DNA sequences downstream of this region, comprising the insulin promoter, displayed negligible enhancer activity (EDLUND et al. 1985), this region displays considerable conservation among the characterized mammalian genes (BELL et al. 1982; CHAN et al. 1984). In order to test whether these sequences are involved in cell specificity, the enhancer region of the insulin gene was substituted with the enhancer of the Moloney sarcoma virus (MSV) which is known to display activity in a wide range of cell types (LAIMINS et al. 1982). In this "enhancer swap" situation, the novel construction retained 10-fold preferential expression in the insulin-producing cell type, indicating that the promoter region per se retains significant ability to direct β cell-specific expression despite the absence of enhancer activity. Thus, cell-specific transcription of the insulin gene appears to result from the combined action of at least two functionally distinct elements, a cell-specific enhancer and a cell-specific promoter. A similar situation has been described for immunoglobulin genes (GROSSCHEDL and BALTIMORE 1985).

III. Positive and Negative Control

In principle, cell-specific gene expression could result from positively acting control mechanisms in producing cells or negative control in nonproducing cells, or from a combination of the two. Indeed, experiments based on fusion of cell types of unlike differentiated phenotype have given evidence for both types of control (reviewed in LEWIN 1980). In general, when such cell fusion experiments are performed the outcome is loss of expression or "extinction" of the differentiated phenotype (KILLARY and FOURNIER 1984). For example, when liver cells are fused to fibroblasts the expression of characteristic liver genes is lost. When such hybrid cells are cultured over time they lose chromosomes and the liver phenotype can be regained. Interestingly, this reacquisition of phenotype correlates with the loss of a specific fibroblast chromosome. These results are consistent with the notion that fibroblasts contain transcriptional repressor molecules which reduce expression of genes normally expressed only in other differentiated cell types. On the other hand, cell fusion can occasionally result in activation of a previously silent locus. This situation is best characterized in cases where muscle cells are a partner in the fusion. Production of heterokaryons can lead to activation of previously silent muscle genes in the fusion partner (BLAU et al. 1985). In the deletion analysis discussed above (Sect. E.I), deletion of sequences led to loss of expression, a result consistent with the loss of a binding site for a positively acting transcription control factor. No increases in expression following such deletions in either expressing or nonexpressing cells were observed. To test directly the possible involvement of negatively acting factors in reducing expression in heterologous cells, an in vivo competition approach was used. The procedure involved introduction to COS fibroblast cells (GLUZMAN 1981) of two plasmids, one containing insulin enhancer sequences linked to the CAT reporter gene (test plasmid) and the other (competitor plasmid) containing the same enhancer sequences without CAT sequences, but containing the SV40 origin of replication (to permit replication in COS cells). Upon introduction of these plasmids to COS cells the competitor plasmid (but not the test plasmid) will replicate to high copy number and titrate out putative repressor molecules, leading to derepression of the test plasmid. Indeed, when such experiments are performed, the presence of insulin enhancer sequences on both plasmids led to a fivefold elevation of expression as compared with a situation where the enhancer is absent from either or both plasmids (NIR et al. 1986). Effectively, the presence of the competitor plasmids reveals cryptic activity of the insulin enhancer in an inappropriate cell type. The role of repressor molecules may be to reduce such "leaky" expression. The putative repressor appears to have a less stringent sequence requirement for binding than the positive factors since the rat amylase enhancer (BOULET et al. 1986), when present at high copy number, was able to activate cryptic enhancer activity in COS cells. If the function of the repressor is to suppress a subset of cell-specific enhancers, it may be present also in insulin-producing cells where its effects are presumably overridden by the dominant activating effects of positive transactivators. Evidence for the existence of additional negatively acting DNA elements in the rat insulin I gene was obtained by LAIMINS et al. (1986), who identified such "silencer" regions 2–4 kb upstream of the rat insulin I gene.

IV. Regulation by Glucose

Insulin plays a central role in integrating carbohydrate metabolism. In order to fulfill this role the output of insulin from the pancreas must respond appropriately to fluctuating levels in the bloodstream of several metabolites, most importantly glucose (HEDESKOV 1980). This is achieved by regulation at multiple levels, including translation (WELSH et al. 1986), mRNA stability (WELSH et al. 1985), and transcription (NIELSEN et al. 1985). To confirm the effect at the level of transcription, and to address the issue of the underlying mechanism, the ability of glucose to increase transcription from a hybrid gene including 410 bp of the rat insulin I gene 5' flanking DNA linked to CAT, was measured following introduction to HIT cells. The presence of 20 mM glucose reproducibly led to a fivefold increase in CAT expression levels using such constructions (L. MOSS, W.J. RUTTER and M.D. WALKER, unpublished work). Plasmids containing unrelated transcriptional control sequences did not respond significantly to glucose. Thus, glucose effects on transcription are mediated at least in part through 5' flanking DNA sequences.

V. Systematic Mutagenesis

Given the complexity of the insulin regulatory region, KARLSSON et al. (1987) set out to define the precise location of the various types of control element. The procedure used was based conceptually on the "linker scanning" approach developed by MCKNIGHT and KINGSBURY (1982). To generate a systematic series of replacement mutants, overlapping oligonucleotides were ligated together into an expression plasmid and the structure of the regulatory region verified by sequencing. For mutant sequences, oligonucleotides were synthesized containing blocks of noncomplementary transversions (i.e., A↔C, G↔T) spanning an average of 10 bp. All plasmids were introduced to cultured HIT cells and expression levels determined by measurement of CAT enzymatic activity and steady state levels of CAT mRNA (KARLSSON et al. 1987). Although mutation of several regions led to marked reduction in activity, relatively few mutations cause dramatic loss of activity (Fig. 3). The most severely affected regions are (−23 to −32) which spans the TATA box, mutant (−104 to −112), and mutant (−233 to −241). The TATA homology, which is found in most *pol II* genes, is unlikely to be contributing to cell-specific expression and is probably involved in interacting with constitutive components of the transcription machinery. However, the two other mutants implicate regions which are conserved well among the various insulin genes (CHAN et al. 1984). Similar though nonidentical sequence motifs

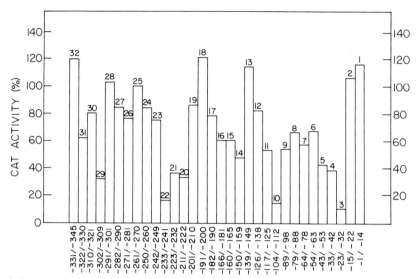

Fig. 3. Systematic mutational analysis of rat insulin I gene 5′ flanking DNA. Plasmids bearing mutations were tranfected into HIT cells and CAT activity was measured. Locations of the mutated sequences are indicated. Activities are expressed relative to that directed by the wild-type 5′ flanking DNA

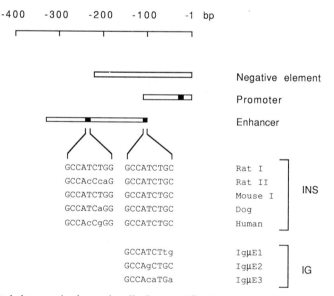

Fig. 4. Control elements in the rat insulin I gene 5′ flanking DNA. The *open bars* show the location of the negative element, promoter, and enhancer. The *filled region* within the promoter indicates the location of the TATA sequence. The *filled region* within the enhancer shows the location of the evolutionarily conserved sequences, mutations within which lead to reductions of activity in transfected HIT cells. The sequences within these regions are shown together with the comparable regions of the rat II, mouse I, dog, and human insulin (*INS*) genes. In addition, the similar regions found in the immunoglobulin (*IG*) heavy chain enhancer (EPHRUSSI et al. 1985; MOSS et al. 1988) are shown. *Uppercase letters* indicate conserved bases whereas *lowercase* indicates altered bases as compared with the rat insulin I gene sequence

(Fig. 4) are implicated in the activity of immunoglobulin enhancers (EPHRUSSI et al. 1985; Moss et al. 1988). Interestingly the -104 to -112 (GCCATCTGC) and -233 to -241 (GCCATCTGG) regions are identical to each other in eight of nine positions (Fig. 4), suggesting that they are involved in interactions with similar or identical transcription factors. Furthermore, these sequences on their own (in the absence of any additional sequence from the rat insulin gene) are able to elevate expression from a heterologous promoter: as a single copy the sequence stimulates expression approximately 2- to 3-fold in insulin-producing cells. When two copies are combined the expression increases to 5- to 10-fold (KARLSSON et al. 1989). Accordingly, it appears that these sequences play an important part in determining cell-specific expression. Although the regulatory region is complex, a measure of cell specificity can be obtained using a very small unit, possibly corresponding to the binding site of a single protein. This observation may significantly simplify analysis of control mechanisms. The locations of the identified control elements in the rat insulin I gene 5' flanking DNA are shown in Fig. 4.

VI. Role of Other Sequences in Cell Specificity

The experiments described above utilizing cultured cells define 5' flanking sequences which play an important role in cell-specific expression. However, these experiments do not directly address the issue of whether other sequences, located elsewhere in the gene, also play a decisive role in vivo. The transgenic mouse experiment involving insulin/T antigen fusion genes (HANAHAN 1985) suggested that expression of the transgene is appropriate in a qualitative sense. However, given the substantial effects of expression of T antigen on cell metabolism, this protein cannot be regarded as a neutral reporter, and hence levels of T antigen mRNA and protein in β cells cannot be used as a quantitative measure of the efficiency of transcription initiation from the insulin control region. In order to address the issue of whether 5' flanking sequences are sufficient to quantitatively direct cell specificity, a construction was designed containing the rat insulin I flanking DNA sequences linked to the human placental lactogen (hPL) gene, a much more neutral reporter function. Preliminary results showed that the levels of pancreatic hPL mRNA in transgenic mice bearing this construction were comparable to those of the endogenous mouse insulin mRNA, (M. D. WALKER, W. J. RUTTER and D. HANAHAN, unpublished work) indicating that the 5' flanking sequences constitute a very significant component of the cell-specific transcription machinery and may represent the only sequences involved in determining cell specificity. Direct measurement of transcription rate will be required to establish this.

VII. Biochemical Approaches

Given the presumed role of sequence-specific transcription factors in activation of promoters and enhancers, much effort has been directed towards purification of these factors biochemically and characterization of their activities. Considerable progress has been made, especially in identification of factors involved in transcription of viral genes (reviewed in JONES et al. 1988).

In vitro analysis of the interaction of transcription factors with DNA is commonly performed using the techniques of DNAse I footprinting (GALAS and SCHMITZ 1978) and gel mobility shift (FRIED and CROTHERS 1981). In both of these methods, radioactively labeled DNA containing the sequence of interest is allowed to interact with cell extracts, and the binding of proteins is monitored by determining the ability of the protein to protect the DNA from nuclease digestion (footprinting assay) or the ability of the protein to alter the mobility of a DNA fragment on polyacrylamide gel electrophoresis (mobility shift assay). In the case of the rat insulin I gene, analysis by DNase I footprinting revealed that three distinct regions, E1 (-290 to -335), E2 (-205 to -222), and E3 (-154 to -169) within the enhancer are protected from DNase I cleavage by nuclear extracts prepared from insulin-producing cells (OHLSSON and EDLUND 1986). The protection of E1 was observed with extracts from insulin-producing cells only, whereas protection of E2 and E3 was also observed with extracts from other cell types. By using the more sensitive gel mobility shift assay, with either whole-cell extracts (Moss et al. 1988) or nuclear extracts (OHLSSON et al. 1988), it was possible to show that the two most important elements of the insulin enhancer, namely -104 to -112 and -233 to -241 also interact in vitro with proteins present in insulin-producing cells. Furthermore, nuclear extracts from other cell types show greatly reduced binding activity (OHLSSON et al. 1988). Since these two sequences compete with one another for binding activity, it seems that they bind to a similar protein(s), designated IEF1, which may play an important role in determining cell-specific expression. In an independent study (SAMPLE and STEINER 1987), gel mobility shift analysis showed interaction of a nuclear factor found preferentially in insulin-producing cells with a fragment encompassing the rat insulin II gene promoter (from -86 to $+47$). HWUNG et al. (1988) have shown that a transcription factor present in HeLa cells and previously shown to bind to promoter sequences of the ovalbumin gene, also interacts with promoter sequences (-53 to -46) of the rat insulin II gene.

F. Expression During Development

Analysis of the expression of hybrid insulin genes during embryonic development in transgenic mice has begun to reveal exciting new insights on both development of pancreatic endocrine cells, and also on the regulatory mechanisms underlying selective gene expression (ALPERT et al. 1988). When transgenes containing rat insulin gene 5′ flanking region are used, islet cells containing glucagon, somatostatin, and pancreatic polypeptide coexpress the transgene when they first appear during development. This unexpected expression pattern is similar, but not identical to that of the endogenous insulin gene. The production of multiple hormones within embryonic β cells is in sharp contrast to the situation postnatally where coexpression is virtually never seen. These findings raise the possibility that the various endocrine cells of the pancreas have a common precursor. This suggestion is consistent with the observation that clonal cell lines of pancreatic endocrine origin are capable of producing more than one hormone (GAZDAR et al. 1980; MADSEN et al. 1986). Whatever mechanisms permit expression of the insulin gene in multiple embryonic cell types, they are clearly operating via a short portion of insulin 5′ flanking DNA, since the behavior of the transgene mimics that of the endogenous gene.

In neuronal tissue, on the other hand, the behavior of the transgene appears to differ from that of the endogenous insulin gene; the transgene is expressed transiently during embryogenesis in cells of the neural tube and neural crest (ALPERT et al. 1988), yet there is no evidence for expression of the insulin gene in these or other cells of the vertebrate brain (BASKIN et al. 1987). ALPERT et al. (1988) suggest that these results could be due to a combinatorial interaction between elements of the insulin regulatory region and cryptic elements within the reporter molecule, as has been suggested in other systems (EVANS et al. 1986). Alternatively, sequence elements outside of the flanking region used may be responsible for negatively controlling expression of the insulin gene in non-pancreatic cell types (LAIMINS et al. 1986). Based on the fact that developing endocrine cells express the characteristic neuronal enzyme tyrosine hydroxylase and that in developing neurons a transferred insulin promoter is active, it is suggested that these cell types may share a common precursor (ALPERT et al. 1988).

The human tyrosine hydroxylase gene has recently been shown to be located adjacent to the 5′ end of the human insulin gene, at a distance of 2.7 kb (O'MALLEY and ROTWEIN 1988). The unexpected expression of the tyrosine hydroxylase in the pancreas may therefore be related to its proximity to the insulin gene enhancer. Whether the insulin gene is influenced by control elements of the tyrosine hydroxylase gene is unknown.

G. Prospects

Given the powerful molecular genetic and biochemical tools currently available one can anticipate relatively rapid progress toward definition of the mechanism underlying the differential expression of the insulin gene. Some of the more promising lines of approach include the following.

I. Biochemical

As described above, biochemical purification of cell extracts can be used to isolate transcription factors. An important technical advance has been the recent development of an effective procedure for purification of sequence-specific DNA binding proteins by affinity chromatography (KADONAGA and TJIAN 1986). The purified material can be assayed for binding activity to the cognate DNA sequence and also for ability to increase transcription activity in in vitro transcription systems. The approach has been applied successfully to several viral transcription factors. However, progress has been slower for cell-specific mammalian genes. GORSKI et al. (1986) have shown that extracts from liver cells are capable of effective in vitro transcription from the mouse albumin promoter, whereas extracts from other cell types are much less effective. BODNER and KARIN (1987) were able to show that partially purified protein fractions isolated from pituitary cells were capable of stimulating the ability of fibroblast extracts to transcribe in vitro the growth hormone gene. One limitation of the approach based on purification by affinity chromatography is the requirement for specific DNA recognition by the purified protein. Several heteromeric transcription factors have recently been isolated (HAHN and GUARENTE 1988; CHODOSH et al. 1988). Since DNA binding activity can be lost upon dissociation of the heteromeric complex, modified isolation procedures will be required in such cases.

II. Genetic

In this approach, DNA from a differentiated cell is used to transfect cultured fibroblast cells and the expression of genes characteristic of the differentiated cell type is either screened for or selected for. An analogous scheme has been successfully applied in the past to isolate oncogenes (SHIH et al. 1981). Recently the approach has been used to isolate two genes which appear to play an important role in the development of differentiated muscle cells (DAVIS et al. 1987; PINNEY et al. 1988). Given the ability of these genes to alter the differentiated phenotype, including the production of dramatic differences in gene expression, the isolated genes may not simply encode transcription factors which activate expression of genes characteristic of the fully differentiated phenotype, but may instead be controlling genes located at a "higher" position in the regulatory hierarchy (BLAU 1988). This highlights a striking advantage of the genetic approach as compared to the biochemical approach, whose utility is limited to isolating transcription factors relevant to the expression of a specific gene. On the other hand, the genetic approach is limited in that an appropriate choice of recipient cell may be critical: 10T1/2 cells (used in the above studies) show an inherent predisposition towards muscle cell differentiation (TAYLOR and JONES 1979). Furthermore, the transfection approach is unlikely to succeed if simultaneous expression of multiple unlinked genes is essential for bringing about the differentiative transition. Finally, transfection efficiency must be high in order to be able to monitor the rare differentiated colony following transfer of DNA, thereby limiting the range of cell types which can be used as recipients. This latter limitation may be overcome if retroviral expression cDNA libraries are used (MURPHY and EFSTRATIADIS 1987); such an approach can be used with any cell type which can be efficiently infected, including derivatives of β cells which have lost the ability to express the insulin gene, possibly because of lack of expression of a key transcription factor (EPISKOPOU et al. 1984).

III. Expression Library Screening

Cloned DNA sequences can be routinely isolated from plasmid or bacteriophage libraries by in situ screening with nucleic acid probes complementary in sequence to the desired gene or with antibody probes directed toward the protein encoded by the desired gene. An additional possibility for genes encoding sequence-specific DNA binding proteins, is to probe the library using a DNA probe corresponding in sequence to the binding site of the particular protein. The feasibility of this approach was demonstrated by SINGH et al. (1988) who identified a recombinant clone from a λGT11 expression library using a DNA sequence corresponding to the binding site of the mammalian transcription factors H2TF1 and NF-κB which are involved in expression of histocompatibility genes and immunoglobulin genes respectively.

The identification of genes encoding transcription factors will allow isolation in quantity of both the wild-type protein and site-directed mutants to permit detailed analysis of the structure and function of the protein. Furthermore, the availability of DNA probes for these genes will permit analysis of the intriguing question of how the genes which encode cell-specific transcription factors are themselves controlled. In the case of the NF-κB factor, considerable experimental evidence has accumulated to suggest that the factor is present in many cell types, but active only in lymphoid cells (SEN and BALTIMORE 1986; BAEUERLE and BALTIMORE 1988). Indeed, the gene identified as encoding the H2TF1/NF-κB factor is transcribed in both lymphoid and nonlymphoid cells (SINGH et al. 1988).

IV. Gene Therapy

In type I diabetes, endogenous insulin production is deficient because of partial or complete destruction of the β cells resulting from autoimmune or viral damage. Although the primary defect is not insulin gene expression per se, the prospect of treating affected individuals by gene therapy is an attractive, if distant, goal (SELDEN et al. 1987a). Some promising steps have been taken: transgenic mice expressing an intact human insulin gene are healthy and show normal regulation of both endogenous and exogenous gene, implying that the DNA fragment used contained all the appropriate signals and that these are capable of responding following gene transfer (SELDEN et al. 1986; BUCCHINI et al. 1986). However, formidable technical problems remain. In particular, an appropriate recipient cell line must be identified: pancreatic β cells would be subject to immune attack in the patient and therefore may be unsuitable. The alternative cell must be capable of efficient expression of the insulin gene and conversion of proinsulin to insulin: perhaps an endocrine cell of nonpancreatic origin (MOORE et al. 1983) would be a candidate. SELDEN et al. (1987a, b) have described a protocol, "transkaryotic implantation," which involves introduction of the desired gene into appropriate cells in culture, followed by characterization of the expression of the introduced gene and implantation of cells to animals. Clearly if such an approach is to be used in humans, it is essential that insulin output by such a modified cell type be regulated appropriately. A thorough understanding of control of expression of the normal gene is a prerequisite for this.

Acknowledgments. I would like to express my deep gratitude to Dr. William J. Rutter under whose guidance I first began to work on insulin gene regulation and whose participation and enthusiastic support were key to the development of my work. It is a pleasure to acknowledge the contributions of several colleagues with whom I have been fortunate to collaborate. In particular Dr. Thomas Edlund was closely involved throughout. In addition I would like to thank Drs. Uri Nir, Anne Boulet, Olof Karlsson, Larry Moss, Jennifer Barnett Moss and Doug Hanahan with whom I enjoyed valuable collaborations. I gratefully acknowledge support from the Juvenile Diabetes Foundation International (Career Development Award). Work carried out in my laboratory was supported by grants from the NIH (GM 38817) and the Israel Academy of Sciences and Humanities.

References

Alpert S, Hanahan D, Teitelman G (1988) Hybrid insulin genes reveal a developmental lineage for pancreatic endocrine cells and imply a relationship with neurons. Cell 53:295–308

Baeuerle PA, Baltimore D (1988) Activation of DNA-binding activity in an apparently cytoplasmic precursor of the NF-κB transcription factor. Cell 53:211–217

Banerji J, Olson L, Schaffner W (1983) A lymphocyte-specific cellular enhancer is located downstream of the joining region in immunoglobulin heavy chain genes. Cell 33:729–740

Baskin DG, Figlewicz DP, Woods SC, Porte D Jr, Dorsa DM (1987) Insulin and the brain. Ann Rev Physiol 49:335–347

Bell GI, Karam JH, Rutter WJ (1981) Polymorphic DNA region adjacent to the 5′ end of the human insulin gene. Proc Natl Acad Sci USA 78:5759–5763

Bell GI, Selby MJ, Rutter WJ (1982) The highly polymorphic region near the human insulin gene is composed of simple tandemly repeating sequences. Nature 295:31–35

Bell GI, Horita S, Karam JH (1984a) A polymorphic locus near the human insulin gene is associated with insulin-dependent diabetes mellitus. Diabetes 33:176–183

Bell GI, Quinto C, Quiroga M, Valenzuela P, Craik CS, Rutter WJ (1984b) Isolation and sequence of a rat chymotrypsin B gene. J Biol Chem 259:14265–14270

Bell GI, Gerhard DS, Fong NM, Sanchez-Pescador R, Rall LB (1985) Isolation of the human insulin-like growth factor genes: insulin-like growth factor II and insulin genes are contiguous. Proc Natl Acad Sci USA 82:6450–6454

Blau HM (1988) Hierarchies of regulatory genes may specify mammalian development. Cell 53:673–674

Blau HM, Pavlath GK, Hardeman EC, Chiu C-P, Dilberstein L, Webster SG, Miller SC, Webster C (1985) Plasticity of the differentiated state. Science 230:758–766

Bodner M, Karin M (1987) A pituitary-specific trans-acting factor can stimulate transcription from the growth hormone promoter in extracts of non-expressing cells. Cell 50:267–275

Boulet AM, Erwin CR, Rutter WJ (1986) Cell specific enhancers in the rat exocrine pancreas. Proc Natl Acad Sci USA 83:3599–3603

Breathnach R, Chambon P (1981) Organization and expression of eukaryotic split genes coding for proteins. Annu Rev Biochem 50:349–585

Bucchini D, Ripoche M-A, Stinnakre M-G, Desbois P, Lores O, Monthioux E, Absil J et al. (1986) Pancreatic expression of human insulin gene in transgenic mice. Proc Natl Acad Sci USA 83:2511–2515

Chan SJ, Episkopou V, Zeitlin S, Karathanasis SK, Mackrell A, Steiner DF, Efstratiadis A (1984) Guinea pig preproinsulin gene: an evolutionary compromise? Proc Natl Acad Sci USA 81:6046–6051

Chodosh LA, Baldwin AS, Carthew RW, Sharp PA (1988) Human CCAAT-binding proteins have heterologous subunits. Cell 53:11–24

Clark JL, Steiner DF (1969) Insulin biosynthesis in the rat: demonstration of two proinsulins. Proc Nat Acad Sci USA 62:278–285

Cordell B, Diamond D, Smith S, Punter J, Schone HH, Goodman HM (1982) Disproportionate expression of the two nonallelic rat insulin genes in a pancreatic tumor is due to translational control. Cell 31:531–542

Darnell JE (1982) Variety in the level of gene control in eukaryotic cells. Nature 297:365–371

Davis RL, Weintraub H, Lassar AB (1987) Expression of a single transfected cDNA converts fibroblasts to myoblasts. Cell 51:987–1000

De Pagter-Holthuizen P, Jansen M, van Shaik FMA, van der Kammen R, Oosterwijk C, van den Brande JL, Sussenbach JS (1987) The human insulin-like growth factor II gene contains two development-specific promoters. FEBS Lett 214:259–264

Edlund T, Walker MD, Barr PJ, Rutter WJ (1985) Cell-specific expression of the rat insulin gene: evidence for role of two distinct 5′ flanking elements. Science 230:912–916

Ephrussi A, Church GM, Tonegawa S, Gilbert W (1985) B lineage-specific interaction of an immunoglobulin enhancer with cellular factors in vivo. Science 227:134–140

Episkopou V, Murphy AJM, Efstratiadis A (1984) Cell-specified expression of a selectable hybrid gene. Proc Natl Acad Sci USA 81:4657–4661

Evans RM (1988) The steroid and thyroid hormone receptor superfamily. Science 240:889–895

Evans RM, Weinberger C, Hollenberg S, Swanson L, Nelson C, Rosenfeld MG (1986) Inducible and developmental control of neuroendocrine genes. Cold Spring Harbor Symp Quant Biol 50:389–397

Fried M, Crothers MD (1981) Equilibria and kinetics of lac repressor operator interactions by polyacrylamide gel electrophoresis. Nucleic Acids Res 9:6505–6525

Galas D, Schmitz A (1978) DNase footprinting: a simple method for detection of protein-DNA binding specificity. Nucleic Acids Res 5:3157–3170

Gazdar AF, Chick WL, Herbert KO, King DL, Gordon CW, Lauris V (1980) Continuous, clonal insulin- and somatostatin-secreting cell lines established from a transplantable rat islet cell tumor. Proc Natl Acad Sci USA 77:3519–3523

Gillies SD, Morrison SL, Oi VT, Tonegawa S (1983) A tissue specific transcription enhancer is located in the major intron of a rearranged immunoglobulin heavy chain gene. Cell 33:717–728

Gluzman Y (1981) SV-40 transformed simian cells support the replication of early SV40 mutants. Cell 23:175–182

Gorman CM (1986) High efficiency gene transfer into mammalian cells. In: Glover DM (ed) DNA cloning: a practical approach. IRL, Oxford, pp 143–190

Gorman CM, Moffat LF, Howard BH (1982a) Recombinant genomes which express chloramphenicol acetyltransferase in mammalian cells. Mol Cell Biol 2:1044–1051

Gorman CM, Merlino GT, Willingham MC, Pastan TR, Howard BH (1982b) The Rous sarcoma virus long terminal repeats is a strong promoter when introduced into a variety of eukaryotic cells by DNA-mediated transfection. Proc Natl Acad Sci USA 79:6777–6781

Gorski K, Carneiro M, Schibler U (1986) Tissue-specific in vitro transcription from the mouse albumin promoter. Cell 47:767–776

Grosschedl R, Baltimore D (1985) Cell type specificity of immunoglobulin gene expression is regulated by at least 3 DNA sequence elements. Cell 41:885–897

Hahn S, Guarente L (1988) Yeast HAP2 and HAP3: transcriptional activators in a heteromeric complex. Science 240:317–320

Hanahan D (1985) Heritable formation of pancreatic β-cell tumours in transgenic mice expressing recombinant insulin/simian virus 40 oncogenes. Nature 315:115–122

Hedeskov CJ (1980) Mechanism of glucose-induced insulin secretion. Physiol Rev 60:442–509

Hwung Y-P, Crowe DT, Wang L-H, Tsai SY, Tsai M-J (1988) The COUP transcription factor binds to an upstream promoter element of the rat insulin II gene. Mol Cell Biol 8:2070–2077

Jacob F, Monod J (1961) Genetic regulatory mechanisms in the synthesis of proteins. J Mol Biol 3:318–356

Jessop NW, Hay RJ (1980) Characteristics of two rat pancreatic exocrine cell lines derived from transplantable tumors. In vitro 16:212

Jones NC, Rigby PWJ, Ziff EB (1988) Trans-acting protein factors and the regulation of eukaryotic transcription: lessons from studies on DNA tumor viruses. Genes Dev 2:267–281

Kadonaga JT, Tjian R (1986) Affinity purification of sequence-specific DNA binding proteins. Proc Natl Acad Sci USA 83:5889–5893

Karlsson O, Edlund T, Moss JB, Rutter WJ, Walker MD (1987) A mutational analysis of the insulin gene transcription control region: expression in beta cells is dependent on two related sequences within the enhancer. Proc Natl Acad Sci USA 84:8819–8823

Karlsson O, Walker MD, Rutter WJ, Edlund T (1989) Individual protein-binding domains of the insulin gene enhancer positively activate β-cell specific transcription. Mol Cell Biol 9:823–827

Keegan L, Gill G, Ptashne M (1986) Separation of DNA binding from the transcription-activating function of a eukaryotic regulatory protein. Science 231:699–704

Kelly JH, Darlington GJ (1985) Hybrid genes: molecular approaches to tissue specific gene regulation. Annu Rev Genet 19:273–296

Khoury G, Gruss P (1983) Enhancer elments. Cell 33:313–314

Killary AM, Fournier REK (1984) A genetic analysis of extinction: trans-dominant loci regulate expression of liver specific traits in hepatoma hybrid cells. Cell 38:523–534

Laimins L, Holmhren-Konig M, Khoury G (1986) Transcriptional "silencer" element in rat repetitive sequences associated with the rat insulin I gene locus. Proc Natl Acad Sci USA 83:3151–3155

Laimins LA, Khoury G, Gorman C, Howard B, Gruss P (1982) Host-specific activation of transcription by tandem repeats from simian virus 40 and Moloney murine sarcoma virus. Proc Natl Acad Sci USA 79:6453–6457

Lewin B (1980) Eucaryotic chromosomes. Wiley, New York (Gene expression, vol 2)

Lomedico P, Rosenthal N, Efstratiadis A, Gilbert W, Kolodner R, Tizard R (1979) The structure and evolution of the two nonallelic rat preproinsulin genes. Cell 18:545–548

Madsen OD, Lersson LI, Rehfeld JF, Schwartz TW, Lernmark A, Labrecque AD, Steiner DF (1986) Cloned cell lines from a transplantable islet cell tumor are heterogeneous and express cholecystokinin in addition to islet hormones. J Cell Biol 103:2025–2034

Maniatis T, Goodbourn S, Fischer JA (1987) Regulation of inducible and tissue-specific gene expression. Science 236:1237–1245

McKnight SL, Kingsbury R (1982) Transcriptional control signals of a eukaryotic protein coding gene. Science 217:316–324

McKnight S, Tjian R (1986) Transcriptional selectivity of viral genes in mammalian cells. Cell 46:795–805

Moore H-P, Walker MD, Lee F, Kelly RB (1983) Expressing a human proinsulin cDNA in a mouse ACTH-secreting cell. Intracellular storage, proteolytic processing, and secretion on stimulation. Cell 35:531–538

Moss L, Moss JB, Rutter WJ (1988) Systematic binding analysis of the insulin gene transcription control region: insulin and immunoglobulin enhancers utilize similar transactivators. Mol Cell Biol 8:2620–2627

Mulligan RD, Berg P (1981) Selection for animal cells that express the E. coli gene coding for xanthine-guanine phosphoribosyltransferase. Proc Natl Acad Sci USA 78:2072–2076

Murphy AJM, Efstratiadis A (1987) Cloning vectors for expression of cDNA libraries in mammalian cells. Proc Natl Acad Sci USA 84:8277–8281

Nagasawa H, Kataoka H, Isogai A, Tamura S, Suzuki A, Mizoguchi A, Fujiwara Y et al. (1986) Amino acid sequence of a prothoracicotropic hormone of the silkworm *Bombyx mori*. Proc Natl Acad Sci USA 83:5840–5843

Nielsen DA, Welsh M, Casadaban MJ, Steiner DF (1985) Control of insulin gene expression in pancreatic β-cells and in an insulin-producing cell line, RIN-5F cells. I. Effects of glucose and cyclic AMP on the transcription of insulin mRNA. J Biol Chem 260:13585–13589

Nir U, Walker MD, Rutter WJ (1986) Regulation of rat insulin 1 gene expression: evidence for negative regulation in non-pancreatic cells. Proc Natl Acad Sci USA 83:3180–3184

Ohlsson H, Edlund T (1986) Sequence-specific interactions of nuclear factors with the insulin gene enhancer. Cell 45:35–44

Ohlsson H, Karlsson O, Edlund T (1988) A beta-cell-specific protein binds to the two major regulatory sequences of the insulin gene enhancer. Proc Natl Acad Sci USA 85:4228–4231

O'Malley KL, Rotwein P (1988) Human tyrosine hydroxylase and insulin genes are contiguous on chromosome 11. Nucleic Acids Res 16:4437–4445

Owerbach D, Bell GI, Rutter WJ, Shows TB (1980) The insulin gene is located on chromosome 11 in humans. Nature 286:82–84

Palmiter RD, Brinster RL (1986) Germ-line transformation of mice. Annu Rev Genet 20:465–499

Pellicer A, Robins D, Wold B, Sweet R, Jackson J, Lowy I, Roberts JM et al. (1980) Altering genotype and phenotype by DNA-mediated gene transfer. Science 209:1414–1422

Pinney DF, Pearson-White SH, Konieczny SF, Lattham KE, Emerson CP Jr (1988) Myogenic lineage determination and differentiation: evidence for a regulatory gene pathway. Cell 53:781–793

Ptashne M (1986) A genetic switch. Blackwell, Palo Alto

Rotwein PS, Chirgwin J, Province M, Knowler WCC, Pettitt DJ, Cordell B, Goodman HM, Permutt MA (1983) Polymorphism in the 5' flanking region of the human insulin gene. A genetic marker for non-insulin-dependent diabetes. N Engl J Med 308:65–71

Sample CE, Steiner DF (1987) Tissue-specific binding of a nuclear factor to the insulin gene promoter. FEBS Lett 222:332–336

Santerre RF, Cook RA, Crisel RMD, Sharp JD, Schmidt RJ, Williams DC, Wilson CP (1981) Insulin synthesis in a clonal cell line of simian virus 40-transformed hamster pancreatic beta cells. Proc Natl Acad Sci USA 78:4339–4343

Santoro C, Mermod N, Andrews PC, Tjian R (1988) A family of human CCAAT-box-binding proteins active in transcription and DNA replication: cloning and expression of multiple cDNAs. Nature 334:218–224

Selden RF, Skoskiewicz MJ, Howie KB, Russell PS, Goodman HM (1986) Regulation of human insulin gene expression in transgenic mice. Nature 32:525–528

Selden RF, Skoskiewicz MJ, Russell PS, Goodman HM (1987a) Regulation of insulin gene expression: implications for gene therapy. N Engl J Med 317:1067–1076

Selden RF, Skoskiewicz MJ, Howie KB, Russell PS, Goodman HM (1987b) Implantation of genetically engineered fibroblasts into mice: implications for gene therapy. Science 236:714–718

Sen R, Baltimore D (1986) Inducibility of κ immunoglobulin enhancer-binding protein NF-κB by a posttranslational mechanism. Cell 47:921–928

Serfling E, Jasin M, Schaffner W (1985) Enhancers and eukaryotic gene transcription. Trends Genet 1:224–230

Shih C, Padhy LC, Murray M, Weinberg RA (1981) Transforming genes of carcinomas and neuroblastomas introduced into mouse fibroblasts. Nature 290:261–264

Sigler PB (1988) Acid blobs and negative noodles. Nature 333:210–212

Singh H, LeBowitz JH, Baldwin AS Jr, Sharp PA (1988) Molecular cloning of an enhancer binding protein: isolation by screening of an expression library with a recognition site DNA. Cell 52:415–423

Smit AB, Vreugdenhil E, Ebberink RHM, Geraerts WPM, Klootwijk J, Joosse J (1988) Growth-controlling molluscan neurons produce the precursor of an insulin-related peptide. Nature 331:535–538

Soares MB, Schon E, Henderson A, Karathanasis SK, Cate R, Zeitlin S, Chirgwin J, Efstratiadis A (1985) RNA-mediated gene duplication: the rat preproinsulin I gene is a functional retroposon. Mol Cell Biol 5:2090–2103

Southern PJ, Berg P (1982) Transformation of mammalian cells to antibiotic resistance with a bacterial gene under control of the SV40 early region promoter. J Mol Appl Genet 1:327–341

Steiner DF, Chan SJ, Welsh JM, Kwok SCM (1985) Structure and evolution of the insulin gene. Annu Rev Genet 19:463–484

Taylor SM, Jones PA (1979) Multiple new phenotypes induced in 10T1/2 and 3T3 cells treated with 5-azacytidine. Cell 17:771–779

Walker MD, Edlund T, Boulet AM, Rutter WJ (1983) Cell-specific expression controlled by the 5' flanking region of insulin and chymotrypsin genes. Nature 306:557–561

Welsh M, Nielsen DA, MecKrell AJ, Steiner DF (1985) Control of insulin gene expression in pancreatic β-cells and in an insulin-producing cell line, RIN-5F cells. II. Regulation of insulin mRNA stability. J Biol Chem 260:13590–13594

Welsh M, Scherberg N, Gilmore R, Steiner DF (1986) Translational control of insulin biosynthesis. Evidence for regulation of elongation, initiation and signal-recognition-particle-mediated translational arrest by glucose. Biochem J 235:459–467

CHAPTER 6

Regulation of Insulin Release by the Intracellular Mediators Cyclic AMP, Ca^{2+}, Inositol 1,4,5-Trisphosphate, and Diacylglycerol

W. J. MALAISSE

A. Introduction

Insulin secretion from the pancreatic B cell is regulated mainly by the concentration of D-glucose in the interstitial fluid surrounding islet cells. However, a number of other agents may modulate the secretory activity of the endocrine pancreas. They include circulating nutrients other than D-glucose and several hormones or neurotransmitters (MALAISSE 1972).[1] All these agents exert immediate and direct effects upon insulin release. In addition, the secretory response of the B cell to D-glucose and other secretagogues may be influenced in a delayed manner by environmental factors of ontogenic, nutritional, and endocrine nature. This complex regulation implies that the B cell is equipped with a number of suitable receptor systems, and organized to integrate the information provided by each sensing device (MALAISSE 1988 a).

The present chapter deals with selected intracellular messengers which couple the identification of regulatory agents to more distal events in the secretory sequence. More precisely, the role of cyclic AMP, cytosolic Ca^{2+}, inositol 1,4,5-trisphosphate, and diacylglycerol as intracellular mediators will be discussed.

It should be underlined, however, that the selection of these messengers, which was apparently motivated by the overall editorial organization of this volume, should not be understood as implying that they are the sole, or even the most important, coupling factors involved in the regulation of insulin secretion. For instance, we believe that, in the process of nutrient-stimulated insulin release, such as that provoked by D-glucose, the coupling of metabolic events (which coincide with the identification of nutrients as insulinotropic agents) to cationic events (which eventually lead to the exocytosis of secretory granules) involves adenine nucleotides (ATP and ADP) as the major intracellular mediators. A change in cytosolic ATP concentration or ATP/ADP ratio is indeed currently held responsible for the closing of a class of K^{+} channels located at the plasma membrane (MALAISSE and SENER 1987; COOK and HALES 1984; RORSMAN and TRUBE 1985). This decrease in K^{+} conductance may then lead to depolarization of the plasma membrane with the subsequent gating of voltage-sensitive Ca^{2+} channels. In the same perspective, the possible role of changes in either redox potential or intracellular pH as coupling factors in the process of nutrient-stimulated insulin release should not be overlooked (MALAISSE et al. 1984c).

---

[1] The references are largely, but not exclusively, restricted to prior review articles.

B. Cyclic AMP

The regulatory role of cyclic AMP in insulin release has been known for more than 20 years since SAMOLS et al. (1965) first reported that glucagon stimulates insulin secretion. Yet, even today, the extent of cyclic AMP participation in the physiological process of glucose-stimulated insulin secretion and the mode of action of the nucleotide in the pancreatic B cell still remain incompletely understood.

I. Cyclic AMP Synthesis and Breakdown

The islet cell content in cyclic AMP reflects the balance between its rate of synthesis, breakdown, and release in the extracellular fluid.

1. Adenylate Cyclase

Adenylate cyclase, which is located at the plasma membrane (HOWELL and WHITFIELD 1972) displays a K_m for ATP close to 0.07 mM (MALAISSE et al. 1984a). Several hormones, such as glucagon, activate the enzyme, whereas α_2-adrenergic agonists inhibit it (GARCIA-MORALES et al. 1984). The modulation of adenylate cyclase by these hormones is mediated by the intervention of the GTP-binding regulatory proteins Ns and Ni. The α subunits of these two regulatory proteins were identified in islet cell membranes by ADP ribosylation in the presence of cholera toxin and the islet-activating protein isolated from the culture medium of *Bordetella pertussis*, respectively (MALAISSE et al. 1984b; SVOBODA et al. 1985). The exposure of intact islets to cholera toxin is associated with activation of adenylate cyclase, enhanced production of cyclic AMP, and an increase in insulin release caused by high concentrations of D-glucose. Inversely, the inhibitory action of adrenergic agents on insulin release is suppressed when the islets are exposed in vivo or in vitro to the *Bordetella pertussis* toxin. This treatment also affects the activity of adenylate cyclase in acellular systems, with an increased responsiveness to GTP relative to its stable analogs and resistance to the inhibitory action of α_2-adrenergic agonists.

Another essential feature of islet adenylate cyclase consists in its activation by Ca-calmodulin (VALVERDE et al. 1979). Thus, whereas Ca^{2+} (0.1–0.7 mM) causes a dose-related inhibition of adenylate cyclase activity, the enzyme is activated, in the presence of Ca^{2+}, by calmodulin. At a fixed concentration of Ca^{2+}, the K_a for calmodulin is close to 0.1 μM (VALVERDE et al. 1981). At a fixed concentration of calmodulin, the apparent K_a for Ca^{2+} is close to 10 μM. Since nutrient secretagogues augment cytosolic Ca^{2+} activity in intact cells, the activation of adenylate cyclase by calmodulin may account for an increase in cyclic AMP content in cells stimulated by exogenous nutrients. The latter increase is indeed abolished when the cells are incubated in the absence of extracellular Ca^{2+}. Incidentally, D-glucose and other nutrient secretagogues fail to exert any obvious effect upon adenylate cyclase activity in islet homogenates.

2. Phosphodiesterase

The nutrient-induced increase in islet cell cyclic AMP content is much more marked, in absolute terms, when the incubation is carried out in the presence of pharmacological inhibitors of cyclic AMP phosphodiesterase such as theophylline or 3-isobutyl-1-methylxanthine. Relative to the paired basal value, the cyclic AMP response to nutrient secretagogues is of comparable magnitude, however, in the absence and presence of the phosphodiesterase inhibitors (VALVERDE et al. 1983). Incidentally, even in islets deprived of exogenous nutrient, the latter agents increase the cyclic AMP content to a much larger extent than that normally seen in response to D-glucose.

The phosphodiesterase activity identified in islet homogenates displays both low and high affinities for the nucleotide and is located mainly in the post-microsomal supernatant (ROSEN et al. 1971; SAMS and MONTAGUE 1972). Ca-calmodulin activates the enzyme (LIPSON and OLDHAM 1983; SUGDEN and ASHCROFT 1981), so that an increase in cytosolic Ca^{2+} activity could, by increasing both the synthesis and breakdown of cyclic AMP, accelerate its turnover rate.

3. Cyclic AMP Outflow from Islet Cells

Several studies have indicated that a fraction of the cyclic AMP generated in the islet cells is released into the extracellular fluid. This release of cyclic AMP occurs independently of, and can be dissociated from the release of insulin. It is inhibited by papaverine or probenecid (CAMPBELL and TAYLOR 1981).

II. Mode of Action of Cyclic AMP

The action of cyclic AMP is thought to be mediated through phosphorylation of specific proteins by a cyclic AMP-dependent protein kinase (HARRISON et al. 1984). The latter kinase and several of its potential polypeptide substrates were found in islet cell homogenates. Agents which increase the cyclic AMP content also affect the phosphorylation of proteins in intact islets. The function of these phosphorylated proteins remains to be established. Several modalities for regulation of insulin secretion by cyclic AMP should be considered.

A first mechanism could consist in an effect of the nucleotide on the metabolism of nutrients in the islet cells. For instance, theophylline, by increasing the cyclic AMP content of islet cells, was shown to stimulate glycogenolysis in glycogen-rich islets (MALAISSE et al. 1977). Cyclic AMP could also provoke lipolysis and hence increase the output of $^{14}CO_2$ from islets prelabeled with [U-^{14}C]palmitate (KAWAZU et al. 1980). The bulk of the evidence suggests, however, that, especially at high glucose concentrations, the enhancing action of cyclic AMP upon insulin release is not attributable to any marked facilitation of nutrient catabolism in the islet cells (MALAISSE and MALAISSE-LAGAE 1984).

A second hypothesis is that cyclic AMP affects Ca^{2+} fluxes in islet cells. For instance, cyclic AMP may facilitate Ca^{2+} inflow into the islet cells (HENQUIN and MEISSNER 1983; HENQUIN et al. 1983). Alternatively, cyclic AMP may cause an intracellular redistribution of Ca^{2+} to favor the cytosolic accumulation of this cation (BRISSON et al. 1972). This could account, in part at least, for the fact that agents such as forskolin, which increase the cyclic AMP content of the islets, allow for a positive secretory response to D-glucose even when the islets are incubated in the absence of extracellular Ca^{2+}. Recent studies suggest, however, that cyclic AMP exerts little effect upon cytosolic Ca^{2+} activity (WOLLHEIM et al. 1984).

In view of the latter finding, it was proposed that cyclic AMP augments the responsiveness to Ca^{2+} of the effector system which controls the access of secretory granules to their site of exocytosis at the plasma membrane (MALAISSE and MALAISSE-LAGAE 1984). In this respect, it is even conceivable that a sufficient concentration of endogenous cyclic AMP plays a permissive role in the process of insulin release (OWEN and MALAISSE 1987). The postulated increased sensitivity of the releasing machinery to Ca^{2+} is compatible with the knowledge that cyclic AMP augments insulin release in permeabilized islets exposed to a suitable concentration of extracellular Ca^{2+} (TAMAGAWA et al. 1985). Incidentally, it was proposed that an appropriate generation of cyclic AMP in the pancreatic B cell may depend on the paracrine influence of adjacent glucagon-producing islet cells (PIPELEERS et al. 1985; SCHUIT and PIPELEERS 1985).

III. Physiological Role of Cyclic AMP

We have already underlined that exogenous nutrients, such as D-glucose, cause a modest increase in the islet cyclic AMP content. It must be underlined, however, that all nonnutrient agents which increase to a much greater extent the cyclic AMP content, e.g., forskolin and theophylline, are unable to cause any marked and sustained increase in insulin output when the islets are deprived of exogenous nutrient or exposed to low concentrations of D-glucose, well below the threshold value for the insulinotropic action of the hexose (MALAISSE et al. 1984a). This situation strongly suggests that cyclic AMP does not represent in itself the key second messenger in the process of nutrient-stimulated insulin release.

An enhanced (or decreased) generation of cyclic AMP may play a more important role in the secretory response of islet cells to those hormonal agents which potentiate (or inhibit) glucose-stimulated insulin output. For instance, the positive insulinotropic action of glucagon (and probably other gastrointestinal hormones) or the inhibitory action upon insulin release of α_2-adrenergic agonists may be mediated through changes in adenylate cyclase activity. It should be underlined, however, that the inhibition of insulin release by adrenergic agents cannot be blamed solely on a decreased production of endogenous cyclic AMP (MALAISSE 1988a). Further work is required to identify novel messengers possibly generated by enzymic systems which would represent targets for the GTP-binding regulatory protein Ni, conceivably involved in the inhibition of insulin release by catecholamines.

C. Calcium

An increase in the cytosolic concentration of ionized Ca^{2+} is currently viewed as a key event in the process of nutrient-stimulated insulin release (MALAISSE

1988 b). The cytosolic accumulation of Ca^{2+} may trigger the release of insulin by activating a microtubular–microfilamentous effector system controlling the intracellular translocation of secretory granules and their extrusion from the cell by exocytosis, in a manner somehow analogous to the process of excitation–contraction coupling in muscle cells (MALAISSE and ORCI 1979). Nevertheless, several agents, which modulate the magnitude of the secretory response to exogenous nutrients apparently do so without causing obvious changes in cytosolic Ca^{2+} activity.

I. Regulation of Ca^{2+} Fluxes in Islet Cells

The cytosolic Ca^{2+} concentration depends on the balance between the rate of Ca^{2+} inflow into and exit from the cell across the plasma membrane. It may also be affected by either an intracellular redistribution of Ca^{2+} between the cytosolic domain and several classes of subcellular organelles susceptible to sequester the cation, or changes in the binding of Ca^{2+} to molecules present in or in contact with the cytoplasm.

1. Calcium Influx into Islet Cells

D-Glucose and other nutrient secretagogues stimulate the influx of Ca^{2+} into the pancreatic B cell, by causing the gating of suitable Ca^{2+} channels (MALAISSE 1988 c). A current view ascribes the facilitation of Ca^{2+} inflow to a primary increase in the cytosolic ATP/ADP ratio, as observed in response to a rise in extracellular D-glucose concentration. This is thought to provoke the closing of ATP-responsive K^+-channels located at the plasma membrane and, hence, to decrease the passive outflow of K^+ along its electrochemical gradient (HENQUIN and MEISSNER 1984). The intracellular accumulation of K^+, by causing a depolarization of the plasma membrane, may lead to the gating of voltage-sensitive Ca^{2+} channels. Other insulinotropic agents, such as hypoglycemic sulfonylureas, may directly act upon the same class of K^+ channels and hence mimic the effect of nutrient secretagogues (MALAISSE 1987).

It should be realized that the entry of Ca^{2+} into the islet cells is not necessarily mediated at the sole intervention of voltage-sensitive Ca^{2+} channels. Thus, careful studies on the concentration–response relationship for the inhibitory action of organic Ca^{2+} antagonists upon both Ca^{2+} inflow and insulin release clearly suggest the existence of distinct modalities for the entry of Ca^{2+} into the B cell (LEBRUN et al. 1982 a, b). The participation of either voltage-insensitive Ca^{2+} channels or native ionophores should, therefore, not be overlooked (MALAISSE 1988 c).

With the latter reservation in mind, there is little doubt that the gating of voltage-sensitive Ca^{2+} channels represents an efficient mechanism for stimulation of insulin release. This concept was recently reinforced by use of organic Ca^{2+} agonists, i.e., agents structurally related to organic Ca^{2+} antagonists, but prolonging the mean open time of Ca^{2+} channels instead of causing their blockade (MALAISSE-LAGAE et al. 1984; MALAISSE and MATHIAS 1985).

2. Calcium Efflux from Islet Cells

The efflux of Ca^{2+} from islet cells is thought to be mediated both at the intervention of a Ca-calmodulin-sensitive ATPase and by a process of Na^+–Ca^{2+} countertransport. Moreover, minute amounts of calcium could be released together with insulin at the exocytotic site. Although D-glucose and other nutrient secretagogues provoke a rapid and sustained decrease in effluent radioactivity from prelabeled islets, perifused in the absence of extracellular Ca^{2+}, this effect is soon masked by a secondary rise in effluent radioactivity when the islets are perifused in the presence of extracellular Ca^{2+} (MALAISSE 1988 b). The reduction of effluent radioactivity is thought to reflect both inhibition of Na^+–Ca^+ countertransport and an accelerated sequestration of Ca^{2+} in cellular organelles. Obviously, these two processes would exert opposite influences on cytosolic Ca^{2+} activity. Under suitable experimental conditions, glucose may indeed cause a transient lowering of cytosolic Ca^{2+} concentration resulting in a paradoxical inhibition of insulin release (HELLMAN 1985). At normal extracellular Ca^{2+} concentration, however, the prevalent effect of the hexose is to augment ^{45}Ca outflow. Such an increase represents the consequence of facilitated Ca^{2+} inflow into the islet cells and could reflect, in part at least, the activation by Ca-calmodulin of the plasma membrane-associated Ca^{2+}-ATPase.

3. Intracellular Calcium Distribution

Several intracellular organelles may play a buffering role in the control of cytosolic Ca^{2+} activity. They include microsomes of the endoplasmic reticulum, mitochondria, and secretory granules (PRENTKI and MATSCHINSKY 1987). In the microsome, the uptake of Ca^{2+} is mediated by a calmodulin-insensitive Ca^{2+}-ATPase, whilst the release of Ca^{2+} is stimulated by inositol 1,4,5-trisphosphate. Likewise, mitochondria isolated from insulin-secreting cells apparently possess independent influx and efflux pathways of Ca^{2+} transport. A rise in either extramitochondrial Na^+ or H^+ concentration seems to favor the mobilization of Ca^{2+} from the mitochondria. Last, secretory granules of pancreatic B cells also contain high amounts of calcium, but their role in the short-term regulation of cytosolic Ca^{2+} activity remains a matter of debate. The interaction between these several organelles in the overall control of Ca^{2+} subcellular distribution may account for changes in cytosolic Ca^{2+} activity independent of any major change in the balance between Ca^{2+} influx and efflux across the plasma membrane. Hence, secretagogue-induced increases in cytosolic inositol 1,4,5-trisphosphate, Na^+, and H^+ concentrations are susceptible to participate in the secretory process by mobilizing Ca^{2+} from intracellular organelles.

II. Target Systems for Cytosolic Ca^{2+}

As already alluded to, an increase in cytosolic Ca^{2+} activity may lead to stimulation of insulin release through activation of Ca-calmodulin-responsive protein kinases, e.g., the myosin light chain kinase. In this way, Ca^{2+} may cause the contraction of the microfilamentous cell web, allowing access of the secretory granules to the exocytotic site at the plasma membrane (MALAISSE and ORCI

1979). The control of motile events in islet cells may also involve trans-glutaminase, which also represents a Ca^{2+}-responsive enzyme and catalyzes the cross-linking of proteins (MALAISSE 1988 b). In addition, the cytosolic Ca^{2+} activity may participate, either directly or at the intervention of Ca-calmodulin, in the activation of several other target systems, including Ca^{2+}-ATPases, adenylate cyclase, cyclic AMP phosphodiesterase, protein kinase C, and, possibly, phospholipase C. In the latter case, the enzymic activation could account for the acceleration of phosphoinositide hydrolysis in nutrient-stimulated islet cells.

D. Inositol 1,4,5-Trisphosphate

It is now firmly established that several insulinotropic agents, especially nutrient and cholinergic secretagogues, activate phospholipase C in pancreatic islet cells (BEST et al. 1984). This coincides with an increased production of inositol phosphates, including inositol 1,4,5-trisphosphate which may act as a second messenger by mobilizing calcium from the endoplasmic reticulum (PRENTKI and WOLLHEIM 1984).

Phosphatidylinositol, phosphatidylinositol 4-phosphate, and phosphatidylinositol 4,5-bisphosphate have been identified in pancreatic islet cells. The kinases involved in the generation of polyphosphoinositides are present in islet cell plasma membrane, in which their activity can be monitored through labeling in the presence of $[\gamma\text{-}^{32}P]$ATP (DUNLOP and MALAISSE 1986). In the same preparation, the activation of phospholipase C by cholinergic agents could be demonstrated. The coupling of muscarinic receptors to the activation of phospholipase C is apparently mediated at the intervention of a GTP-binding regulatory protein (BLACHIER and MALAISSE 1987) distinct from both Ns and Ni (BLACHIER et al. 1987). The activation of phospholipase C is also modulated by the ambient Ca^{2+} concentration (VALLAR et al. 1987) and this may account, as already mentioned, for the stimulation of phosphoinositide hydrolysis in nutrient-stimulated islet cells (BEST 1986). The purinergic pathway for stimulation of insulin release also apparently involves activation of phospholipase C (BLACHIER and MALAISSE 1988). However, secretagogues such as hypoglycemic sulfonylureas, glucagon, and the cationic amino acids L-arginine and L-ornithine apparently fail to affect the turnover of inositol-containing phospholipids in islet cells (BLACHIER et al. 1987; MATHIAS et al. 1985a).

Whatever the precise mechanism responsible for the activation of phospholipase C, the mobilization of inositol 1,4,5-trisphosphate from phosphatidylinositol 4,5-bisphosphate may lead to an increased release of Ca^{2+} from the endoplasmic reticulum (PRENTKI et al. 1984). This is documented by the increase in effluent radioactivity evoked by cholinergic agents in islets prelabeled with ^{45}Ca. At variance with the process of glucose-stimulated ^{45}Ca efflux, the cationic response to cholinergic agents persists even when the islets are perifused in the absence of extracellular Ca^{2+} (MATHIAS et al. 1985b).

The production of inositol 1,4,5-trisphosphate may lead to the subsequent generation, by either phosphorylation or through the action of phosphomonoesterases, of inositol 1,3,4-5-tetrakisphosphate, inositol 1,3,4-

trisphosphate, inositol 1,4-bisphosphate, and inositol 1-phosphate. There is yet no evidence to ascribe a physiological role to the latter four phosphate esters. Nevertheless, the eventual production of free myoinositol, which is in part released into the extracellular medium, allows for resynthesis of the inositol-containing phospholipids. Incidentally, the measurement of effluent radioactivity from islets prelabeled with tritiated myoinositol represents a sensitive procedure to assess the time course and magnitude of phospholipase C activation by distinct secretagogues (MATHIAS et al. 1985 a).

E. Diacylglycerol

Diacylglycerol represents the last intracellular mediator of insulin release considered in this chapter.

I. Generation of Diacylglycerol

There exist several theoretical modalities for the production of diacylglycerol in islet cells (PRENTKI and MATSCHINSKY 1987). First, it may be generated through the breakdown of inositol phospholipids by phospholipase C. This process was documented in membranes prepared from cells prelabeled with [U-^{14}C]arachidonate (DUNLOP and MALAISSE 1986). Other speculative mechanisms for diacylglycerol production consist in the hydrolysis of triglycerides (MALAISSE et al. 1983), the de novo synthesis from glycolytic intermediates, the activation of phosphatidate phosphohydrolase, and the esterification of fatty acyl-CoA residues (BEST and MALAISSE 1984 a). Only scanty information is so far available on the quantitative aspects of diacylglycerol metabolism in islet cells.

II. Activation of Protein Kinase C

Pancreatic islet homogenates display protein kinase C activity (HUBINONT et al. 1984). This phospholipid-dependent and Ca^{2+}-responsive enzyme is activated by diacylglycerol. Tumor-promoting phorbol esters such as 12-O-tetradecanoylphorbol-13-acetate (TPA) may substitute for diacylglycerol as enzyme activator. At high concentrations of TPA, the enzyme can even be activated in the absence of Ca^{2+} (HUBINONT and MALAISSE 1985). Hence, the phosphorylation of substrate proteins could occur at the basal cytosolic Ca^{2+} concentration ($\leq 0.1\ \mu M$) found in resting islet cells. This process probably accounts for stimulation of insulin release by either TPA or the synthetic diglyceride 1-oleoyl-2-acetyl-sn-glycerol. The action of protein kinase C in secretion is apparently mediated through phosphorylation of specific proteins, which remain to be fully identified. Incidentally, in intact islets, TPA fails to affect ^{86}Rb efflux and phosphoinositide turnover, suggesting that the phorbol ester does not reproduce the effect of nutrient secretagogues to decrease K^+ conductance with the subsequent increase in cytosolic Ca^{2+} activity and activation of phospholipase C (MALAISSE et al. 1980).

III. Functional Implications

Although the findings so far reviewed indicate that activation of protein kinase C represents an efficient modality for stimulation of insulin release, the role played by this enzyme in the normal process of glucose-induced insulin release remains ill defined. Indirect information based on the use of a presumably specific inhibitor of protein kinase C, 1-(5-isoquinolinesulfonyl)-2-methylpiperazine (H-7), suggests that activation of the enzyme only plays an amplifying role when the B cell is exposed to very high concentrations of the hexose (MALAISSE and SENER 1985). However, the increment in insulin release attributable to carbamylcholine appears more sensitive to inhibition by H-7.

F. Concluding Remarks

The present chapter deals with the possible role of cyclic AMP, Ca^{2+}, inositol 1,4,5-trisphosphate, and diacylglycerol as intracellular mediators in the regulation of insulin release. It should again be emphasized that these are not the sole messengers involved in the coupling process between the identification of secretagogues by the B cell and the eventual activation of the effector system for exocytosis of secretory granules from this cell. Moreover, it would be unwise to consider that each of the intracellular mediators under consideration acts independently of the others in response to stimulation of the B cell by distinct types of secretagogues. On the contrary, several modalities for the simultaneous generation of two or more mediators or for reciprocal interactions between them can be visualized.

In the latter respect, the physiologically essential process of glucose-stimulated insulin release could be considered as the most illustrative example for such a multifactorial mediation of stimulus–secretion coupling. It is indeed remarkable that the accelerated catabolism of D-glucose seen in islet cells exposed to a rise in the extracellular concentration of the hexose, and the concomitant increase in both O_2 uptake and ATP generation rate, which coincide with the identification of D-glucose as an insulinotropic agent, may eventually lead to an increased production of all four mediators here under review. Thus, exposure of islet cells to a high concentration of D-glucose indeed results in an increase in cyclic AMP content (VALVERDE et al. 1983), Ca^{2+} cytosolic activity (DELEERS et al. 1985), inositol 1,4,5-trisphosphate formation (BEST and MALAISSE 1984b), and, presumably, diacylglycerol production in the islet cells. Needless to say, the simultaneous increase in the generation or concentration of all these mediators is probably quite appropriate to ensure an optimal magnitude of the secretory response to the hexose. This is not meant to deny, however, that other insulinotropic agents may increase, to a greater extent than D-glucose, the cell content or production rate of one or more of these four messengers.

Further progress in understanding the functional organization of the insulin-producing B cell may precisely depend on a more accurate and comparative quantification of the respective role played by each mediator in response to distinct secretagogues. It may also depend on the identification of novel messengers (MALAISSE 1989).

Acknowledgments. I am grateful to F. Malaisse-Lagae and C. Demesmaeker for their help in the preparation of this chapter.

References

Best L (1986) A role for calcium in the breakdown of inositol phospholipids in intact and digitonin-permeabilized pancreatic islets. Biochem J 238:773–779

Best L, Malaisse WJ (1984a) Enhanced de novo synthesis of phosphatidic acid and phosphatidylinositol in rat pancreatic islets exposed to nutrient or neurotransmitter stimuli. Arch Biochem Biophys 234:253–257

Best L, Malaisse WJ (1984b) Nutrient and hormone-neurotransmitter stimuli induce hydrolysis of polyphosphoinositides in rat pancreatic islets. Endocrinology 115:1814–1820

Best L, Dunlop M, Malaisse WJ (1984) Phospholipid metabolism in pancreatic islets. Experientia 40:1085–1091

Blachier F, Malaisse WJ (1987) Possible role of a GTP-binding protein in the activation of phospholipase C by carbamylcholine in tumoral insulin-producing cells. Res Commun Chem Pathol Pharmacol 58:237–255

Blachier F, Malaisse WJ (1988) Effect of exogenous ATP upon inositol phosphate production, cationic fluxes and insulin release in pancreatic islet cells. Biochim Biophys Acta 970:222–229

Blachier F, Segura MC, Malaisse WJ (1987) Unresponsiveness of phospholipase C to the regulatory proteins Ns and Ni in pancreatic islets. Res Commun Chem Pathol Pharmacol 55:335–355

Brisson GR, Malaisse-Lagae F, Malaisse WJ (1972) The stimulus-secretion coupling of glucose-induced insulin release. VII. A proposed site of action for adenosine-3',5'-cyclic monophosphate. J Clin Invest 51:232–241

Campbell IL, Taylor KW (1981) The effect of metabolites, papaverine, and probenecid on cyclic AMP efflux from isolated rat islets of Langerhans. Biochim Biophys Acta 677:357–364

Cook DL, Hales CN (1984) Intracellular ATP directly blocks K^+ channels in pancreatic B-cells. Nature 311:271–273

Deleers M, Mahy M, Malaisse WJ (1985) Glucose increases cytosolic Ca^{2+} activity in pancreatic islet cells. Biochem Int 10:97–103

Dunlop ME, Malaisse WJ (1986) Phosphoinositide phosphorylation and hydrolysis in pancreatic islet cell membrane. Arch Biochem Biophys 244:421–429

Garcia-Morales P, Dufrane SP, Sener A, Valverde I, Malaisse WJ (1984) Inhibitory effect of clonidine upon adenylate cyclase activity, cyclic AMP production and insulin release in rat pancreatic islets. Biosci Rep 4:511–521

Harrison DE, Ashcroft SJH, Christie MR, Lord JM (1984) Protein phosphorylation in the pancreatic B-cell. Experientia 40:1075–1084

Hellman B (1985) β-Cell cytoplasmic Ca^{2+} balance as a determinant for glucose-stimulated insulin release. Diabetologia 28:494–501

Henquin JC, Meissner HP (1983) Dibutyryl cyclic AMP triggers Ca^{2+} influx and Ca^{2+}-dependent electrical activity in pancreatic B-cell. Biochem Biophys Res Commun 112:614–620

Henquin JC, Meissner HP (1984) Significance of ionic fluxes and changes in membrane potential for stimulus-secretion coupling in pancreatic B-cells. Experientia 40:1043–1052

Henquin JC, Schmeer W, Meissner HP (1983) Forskolin, an activator of adenylate cyclase, increases Ca^{2+}-dependent electrical activity induced by glucose in mouse pancreatic B-cells. Endocrinology 112:2218–2220

Howell SL, Whitfield M (1972) Cytochemical localization of adenyl cyclase activity in rat islets of Langerhans. J Histochem Cytochem 20:873–879

Hubinont CJ, Malaisse WJ (1985) Protein kinase C activity in pancreatic islets: effects of Ca^{2+}, calmodulin and retinoic acid. Biochem Int 10:577–584

Hubinont CJ, Best L, Sener A, Malaisse WJ (1984) Activation of protein kinase C by a tumor-promoting phorbol ester in pancreatic islets. FEBS Lett 170:247–253

Kawazu S, Sener A, Couturier E, Malaisse WJ (1980) Metabolic, cationic and secretory effects of hypoglycemic sulfonylureas in pancreatic islets. Naunyn Schmiedebergs Arch Pharmacol 312:277–283

Lebrun P, Malaisse A, Herchuelz A (1982a) Evidence for two distinct modalities of Ca^{2+} influx into the pancreatic B-cell. J Physiol (Lond) 242:E59–E66

Lebrun P, Malaisse WJ, Herchuelz A (1982b) Modalities of gliclazide-induced Ca^{2+} influx into the pancreatic B-cell. Diabetes 31:1010–1015

Lipson LG, Oldham SB (1983) The role of calmodulin in insulin secretion: the presence of a calmodulin-stimulatable phosphodiesterase in pancreatic islets of normal and pregnant rats. Life Sci 32:775–780

Malaisse WJ (1972) Hormonal and environmental modification of islet activity. In: Steiner DF, Freinkel N (eds) Handbook of physiology, sect 7, vol 1. American Physiological Society, Washington, pp 237–260

Malaisse WJ (1987) Mechanism of the insulinotropic effect of hypoglycemic sulfonylureas. Excerpta Medica Int Congr Ser 754:3–8

Malaisse WJ (1988a) Stimulation-secretion coupling in the pancreatic B-cell. In: Ganten D, Pfaff D (eds) Stimulus-secretion coupling in neuroendocrine systems. Springer, Berlin Heidelberg New York (Current topics in neuroendocrinology, vol 9), pp 231–251

Malaisse WJ (1988b) Cellular calcium: secretion of hormones. In: Nordin BEC (ed) Calcium in human biology. Springer, Berlin Heidelberg New York pp 367–384

Malaisse WJ (1988c) Calcium entry and activation of endocrine cells. Ann NY Acad Sci 552:284–295

Malaisse WJ (1989) Dual role of lipids in the stimulus-secretion coupling for insulin release. Biochem Soc Trans 17:59–60

Malaisse WJ, Malaisse-Lagae F (1984) The role of cyclic AMP in insulin release. Experientia 40:1068–1075

Malaisse WJ, Mathias PCF (1985) Stimulation of insulin release by an organic calcium agonist. Diabetologia 28:153–156

Malaisse WJ, Orci L (1979) The role of the cytoskeleton in pancreatic B-cell function. In: Gabbiani E (ed) Methods of achievements in experimental pathology, vol 9. Karger, Basel, pp 112–136

Malaisse WJ, Sener A (1985) Inhibition by 1-(5-isoquinolinesulfonyl)-2-methylpiperazine (H-7) of protein kinase C activity and insulin release in pancreatic islets. IRCS Med Sci 13:1239–1240

Malaisse WJ, Sener A (1987) Glucose-induced changes in cytosolic ATP content in pancreatic islets. Biochim Biophys Acta 927:190–195

Malaisse WJ, Sener A, Koser M, Ravazzola M, Malaisse-Lagae F (1977) The stimulus-secretion coupling of glucose-induced insulin release. XXV. Insulin release due to glycogenolysis in glucose-deprived islets. Biochem J 164:447–454

Malaisse WJ, Sener A, Herchuelz A, Carpinelli AR, Poloczek P, Winand A, Castagna M (1980) Insulinotropic effect of the tumor promoter 12-O-tetradecanoylphorbol-13-acetate in rat pancreatic islets. Cancer Res 40:3827–3831

Malaisse WJ, Best L, Kawazu S, Malaisse-Lagae F, Sener A (1983) The stimulus-secretion coupling of glucose-induced insulin release. LV. Fuel metabolism in islets deprived of exogenous nutrient. Arch Biochem Biophys 224:102–110

Malaisse WJ, Garcia-Morales P, Dufrane SP, Sener A, Valverde I (1984a) Forskolin-induced activation of adenylate cyclase, cyclic adenosine monophosphate production and insulin release in rat pancreatic islets. Endocrinology 115:2015–2020

Malaisse WJ, Svoboda M, Dufrane SP, Malaisse-Lagae F, Christophe J (1984b) Effect of Bordetella pertussis toxin on ADP-ribosylation of membrane proteins, adenylate cyclase activity and insulin release in rat pancreatic islets. Biochim Biophys Res Commun 124:190–196

Malaisse WJ, Malaisse-Lagae F, Sener A (1984c) Coupling factors in nutrient-induced insulin release. Experientia 40:1035–1043

Malaisse-Lagae F, Mathias PFC, Malaisse WJ (1984) Gating and blocking of calcium channels by dihydropyridines in the pancreatic B-cell. Biochem Biophys Res Commun 23:1062–1068

Mathias PCF, Best L, Malaisse WJ (1985a) Stimulation by glucose and carbamylcholine of phospholipase C in pancreatic islets. Cell Biochem Funct 3:173–177

Mathias PCF, Carpinelli AR, Billaudel B, Garcia-Morales P, Valverde I, Malaisse WJ (1985b) Cholinergic stimulation of ion fluxes in pancreatic islets. Biochem Pharmacol 34:3451–3457

Owen A, Malaisse WJ (1987) Mathematical modelling of stimulus-secretion coupling in the pancreatic B-cell. V. Threshold phenomenon for the response to cyclic AMP. Diabete Metab 13:514–519

Pipeleers DG, Schuit FC, In't Veld PA, Maes E, Hooghe-Peters EL, van de Winkel M, Gepts W (1985) Interplay of nutrients and hormones in the regulation of insulin release. Endocrinology 117:824–833

Prentki M, Matschinsky FM (1987) Ca^{2+}, cAMP, and phospholipid-derived messengers in coupling mechanisms of insulin secretion. Physiol Rev 67:1185–1248

Prentki M, Wollheim CB (1984) Cytosolic free Ca^{2+} in insulin secreting cells and its regulation by isolated organelles. Experientia 40:1052–1060

Prentki M, Biden TJ, Janjic D, Irvine RF, Berridge MJ, Wollheim CB (1984) Rapid mobilization of Ca^{2+} from rat insulinoma microsomes by inositol 1,4,5-trisphosphate. Nature 309:562–564

Rorsman P, Trube G (1985) Glucose dependent K^+-channels in pancreatic B-cells are regulated by intracellular ATP. Pflugers Arch 405:305–309

Rosen OM, Hirsch AH, Goren EN (1971) Factors which influence cyclic AMP formation and degradation in an islet cell tumor of the Syrian hamster. Arch Biochem 146:600–603

Samols E, Marri G, Marks V (1965) Promotion of insulin secretion by glucagon. Lancet 2:415–416

Sams DJ, Montague W (1972) The role of adenosine-3':5'-cyclic monophosphate in the regulation of insulin release. Properties of islet-cell adenosine-3':5'-cyclic monophosphate phosphodiesterase. Biochem J 129:945–952

Schuit FC, Pipeleers DG (1985) Regulation of adenosine 3',5'-monophosphate levels in the pancreatic B-cell. Endocrinology 117:834–840

Sugden MC, Ashcroft SJH (1981) Cyclic nucleotide phosphodiesterase of rat pancreatic islets. Effects of Ca^{2+}, calmodulin and trifluoperazine. Biochem J 197:459–464

Svoboda M, Garcia-Morales P, Dufrane SP, Sener A, Valverde I, Christophe J, Malaisse WJ (1985) Stimulation by cholera toxin of ADP-ribosylation of membrane proteins, adenylate cyclase and insulin release in pancreatic islets. Cell Biochem Funct 3:25–32

Tamagawa T, Niki H, Niki A (1985) Insulin release independent of a rise in cytosolic free Ca^{2+} by forskolin and phorbol ester. FEBS Lett 183:430–432

Vallar L, Biden TJ, Wollheim CB (1987) Guanine nucleotides induce Ca^{2+}-independent insulin secretion from permeabilized RINm5F cells. J Biol Chem 262:5049–5056

Valverde I, Vandermeers A, Anjaneyulu R, Malaisse WJ (1979) Calmodulin activation of adenylate cyclase in pancreatic islets. Science 206:225–227

Valverde I, Sener A, Herchuelz A, Malaisse WJ (1981) The stimulus-secretion coupling of glucose-induced insulin release. XLVII. The possible role of calmodulin. Endocrinology 108:1305–1312

Valverde I, Garcia-Morales P, Ghiglione M, Malaisse WJ (1983) The stimulus-secretion coupling of glucose-induced insulin release. LIII. Calcium dependency of the cyclic AMP response to nutrient secretagogues. Horm Metab Res 15:62–68

Wollheim CB, Ullrich S, Pozzan T (1984) Glyceraldehyde, but not cyclic AMP-stimulated insulin release is preceded by a rise in cytosolic free Ca^{2+}. FEBS Lett 177:17–22

The Role of Cholecystokinin and Other Gut Peptides on Regulation of Postprandial Glucose and Insulin Levels

R. J. RUSHAKOFF, R. A. LIDDLE, J. A. WILLIAMS, and I. D. GOLDFINE

A. Summary

The administration of glucose and amino acids into the gastrointestinal tract produces higher insulin levels and lower glucose levels than when these substances are administered intravenously. There is considerable evidence that gut peptides released from the gastrointestinal tract potentiate glucose- and amino acid-induced insulin release from pancreatic β cells. One of the best characterized peptides is the hormone cholecystokinin (CCK). CCK is released from the gut in response to feeding either amino acids, protein, or fat. CCK has been clearly documented to play a role in gastrointestinal motility and secretion. Recent studies now demonstrate that CCK plays a significant role in regulating postprandial glucose and insulin levels. CCK has two major actions. First, CCK delays the gastric emptying of glucose and other nutrients, thus blunting postprandial hyperglycemia. Second, CCK potentiates amino acid-stimulated insulin release.

B. Background

Insulin secretion by β cells of the pancreas is regulated by circulating nutrients, neurotransmitters, paracrine agents, and gut hormones. Since 1906 there has been an interest in the gut hormones that influence carbohydrate metabolism (MOORE et al. 1906). In the late 1920s and early 1930s LaBarre and colleagues presented evidence that an endocrine factor, secreted from the gut, enhanced insulin secretion from the pancreas (LABARRE and STILL 1930; ZUNZ and LABARRE 1929; LABARRE 1936). This factor was not the hormone secretin, which at that time had been identified as a substance which stimulated the exocrine secretion of the pancreas (LABARRE and STILL 1930). Based on this work, LaBarre introduced the term "incretin" to describe a gut factor(s) that stimulate the endocrine pancreas (LABARRE 1932). Concurrent with LaBarre's observations, Heller described a duodenal factor that was able to reduce postprandial hyperglycemia which he named "duodenin" (HELLER 1935).

Further progress in incretin research was not made until the development of the radioimmunoassay (RIA) for insulin in the 1960s. Investigators were then able to measure insulin levels after the administration of glucose and other nutrients. Studies demonstrated that when glucose was administered by the oral (or jejunal) route, higher plasma insulin and lower plasma glucose levels were observed than when glucose was administered by the intravenous route (McINTYRE et al. 1964; ELRICK et al. 1964). Similar results were observed when comparing the intravenous and intraduodenal administration of mixed amino acids (RAPTIS et al. 1973). It was also demonstrated that this phenomenon was a result of enhanced insulin production and not decreased insulin turnover (LICKLEY et al. 1975; McINTYRE et al. 1970). A number of candidate peptide hormones were then studied as possible incretins (CREUTZFELDT 1979; DUPRE 1980; MARKS and TURNER 1977). At that time, however, work was hampered by the lack of pure peptides and the absence of assays capable of quantitating these peptides in plasma. Now, in the 1980s, evidence of incretin action remains (HAMPTON et al. 1986) and the problems with impure peptide preparations and inadequate assays have been largely overcome. Moreover, a number of new gut peptides and chemical variants of established peptides have been discovered. Thus, old factors have been reevaluated and new factors studied for the first time. Evidence for and against these factors as incretins is presented in the following sections.

C. Incretin Criteria

We believe the following criteria must be met before a hormone may be considered as an incretin:

1. The hormone should be produced by gut epithelial cells and released by oral nutrients.
2. Exogenous administration of the pure hormone to achieve physiological plasma concentrations should augment insulin release.
3. Inhibition of hormone action by use of an antagonistic or a blocking antibody should reduce insulin secretion in response to oral nutrients.
4. Receptors for the hormone should be present on pancreatic β cells.

D. Candidate Incretins

I. Glucose Insulinotropic Peptide

Glucose insulinotropic peptide (GIP), a hormone in the glucagen–secretin family, was discovered by BROWN and PEDERSON (1970) who observed that a 10% CCK preparation inhibited gastric acid secretion more strongly than a 40% pure CCK preparation, and speculated that another gut hormone was present along with CCK. The hormone that was ultimately isolated, GIP, is a 42 amino acid peptide with a molecular weight of 5104 (BROWN 1971; JORNVALL et al. 1981). Subsequently, GIP was identified in specific intestinal endocrine cells of the mucosa of the duodenum, jejunum, and ileum termed the K cells (BUCHAN et al. 1978; POLAK et al. 1973; USELLI et al. 1984). The gastric inhibitory effect of GIP is now thought to be of little physiologic significance since supraphysiologic doses of GIP are required for this inhibition to occur (BROWN and PEDERSON 1977).

GIP is released by oral glucose in a dose-related manner (PEDERSON et al. 1975), whereas intravenous glucose does not stimulate GIP release (BROWN et al. 1975; CATALAND et al. 1974). Fat ingestion also stimulates GIP release, but there is no concomitant rise in insulin levels (BROWN et al. 1975; CLEATOR and GOUR-LAY 1975; FALKO et al. 1975; ROSS and DUPRE 1978; EBERT et al. 1979a). Amino acids weakly stimulate GIP release (O'DORISIO et al. 1976; THOMAS et al. 1978).

The insulinotropic action of GIP has been demonstrated in humans (BROWN et al. 1975; ELAHI et al. 1979; DUPRE et al. 1973; CROCKETT et al. 1976), dogs (PEDERSON et al. 1975), and rats (EBERT and CREUTZFELDT 1982; RABINOVITCH and DUPRE 1974; SZECOWKA et al. 1982). In vitro, GIP stimulates insulin secretion in the perfused pancreas (PEDERSON and BROWN 1976; BATAILIE et al. 1977) and in isolated pancreatic islets (MAZZAFERRI et al. 1983; SCHAFER and SCHATZ 1979; SCHAUDER et al. 1975). In all situations, stimulation of insulin release by GIP only occurs in the presence of hyperglycemia (blood glucose levels >115 mg/dl; ANDERSEN et al. 1978; CREUTZFELDT 1979; MCCULLOUGH et al. 1983).

GIP has been a leading candidate for an incretin (CREUTZFELDT 1979). However, analysis of plasma GIP by RIA has been difficult and recent evidence indicates that GIP may exert its insulinotropic effect only at supraphysiologic concentrations (SARSON et al. 1984). In addition, enhanced insulin release by oral glucose occurs even in the presence of excess circulating anti-GIP antibodies (EBERT et al. 1979b). At present there are no known GIP antagonists. The physiological role of GIP as an incretin therefore needs further investigation.

II. Secretin

Secretin was discovered by Bayliss and Starling in 1902, when they observed that an intravenous injection of jejunal mucosal extract stimulated pancreatic exocrine secretion. Secretin is a 27 amino acid polypeptide with a molecular weight of 3055. This hormone is produced by the mucosal S cells of the small intestine (POLAK et al. 1971). Intestinal acidification is the most potent stimulus for secretin secretion (DAVENPORT 1971). The most important physiologic action of secretin is stimulation of pancreatic bicarbonate and water secretion.

In early studies, crude secretin preparations stimulated insulin release *both* in vitro and in vivo (DUPRE 1964; PFEIFFER et al. 1965). These same preparations, when administered intravenously to human subjects, increased insulin levels and improved glucose tolerance (DUPRE et al. 1966; DUPRE 1964). However, these effects are now believed to be pharmacologic phenomena (BUCHANAN et al. 1968). In addition, specific assays for secretin demonstrate that oral glucose does not stimulate secretin secretion (BLOOM and WARD 1975; BODEN et al. 1974; BLOOM 1974; BODEN et al. 1975). Finally, infusion of secretin to levels reproducing physiologic blood levels of the hormone do not stimulate insulin secretion, even in the presence of hyperglycemia (FAHRENKRUG 1978).

III. Gastrin

Gastrin is produced in the G cell of the antral portion of the gastric mucosa. The three main forms of gastrin contain 34, 17, and 14 amino acids. Sulfated and non-sulfated forms exist. All forms have the same COOH terminal configuration. In addition, all forms share the same five COOH terminal amino acids with CCK. While pharmacological amounts of gastrin have been demonstrated to increase glucose-stimulated insulin release, no release of insulin has been demonstrated using physiological amounts of gastrin (Unger et al. 1967; Dupre et al. 1969; Creutzfeldt et al. 1970; Rehfeld 1972; Ipp et al. 1977b). Moreover, low dose gastrin infusions have been reported to inhibit glucose-induced insulin release (Creutzfeldt et al. 1970). Thus, it is unlikely that gastrin is an incretin.

IV. Vasoactive Inhibitory Peptide

Vasoactive inhibitory peptide (VIP) is a 28 amino acid polypeptide with a molecular weight of 3326. VIP has sequence homology to secretin. Initially, VIP was thought to be present in endocrine cells of the gastrointestinal tract, but more recent studies have identifed the peptide only in nerve fibers of the gut (Larsson et al. 1984). VIP has also been isolated in the brain, specifically in the cerebellar cortex, hypothalamus, amygdaloid nucleus, and corpus striatum.

VIP is released secondary to neural stimulation, including that of the vagus nerve. Normally after a meal there is a small increase in VIP in portal venous plasma, but no increase in peripheral blood. The physiologic role of VIP has not yet been accurately defined, but it is believed to play a role in local regulation of blood flow, smooth muscle relaxation, and exocrine secretion. In the exocrine pancreas, VIP can interact with both its own receptors and receptors for secretin, and thus stimulates the secretion of bicarbonate-rich pancreatic juice.

The injection of pharmacological amounts of VIP augments glucose-stimulated insulin secretion (Schebalin et al. 1977; Szecowka 1983; Lindkaer Jensen et al. 1978). Moreover, ingestion of protein and fat, but not carbohydrate, stimulate VIP release (Shuster et al. 1988; Hill et al. 1986; Andrews et al. 1981). Thus, the data suggest that VIP is not an incretin for glucose, but that further studies are needed to define its role as an incretin for amino acids.

V. Peptide Histidine Isoleucine and Peptide YY

Several new peptides have been tested for incretin activity. Peptide histidine isoleucine (PHI) is a neurotransmitter that has sequence homology to VIP and secretin (Miller 1984). It is a 27 amino acid neuropeptide that has been found in the same neurons of the brain and gut that secrete VIP (Christofides et al. 1982). Under hyperglycemic conditions, PHI has been reported to potentiate insulin release in the perfused rat pancreas (Szecowka et al. 1983; Yanaihara et al. 1986).

Peptide YY (PYY) is a 36 amino acid peptide that has been localized to endocrine cells of the mucosa of the ileum, colon, and rectum (Adrian et al. 1985). Levels of PYY increase after ingestion of either meat or fatty acid meals. To date, PYY has not been demonstrated to influence insulin levels (Adrian et al. 1986). Although PYY has been shown to increase blood pressure, inhibit the secretion of gastric acid and pepsin, and delay intestinal transit time (ADRIAN et al. 1986; SAVAGE et al. 1987), the physiologic role of PYY has not been defined.

VI. Enteroglucagon

Glicentin contains 69 amino acids with residues 33-61 being identical to pancreatic glucagon (THIM and MOODY 1981). The COOH terminal portion of glicentin consists of the glucagon molecule with a COOH terminal extension. Oxyntomodulin is the 33-69 sequence and hence is glucagon plus a COOH terminal octapeptide extension (BATAILLE et al. 1982). Recently two other additional peptides have been identified which are derived from the 180 residue preproglucagon. These two glucagon-like peptides are termed GLP-I and GLP-II (BELL et al. 1983b).

The intestine contains both pancreatic glucagon, oxyntomodulin, and glicentin. These latter two gut glucagon peptides are derived from the same messenger RNA as glucagon, but are produced in different cells by differential processing of the preprohormone (BELL et al. 1983a). In the intestine the distribution of glicentin is different from glucagon (SUNDBY et al. 1976). Glicentin has some glucagon-like activity in the liver, and it has larger molecular weight (approximately 11000 for glicentin vs 3485 for glucagon).

Studies have demonstrated that glicentin levels increase after glucose and fat ingestion (OHNEDA et al. 1987). Moreover, studies with the perfused pancreas indicate that glicentin, at physiological concentrations, may stimulate insulin release (OHNEDA et al. 1986). However, further studies will be necessary to define its role as an incretin.

E. An Established Incretin: Cholecystokinin

CCK is a gut hormone that is released after the ingestion of protein, amino acids, and fat. CCK is a major physiologic regulator of pancreatic exocrine secretion, gallbladder contraction, and bowel motility (MUTT 1980; WILLIAMS 1982). In addition, CCK may play a role in mediating satiety. Recent studies, employing pure synthetic CCK and a sensitive bioassay for the hormone, have demonstrated that CCK meets all criteria for an incretin.

Table 1. The amino acid sequence of cholecystokinin and gastrin

Chole-cysto-kinin$_{33}$	Lys-Ala-Pro-Ser-Gly-Arg-Val- Ser-Met-Ile-Lys-Asn-Leu-Glu-Ser- SO_3H Leu-Asp-Pro-Ser-His-Arg-Ile-Ser-Asp-Arg-Asp-Tyr-Met-Gly-Trp-Met-Asp-Phe-NH$_2$
Gastrin$_{17}$	Glp-Gly-Pro-Trp-Met-Glu-Glu-Glu-Glu-Glu-Ala-Tyr[a]-Gly-Trp-Met-Asp-Phe-NH$_2$

[a] Gastrin exists in two forms, I (nonsulfated tyrosine) and II (sulfated tyrosine). Glp, pyroglutamic acid.

CCK is produced in the I cells of the intestine, but is most highly concentrated in the proximal small intestine. Although CCK was first identified in intestinal extracts as a 33 amino acid polypeptide extracted from the intestine, other molecular forms of CCK originating from a common precursor have been identified in intestine, brain, and plasma. These forms include larger and smaller molecules CCK-58, CCK-39, CCK-22, CCK-12, CCK-8, and CCK-4 (Table 1; EYSSELEIN et al. 1982; MUTT and JORPES 1968; ENG et al. 1984; REHFELD 1978). In human plasma, several CCK molecules have been identified including CCK-33, CCK-8, and a third form intermediate in size between CCK-33 and CCK-8 (LIDDLE et al. 1985). The full biological activity of the molecule is contained in CCK-8, the eight COOH terminal amino acids. Unlike gastrin, all CCK molecules contain a sulfated tyrosine; in CCK this sulfation occurs at position 7 from the COOH terminus. The relative biological activities of CCK-8 and CCK-33 are nearly identical (LIDDLE et al. 1986; SOLOMON et al. 1984). The relative biological activity of CCK-58 has not been fully analyzed, and CCK-4 has low bioactivity in the pancreas and gallbladder.

Studies in vitro demonstrate that CCK binds to specific receptors on islet cells and stimulates the release of insulin (UNGER et al. 1967; VERSPOHL et al. 1986) and other islet hormones (UNGER et al. 1967; SZECOWKA et al. 1982; IPP et al. 1977a). Employing isolated rat islets and radiolabeled CCK, the presence of specific, high affinity CCK receptors was demonstrated (Fig. 1). In the same preparations, CCK analogs stimulated insulin release in direct proportion to their receptor binding ability (Fig. 2). Moreover, both the binding of CCK to its receptor and stimulation of insulin release were blocked by the CCK antagonist N-CBZ-L-tryptophan (VERSPOHL et al. 1986).

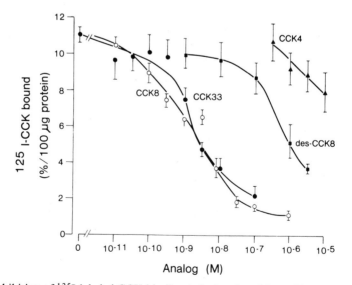

Fig. 1. Inhibition of ^{125}I-labeled CCK binding to isolated rat islets of Langerhans by CCK analogs. (VERSPOHL et al. 1986)

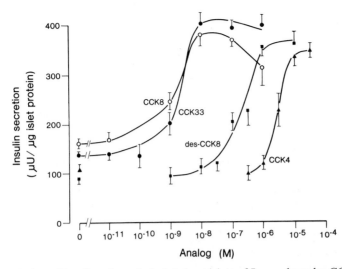

Fig. 2. Stimulation of insulin release in isolated rat islets of Langerhans by CCK analogs. (VERSPOHL et al. 1986)

Employing the isolated perfused rat pancreas, electron microscope autoradiographic studies with radioiodinated CCK localized this hormone over β, and other islets cells (Fig. 3; SAKAMOTO et al. 1985). CCK was most highly accumulated over β cells; this accumulation was similar to that observed for CCK over pancreatic acinar cells. In the isolated perfused rat pancreas, physiological concentrations of CCK stimulate insulin release (SZECOWKA et al. 1982). In both isolated islets and perfused pancreas, CCK augmented glucose-induced insulin release. These in vitro studies indicate, therefore, that islet β cells have receptors for CCK that are linked to stimulation of insulin release.

With the recent development of potent and specific CCK receptor antagonists, it has been possible to determine the effects of CCK receptor blockade on insulin secretion in an experimental model (CHANG and LOTTI 1986; ROSSETTI et al. 1987). In rats, treatment with L-364,718, a specific antagonist of CCK binding to its membrane receptor, significantly attenuated the increase in plasma insulin and glucagon levels seen after duodenal feeding of a protein meal (ROSSETTI et al. 1987). These results support the concept that CCK plays an important role in the in vivo regulation of plasma insulin after protein ingestion.

Prior studies in humans, employing CCK-rich extracts of porcine duodenum, suggested that CCK potentiated the release of insulin and glucagon (DUPRE et al. 1969). The interpretation of these studies was complicated, however, because the extracts used contained, in addition to CCK, other gut peptides that had the ability to stimulate insulin release (DUPRE et al. 1973). Furthermore, technical difficulties in measuring circulating CCK levels limited research efforts in this area.

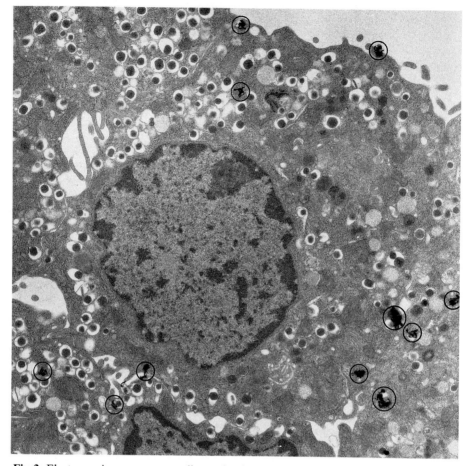

Fig. 3. Electron microscope autoradiograph of rat pancreatic islets after incubation with ^{125}I-labeled CCK. Silver grains over a β cell are shown. For clarity, the silver grains are *circled*. (VERSPOHL et al. 1986) $\times 10\,750$

Recently a specific and sensitive bioassay for measuring plasma CCK levels has been developed (LIDDLE et al. 1984, 1985). This assay is sufficiently sensitive to allow detection of basal CCK levels in humans, and is specific for CCK, with no detectable interference from other hormones and neurotransmitters (LIDDLE et al. 1985). The CCK levels following a mixed liquid meal are shown in Fig. 4. Basal CCK levels are very low and average approximately 1 pM. Some 10 min after the ingestion of a meal containing 60 g glucose, 27 g lipid, and 30 g protein, CCK levels increase sevenfold.

In order to reproduce the plasma CCK levels seen after a meal, pure synthetic CCK-8 was infused into normal volunteers. Infusion of CCK did not alter insulin levels either alone or in combination with intravenous glucose (RUSHAKOFF et al. 1987). Thus, the β cell in the human is different from the β cell in the rat, where CCK does potentiate glucose-induced insulin release. CCK however, markedly

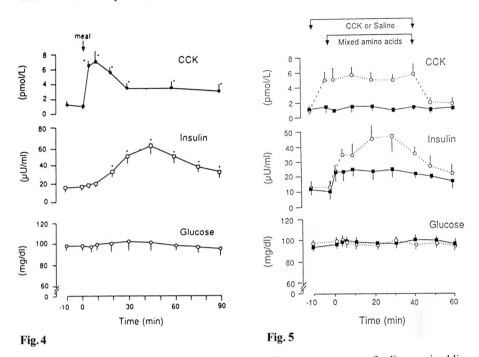

Fig. 4. Mean plasma CCK, insulin, glucagon, and glucose responses to feeding a mixed liquid meal containing 60 g glucose, 27 g lipid, and 30 g protein. The *asterisk* signifies values statistically different from basal ($p < 0.05$). (RUSHAKOFF et al. 1987)

Fig. 5. Mean plasma CCK, insulin, glucagon, and glucose responses to intravenous infusions of mixed amino acids and CCK. The insulin and glucagon levels were significantly elevated during the CCK infusion ($p < 0.005$). (RUSHAKOFF et al. 1987)

potentiated insulin release in response to mixed amino acids (Fig. 5). Integration of the insulin secreted during the infusion indicated that CCK doubled insulin secretion (RUSHAKOFF et al. 1987).

Interestingly, after ingestion of the mixed liquid meal containing glucose, protein, and fat, there was very little, if any, rise in blood glucose levels (see Fig. 4; RABINOWITZ et al. 1966; TASAKA et al. 1975). In contrast, when the same amount of glucose was given alone, hyperglycemia resulted (Fig. 6; RABINOWITZ et al. 1966; BANTLE et al. 1983). There is evidence that CCK, together with the noncarbohydrate elements of a meal, are responsible for preventing postprandial hyperglycemia. First, when a mixed meal is given, ingested protein and amino acids stimulate insulin release. Second, the protein, amino acids, and fat in a mixed meal are potent stimulants of CCK release (LIDDLE et al. 1985); glucose, in contrast, is a weaker stimulant (LIDDLE et al. 1985). Thus, the CCK released by protein, amino acids, and fat then serves to prevent postprandial hyperglycemia by potentiating amino acid-stimulated insulin release.

Physiological levels of CCK inhibit gastric emptying (LIDDLE et al. 1986; DEBAS et al. 1975; YAMAGISHI and DEBAS 1978). Figure 7 shows the gastric

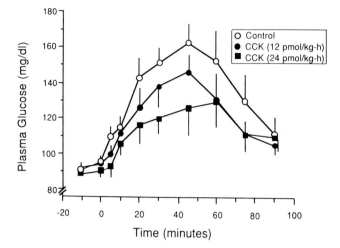

Fig. 6. Effect of CCK infusions on plasma glucose levels after the oral administration of 60 g glucose. Each value is the mean of eight subjects. (LIDDLE et al. 1988)

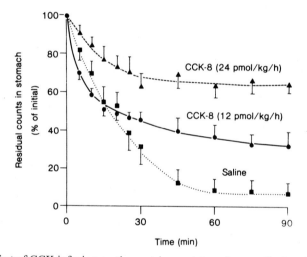

Fig. 7. The effect of CCK infusion on the gastric emptying of water. Each value is the mean of five subjects. (LIDDLE et al. 1986)

emptying of saline, which is rapid with a $t_{1/2}$ of approximately 15 min. A mixed liquid meal, in contrast, empties with a $t_{1/2}$ of >90 min (LIDDLE et al. 1988). When CCK was infused to produce either a low (4 pM) or high (8 pM) physiological level, gastric emptying of saline was markedly delayed. These data suggested, therefore, that the effect of CCK on gastric emptying also served to prevent postprandial hyperglycemia. Since as previously noted, glucose is not a

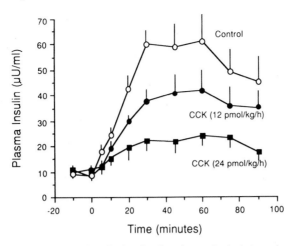

Fig. 8. Effect of CCK on plasma insulin levels after the oral administration of 60 g glucose. Each value is the mean of eight subjects. (LIDDLE et al. 1988)

major regulator of CCK secretion, it is possible that the postprandial hyperglycemia seen after the ingestion of carbohydrate alone is due, in part, to the lack of CCK release. If CCK were present, it would have delayed gastric emptying of glucose and reduced hyperglycemia.

To investigate the role of CCK in the regulation of gastric emptying during oral glucose administration, normal human volunteers were given 60 g glucose orally in the presence and absence of a CCK infusion. In these studies plasma levels of CCK, glucose, and insulin, and gastric emptying rates, were measured. CCK infusion significantly lowered postingestion plasma glucose levels (see Fig. 6). A concomitant diminution of plasma insulin levels was also observed (Fig. 8). Because CCK infusions lowered plasma glucose levels without increasing insulin levels, gastric emptying rates were measured. CCK infusions delayed gastric emptying in a dose-related manner (Fig. 9). When glucose was infused intraduodenally, no effect of CCK was observed.

To determine if endogenously released CCK exhibited effects similar to those produced by infused CCK, lipid was added to the oral glucose solution (Fig. 10). Ingestion of glucose alone did not increase CCK levels and ingestion of lipid alone did not alter insulin levels (LIDDLE et al. 1988). However, addition of lipid to the glucose increased CCK levels. Coincident with this lipid-induced increase in CCK levels, the rate of gastric emptying of glucose and lipid was significantly slower than that of glucose alone. In addition, with the added lipid, plasma glucose and insulin levels were lower than with the glucose alone. These experiments indicated, therefore, that after the ingestion of glucose, CCK in physiological concentrations delays gastric emptying, slows the delivery of glucose to the duodenum, and reduces hyperglycemia.

The findings that endogenous and exogenous CCK reduces postprandial hyperglycemia by both delaying gastric emptying and potentiating amino acid-

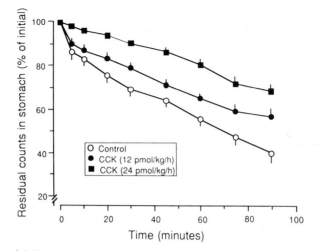

Fig. 9. Effect of CCK on the gastric emptying of glucose. *Open circles*, glucose alone. *Full circles*, glucose plus CCK at 12 pmol kg^{-1} h^{-1}. *Full squares*, glucose plus CCK at 24 pmol kg^{-1} h^{-1}. Each value is the mean of eight subjects. (Liddle et al. 1988)

induced insulin release extends the physiologic role of the integrative actions of this hormone. After a meal, in a highly coordinated fashion, CCK: (a) regulates the movement of nutrients through the gastrointestinal tract (Liddle et al. 1986, 1988; Debas et al. 1975; Yamagishi and Debas 1978); (b) contracts the gallbladder (Liddle et al. 1985; Wiener et al. 1981; Byrnes et al. 1981) and stimulates pancreatic exocrine secretion to facilitate digestion (Mutt 1980; Beglinger et al. 1985; Grossman 1971; Go et al. 1970); and (c) to maintain euglycemia, potentiates amino acid-induced insulin secretion (Rushakoff et al. 1987) and delays gastric emptying (Liddle et al. 1988). CCK thus has an essential role in regulating the intake, processing, and distribution of essential nutrients.

F. Conclusion

Hormones released from the gut play a major role in regulating postprandial glucose and insulin levels. In the case of one hormone, CCK, incretin activity has been demonstrated. In humans CCK potentiates insulin release in response to amino acids. Therefore, another gut hormone(s) must exist that potentiates insulin release in response to glucose. The two leading candidates for this other incretin are GIP and glicentin. Further studies, therefore, will be necessary to elucidate the individual physiologic roles of these latter two hormones. Moreover, it is possible that gut peptides act synergistically since they are released by distinct nutrients. Further studies in this field should clarify the role of incretins in maintaining nutritional homeostasis.

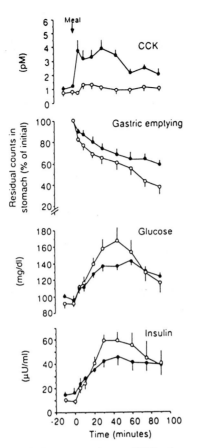

Fig. 10. Effect of oral lipid and glucose on plasma CCK, glucose, and insulin levels, and gastric emptying. *Open circles*, 60 g gluco > e alone. *Full circles*, 60 g glucose plus 50 g lipid. Each value is the mean of four subjects. (LIDDLE et al. 1988)

References

Adrian TE, Ferri GL, Bacarese-Hamilton AJ, Fuessl HS, Polak JM, Bloom SR (1985) Human distribution and release of a putative new gut hormone, peptide YY. Gastroenterology 89:1070–1077

Adrian TE, Sagor GR, Savage AP, Bacarese-Hamilton AJ, Hall GM, Bloom SR (1986) Peptide YY kinetics and effects on blood pressure and circulating pancreatic and gastrointestinal hormones and metabolites in man. J Clin Endocrinol Metab 63:803–807

Andersen DK, Elahi D, Brown JC, Tobin JD, Andres R (1978) Oral glucose augmentation of insulin secretion: interactions of gastric inhibitory polypeptide with ambient glucose and insulin levels. J Clin Invest 62:152–161

Andrews WJ, Henry RW, Alverti KGMM, Buchanan KD (1981) The gastro-entero-pancreatic hormone response to fasting in obesity. Diabetologia 21:440–445

Bantle JP, Laine DC, Castle GW, Thomas JW, Hoogwerf BJ, Goetz FC (1983) Postprandial glucose and insulin responses to meals containing different carbohydrates in normal and diabetic subjects. N Engl J Med 309:7–12

Batailie D, Jarrousse C, Vauclin N, Gespach C, Rosselin G (1977) Effect of vasoactive intestinal peptide (VIP) and gastric inhibitory polypeptide (GIP) on insulin and glucagon release by perifused newborn rat pancreas. In: Goa PP, Bajaj JS, Foa NL (eds) Glucagon. Its role in physiology and clinical medicine. Springer, Berlin Heidelberg New York, pp 255–269

Bataillie D, Coudray A, Carlqvist M, Rosselin G, Mutt V (1982) Isolation of glucagon-27 (bioactive enteroglucagon/oxyntomodulin) from porcine jejuno-ileum. FEBS Lett 146:73–78

Beglinger C, Fried M, Whitehouse I, Jansen JB, Lamers CB, Gyr K (1985) Pancreatic enzyme response to a liquid meal and to hormonal stimulation: correlation with plasma secretin and cholecystokinin levels. J Clin Invest 75:1471–1476

Bell G, Sanchez-Pescador R, Laybourn P, Najarian R (1983 a) Exon duplication and divergence in the human preproglucagon gene. Nature 304:368–371

Bell G, Sanierre R, Mullenbach G (1983 b) Hamster preproglucagon contains the sequence of glucagon and two related peptides. Nature 302:716–718

Bloom SR (1974) Hormones of the gastrointestinal tract. Br Med Bull 30:62–71

Bloom SR, Ward AS (1975) Failure of secretin release in patients with duodenal ulcer. Br Med J 1:126–127

Boden G, Essa L, Owen OE, Reichle FA (1974) Effects of intraduodenal administration of HCl and glucose on circulating immunoreactive secretin and insulin concentrations. J Clin Invest 53:1185

Boden G, Essa N, Owen OE (1975) Effects of intraduodenal amino acids, fatty acids and sugars on secretin concentrations. Gastroenterology 68:722–728

Brown JC (1971) A gastric inhibitory polypeptide. 1. The amino acid composition and the tryptic peptides. Can J Bio Chem 49:255–261

Brown JC, Pederson RA (1970) A multiparameter study on the action of preparations containing cholecystokinin pancreozymin. Scand J Gastroenterol 5:537–541

Brown JC, Pederson RA (1977) GI hormones and insulin secretion. In: Endocrinology. Proceedings of the Vth international congress of endocrinology, vol 2. Excerpta Medica, Amsterdam, pp 568–570

Brown JC, Dryburgh JR, Ross SA, Dupre J (1975) Identification and actions of gastric inhibitory polypeptide. Reidentification and actions of gastric inhibitory polypeptide. Recent Prog Horm Res 31:487–532

Buchan AMJ, Polak JM, Solcia E, Capella C, Pearse AGE (1978) Electroimmunocytochemical evidence for the K cell localization of gastric inhibitory polypeptide (GIP) in man. Histochemistry 56:37–44

Buchanan KD, Vance JE, Morgan A, Williams RH (1968) Effect of pancreozymin on insulin and glucagon levels in blood and bile. Am J Physiol 215:1293–1299

Byrnes DJ, Borody T, Daskalopoulos G, Boyle M, Benn I (1981) Cholecystokinin and gall bladder contraction: effect of CCK infusion. Peptides 2:259–262

Cataland S, Crockett SE, Brown JC, Mazzaferri EL (1974) Gastric inhibitory polypeptide (GIP) stimulation by oral glucose in man. J Clin Endocrinol Metab 39:223–228

Chang RS, Lotti VJ (1986) Biochemical and pharmacological characterization of an extremely potent and selective nonpeptide cholecystokinin antagonist. Proc Natl Acad Sci USA 83:4923–4923

Christofides ND, Adalaja AB, Tatemoto K, Ferri GL, Polak JM, Bloom SR (1982) Presence of PHI in the human intestine. Regul Pept 3:69–72

Cleator JCM, Gourlay RH (1975) Release of immunoreactive gastric inhibitory polypeptide (IR-GIP) by oral ingestion of food substances. Am J Surg 130:128–135

Creutzfeldt W (1979) The incretin concept today. Diabetologica 16:75–85

Creutzfeldt W, Feurle G, Ketterer H (1970) Effect of gastrointestinal hormones on insulin and glucagon secretin. N Engl J Med 282:1139–1141

Crockett SR, Cataland S, Falko JM, Mazzaferri EL (1976) The insulinotropic effect of endogenous gastric inhibitory polypeptide in normal subjects. J Clin Endocrinol Metab 42:1090–1103

Davenport HW (1971) Physiology of the digestive tract, 3rd edn. Year Book Medical, Chicago

Debas HT, Farooq O, Grossman MI (1975) Inhibition of gastric emptying is a physiological action of cholecystokinin. Gastroenterology 68:1211–1217

Dupre J (1964) An intestinal hormone effecting glucose disposal in man. Lancet 2:672–673

Dupre J (1980) The entero-insular axis and the metabolic effects of gastroenteropancreatic polypeptides. Clin Gastroenterol 9:711–732

Dupre J, Rojas L, White JJ, Unger RH, Beck JC (1966) Effects of secretin on insulin and glucagon in portal and peripheral blood in man. Lancet 2:26–27

Dupre J, Curtis JD, Unger RH, Waddell RW, Beck JC (1969) Effects of secretion, pancreozymin, or gastrin on the response of the endocrine pancreas to administration of glucose or arginine in man. J Clin Invest 48:747–758

Dupre J, Ross SA, Watson D, Brown JC (1973) Stimulation of insulin secretion by gastric inhibitory polypeptide in man. J Clin Endocrinol Metab 37:826–828

Ebert R, Creutzfeldt W (1982) Influence of gastric inhibitory polypeptide antiserum on glucose-induced insulin secretion in rats. Endocrinology 111:1601–1606

Ebert R, Frerichs H, Creutzfeldt W (1979a) Impaired feedback control of fat induced gastric inhibitory polypeptide (GIP) secretion by insulin in obesity and glucose tolerance. Eur J Clin Invest 9:129–135

Ebert R, Illmer K, Creutzfeldt W (1979b) Release of gastric inhibiting polypeptide (GIP) by intraduodenal acidification in rats and humans and abolishment of the incretin effect of acid by GIP-antiserum in rats. Gastroenterology 76:515–523

Elahi D, Andersen DK, Brown JC, Debas H, Hershcopf RJ, Raizers GS, Tobin JD, Andres R (1979) Pancreatic alpha- and beta-cell responses to GIP infusion in normal man. Am J Physiol 237:E185–E191

Elrick H, Stimmler L, Hlad CJ, Arai J (1964) Plasma insulin responses to oral and intravenous glucose administration. J Clin Endocrinol Metab 24:1076–1082

Eng J, Du B-H, Pan Y-C, Chang M, Hulmes J, Yallow R (1984) Purification and sequencing of a rat intestinal 22 amino acid C-terminal CCK fragment. Peptides 5:1203–1206

Eysselein VE, Reeve JR, Shively J, Hawke D, Walsh J (1982) Partial structure of a large canine cholecystokinin (CCK58): amino acid sequence. Peptides 3:687–691

Fahrenkrug J, Schaffalitzky de Muckadell OB, Kohl C (1978) Effect of secretin on basal and glucose-stimulated insulin secretion in man. Diabetologia 14:229–234

Falko JM, Crockett SE, Cataland S, Mazzaferri EL (1975) Gastric inhibitory polypeptide (GIP) stimulated by fat ingestion in man. J Clin Endocrinol Metab 41:260–265

Go VLW, Hoffman AF, Summerskill WHJ (1970) Pancreozymin bioassay in man based on pancreatic enzyme secretion: potency of specific amino acid and other digestive products. J Clin Invest 49:1558–1564

Grossman MI (1971) Control of pancreatic secretion. In: Beck IT, Sinclair DG (eds) The exocrine pancreas. Churchill Livingstone, London, p 59

Hampton SM, Morgan JM, Tredger JA, Cramb R, Marks V (1986) Insulin and C-peptide levels after oral and intravenous glucose. Diabetes 35:612–616

Heller H (1935) Über das insulinotrope Hormon der Darmschleimhaut (Duodenin). Nauyn Schmiedebergs Arch Pharmacol 147:127–133

Hill P, Thijssen JHH, Garbaczewski L, Koppewchaar HPF, de Waard F (1986) VIP and prolactin release in response to meals. Scand J Gastroenterol 21:958–960

Ipp E, Dobbs RE, Arimura A, Vale W, Harris V, Unger RH (1977a) Release of immunoreactive somatostatin from the pancreas in response to glucose, amino acids, pancreozymin-cholecystokinin, and tolbutamide. J Clin Invest 60:760–771

Ipp E, Dobbs RE, Harris V, Arimura A, Vale W, Unger RH (1977b) The effects of gastrin, gastric inhibitory polypeptide, secretin and the octapeptide of cholecystokinin upon immunoreactive somatostatin release by the perfused canine pancreas. J Clin Invest 60:1216–1219

Jornvall H, Carlquist M, Kwauk S, Otte SC, Mcintosh CHS, Brown JC, Mutt V (1981) Amino acid sequence and heterogeneity of gastric inhibitory polypeptide (GIP). FEBS Lett 123:205–210

LaBarre (1932) Sur les possibilités d'un traitement du diabetic par l'incretin. Bell Acad Voy Med Belg 12:620–634

LaBarre J (1936) La secretine: son rôle physiologique, ses propriétés thérapeutiques. Masson, Paris

LaBarre J, Still EU (1930) Studies on the physiology of secretin. Am J Physiol 91:649–653

Larsson LI, Fahrenkrug J, Holst JJ, Schaffalitzky de Schusdziarra V, Lenz N, Rewes B, Pfeiffer EF (1984) Endogenous opioids modulate the effect of cholecystokinin on insulin release in dogs. Neuropeptides 4:507–513

Lickley HLA, Chisholm DJ, Rabinovitch A, Wexler M, Dupre J (1975) Effects of portacaval anastomosis on glucose tolerance in the dog: evidence of an interaction between the gut and the liver in oral glucose disposal. Metabolism 24:1157–1168

Liddle RA, Goldfine ID, Williams JA (1984) Bioassay of plasma cholecystokinin in rats: effects of food, trypsin inhibitor, and alcohol. Gastroenterology 87:542–549

Liddle RA, Goldfine ID, Rosen MS, Tapliz RA, Williams JA (1985) Cholecystokinin bioactivity in human plasma: molecular forms, responses to feeding, and relationship to gallbladder contraction. J Clin Invest 75:1144–1152

Liddle RA, Morita ET, Conrad CK, Williams JA (1986) Regulation of gastric emptying in humans by cholecystokinin. J Clin Invest 77:992–996

Liddle RA, Rushakoff RJ, Morita ET, Beccaria L, Carter JD, Goldfine ID (1988) Physiological role for cholecystokinin in reducing postprandial hyperglycemia in humans. J Clin Invest 81:1675–1681

Lindkaer Jensen S, Fahrenkrug J, Holst JJ, van Nielsen O, Schaffalitzky de Muckadell OB (1978) Secretory effects of VIP on isolated perfused procine pancreas. Am J Physiol 235:E387–E391

Loew ER, Gray JS, Ivy AC (1940) Is a duodenal hormone involved in carbohydrate metabolism? Am J Physiol 129:659–663

Marks V, Turner DS (1977) The gastrointestinal hormones with particular reference to their role in the regulation of insulin secretion. Essays Biochem 3:109–152

Mazzaferri EL, Ciofalo L, Waters LA, Starich GH, Groshong JC, DePalma L (1983) Effects of gastric inhibitory polypeptide on leucine and arginine-stimulated insulin release. Am J Physiol 245:E114–E119

McCullough AJ, Miller CJ, Service FJ, Go VLW (1983) Effect of graded intraduodenal glucose infusion on the release and physiological action of gastric inhibitory polypeptide. J Clin Endocrinol Metab 56:234–241

McDonald TJ, Jornvall H, Nilsson G, Vagne M, Ghatei M, Bloom SR, Mutt V (1979) Characterization of a gastrin releasing peptide from porcine non-antral gastric tissue. Biochem Biophys Res Commun 90:227–238

McDonald TJ, Ghatei MA, Bloom SR, Track ND, Radziuk J, Dupre J, Mutt V (1981) A qualitative comparison of canine plasma gastroenteropancreatic hormone responses to bombesin and the porcine gastrin-releasing peptide. Regul Pept 2:293–305

McIntyre N, Holdsworth DC, Turner DS (1964) New interpretation of oral glucose tolerance. Lancet 2:20–21

McIntyre N, Turner DS, Holdsworth CD (1970) The role of the portal circulation in glucose and fructose tolerance. Diabetologia 6:593–596

Miller RJ (1984) New perspectives on gut peptides. J Med Chem 27:1239–1245

Moore B, Edie ES, Abran JH (1906) On the treatment of diabetes mellitus by acid extract of duodenal mucous membrane. Biochem J 1:28

Muckadell OB (1978) Innervation of the pancreas by vasoactive intestinal polypeptide (VIP) immunoreactive nerves. Life Sci 22:773–780

Mutt V (1980) Cholecystokinin: isolation, structure and functions. In: Glass GB (ed) Gastrointestinal hormones. Raven, New York, p 169

Mutt V, Jorpes J (1968) Structure of porcine cholecystokinin-pancreozymin. Eur J Biochem 6:156–162

O'Dorisio TM, Cataland S, Stevenson M, Mazzaferri EL (1976) Gastric inhibitory polypeptide (GIP). Intestinal distribution and stimulation by amino acids and medium-chain triglycerides. Am J Dig Dis 21:761–765

Ohneda A, Kobayashi T, Nihei J (1986) Effect of glicentin-related peptides on glucagon secretion in anaesthetized dogs. Diabetologia 29:397–401

Ohneda A, Takahashi H, Maruyama Y (1987) Response of plasma glicentin to fat ingestion in piglets. Diabetes Res Clin Pract 3:103–109

Pederson RA, Brown JC (1976) The insulinotropic action of gastric inhibitory polypeptide in the perfused isolated rat pancreas. Endocrinology 99:780–785

Pederson RA, Schubert HE, Brown JC (1975) Gastric inhibitory polypeptide. Its physiological release and insulinotropic action in the dog. Diabetes 24:1050–1056

Pfeiffer EF, Teib M, Ammon J, Melani F, Ditschuneit H (1965) Direkte Stimulierung der Insulinseketion in vitro durch Sekretin. Dtsch Med Wochenschr 90:1663–1667

Polak JM, Coulling I, Bloom S, Pearse AGE (1971) Immunofluorescent localization of secretin and enteroglucagon in human intestinal mucosa. Scand J Gastroenterol 6:739

Polak JM, Bloom SR, Kuzio M, Brown JC, Pearse AGE (1973) Cellular localization of gastric inhibitory polypeptide in the duodenum and jejunum. Gut 14:284–288

Rabinovitch A, Dupre J (1974) Effects of gastric inhibitory polypeptide present in impure pancreozymin-cholecystokinin on plasma insulin and glucagon in the rat. Endocrinology 94:1139–1144

Rabinowitz D, Merimee TJ, Maffessol R, Burgess JA (1966) Patterns of hormonal release after glucose, protein and glucose plus protein. Lancet ii:454–462

Raptis S, Dollinger HC, Schroder KE, Schleyer M, Rothenbuchner G, Pfeiffer EF (1973) Differences in insulin, growth hormone and pancreatic enzyme secretion after intravenous and intraduodenal administration of mixed amino acids. N Engl J Med 228:1199–1202

Rehfeld JF (1972) Gastrointestinal hormones and insulin secretion. Scand J Gastroenterol 7:289–301

Rehfeld JF (1978) Immunochemical studies on cholecystokinin. II. Distribution and molecular heterogeneity in the central nervous system and small intestine of man and hog. J Biol Chem 253:4016–4021

Ross SA, Dupre J (1978) Effects of ingestion of triglyceride or galactose on secretion of gastric inhibitory polypeptide and on responses to intravenous glucose in normal and diabetic subjects. Diabetes 27:327–333

Rossetti L, Shulman GI, Zawalich WS (1987) Physiological role of cholecystokinin in meal-induced insulin secretion in conscious rats. Studies with L 364718, a specific inhibitor of CCK-receptor binding. Diabetes 36:1212–1215

Rushakoff RJ, Goldfine ID, Carter JD, Liddle RA (1987) Physiological concentrations of cholecystokinin stimulate amino acid-induced release in humans. J Clin Endocrinol Metab 65:395–401

Sakamoto C, Goldfine ID, Roach E, Williams JA (1985) Localization of CCK receptors in rat pancreatic islets by light and electron microscope autoradiography. Diabetes 34:390–394

Sarson PL, Wood SM, Kansal PC, Bloom SR (1984) Glucose-dependent insulinotropic polypeptide augmentation of insulin physiology and pharmacology? Diabetes 33:398

Savage AP, Adrian TE, Carolan G, Chatterjee VK, Bloom SR (1987) Effects of peptide YY (PYY) on mouth to caecum intestinal transit time and on the rate of gastric emptying in healthy volunteers. Gut 28:L166–170

Schafer R, Schatz M (1979) Stimulation of (pro)insulin biosynthesis and release by gastric inhibitory polypeptide in isolated islets of rat pancreas. Acta Endocrinol 91:493–500

Schauder P, Brown JC, Frerichs H, Creutzfeldt W (1975) Gastric inhibitory polypeptide: effect on glucose induced insulin release from isolated rat pancreatic islets in vitro. Diabetologia 11:483–484

Schebalin M, Said SI, Makhlouf FM (1977) Stimulation of insulin and glucagon secretion by vasoactive intestinal peptide. Am J Physiol 232:E197–E200

Shuster LT, Go VLW, Rizza RA, O'Brien PC, Service ED (1988) Potential incretins. Mayo Clin Proc 63:794–800

Solomon TET, Yamada T, Elashoff J, Wood J, Beglinger C (1984) Bioactivity of choleystokinin analogues: CCK-8 is not more potent than CCK-33. Am J Physiol 247:G105–111

Sundby F, Jacobson H, Moody A (1976) Purification and characterization of a protein from porcine gut with glucagon-like immunoreactivity. Horm Metab Res 8:366–371

Szecowka J, Lins PE, Efendic S (1982) Effects of cholecystokinin, gastric inhibitory polypeptide and secretin on insulin and glucagon secretion in rats. Endocrinology 110:1268–1272

Szecowka J, Lins PE, Tatemoto K, Efendic S (1983) Effects of porcine intestinal heptacosapeptide and vasoactive intestinal polypeptide on insulin and glucagon secretion in rats. Endocrinology 112:1469–1473

Tasaka Y, Sekine M, Wakatuki M, Obgawasa H, Shizume K (1975) Levels of pancreatic glucagon, insulin and glucose during twenty-four hours of the day in normal subjects. Horm Metab Res 7:205–213

Thim L, Moody A (1981) The amino acid sequence of porcine glicentin. Peptides 2:37–39

Thomas FB, Sinar D, Mazzaferri EL, Cataland S, Mekhjian HS, Caldwell JH, Fromkes JJ (1978) Selective release of gastric inhibitory polypeptide by intraduodenal amino acid perfusion in man. Gastroenterology 74:1261–1265

Unger RH, Ketterer H, Dupre J, Eisentraut AM (1967) The effects of secretin, pancreozymin, and gastrin on insulin and glucagon secretion in anesthetised dogs. J Clin Invest 46:1630–1642

Usellini L, Capella C, Solcia E, Buchan AMJ, Brown JC (1984) Ultrastructural localization of gastric inhibitory polypeptide (GIP) in a well characterized endocrine cell of canine duodenal mucosa. Histochemistry 80:85–89

Verspohl EJ, Ammon HPT, Williams JA, Goldfine ID (1986) Evidence that cholecystokinin interacts with specific receptors and regulates insulin release in isolated rat islets of Langerhans. Diabetes 35:38–43

Wiener I, Inoue K, Fagan CJ, Lija P, Watson LC, Thompson JC (1981) Release of cholecystokinin in man: correlation of blood levels with gallbladder contraction. Ann Surg 191:321–327

Williams JA (1982) Cholecystokinin: a hormone and a neurotransmitter. Biomed Res 3:107–119

Yamagishi T, Debas HT (1978) Cholecystokinin inhibits gastric emptying by action on both proximal stomach and pylorus. Am J Physiol 234:E375–378

Yanaihara C, Hashimoto Y, Takeda Y, Kato I, Track NS, Nokihara K, Manaka H, Iwanaga T, Fujita T, Okamoto H, Yanaihara N (1986) PHI structural requirements for potentiation of glucose-induced insulin release. Peptides 7 [Suppl 1]:83–88

Zunz E, LaBarre J (1929) Contributions à l'étude des variations physiologiques de la sécrétion interne du pancréas: relations entre les sécrétions externe et interne du pancréas. Arch Int Physiol Biochim 31:20–44

CHAPTER 8

Insulin-Degrading Enzyme

W. C. DUCKWORTH

A. Introduction

Numerous studies have reported that a relatively specific insulin-degrading protease can be found in most tissues. Several different laboratories have examined this enzyme and many of its characteristics have been established, but many contradictory findings have been reported and much confusion and controversy continues to exist. Recent work from Dr. Richard Roth's laboratory has clarified some issues and established a reference preparation with which to compare other studies, but a number of issues remain unsettled. Some of the current confusion is due to experimental variations among different studies, but other literature discrepancies are not easily reconciled. Some possible explanations will be discussed below, but complete clarification will require additional investigation.

The first systematic studies of tissue insulin-degrading activity was by Mirsky who identified a proteolytic activity he termed insulinase (MIRSKY 1957, 1964; MIRSKY and BROH-KAHN 1949; MIRSKY and PERISUTTI 1957; BROH-KAHN and MIRSKY 1949). He characterized this activity in a number of studies which have been largely confirmed by subsequent studies. His overall conclusion was that a relatively specific insulin-degrading protease is present in multiple tissues of the body. Two properties of this enzymatic activity which are particularly relevant to current studies were that it was difficult to purify and that the activity was unstable.

The failure to obtain a purified preparation and the discovery of an insulin-degrading enzyme, glutathione insulin transhydrogenase (GIT), which acted by reductive cleavage of the disulfide bonds of insulin (KATZEN and TIETZE 1966) led to the conclusion that "insulinase" was actually a mixture of GIT and nonspecific proteases which further degraded the separate insulin chains. This led to the sequential theory of insulin degradation in which the obligatory initial step in insulin degradation by tissues was reductive cleavage by GIT followed by nonspecific proteolysis (VARANDANI et al. 1972). A number of studies on this process followed and this became widely accepted for some time as the physiological mechanism of insulin degradation. Although current evidence does not support a major role for GIT in cellular insulin degradation, the evidence accumulated for this theory illustrates two of the continuing problems in examining insulin degradation, namely nonspecific assays and contaminating nonspecific proteases. Most studies on tissue insulin degradation use the trichloracetic acid (TCA) method for assay. This assay measures the solubilization of labeled insulin in TCA to determine degradation. This does not differentiate between disulfide bond reduction and direct proteolysis and is influenced by the presence of non-specific proteases which do not degrade intact insulin, but do solubilize fragments of insulin which are partially TCA precipitable. This is discussed further below.

Several studies of cells and tissue extracts suggested a direct proteolytic degradation of insulin (BRUSH and KITABCHI 1970; RUDMAN et al. 1966), but were largely discounted because of the problems discussed above. Attention was refocused on a proteolytic enzyme by the studies of BRUSH (1971) who partially purified a protease from rat skeletal muscle which appeared to have specificity for insulin. Additional studies on more purified preparations ensued, but the lack of complete purification and discrepancies in various reports were taken to be evidence that the studies were on mixtures of GIT and various proteases (THOMAS and VARANDANI 1979; VARANDANI 1973).

Although it perhaps cannot be said that this controversy has been established with total finality, it is now clear: that a single proteolytic enzyme excists which can degrade insulin directly without requiring disulfide cleavage; that this enzyme is the primary insulin-degrading activity in tissues; that, from numerous cell studies, much evidence has accumulated that GIT plays little, if any, role in insulin degradation in tissues; and that the proteolytic enzyme is the primary mechanism for cellular insulin degradation.

A number of different laboratories have studied this insulin-degrading activity and have given it a variety of names, including insulinase, insulin-specific protease, insulin–glucagon protease, insulin protease, insulin-degrading enzyme, neutral thiolpeptidase, and metalloendoprotease (BURGHEN et al. 1972; DUCKWORTH and KITABCHI 1974; DUCKWORTH et al. 1972; KIRSCHNER and GOLDBERG 1983; ROTH et al. 1985; SHII et al. 1986; SHROYER and VARANDANI 1985). These all appear to be the same enzyme, although a number of differences have been reported. We have studied extensively an insulin-degrading enzyme from rat skeletal muscle which we named insulin protease (IP) (DUCKWORTH et al. 1972). Roth's group has purified to homogeneity an enzyme from red blood cells with similar characteristics which they refer to as insulin-degrading enzyme (IDE) (SHII et al. 1986). In unpublished studies, we have found that monoclonal antibodies to their enzyme recognize IP and that IDE and IP generate the same degradation products from insulin. These enzymes therefore appear to be identical and IP and IDE will be used interchangeably throughout this discussion.

The insulin-degrading protease is a neutral metalloproteinase with a thiol group required for activity. Most studies have found a pH optimum in the 7.0–8.0 range (see below for exceptions) and in spite of early reports to the contrary have found inhibition by chelators (EDTA, EGTA, phenanthroline). Sulfhydryl inhibitors such as N-ethylmaleimide (NEM) and p-chloromercuribenzenesulfonic acid (PCMBS) inactivate the enzyme, but other thiol active materials such as E64 (and related compounds), leupeptin, and iodoacetate do not (DUCKWORTH 1976a, b; DUCKWORTH et al. 1972; DUCKWORTH and KITABCHI 1974; BURGHEN et al. 1972; KIRSCHNER and GOLDBERG 1983; ROTH et al. 1985; SHII et al. 1986; SHROYER and VARANDANI 1985; ANSORGE et al. 1984; BRUSH and NASCIMENTO 1982; CHOWDHARI et al. 1985; KOLB and

STANDLE 1980; MANNOR et al. 1984; MCKENZIE and BURGHEN 1984; RYAN et al. 1985; YOKONO et al. 1981; POSNER 1973).

The metal required by the enzyme is not clear. We find reactivation of chelator-treated enzyme with Mn^{2+} or low concentrations of Zn^{2+}, with higher concentrations of Zn inhibitory, as have others, but unlike others, we can also reactivate with Ca^{2+} under certain conditions (RYAN et al. 1985). The effect of Ca^{2+} is pH dependent with reactivation above pH 7, but at lower pH Ca^{2+} is ineffective although Mn^{2+} or Zn^{2+} will reactivate. One group found irreversible inactivation of their enzyme with chelators, although Zn^{2+}, Co^{2+}, and Mn^{2+} protected against inactivation (ANSORGE et al. 1984).

Other enzyme properties are even less established. The molecular weight of Roth's enzyme is 300000 under nonreducing conditions and 110000 on reduced SDS gels (SHII et al. 1986). This is in agreement with the results of KIRSCHNER and GOLDBERG (1983) for their enzyme (see further below), but many other results have been reported (Table 1). Some of these discrepancies are due to the method used for determination of molecular weight and some are due to impure preparations. Apparent molecular weight determined by molecular sieve chromatography depends on the conditions used and the type of molecular sieve material. Molecular weight of the enzyme as determined by Sepharose 4B was over 300000, but by Sephadex G-200 was 170000 in one study (SHII et al. 1986). Under some conditons, elution of enzymatic activity from molecular sieve is asymmetrical, making accurate estimation of molecular weight impossible. In studies of less pure preparations by SDS-PAGE, the protein band seen and used for molecular weight calculation may not be the enzyme, but may be a contaminating protein. Smaller proteins may also result from autolysis of the enzyme. Similar considerations may apply to the various isoelectric points reported (Table 1).

Table 1. Properties of insulin-degrading enzyme reported from various studies

Study	Enzyme source	Approximate K_m (nM)	Molecular weight	Isoelectric point	pH optimum	Purification (fold)
BRUSH 1971	Muscle	180			7.5–7.7	50
BURGHEN et al. 1972	Muscle	100	80000		7.5–7.7	97
DUCKWORTH et al. 1972	Muscle	22			7.4	1013
YOKONO et al. 1980	Muscle	70	135000[a]	5.3		741
KOLB and STANDL 1980	RBC		160000[c]	5.8		5800
ROTH et al. 1985	IM9 lymphocytes	30	130000[a]	4.7, 5.0[a]		90
KIRSCHNER and GOLDBERG 1983	RBC		300000[b]		8.6	40600
SHII et al. 1986	RBC		300000[b]	5.2	7.0	50729
SHROYER and VARANDANI	Liver	34	180000[a]	5.9, 6.3		83

[a] Two subunits;   [b] three subunits;   [c] four subunits

The K_m of the enzyme for insulin is in somewhat better agreement. Most reports of more purified preparations have been in the 10–70 nM range with less purified preparations an order of magnitude higher (Burghen et al. 1972; Duckworth et al. 1972). The majority of these studies, however, have used TCA for the assay and the K_m using a more specific and sensitive assay is not known.

Purification methods, fold purification results, and the specific activity of purified enzyme have been major areas of controversy. The enzyme reported by Shii et al. (1986) required a 50 000-fold purification from red blood cell lysates to obtain homogeneous enzyme. The purification results are shown in Table 2. The specific activity of the final preparation was 1.37 pmol mg^{-1} min^{-1}. Eight steps were required for purification. The purification approach was derived from a combination of steps used previously by these investigators and the approach used by Kirschner and Goldberg (1983). The fold purification achieved by Kirschner and Goldberg was similar (40 600) as were recoveries (1–2%) of activity, but the specific activity was approximately 500 nmol mg^{-1} min^{-1} (see further below).

Many other purification approaches have been used with variable, but lesser success. Most have employed ammonium sulfate fractionation, molecular sieve and ion exchange chromatography with or without additional steps, although one report claimed little or no purification by these methods (Brush and Nascimento 1982). Our original purification included an affinity step using insulin–agarose (or later glucagon–agarose; Duckworth et al. 1972). Other investigators have also used this affinity step with success (McKenzie and Burghen 1984; Yokono et al. 1981), but other reports have not found it useful (Brush

Table 2. Purification of IDE from human erythrocytes

Step	Volume (ml)	Total protein (mg)	Total activity[a] (pmol/ min)	Yield (%)	Specific activity[a] (pmol mg^{-1} min^{-1})	Relative purification (fold)
1. Lysate supernatant	2000	1.5×10^5	4.0	100	2.7×10^{-5}	1
2. DEAE–Sephadex (batch)	900	2.0×10^3	2.8	72	1.4×10^{-3}	52
3. 45–60% $(NH_4)_2SO_4$ cut	90	7.0×10^2	2.3	58	3.2×10^{-3}	118
4. DEAE–Sephadex column	119	10^2	0.72	18	6.9×10^{-3}	257
5. Pentylagarose	104	6.2	0.68	17	0.1	3878
6. Hydroxylapatine	245	4.9	0.56	14	0.11	4173
7. Chromatofocusing	45	0.18	0.16	4	0.75	28 000
8. Ultrogel AcA-34 chromatography	9	0.04	0.056	1.4	1.37	50 729

[a] The activity of the enzyme was determined by the TCA precipitation assay described in Sect. F, which uses 16 pM insulin. Only values that were within the linear range of this assay ($\sim$25% of the insulin degraded) Were used to calculate these values. Assays with higher concentrations of insulin give higher specific activities because the K_m of this enzyme for insulin is $\sim$100 nM. Thus, the specific activity of the enzyme after step 8 would be $\sim$500 times greater if 100 nM insulin is used in the assay.

and NASCIMENTO 1982; KOLB and STANDLE 1980). We have similarly had mixed results with this approach over the years. Some preparations of insulin– (or glucagon)–agarose do not bind the enzyme well. In addition, we have found that altering previous steps in the purification (e.g., substituting DEAE for QAE or including EDTA) results in a preparation which does not bind well to insulin–agarose. Nevertheless, following the original description rigidly gives reproducible purification although the final preparation is not homogeneous. Adding preparative electrophoresis to the originally described procedure does produce a homogeneous enzyme, but the preparation is very unstable.

A number of other purification steps have been used as well. The original partial purification of insulin-specific protease involved a calcium phosphate gel adsorption (BRUSH 1971) and elution step and several current approaches use hydroxylapatite columns. A hydrophobic step also is of considerable use with pentyl-Sepharose and phenyl-Sepharose both effective. KOLB and STANDL (1980) used an organomercuri-Sepharose step effectively, but we have not had success with this method.

Other factors confound purification results. These include assay conditions, contaminating proteases, endogenous inhibitors, and enzyme instability. The common use of TCA solubility as an assay means that nonspecific proteases present in less pure preparations increase apparent activity by further degrading TCA-precipitable fragments generated by the specific protease (DUCKWORTH 1976a, b; SHII et al. 1986). Removal of these proteases during purification decreases recoveries and alters specific activity results. Loss of enzyme catalytic activity as purification progresses also alters recoveries and specific activity calculations. The conditions of the assay affect results (see below). Thus, definitive conclusions about enzyme purification are difficult and comparisons among studies frequently impossible.

B. Substrate Specificity

While this enzyme has a high specificity for insulin, the specificity is by no means complete. In fact, one study concluded that this enzyme has relatively little activity toward insulin, but primarily attacks chains (KIRSCHNER and GOLDBERG 1983). While this conclusion is apparently due to the assay conditions used (see below), the enzyme does have activity against insulin B chain. In early studies using relatively insensitive assays, several groups reported no apparent activity toward A or B chains (DUCKWORTH et al. 1975; YOKONO et al. 1980), but using iodinated B chain, considerable activity can be seen. In fact, if substrate concentrations above the K_m for insulin are used, the apparent activity for B chain is greater than for insulin, but at lower concentrations insulin is a better substrate.

Proinsulin is a poor substrate for the enzyme, although it is an effective inhibitor with a K_i of 40 nM (BRUSH 1971; DUCKWORTH et al. 1972; THOMAS and VARANDANI 1979; VARANDANI 1973). Proinsulin intermediates have variable susceptibilities to the enzyme and, interestingly, the susceptibility correlates well with the biological activity of the derivatives (BASKIN and KITABCHI 1973; PEAVY et al. 1985; YU and KITABCHI 1973).

Glucagon is also a substrate for IP, but the K_m is much higher than for insulin (DUCKWORTH and KITABCHI 1974). Other glucagon-related peptides, with the possible exception of secretin, are not substrates and other peptide hormones such as GH, PTH, ACTH, FSH, LH, PP, and somatostatin are also not substrates (DUCKWORTH and KITABCHI 1981).

The interaction of growth factors with IP is an area of increasing interest. The insulin-like growth factors were early shown to be inhibitors of insulin degradation (BURGHEN et al. 1976; KAHN et al. 1976) and more recent studies have shown that IGF-II is a good substrate whereas IGF-I inhibits, but is a poor substrate (MISBIN et al. 1983; ROTH et al. 1984). The status of the other growth factors is less clear, but EGF binds to the homologous IDE in *Drosophila* with a very high affinity such that the enzyme was originally described as an EGF receptor, but the peptide does not appear to be degraded (THOMPSON et al. 1985; GARCIA et al. 1989). The suggestion has been made that the cytoplasmic IDE may be an intracellular growth factor receptor.

C. Inhibition Studies

The effect of various inhibitors on the enzyme is also an area of some confusion. As discussed above, chelators and some, but not all, sulfhydryl inhibitors affect the enzyme. Bacitracin is also an effective inhibitor in essentially all studies, but variable results have occurred with other agents. The microbial inhibitors leupeptin, antipain, pepstatin, and chymostatin in general have little or no effect (Table 3). Endogenous inhibitors of the enzyme have also been reported and partially purified and characterized. The potential physiological role of these inhibitors is not currently known, but the presence of these in tissues further confounds determinations of specific activity, purification results, etc. (RYAN and DUCKWORTH 1983; McKENZIE and BURGHEN 1984).

D. Subcellular Location

The majority of the cellular IDE is cytoplasmic as defined by being found in the $10000\,g$ supernatant after homogenization and differential centrifugation (BRUSH 1971; DUCKWORTH et al. 1972). Immunochemical studies have confirmed the enzyme to be predominantly cytoplasmic (DORN et al. 1986). Insulin-degrading activity can also be found in other subcellular compartments. Plasma membrane preparations typically contain variable amounts of insulin-degrading activity (DUCKWORTH 1978, 1979; YOKONO et al. 1979). Most of this activity has characteristics typical of IP, but it is not clear whether or not the membrane IDE is from adsorbed cytoplasmic activity or is an intrinsic membrane enzyme. We have found that extensive washing of frozen-thawed membrane preparations will remove most of the activity, but this does not exclude a functional association in the cell. A portion of the degrading activity may be located on the exofacial side of the membrane (GOLDFINE et al. 1984; YASO et al. 1987; YOKONO et al. 1982) which would explain membrane-mediated degradation and also potentially localize the enzyme to the inside of the endosome.

Table 3. The effect of inhibitors on insulin-degrading activity of various purified enzyme preparations

Inhibitor	Concentration	Percent inhibition KIRSCHNER and GOLDBERG (1983)	Percent inhibition SHROYER and VARANDANI (1985)	Percent inhibition ROTH et al. (1985)	Percent inhibition ANSORGE et al. (1984)	Percent inhibition SHII et al. (1986)	Percent inhibition DUCKWORTH and KITABCHI (1974)
NEM	0.1 mM	84	56		48		
	0.2 mM			99			
	1.0 mM						
PCMBS	0.01 mM	85	90		85	100	100
	0.1 mM						
EDTA	0.01 mM	83	98		98		100
	1.0 mM		30			32	
	5.0 mM		44			100	
Phenanthroline	1.0 mM	81	40	10		100	100
PMFS	1.0 mM					10	5–10
Bacitracin	0.1 mg/ml		58	99		19	20
	1.0 mg/ml			0		100	100
Phosphoramidon	10 μg/ml	0	0		7	0	0
	50 μg/ml	0	0				
Aprotinin	1 mg/ml		0	60		18	
Iodoacetate	1 mM	0	5				
Iodoacetamide	1 mM		84				0
	5 mM		99				
Antipain	10 μg/ml	2		10	0	0	0
	100 μg/ml						
Pepstatin	10 μg/ml	0	8		0		0
	100 μg/ml		8				
Leupeptin	10 μg/ml	1	0	10	4	8	0
	100 μg/ml		0		1		
Bestatin	10 μg/ml	0					0
TPCK	0.5 mM	63	98				0
	1.0 mM		100				

Table 3 (continued)

Inhibitor	Concentration		Percent inhibition KIRSCHNER and GOLDBERG (1983)	Percent inhibition SHROYER and VARANDANI (1985)	Percent inhibition ROTH et al. (1985)	Percent inhibition ANSORGE et al. (1984)	Percent inhibition SHII et al. (1986)	Percent inhibition DUCKWORTH and KITABCHI (1974)
Chymostatin	10	μg/ml	11					0
	50	μg/ml						0
TLCK	0.5	mM	15	0		0		
	1.0	mM		12				
Ep475	90	μg/ml	9					0
Soybean Try inh.	50	μg/ml			95			
Pancreas Try inh.	1	mg/ml			95			
Chlorquin	0.2	mM			0			0
	1.0	mM						0
Thiorphan	100	μg/ml					0	
ACTH	0.1	μg/ml					0	0
	1.0	μg/ml					67	40
DFP	5	mM					5	
Diamide	1.0	mM				57		0

Endosomes contain insulin-degrading activity and several studies have shown insulin degradation occurring in early endosomes (HAMEL et al. 1989; PEASE et al. 1985). The insulin-degrading activity in disrupted endosomes has characteristics consistent with IP, but the activity in intact endosomes has been considered not to be IP, based on a lack of inhibition by NEM (PEASE et al. 1987). We have confirmed this finding (unpublished observation), but this does not exclude a role for IP in endosomal degradation, perhaps with a protected sulfhydryl group within the intact endosome.

A mitochondrial insulin-degrading enzyme has been reported, but relatively little is known about this enzyme. It is inhibited by NEM, but stimulated by EDTA, making it unlike IP. Its possible cellular role is unclear (HARE 1978).

E. Tissue Distribution

Insulin-degrading activity with characteristics of IP has been found in essentially all tissues examined. Since in crude preparations, other enzymes participate in the degradation of insulin, the tissue distribution of this enzyme has not been precisely determined, but liver, kidney, muscle, brain, fibroblasts, and red blood cells all contain large amounts of insulin-degrading activity (KITABCHI et al. 1972; STENTZ et al. 1985).

F. Insulin Degradation Assays

Many different methods have been used to assay insulin degradation (MISBIN et al. 1984; RYAN and DUCKWORTH 1989). The choice of the assay and the conditions used for the assay can have profound effects on the apparent results. If insulin degradation is defined as the initial event altering the structure or function of the molecule, then an ideal assay would be one which detects this alteration specifically and sensitively and which has a large throughput to be used for large numbers of samples. Such an assay does not currently exist. The only current assay which can detect early changes, such as a single peptide bond cleavage, is high performance liquid chromatography (HPLC) (HAMEL et al. 1986). This procedure is highly sensitive and specific, but impractical at present for assaying large numbers of samples and thus has been used primarily for studies of degradation products.

Most studies of insulin degradation have used the TCA precipitation assay. This assay depends on the conversion of ^{125}I-labeled insulin, which is precipitable in TCA, to fragments which are soluble. This assay underestimates degradation to variable degrees, depending on the type of iodinated insulin used, the concentration of TCA, and the incubation conditions. Insulin degradation occurs through the formation of a series of intermediates which are partially or even (single bond cleaved) completely precipitable in TCA. In comparisons of TCA with HPLC, we have found that complete degradation (no intact insulin by HPLC) can be seen with TCA solubility of as little as 30%–40%, and as much as 20%–40% degradation by HPLC with little or no change (<2%) in TCA solubility. These findings, however, can vary with the location of the iodine, the

final TCA concentration, the purity of the enzyme preparation, and the incubation conditions (MISBIN et al. 1984; RYAN and DUCKWORTH 1989).

The extent of iodination and the location of the iodine on the insulin molecule
can affect degradation results. Early studies emphasized that heavily iodinated
insulin (average of more than 1 mol iodine/mol insulin) was relatively resistant to
degradation (SODOYEZ et al. 1975). These findings emphasized the importance of
using monoiodinated insulin (less than 1 mol iodine/mol insulin) for studying
degradation as well as for studies of receptor binding. Insulin, however, has four
tyrosines (A14, A19, B16, and B26) and thus even monoiodinated insulin can be
heterogeneous.

We developed a method for preparing and separating the four specifically
labeled monoiodoinsulin isomers and characterized their properties both in cells
(binding, degradation, and action; FRANK et al. 1983; PEAVY et al. 1984) and with
enzyme (degradation; RYAN et al. 1984). The four isomers have different biological activities, and different susceptibilities to degradation, but, not surprisingly,
the relative susceptibility to degradation varies depending on the assay used.
Using TCA, the A14-labeled isomer had the highest initial susceptibility and the
A19 isomer the lowest with the two B chain isomers intermediate in their
degradation (RYAN et al. 1984). With prolonged incubation, however, the B26
isomer showed the most degradation and the B16 the least (Fig. 1). Subsequent studies and identification of the cleavage sites (see below) have clarified the

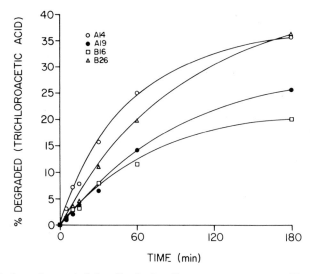

Fig. 1. Degradation of monoiodoinsulins by insulin protease as measured by trichloracetic
acid solubility. Each of the four monoiodoinsulins ($10^{-11}M$) was incubated with insulin
protease (0.12 µg/ml) at 37° C in 0.1 M tris-C1 buffer (pH 7.5), containing 0.5% BSA. At
each of the indicated times, an aliquot of incubation mixture was removed and added to an
equal volume of 10% trichloracetic acid and the sample was centrifuged. Degradation is
shown as the percent change in solubility. Each data point represents the mean of three experiments, each done in triplicate. The relative order of susceptibility of the four isomers to
hydrolysis was identical in the three experiments. (RYAN et al. 1984)

reasons for this. Initial cleavages occur within the disulfide bonds, producing large fragments with variable TCA solubility. Later cleavages occur at B25-B26, resulting in a fragment (B26-B30) which is TCA soluble. The kinetics of degradation also varied, with B16 having the highest K_m and the lowest V_{max}, and A14 and B26 the lowest K_m and the highest V_{max}. Thus, the conclusions from degradation studies could vary both with the isomer used and the extent of the degradation. In general, TCA assays should be used with final solubilities of less than 15% because, above this, linearity is lost and multiple processes are occurring.

The amount of TCA used can also alter apparent results. With higher concentrations (over 5% final) some of the partially degraded, soluble fragments will precipitate and thus cause a further underestimation of degradation. The purity of the enzyme preparations alters the TCA results. If nonspecific proteases are present, they can degrade the TCA-precipitable fragments and increase TCA solubility. Removal of these proteases will therefore decrease apparent activity (by TCA) and alter results such as recoveries, fold purifications, etc. The presence of these nonspecific proteases can therefore increase the sensitivity of the TCA assay. We have used an assay combining IP degradation followed by addition of a peptidase which attacks fragments, but not intact insulin. This greatly increases TCA assay sensitivity (RYAN and DUCKWORTH 1989). The use of this approach with the different isomers prevented the apparent change in relative susceptibility of the isomers during prolonged incubation.

Assay conditions can also alter TCA results. Some alteration (pH, buffer type, etc.), may affect the enzyme rather than the assay, but others may alter the assay directly. The primary example of this is the use of reducing reagents such as dithiothreitol (DTT). With low concentrations of substrate the addition of DTT will result in reductive cleavage of variable amounts of the insulin to separate chains and the assay then becomes a chain-degrading assay rather than an insulin-degrading assay. On the other hand, addition of DTT at the end of the assay (after inactivating the enzyme) can increase the sensitivity of the TCA assay by cleaving the disulfide bonds of TCA-precipitable fragments and producing smaller TCA-soluble chain fragments. With this approach, the TCA assay can be made almost as sensitive as the HPLC assay for degradation in the 5%–50% range.

Molecular sieve chromatography can also be used to assay insulin degradation, as a function of the smaller size of the degradation products, but again, will underestimate early degradation and the results will depend on the location of the iodine. Using Sephadex G-50 chromatography, we found results with the four isomers similar to those with TCA and demonstrated that each of the isomers was converted to a complex mixture of intermediate and low molecular weight materials (RYAN et al. 1984).

Changes in other properties of the insulin molecule can also be used to assay degradation. The immunoreactivity of insulin is a property which can be used as a sensitive assay. Immunoassay can be used to measure degradation of unlabeled insulin, but most frequently immunoprecipitation of labeled insulin is the approach used. This again depends on a number of variables, including the antisera used and the location of the iodine. In general, however, this approach is at least twice as sensitive as the TCA assay.

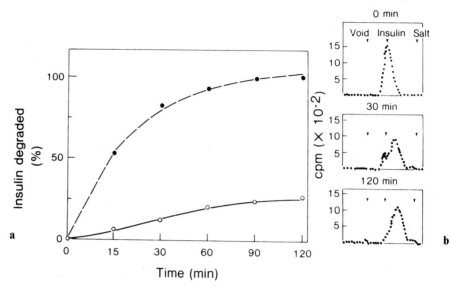

Fig. 2. Measurement of ^{125}I-labeled insulin degradation by IDE assessed by (**a**) TCA precipitation and receptor binding, and (**b**) gel filtration. IDE was incubated for different periods of time with ^{125}I-labeled insulin, and the extent of degradation was assessed by TCA precipitation (*open circles*), binding to receptor (*full circles*), or gel filtration. (SHII et al. 1986)

The most sensitive assay (outside of HPLC) may be receptor binding. Relatively small changes in the insulin molecule alter its affinity for the insulin receptor and thus this property can be used as a sensitive approach for detecting insulin degradation. With purified enzyme receptor binding may be tenfold more sensitive than TCA (Fig. 2).

Given these considerations, it is not surprising that discrepancies have been reported in different studies and any comparison of studies must take into account the different conditions and assays used. A particular and interesting case in point is the enzyme described by KIRSCHNER and GOLDBERG (1983). The overall characteristics of this enzyme suggest strongly that it is the same IDE as described by SHII et al. (1986). Two major differences were reported. First of all, the Kirschner and Goldberg enzyme had a much higher pH optimum and secondly, their enzyme degraded insulin very poorly. To see appreciable activity with ^{125}I-labeled insulin sufficient DTT had to be added to reduce the disulfide bonds so that their assay was actually for chains rather than insulin. The apparent explanation for this difference from SHII et al. (1986) was that the assay used by KIRSCHNER and GOLDBERG (1983) had an insulin concentration of 40 µg/ml, far above the K_m for insulin of IP (IDE). Since the K_m of the enzyme for chains is much higher than for insulin, chain-degrading, but not intact insulin-degrading activity could be detected. While this does not apparently explain the different pH optimum, this or other assay variation may be the reason.

G. Enzyme Products

Although there has been much interest in and many studies on the degradation products of insulin generated by this enzyme, only recently have the degradation products been identified. Several different approaches have been used for studying the degradation products. Our approach has been to use the four specifically labeled monoiodoinsulin isomers, incubating each of the isomers with the enzyme, separating the degradation products on HPLC, and then sequencing each of them. The sequenator step in which radioactivity appears identifies the number of amino acids amino terminal to the labeled tyrosine and thus, with the known position of the radioactivity, identifies cleavages amino terminal to each of the labeled tyrosines. This approach was chosen for identifying enzyme products so that low concentrations of substrate can be used in order to compare enzyme products and cell products at the same substrate concentrations. With this approach, we have identified two A chain cleavages produced by insulin protease; one at A13-A14 and one at A14-A15 (DUCKWORTH et al. 1987). Seven B chain cleavages are produced by insulin protease in intact insulin. The HPLC separation of these seven peptides after sulfitolysis of the B26-labeled insulin digestion by insulin protease is shown in Fig. 3. Four of these cleavages are major and three are minor. The four major cleavages occur at B9-B10, B13-B14, B16-

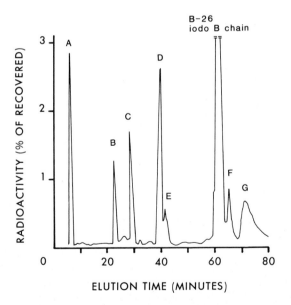

Fig. 3. Elution pattern from reverse-phase HPLC of sulfitolized B chain peptides obtained from insulin-sized products after incubation of [^{125}I]iodo (B26) insulin with insulin protease. A portion of the Sephadex G-50 insulin-sized products was sulfitolized, and labeled sulfitolized peptides were injected on reverse-phase HPLC. The distinct and reproducible peptide peaks are labeled *A–G* and the intact chain is identified. (DUCKWORTH et al. 1988)

Table 4. Sequence analysis and deduced peptide compositions of [^{125}I]iodo (B26) insulin degraded by insulin protease

B25-labeled peptide	Sequence cycle	Bond cleaved	Peptide composition [a]
A	1	Phe-Tyr	B26+
B	2	Phe-Phe	B25-B26+
C	10	Tyr-Leu	B17-B26+
D	13	Glu-Ala	B24-B26+
E	12	Ala-Leu	B15-B26+
F	16	His-Leu	B11-B26+
G	17	Ser-His	B10-B26+
B chain	26		B1-B26+

[a] The peptide compositions presented are minimal sequences that could be deduced. Using [^{125}I]iodo (B26) insulin only, the amino terminus through the B26 tyrosine could be determined, as indicated by B26+.

B17, and B25-B26 (DUCKWORTH et al. 1988). The minor cleavage sites are at B10-B11, B14-B15, and B24-B25 (Table 4). The nature of the bonds cleaved does not allow a simple classification of peptide bond specificity for the enzyme, but all bonds except B24-B25 and B25-B26 are in close proximity in the three-dimensional structure of insulin, suggesting that the specificity is to the molecule itself rather than to specific amino acid residues. We have not yet determined whether the bonds are cleaved randomly or whether there is a sequence to the cleavage, except that the B24-B25 and B25-B26 seem to occur later. In a previous study using very different conditions (high concentrations of unlabeled insulin and very short incubation times), we found a single, early product with only one bond cleaved (B16-B17) and postulated that this was the initial site of cleavage (DUCKWORTH et al. 1979). While we now find products with this bond cleavage, we also have products with it intact and thus it is not an obligatory initial site. It is obviously an important and early cleavage site in intact cells, however (see below). VARANDANI and SCHROYER (1987) identified a product of their neutral thiol peptidase, which appears to be identical with insulin protease, as a fragment produced by cleavages at A13-A14 and B9-B10. This would be generated by the cleavages described above.

MUIR et al. (1986) and DAVIES et al. (1986, 1988) have examined degradation products of insulin protease extensively using a somewhat different approach (MUIR et al. 1986; DAVIES et al. 1986, 1988). They have used semisynthetic insulin labeled with tritium at the ends of the A and B chain and have characterized the products generated by the degradation of insulin protease by various biochemical approaches. In general, their conclusions are consistent with ours. They found a fragment cleaved at the A13-A14 and B9-B10 sites, with another minor B chain cleavage at B10-B11 (or B11-B12). In more recent studies using fast atom bombardment mass spectrometry, they have found cleavages at A13-A14, A14-A15, B9-B10, B13-B14, B24-B25, and B25-B26 (SAVOY et al. 1988). Thus, their results agree with all our cleavage sites except for a minor B14-B15 and a major B16-B17. The minor B14-B15 cleaved peptide is difficult to separate from the B13-

B14 cleavage product and is present in very small amounts. The failure to find a B16-B17 cleaved product is more difficult to explain. The most obvious explanation is that they used insulin labeled on the ends of the chains and studied extensively degraded material. The cleavage in the midportion of the chain (B16-B17) would be obscured by additional cleavages (B9-B10, B13-B14, B24-B25, B25-B26) on each side.

YONEZAWA et al. (1986) have also examined insulin degradation by their insulin-degrading enzyme and have separated the degradation products on HPLC. While they have not yet identified cleavage sites, the HPLC elution profile is similar to that seen in our studies. They have examined some of the biological properties of their degradation products. Some of the more hydrophobic products had binding to insulin receptors and to anti-insulin antibodies, suggesting the retention of some insulin-like properties. They also found evidence for an initial product, their peak 6, which preceded the formation of more extensively degraded products. Since their conditions, high unlabeled insulin concentrations and short incubation times, were similar to our original studies, this peak 6 could well be a product with a single cleavage at B16-B17.

H. Physiological Role of Insulin Protease

The initial step in insulin degradation by intact cells is binding of the hormone to a specific receptor on the cell membrane (TERRIS and STEINER 1975). The receptor binding step initiates the cellular response to insulin and also is the first in a series of steps resulting in insulin degradation by the cell. Some insulin is degraded on the cell membrane, apparently after receptor binding. The insulin receptor itself does not degrade the hormone, but allows exposure to an insulin-degrading enzyme. Much evidence has accumulated that the cell surface degrading activity is due to insulin protease. Isolated cell membranes have degrading activity which is apparently due to IP and several reports have now demonstrated that IDE, which is identical with IP, may be present on the exofacial surface of the cell membrane. We have recently demonstrated that insulin degradation products found in the medium from incubated hepatocytes are identical with those produced by IP (DUCKWORTH et al. 1988).

In addition to some receptor-bound insulin being degraded on the membrane, some insulin is released intact from the receptor, and some receptor-bound insulin is internalized into the cell. The internalization process results in insulin–receptor complexes being internalized into endosomes. Although conventional wisdom has it that endosomes containing insulin fuse with lysosomes where the insulin is degraded, it is now clear that insulin degradation is initiated in early endosomes (HAMEL et al. 1989; PEASE et al. 1985). Several groups have found insulin degradation products in prelysosomal endosomes and we have demonstrated that typical intracellular insulin degradation products can be isolated from early or light endosomes as rapidly as 1 min after intrajugular insulin injection in the rat. These degradation products consist of insulin with an intact A chain, and one or more cleavages in the B chain with the earliest apparent cleavage at B16-B17 consistent with an early insulin protease cleavage. It has not been shown with finality that the insulin-degrading activity of the endosome is IP, but the characteristics of insulin-degrading activity in disrupted endosomes and the products generated within endosomes strongly suggest that this enzyme is the endosomal insulin-degrading enzyme.

A physiological role for IDE in cellular insulin metabolism has been shown clearly by the studies of Roth's group. These investigators demonstrated that labeled insulin added to tissue homogenates could be cross-linked to a protein of 110 000 molecular weight (SHII et al. 1985). Monoclonal antibodies raised against their insulin-degrading enzyme precipitated this cross-linked protein, showing that IDE interacts with and can be cross-linked to insulin. They next added labeled insulin to intact cells and after a period of time, added cross-linking reagents to the cells and demonstrated that the same 110 000 molecular weight cytoplasmic protein was bound to the labeled insulin (HARI et al. 1987). Again, this protein cross-reacted with monoclonal antibodies to the insulin-degrading enzyme. Additional studies with the monoclonal antibodies to IDE were done by microinjecting a pool of these antibodies into the cytoplasm of cultured hepatocytes which resulted (after a 24-h delay) in an 18%–54% inhibition of insulin degradation by the cells, suggesting an important although perhaps not total role for the enzyme (SHII and ROTH 1986). Thus, under normal conditions, this enzyme interacts with intracellular insulin and, as shown by the microinjection of antibodies, degrades insulin. The failure of the microinjected antibodies to inhibit totally insulin degradation by the cells is of interest. There are a number of possible explanations for this partial effect. First of all is that the antibody concentration intracellularly was not high enough to reduce enzyme levels further. Additionally, subcellular compartmentalization of the enzyme could have prevented access of the cytoplasmically injected antibody. Finally, the lack of total inhibition could have resulted from the contribution of another enzyme to cellular degradation. At the present time, however, most cellular insulin degradation can be attributed to IP (IDE). Membrane degradation clearly appears to be due to this enzyme and at least part of the intracellular degrading activity similarly appears due to this enzyme.

Factors affecting the control of cellular enzyme activity are not clear. Although many studies of tissue insulin-degrading activity in various physiological and pathological states have been performed, these must be interpreted with care. The difficulty in determining true enzyme activity in crude tissues and the multistep and multipathway process of cellular insulin degradation make conclusion about alterations in enzyme activity difficult.

J. Other Insulin-Degrading Enzymes

I. Glutathione Insulin Transhydrogenase

Glutathione insulin transhydrogenase, or GIT, was at one time hypothesized to be the obligatory initial event in the degradation of insulin by cells. GIT degrades insulin by cleaving the disulfide bonds of insulin in the presence of glutathione or other reducing agents, resulting in separate and inactive A and B chains. These separate chains are then susceptible to degradation by nonspecific peptidases in the cell. While a number of observations made this hypothesis attractive, subsequent studies have strongly suggested that GIT plays little if any role in the cellular degradation of insulin (ASSOIAN and TAGER 1981, 1982; BRUSH and JERING 1979; POOLE et al. 1982; MISBIN et al. 1983). The general characteristics of subcellular degradation are not consistent with the properties of GIT and it has been clearly shown that partially degraded disulfide-linked fragments of insulin are present in the cell, thus conclusively showing that GIT action is not an obligatory initial step in the degradation of insulin. A number of studies have concluded that GIT is also not a major insulin-degrading enzyme in cells. The microinjection study from Roth's laboratory actually used antibodies to GIT as a negative control, showing that the microinjection of these antibodies did not affect cellular degradation.

The primary cellular role for GIT appears to be in the synthesis of disulfide-containing proteins. In the presence of GIT, these proteins assume their normal three-dimensional structures with correct oxidation of sulfhydryls to form disulfide bonds. Although controversial for some time, it is now clear that GIT is the same enzyme as protein disulfide isomerase (PDI) (BJELLAND et al. 1983). This enzyme was originally shown to reactivate "scrambled" ribonuclease (GOLDBERGER et al. 1963). GIT (or PDI) has now been shown to be involved in the synthesis of disulfide-containing proteins such as immunoglobulin and prolactin (FREEDMAN et al. 1984; KADERBHAI and AUSTEN 1985; ROTH and KOSHLAND 1981).

GIT also has other functions in the cell. One subunit of the enzyme proline hydroxylase is identical with GIT (PDI) (KOIVU et al. 1987). In addition, another protein has been isolated from cells which appears to serve as a thyroid hormone receptor, which also contains as part of its structure the amino acid sequence of PDI (OBATA et al. 1988). Thus, this enzyme may have multiple functions in the cell and may be involved in a number of processes which require disulfide interchange interactions. One potential role for the enzyme is in insulin–insulin receptor interaction (JACOBS and CUATRECASAS 1980; MORGAN et al. 1985), since disulfide interchange events occur after insulin binds to its receptor, resulting in changes in receptor affinity, as well as some insulin receptor covalently bound complexes, apparently bound through disulfide interchange (CLARK and HARRISON 1982, 1985). A role for GIT in the degradation of insulin has not been totally excluded, but if this occurs, it must be under special conditions. GIT might also be involved in reductive cleavage of some of the partially degraded fragments of insulin within the cell.

II. High Molecular Weight Insulin-Degrading Enzyme

We have isolated another insulin-degrading enzyme from cells and purified it to homogeneity (RYAN and DUCKWORTH 1989). This enzyme accounts for only 5%–10% of the total cellular degrading activity and is quite distinct from IP. In addition to a lack of metal dependency and a different pH curve, the substrate specificity of this enzyme is different. It degrades proinsulin and insulin with roughly equal potency, whereas insulin protease is 20–25 times more active against insulin than proinsulin. This enzyme can be separated from insulin protease on ion exchange and can be purified to homogeneity by a combination of ammonium sulfate, ion exchange, hydrophobic chromatography, and molecular sieve chromatography. After purification, it migrates as a single band on native PAGE, but on SDS migrates as a series of bands ranging from 20 to 40 000 kilodalton. On molecular sieve the apparent molecular weight is over 300 000, making this enzyme very similar to a high molecular weight, apparently multicatalytic enzyme isolated from brain, kidney, and other tissues (Dahlmann et al. 1985, 1986; Ishiura et al. 1986; Nojima et al. 1986; Orlowski and Wilk 1981; Orlowski et al. 1983; Tanaki et al. 1986; Yamamoto et al. 1986; Zolfaghari et al. 1987 a, b).

Although it is not yet totally established that all of these studies are on the same enzyme, a high molecular weight proteinase composed of multiple small subunits has been described which has three separate catalytic activities, a trypsin-like, a chymotrypsin-like, and a cucumsin-like activity. These sites have different substrate and inhibitor specificities and can be assayed independently. The recommended name for this enzyme is multicatalytic endopeptidase complex until the current uncertainties are settled, as discussed in the *ICOP Newsletter* (January 1988).

The high molecular weight multicatalytic complex has been implicated in the nonlysosomal pathway for intracellular protein degradation. Recent reports have shown homologies between this enzyme and the 19S ribonucleoprotein in *Drosophila* (Falkenburg et al. 1988) and evidence that this enzyme is the same as the "prosome" particle in HeLa cells (Arrigo et al. 1988). The cellular function of this enzyme is of major interest.

K. Future Studies

The cellular degradation of insulin is a complex process involving multiple steps and pathways. We need additional information on the mechanisms and enzymes involved and on physiological and pathological changes in the process (Duckworth 1989). Although many studies have been done on the control of insulin degradation and on changes with various pathological states, the complexity of the process and problems with assays, different approaches, etc., have limited our understanding of the events. With the purification of IDE and the development of specific antibodies, it is expected that rapid progress in the isolation and cloning of the gene will occur. With these tools, a better understanding of the role of this enzyme and of insulin degradation in cells should be achieved.

Acknowledgments. The author thanks Kimberley Dempsey for excellent secretarial assistance. Portions of this work were supported by Veterans Administration research funds.

References

Ansorge S, Bohley P, Kirschke J, Langner J, Weideranders B (1984) The insulin and glucagon degrading proteinase of rat liver: a metal dependent enzyme. Biomed Biochim Acta 1:39–46

Arrigo AP, Tanaka K, Goldberg AL, Welch WL (1988) Identity of the 19 S "prosome" particle with the large multifunctional protease complex of mammalian cells (the proteasome). Nature 331:192–194

Assoian RK, Tager HS (1981) [^{125}I] Iodotyrosyl insulin: semisynthesis receptor binding and cell mediated degradation of a B chain labeled insulin. J Biol Chem 256:4042–4049

Assoian RK, Tager HS (1982) Peptide intermediates in the cellular metabolism of insulin. J Biol Chem 257:9078–9085

Baskin FK, Kitabchi AE (1973) Substrate studies for insulin specific protease. Eur J Biochem 37:489

Bjelland S, Wallevik K, Kroll J, Dixon JE, Morin JE, Freedman RB, Lambert N et al. (1983) Immunological identity between bovine preparations of thiol: protein-disulphide oxidoreductase, glutathion-insulin transhydrogenase, and protein di-sulphide isomerase. Biochim Biophys Acta 747:197–199

Bond J, Orlowski M (1988) ICOP Newsletter (January) 1:3

Broh-Kahn RH, Mirsky IA (1949) The inactivation of insulin by tissue extracts. II. The effects of fasting on the insulinase content of rat liver. Arch Biochem Biophys 20:10–21

Brush JS (1971) Purification and characterization of a protease with specificity for insulin from rat muscle. Diabetes 20:151–155

Brush JS, Jering H (1979) The importance of proteolysis as the initial step of insulin degradation in rat liver homogenates. Endocrinology 104:1639–1643

Brush JS, Kitabchi AE (1970) Metabolic disposition of I-131-iodoinsulin within the rat diaphragm. Biochim Biophys Acta 215:134–144

Brush JS, Nascimento CG (1982) Studies of the properties of the insulin protease of rat liver. Biochim Biophys Acta 704:398–4021

Burghen GA, Kitabchi AE, Brush JS (1972) Characterization of a rat liver protease with specificity for insulin. Endocrinology 91:633–642

Burghen GA, Duckworth WC, Kitabchi AE, Solomon SS, Poffenbarger PL (1976) Inhibition of insulin degradation by nonsuppressible insulin-like activity. J Clin Invest 57:1089–1093

Chowdhary BK, Smith GD, Peters TJ (1985) Subcellular localization and partial characterization of insulin proteolytic activity in rat liver. Biochim Biophys Acta 840:180–186

Clark S, Harrison LC (1982) Insulin binding leads to the formation of covalent (S-S) hormone receptor complexes. J Biol Chem 257:12239–12244

Clark S, Harrison LC (1985) Structure of covalent insulin-receptor complexes (I-S-S-R) in isolated rat adipocytes and human placental membranes. Biochem J 229:513–519

Dahlmann B, Huehn L, Rutschmann M, Reinauer H (1985) Purification and characterization of a multicatalytic high-molecular-mass proteinase from rat skeletal muscle. Biochem J 228:161–170

Dahlmann B, Kopp F, Huehn L, Reinauer H, Schwenen M (1986) Studies on the multi-catalytic proteinase from rat skeletal muscle. Biomed Biochim Acta 45:1493–1501

Davies JG, Muir AV, Offord RE (1986) Identification of some cleavage sites of insulin by insulin proteinase. Biochem J 240:609–612

Davies JG, Muir AV, Rose K, Offord RE (1988) Identification of radioactive insulin fragments liberated during the degradation of semisynthetic [^{3}H]Gly^{A1} insulin and [^{3}H]Phe^{B1} insulin. Biochem J 249:209–214

Dorn A, Bernstein HG, Reiser M, Rinne A, Ansorge S (1986) Degradation of insulin and glucagon in developing rat kidney: immunolocalization of insulin-glucagon-specific protease and quantitative estimation. J Histochem Cytochem 34(3):411–412

Duckworth WC (1976a) Insulin and glucagon degradation in the kidney. I. Subcellular distribution under different assay conditions. Biochem Biophys Acta 437:518–530

Duckworth WC (1976b) Insulin and glucagon degradation by the kidney. II. Characterization of the mechanisms at neutral pH. Biochem Biophys Acta 437:531–542

Duckworth WC (1978) Insulin and glucagon binding and degradation by the kidney cell membrane. Endocrinology 102:1766–1774

Duckworth WC (1979) Insulin degradation by liver cell membranes. Endocrinology 104:1758–1764

Duckworth WC (1989) Insulin degradation: mechanisms, products, and significance. Endocr Rev 9(3):319–345

Duckworth WC, Kitabchi AE (1974) Insulin and glucagon degradation by the same enzyme. Diabetes 23:536–543

Duckworth WC, Kitabchi AE (1981) Insulin metabolism and degradation. Endocr Rev 2:210–233

Duckworth WC, Heinemann MA, Kitabchi AE (1972) Purification of insulin specific protease by affinity chromatography. Proc Natl Acad Sci USA 69:3698–3702

Duckworth WC, Heinemann MA, Kitabchi AE (1975) Proteolytic degradation of insulin and glucagon. Biochim Biophys Acta 377:421–430

Duckworth WC, Stentz FB, Heinemann M, Kitabchi AE (1979) Initial site of cleavage of insulin by insulin protease. Proc Natl Acad Sci USA 76:635–639

Duckworth WC, Hamel FG, Liepnieks JJ, Peavy DE, Ryan MP, Hermodson MA, Frank BH (1987) Identification of A chain cleavage sites in intact insulin produced by insulin protease and isolated hepatocytes. Biochem Biophys Res Commun 147(2):615–621

Duckworth WC, Hamel FG, Peavy DE, Liepnieks JJ, Ryan MP, Hermodson MA, Frank BH (1988) Degradation products of insulin generated by hepatocytes and by insulin protease. J Biol Chem 263(4):1826–1831

Falkenburg PE, Haass C, Kloetzel PM, Niedel B, Kopp F, Luehn L, Dahlmann B (1988) Drosophila small cytoplasmic 19S ribonucleoprotein is homologous to the rat multicatalytic proteinase. Nature 331:190–192

Frank BH, Peavy DE, Hooker CS, Duckworth WC (1983) Receptor binding properties of monoiodotyrosyl insulin isomers purified by high performance liquid chromatography. Diabetes 32:705–711

Freedman RB, Brockway BE, Lambert N (1984) Protein disulphide-isomerase and the formation of native disulphide bonds. Biochem Soc Trans 12:929–932

Garcia JV, Fenton BW, Rosner MRR (1989) Isolation and characterization of an insulin-degrading-enzyme from Drosophila melanogaster. Biochemistry (in press)

Goldberger RF, Epstein CJ, Anfinsen CB (1963) Acceleration of reactivation of reduced bovine pancreatic ribonuclease by a microsomal system from rat liver. J Biol Chem 238:628

Goldfine ID, Williams JA, Bailey AC, Wong KY, Iwamoto Y, Yokono K, Baba S, Roth RA (1984) Degradation of insulin by isolated mouse pancreatic acini: evidence for cell surface protease activity. Diabetes 33:64–72

Hamel FG, Peavy DE, Ryan MP, Duckworth WC (1986) High performance liquid chromatographic analysis of insulin degradation by rat skeletal muscle insulin protease. Endocrinology 118(1):328–333

Hamel FG, Posner BI, Bergeron JJM, Frank BH, Duckworth WC (1989) Isolation of insulin degradation products from endosomes derived from intact rat liver. J Biol Chem (in press)

Hare JF (1978) A novel proteinase associated with mitochondrial membranes. Biochim Biophys Res Commun 83:1206–1215

Hari J, Shii K, Roth RA (1987) In vivo association of ^{125}I insulin with a cytosolic insulin-degrading enzyme: detection by covalent cross-linking and immunoprecipitation with a monoclonal antibody. Endocrinology 120(2):829–831

Ishiura S, Yamamoto T, Nojima M, Sugita H (1986) Ingensin, a fatty acid-activated serine protease from rat liver cytosol. Biochem Biophys Acta 882:305–310

Jacobs S, Cuatrecases P (1980) Disulfide reduction converts the insulin receptor of human placenta to a low affinity form. J Clin Invest 66:1424–1427

Kaderbhai MA, Austen BM (1985) Studies on the formation of interchain disulfide bonds in newly synthesized bovine prolactin. Eur J Biochem 153:167–178

Kahn CR, Megyesi K, Roth J (1976) Nonsuppressible insulin-like activity of human serum. A potent inhibitor of insulin degradation. J Clin Invest 57:526–529

Katzen HM, Tietze F (1966) Studies on the specificity and mechanism of action of hepatic glutathione-insulin transhydrogenase. J Biol Chem 241:3561–3570

Kirschner RG, Goldberg AL (1983) A high molecular weight metalloendoprotease from the cytosol of mammalian cells. J Biol Chem 258:967–976

Kitabchi AE, Stentz FB, Bobal MA (1972) Degradation of insulin and proinsulin by various organ homogenates of rat. Diabetes 21(11):1091–1101

Koivu J, Myllyla R, Helaakuski T, Pihlajaniemi T, Tasanen K, Kivirikko KI (1987) A single polypeptide acts both as the B subunit of prolyl 4-hydroxylase and as a protein disulfide isomerase. J Biol Chem 262:6447–6449

Kolb JH, Standl E (1980) Purification to homogeneity of an insulin-degrading enzyme from human erythrocytes. Hoppe Seylers Z Physiol Chem 361:1029–1039

Mannor G, Movsas B, Yalow RS (1984) Characterization of insulinase from mammalian and non-mammalian livers. Life Sci 34:1341–1345

McKenzie RA, Burghen GA (1984) Partial purification and characterization of insulin protease and its intracellular inhibitor from rat liver. Arch Biochem Biophys 229:604–611

Mirsky IA (1957) Insulinase, insulinase inhibitors, and diabetes mellitus. Recent Prog Horm Res 13:429–471

Mirsky IA (1964) The metabolism of insulin. Diabetes 13:225–229

Mirsky IA, Broh-Kahn RH (1949) The inactivation of insulin by tissue extracts. I. The distribution and properties of insulin inactivating extracts (insulinase). Arch Biochem Biophys 20:1–9

Mirsky IA, Perisutti G (1957) The relative specificity of the insulinase activity of rat liver extracts. J Biol Chem 228:77–83

Misbin RI, Almira EC, Duckworth WC, Mehl TD (1983) Inhibition of insulin degradation by insulin-like growth factors. Endocrinology 113:1525–1527

Misbin RI, Almira EC, Buynitzky SJ (1983) Insulin metabolism in rat hepatocytes. J Biol Chem 258:2157–2162

Misbin RI, Ryan MP, Duckworth WC (1984) Methods for assaying insulin degradation. In: Larner J, Pohl S (eds) Methods in diabetic research, vol 1, part A. Wiley, New York, pp 389–401

Morgan MS, Darrow RM, Nafz MA, Varandani PT (1985) Participation of cellular thiol/disulphide groups in the uptake, degradation, and bioactivity of insulin in primary cultures of rat hepatocytes. Biochem J 225:349–356

Muir A, Offord RE, Davies JG (1986) The identification of a major product of the degradation of insulin by insulin proteinase (EC 3.4.22.11). Biochem J 237:631–637

Nojima M, Ishiura S, Yamamoto T, Okayama T, Furuya H, Sugita H (1986) Purification and characterization of a high-molecular-weight protease, ingensin, from human placenta. J Biochem 99:1605–1611

Obata T, Kitagawa S, Gong Q-H, Pastan I, Cheng S-Y (1988) Thyroid hormone down-regulates p55, a thyroid hormone-binding protein that is homologous to protein disulfide isomerase and the B subunit of prolyl-4-hydroxylase. J Biol Chem 263(2):782–785

Orlowski M, Wilk S (1981) A multi-catalytical protease complex from pituitary that forms enkephalin and enkephalin-like containing peptides. Biochem Biophys Res Commun 101(3):814–822

Orlowski M, Michaud C, Chu TG (1983) A soluble metalloendopeptidase from rat brain. Eur J Biochem 135:81–88

Pease RJ, Smith GD, Peters TJ (1985) Degradation of endocytosed insulin in rat liver is mediated by low density vesicles. Biochem J 228:137–146

Pease RJ, Smith GD, Peters TJ (1987) Characterization of insulin degradation by rat liver low density vesicles. Eur J Biochem 164:251–257

Peavy DE, Abram JD, Frank BH, Duckworth WC (1984) Receptor binding and biological activity of specifically labeled ^{125}I and ^{127}I monoiodoinsulin isomers in isolated rat adipocytes. Endocrinology 114:1818–1824

Peavy DE, Brunner MR, Duckworth WC, Hooker CS, Frank BH (1985) Receptor binding and biological potency of several split forms (conversion intermediates) of human proinsulin: studies in cultured IM-9 lymphocytes and in vitro in rats. J Biol Chem 260(26):13989–13994

Poole GP, O'Connor KJ, Lazarus NR, Pogson CI (1982) ^{125}I labeled insulin degradation by isolated rat hepatocytes: the roles of glutathione-insulin transhydrogenase and insulin-specific protease. Diabetologia 23:49–53

Posner BI (1973) Insulin metabolizing enzyme activities in human placental tissue. Diabetes 22:552–563

Roth RA, Koshland ME (1981) Role of disulfide interchange enzyme in immunoglobulin synthesis. Biochemistry 20:6594–6599

Roth RA, Mesirow ML, Yokono K, Baba S (1984) Degradation of insulin-like growth factors I and II by a human insulin degrading enzyme. Endocr Res 10:101–112

Roth RA, Mesirow ML, Cassell DJ, Yokono K, Baba S (1985) Characterization of an insulin degrading enzyme from cultured human lymphocytes. Diabetes Res Clin Pract 1:31–39

Rudman D, Garcia L, DiGirolamo M, Shank PW (1966) Cleavage of bovine insulin by rat adipose tissue. Endocrinology 78:169–185

Ryan MP, Duckworth WC (1983) Partial characterization of an endogenous inhibitor of a calcium-dependent form of insulin protease. Biochem Biophys Res Commun 114: 195–203

Ryan MP, Duckworth WC (1989) Insulin degradation: assays and enzymes. In: Kahn CR, Harrison LC (eds) The insulin receptor, vols I, II. Liss, New York (in press)

Ryan MP, Peavy DE, Frank BH, Duckworth WC (1984) The degradation of monotyrosyl insulin isomers by insulin protease. Endocrinology 115:591–599

Ryan MP, Gifford JD, Solomon SS, Duckworth WC (1985) The calcium dependence of insulin degradation by rat skeletal muscle. Endocrinology 117(4):1693–1698

Savoy LA, Jones RML, Pochon S, Davies JG, Muir AV, Offord RE, Rose K (1988) Identification by fast atom bombardment mass spectrometry of insulin fragments produced by insulin proteinase. Biochem J 249:215–222

Shii K, Roth TA (1986) Inhibition of insulin degradation by hepatoma cells after microinjections of monoclonal antibodies to a specific cytosolic protease. Proc Natl Acad Sci USA 83:4147–4151

Shii K, Baba S, Yokono K, Roth RA (1985) Covalent linkage of ^{125}I insulin to a cytosolic insulin degrading enzyme. J Biol Chem 260:6503–6506

Shii K, Yokono K, Baba S, Roth RA (1986) Purification and characterization of insulin-degrading enzyme from human erythrocytes. Diabetes 35:675–683

Shroyer LA, Varandani PT (1985) Purification and characterization of a rat liver cytosol neutral thiol peptidase that degrades glucagon, insulin, and isolated insulin A and B chains. Arch Biochem Biophys 236:205–219

Sodoyez JC, Sodoyez-Goffaux F, Goff MM, Zimmerman AE, Arquilla ER (1975) ^{125}I or carrier free ^{125}I monoiodoinsulin. Preparation, physical, immunological, and biological properties, and susceptibility to insulinase degradation. J Biol Chem 250:4268–4277

Stentz FB, Harris HL, Kitabchi AE (1985) Characterization of insulin-degrading activity of intact and subcellular components of human fibroblasts. Endocrinology 116(3):926–934

Tanaki KI, Ichihara A, Waxman L, Goldberg A (1986) A high molecular weight protease in the cytosol of rat liver. I. Purification, enzymological properties, and tissue distribution. J Biol Chem 261(3):15197–15203

Terris S, Steiner DG (1975) Binding and degradation of ^{125}I insulin by rat hepatocytes. J Biol Chem 250:8389–8398

Thomas JH, Varandani PT (1979) Insulin degradation. XXV. Glutathione insulin transhydrogenase activity of rat liver and kidney during development of streptozotocin-diabetes. Biochim Biophys Acta 567:88–95

Thompson KL, Decker SJ, Rosner MR (1985) Identification of a novel receptor in *Drosophila* for both epidermal growth factor and insulin. Proc Natl Acad Sci USA 82:8443–8447

Varandani PT (1973) Insulin degradation S. Identification of insulin degrading activity of rat liver plasma membrane as glutathione insulin transhydrogenase. Biochem Biophys Res Commun 55:689–696

Varandani PT, Shroyer LA (1987) Identification of an insulin fragment produced by an insulin degrading enzyme neutral thiopeptidase. Mol Cell Endocrinol 50:171–175

Varandani PT, Shroyer LA, Nafz MA (1972) Sequential degradation of insulin by rat liver homogenates. Proc Natl Acad Sci USA 69(7):1681–1684

Yamamoto T, Nojima M, Ishiura S, Sugita H (1986) Purification of the two forms of the high-molecular-weight neutral proteinase ingensin from rat liver. Biochem Biophys Acta 882:297–304

Yaso S, Yokono K, Hari J, Yonezawa K, Shii K, Baba S (1987) Possible role of cell surface insulin degrading enzyme in cultured human lymphocytes. Diabetologia 30:27–32

Yokono K, Imamura Y, Sakai H, Baba S (1979) Insulin-degrading activity of plasma membranes from rat skeletal muscle. Diabetes 28:810–817

Yokono K, Imamura Y, Shii K, Mizuno N, Sakai H, Baba S (1980) Immunochemical studies on the insulin degrading enzyme from pig and rat skeletal muscle. Diabetes 29:856–859

Yokono K, Imimura Y, Shii K, Sakai J, Baba S (1981) Purification and characterization of insulin degrading enzyme from pig skeletal muscle. Endocrinology 108:1527–1532

Yokono K, Roth RA, Baba S (1982) Identification of insulin degrading enzyme on the surface of cultured human lymphocytes, rat hepatoma cells, and primary cultures of rat hepatocytes. Endocrinology 111:1102–1108

Yonezawa K, Yokono K, Yaso S, Hari J, Amano K, Kawase Y, Sakamoto T et al. (1986) Degradation of insulin by insulin degrading enzyme and biological characterization of its fragments. Endocrinology 118:1989–1996

Yu SS, Kitabchi AE (1973) Biological activity of proinsulin and related polypeptides in the fat tissue. J Biol Chem 248:3753

Zolfaghari R, Baker CRF, Amirgholami A, Canizaro PC, Behal FJ (1987a) A multi-catalytic high molecular weight neutral endopeptidase from human kidney. Arch Biochem Biophys 258:42–50

Zolfaghari R, Baker CRF Jr, Cauizaro PC, Amirgholami A, Behal FG (1987b) A high-molecular-mass neutral endopeptidase – 24.5 from human lung. Biochem J 241:129–135

Part III   Insulin Receptor

CHAPTER 9

Insulin Receptor Structure

R. A. ROTH

A. Background

The insulin receptor serves to not only concentrate insulin on the appropriate target cells, but also to initiate the responses of these cells to the hormone. Consequently, a great deal of research has been focused on this molecule. Early efforts were directed at quantitating the interaction of radioactively labeled insulin with its receptor (for reviews of these studies see GAMMELTOFT 1984; ROTH and GRUNFELD 1981; KAHN 1976). From these studies it was possible to

1. Estimate the number of insulin receptors on different cells under various physiological conditions;
2. Show that the interaction of insulin with its receptor results in a subsequent decrease in the number of cell surface receptors (the phenomenon called "downregulation");
3. Show that the interaction of insulin with its receptors is not a simple one since Scatchard plots of binding data were curvilinear, a phenomenon which has been interpreted to mean that insulin was interacting with two populations of receptors or that there was one class of receptors which exhibits negative cooperativity;
4. Show that the rank order of potency of different insulin analogs varies in their binding to insulin receptors in different tissues, suggesting that the insulin receptor might differ in these tissues.

The goal of the structural studies will, in part, be to explain these phenomena in molecular terms. The present chapter will focus on our current understanding of the structure of the insulin receptor.

B. Early Glimpses of the Structure of the Insulin Receptor

Several approaches were applied to decipher the basic structure of the insulin receptor. One approach was to cross-link radioactive insulin to the receptor and analyze the insulin–receptor complex by electrophoresis on SDS polyacrylamide gels (reviewed in CZECH and MASSAGUE 1982). Another approach was to purify the insulin receptor and characterize the purified protein (reviewed in JACOBS and CUATRECASAS 1981). A third approach was to utilize antibodies to the receptor to immunoprecipitate the receptor from metabolically labeled cells and analyze these immunoprecipitates by electrophoresis on SDS polyacrylamide gels (KAHN and MARON 1984). Based on all the data from these different approaches, a pic-

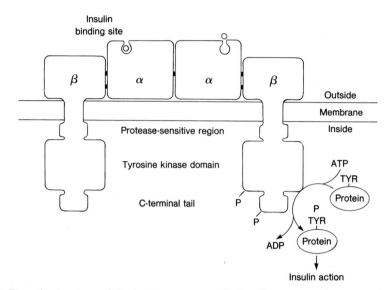

Fig. 1. Domain structure of the insulin receptor. The insulin receptor is composed of two different polypeptides, called α ($M_r \sim 135\,000$) and ($M_r \sim 95\,000$). The α subunit is completely extracellular and is primarily involved in binding insulin. The β subunit is partly extracellular and partly intracellular. The cytoplasmic portion of the β subunit has an intrinsic tyrosine-specific protein kinase activity. The β subunit is also very susceptible to proteolytic cleavage, sometimes resulting in the formation of a fragment ($M_r \sim 45\,000$, called β') which consists primarily of the extracellular domain

ture of the receptor emerged (Fig. 1). This model consisted of an oligomer of two α and two β subunits linked via disulfide bonds. The α and β subunits were found to have apparent M_r of $\sim 130\,000$ and $95\,000$, respectively. Both subunits were found to be glycosylated, indicating that both were at least partly facing the extracellular milieu (HERZBERG et al. 1985).

The α subunit appeared to be primarily involved in binding insulin since it was most strongly labeled in cross-linking studies with ^{125}I-labeled insulin (JACOBS et al. 1979; PILCH and CZECH 1979; YIP et al. 1980). In addition, proteolysis of the β subunit did not appear to affect insulin binding (SHIA et al. 1983; ROTH et al. 1983). These studies did not, however, rule out a contribution of the β subunit in the formation of the insulin binding site. Labeling studies with right-side-out and inside-out vesicles indicated that the α subunit was primarily extracellular (HEDO and SIMPSON 1984).

These same labeling studies also indicated that the β subunit had both an extracellular and intracellular domain (HEDO and SIMPSON 1984). In addition, insulin was found to stimulate the phosphorylation of the β subunit of the receptor in both intact cells and with isolated receptor (KASUGA et al. 1982a, b). Subsequent studies demonstrated that the receptor was itself a tyrosine-specific kinase which became activated after binding insulin (reviewed in ROSEN 1987). This activity appeared to be localized primarily in the β subunit since the β subunit was predominantly labeled when radioactive ATP was cross-linked to the receptor (SHIA and PILCH 1983; ROTH and CASSELL 1983; VAN OBBERGHEN et al. 1983). Also, proteolysis of the β subunit readily eliminated the kinase activity of the receptor without affecting the insulin binding activity (SHIA et al. 1983; ROTH et al. 1983).

Studies of the synthesis of the insulin receptor indicated that the α and β subunits were actually synthesized as a single polypeptide (HEDO et al. 1983; JACOBS et al. 1983). This proreceptor ($M_r \sim 210\,000$) could be shown to both bind insulin and to be phosphorylated on tyrosine residues (HEDO et al. 1983; JACOBS et al. 1983; BLACKSHEAR et al. 1983; REES-JONES et al. 1983). This polypeptide was found to be converted into the mature receptor by a proteolytic step several hours after synthesis. Recent studies indicate that the correct formation of disulfide bonds in the proreceptor may be required before the proreceptor can bind insulin (OLSON et al. 1986).

C. Sequence of the Insulin Receptor

I. The α Subunit

The availability of purified insulin receptor preparations allowed the determination of the amino acid sequence of the NH_2 termini of the α and β subunits of the receptor as well as various tryptic peptide fragments of the receptor (ULLRICH et al. 1985; EBINA et al. 1985). These protein sequences were used to design oligonucleotide probes which could be used to isolate a cDNA which encodes for the insulin receptor (ULLRICH et al. 1985; EBINA et al. 1985). The nucleotide sequence of this cDNA allows one to deduce the complete amino acid sequence of the receptor. This amino acid sequence of the receptor confirmed and extended the prior biochemical information on the structure of the molecule (Fig. 2).

First, the NH_2 terminal amino acid sequence of the α subunit was preceded by a stretch of 27 hydrophobic residues beginning with a methionine. This sequence is characteristic of the signal sequence which is required to direct proteins into the membrane. The remainder of the sequence predicts a protein of 1355 amino acids with a calculated M_r of 153917 which includes both the α and β subunits. [1] These results thereby confirm the pulse–chase experiments indicating that the receptor is synthesized as a single precursor polypeptide.

After cleavage of the signal peptide, the predicted sequence of the α subunit deduced from the cDNA agrees with the actual NH_2 terminal sequence of the isolated α subunit, indicating that there is no additional processing at this end of the molecule. At the COOH terminus of the α subunit there are four dibasic residues (Arg-Lys-Arg-Arg, residues 732–735) which serve as a cleavage site to yield the α (731 amino acids) and β (620 amino acids) subunits. The resulting α subunit sequence contains 15 consensus sequences (Asn-X-Ser/Thr) for asparagine-linked glycosylation which are scattered throughout the α subunit. The α subunit sequence does not contain a stretch of hydrophobic residues of sufficient length to span the plasma membrane, suggesting that this subunit is completely extracellular. In support of this hypothesis is the finding that the α subunit

---

[1] The amino acid and residue numbers used throughout this chapter refer to those of EBINA et al. (1985).

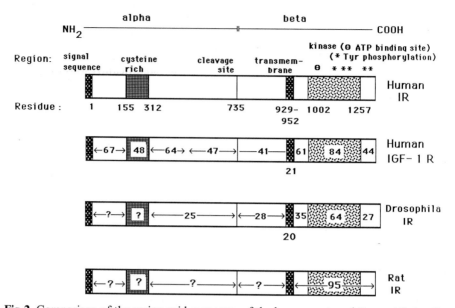

Fig. 2. Comparison of the amino acid sequences of the human, rat, and *Drosophila* insulin receptors (*IR*) and the IGF-I receptor (*IGF-1 R*). The amino acid sequences of these four receptors deduced from their respective cDNA clones are compared for the different regions. The percent sequence identity with the human receptor is given. Regions for which the sequences have not yet been described are indicated by *question marks*

can be released from cells after reduction of disulfide bonds and the disruption of noncovalent interactions, but without solubilization of the membrane (GRUNFELD et al. 1985).

Interestingly, a recent study has identified an insulin-resistant patient with a mutation in the receptor such that the proreceptor molecule is not cleaved to the α and β subunits (YOSHIMASA et al. 1988). This patient was found to have a single mutation in the cleavage site of the proreceptor, changing the tetrabasic cleavage site from Arg-Lys-Arg-Arg to Arg-Lys-Arg-Ser. This mutation resulted in the inability of the cells to convert the proreceptor into mature receptor. The proreceptor had a decreased ability to bind insulin that could be activated by mild trypsin treatment.

One notable feature in the sequence of the α subunit is the presence of 26 cysteine residues between residues 155 and 312. This "cysteine-rich" region of the α subunit is hydrophilic. These cysteine residues may be involved in forming disulfide bonds with the β subunit as well as with another α chain. A similar cysteine-rich region is present in the receptors for insulin-like growth factor I and epidermal growth factor, whereas the receptors for platelet-derived growth factor and colony-stimulating factor do not have an equivalent cysteine-rich region (YARDEN and ULLRICH 1988). Since all of these receptors have a ligand-activated intrinsic tyrosine kinase activity, this region either must not play a critical role in the activation mechanism or these receptors must use different mechanisms for inducing the activation state.

One could also hypothesize that the cysteine-rich region of the α subunit forms the insulin-binding pocket of the receptor. Consistent with this hypothesis is the finding that the cysteine-rich region shares only 48% sequence identity with the homologous region of the IGF-I receptor, a value significantly lower than the surrounding region (ULLRICH et al. 1986). If one excludes the cysteines, the percent sequence identity of these two regions falls to 25%. This degree of sequence identity is much lower than any other region of these two receptors. One could therefore interpret these data to indicate that the cysteine-rich region is involved in ligand binding and is therefore more divergent between these two receptors. In addition, studies with insulin cross-linked to the receptor have indicated that insulin is linked to a region of the α subunit which has extensive disulfide bonds, i.e., the cysteine-rich region (BÖNI-SCHNETZLER et al. 1987).

Other data, however, do not support the localization of the insulin-binding region to the cysteine-rich region of the α subunit. Deletion mutants of the receptor and cross-linking of azidoinsulin to the receptor have implicated the NH_2 terminus of the α subunit in binding insulin (BODSCH et al. 1988). In particular, residues 60–90 have been suggested to be involved in insulin binding (BODSCH et al. 1988). A recent study of the insulin receptor of an insulin-resistant patient has indicated that Lys-460 may also contribute to the formation of the binding site since changing this residue to a Glu affects the pH sensitivity of insulin dissociation (KADOWAKI et al. 1988). Additional studies are therefore required to further delineate the regions of the receptor involved in binding insulin.

II. The β Subunit

The deduced NH_2 terminal sequence of the β subunit of the receptor agrees with that determined from the isolated subunit, confirming the predicted cleavage of the prorecceptor after residue 735 (ULLRICH et al. 1985; EBINA et al. 1985). The β subunit sequence contains a stretch of 23 hydrophobic residues (930–952) which are of sufficient length to cross the membrane one time. Thus, the β subunit is predicted to have both an extracellular (196 amino acids) and an intracellular domain (403 amino acids). This proposed structure is consistent with labeling studies indicating that the β subunit can be radioiodinated in both inside-out and right-side-out membrane vesicles (HEDO and SIMPSON 1984). The resulting extracellular β subunit sequence contains four potential asparagine-linked glycosylation sites.

The cytoplasmic portion of the β subunit contains a region (residues 1002–1257) which is homologous to other kinases (ULLRICH et al. 1985; EBINA et al. 1985). This region includes the consensus sequence for an ATP-binding site (Gly-X-Gly-XX-Gly) followed by a lysine (residue 1030) 22 amino acids later. That this lysine is critical for the kinase activity of the receptor has been confirmed by site-directed mutagenesis studies (EBINA et al. 1987; CHOU et al. 1987; McCLAIN et al. 1987). Replacement of this lysine with either an arginine, alanine, or a methionine results in the formation of a receptor which binds insulin normally, but is totally devoid of kinase activity. The overall structure of the receptor appeared unaffected since the receptor reacted normally with several conformation-specific monoclonal antibodies (EBINA et al. 1987).

The kinase domain of the insulin receptor has highest homology to the kinase domain of the IGF-I receptor, being 84% identical in sequence (ULLRICH et al. 1986). This region of these two receptors is more homologous than any other region of the two receptors (Fig. 2). This is in agreement with immunological studies showing that antibodies to the kinase region of the insulin receptor can cross-react with almost identical potency with the kinase region of the IGF-I receptor (MORGAN and ROTH 1986). This high homology of the kinase domains of these two receptors is also consistent with the finding that the same endogenous proteins are phosphorylated in response to insulin and IGF-I (KADOWAKI et al. 1987). In addition, this high homology is consistent with the ability of both receptors to mediate similar responses (STEELE-PERKINS et al. 1988).

The kinase domain of the insulin receptor also has 52% sequence identity with the protein product of the v-ros oncogene of the UR2 avian sarcoma virus (ULLRICH et al. 1985; EBINA et al. 1985). This suggested that the cellular homolog of this oncogene is the chicken insulin receptor. However, monoclonal antibodies which recognize the avian insulin receptor kinase domain do not recognize the protein product of v-ros (ELLIS et al. 1987b). In addition, outside the kinase domains these two proteins do not appear to be related although the cellular homolog of v-ros appears to encode a receptor-like molecule (MATSUSHIME et al. 1986).

The kinase domain of the insulin receptor also contains several autophosphorylation sites. Tyr-1162 is in an equivalent position to a major autophosphorylation site of the protein product of the viral src oncogene and has been found to be autophosphorylated in the receptor and to serve a regulatory function (WHITE et al. 1988; TORNQVIST et al. 1988; HERRERA and ROSEN 1986; ELLIS et al. 1986a). In addition, Tyr-1163 and -1158 also appear to be autophosphorylated and to play a role in the activation of the receptor kinase (WHITE et al. 1988; TORNQVIST et al. 1988; HERRERA and ROSEN 1986; ELLIS et al. 1986a). One could envision that the phosphorylation of this cluster of tyrosine residues adds enough of a negative charge to this region of the receptor molecule such that the active site of the receptor becomes more accessible to substrates (Fig. 3). Indeed, polyclonal anti-peptide antibodies to this region of the receptor (residues 1152–1174) recognize the native receptor after autophosphorylation, but not before (HERRERA and ROSEN 1986). Thus, in analogy with other kinases, one could propose that these residues are generally sitting in the active site of the receptor, thereby blocking access to substrates. Once the receptor binds insulin, these sites become phosphorylated and move out of the active site of the receptor.

The COOH terminal tail (residues 1260–1355) of the β subunit shows less homology to other kinases (ULLRICH et al. 1985; EBINA et al. 1985). For example, this region exhibits only 44% sequence identity with the IGF-I receptor (see Fig. 2). This region is also highly hydrophilic. One could propose that this region confers specificity to the insulin receptor kinase. This region also contains two autophosphorylation sites, Tyr-1328 and -1334. Although these sites appear to be phosphorylated (WHITE et al. 1988; TORNQVIST et al. 1988; HERRERA and ROSEN 1986), they do not appear to serve a regulatory role. Intriguingly, the expression

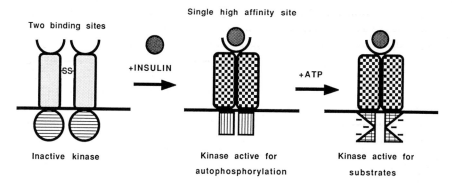

Two binding sites

Single high affinity site

+INSULIN

+ATP

Inactive kinase

Kinase active for
autophosphorylation

Kinase active for
substrates

Fig. 3. Schematic of the activation of the insulin receptor. The binding of insulin to its receptor appears to induce a substantial conformational change in the receptor such that: (1) the receptor loses one of its high affinity insulin-binding sites, and (2) the receptor autophosphorylates. The latter effect causes a subsequent increase in the ability of the receptor kinase to phosphorylate exogenous substrates, possibly as a result of the increase in negative charge in the kinase domain

of a mutated receptor lacking the COOH terminal 43 residues was found incapable of mediating a rapid biological response, stimulation of glucose uptake, or to cause stimulation of glycogen synthesis (MAEGAWA et al. 1988). However, this mutated receptor did appear to autophosphorylate normally and to phosphorylate exogenous substrates (MAEGAWA et al. 1988). It will be important to determine whether this mutated receptor can cause the phosphorylation of the same endogenous substrates as observed with the wild-type receptor in the intact cell.

III. Domain Structure of the Insulin Receptor

From the sequence of the insulin receptor, one can subdivide the receptor into an extracellular domain (containing the complete α subunit and part of the β subunit, a total of 929 residues), the transmembrane domain (residues 930–952), and the cytoplasmic domain (residues 953–1355 of the β subunit). The cytoplasmic domain appears to fold independently since cDNA constructs encoding for only this portion of the molecule can be expressed and yield an active tyrosine-specific kinase (ELLIS et al. 1987a; HERRERA et al. 1988; ELLIS et al. 1988a). Interestingly, this domain appears to be fully active as a kinase, suggesting that the extracellular domain normally acts to inhibit this domain (ELLIS et al. 1987a; HERRERA et al. 1988; ELLIS et al. 1988a). Additional support for this hypothesis comes from the finding that removal of the extracellular domain of the receptor by proteolysis also activates the kinase domain (SHOELSON et al. 1988).

The extracellular domain also appears to be capable of forming independently. Expression of cDNA clones encoding this region of the molecule have also been expressed either alone or attached to the transmembrane and cytoplasmic portion of other receptors (ELLIS et al. 1986b; WHITAKER and OKAMOTO 1988; JOHNSON et al. 1989; ELLIS et al. 1988b). These forms of the molecule appear to fold normally since they can bind insulin with the same affinity as the intact receptor, are recognized by conformation-specific monoclonal antibodies, are glycosylated, and are properly processed into heterotetramers (ELLIS et al. 1986b; WHITAKER and OKAMOTO 1988; JOHNSON et al. 1989; ELLIS et al. 1988b). However, the expression of only the extracellular domain of the α subunit has, so far, not yielded a product which can bind insulin (JOHNSON et al. 1989). Thus, the extracellular domain of the β subunit contributes either to the stability of the extracellular domain or is required for its productive synthesis.

The heterotetrameric structure ($\alpha_2\beta_2$) of the receptor appears to be required to duplicate the interactions of the wild-type receptor with insulin. Isolated heterodimeric complexes ($\alpha\beta$) have been found to exhibit linear binding curves in contrast to the curvilinear Scatchard plots observed with intact receptor (DEGER et al. 1986; SWEET et al. 1987). In addition, insulin enhanced the dissociation of labeled insulin from the tetramer, but not the dimer (DEGER et al. 1986). These results support the hypothesis that the intact receptor has two binding sites which exhibit negative cooperativity (GU et al. 1988) and indicate that the intact heterotetramer is required for this conformational change. A similar requirement has been observed in the insulin stimulation of the receptor kinase (BÖNI-SCHNETZLER et al. 1988). These two events may be related. That is, the binding of insulin to the intact heterotetramer may induce substantial conformational changes in the extracellular domain of the receptor which thereby renders the second insulin-binding site inactive and removes the inhibitory effect of the extracellular domain on the kinase activity of the cytoplasmic domain of the receptor (Fig. 3). Evidence for such an insulin-induced conformational change in the receptor has been obtained in studies in which an increased susceptibility to proteolysis was observed in the α subunit after insulin binding (PILCH and CZECH 1980; DONNER and YONKERS 1983).

D. Insulin Receptors in Different Tissues and Species

The insulin receptor in different tissues has been reported to vary in molecular weight, affinity for various ligands, state of aggregation, and interaction with antibodies (HEIDENREICH et al. 1983; HENDRICKS et al. 1984; JONAS et al. 1986; ROTH et al. 1986; SCHWEITZER et al. 1980; BURANT et al. 1987; MCELDUFF et al. 1985; CARO et al. 1988). Some of these differences have been attributed to variation in carbohydrate content or disulfide bonds (HEIDENREICH et al. 1983; HENDRICKS et al. 1984; POTTICK et al. 1981; JACOBS and CUATRECASAS 1980). However, it is also possible that the amino acid sequence of the insulin receptor differs in various tissues. The two cDNA clones which were originally isolated differ by the presence of 12 amino acids at the COOH terminal of the α subunit (718–729) of one clone, but not the other (ULLRICH et al. 1985; EBINA et al. 1985). It has been reported that this difference arises from the alternative splicing of the mRNA and that this is regulated in a tissue-specific manner (YARDEN and ULLRICH 1988). Thus, even though there appears to be only a single gene for the insulin receptor (ULLRICH et al. 1985; EBINA et al. 1985), alternative processing of the mRNA can yield receptors with different amino acid sequences and consequently different characteristics. Indeed, multiple insulin receptor mRNA species have been observed in various human and rat tissues (ULLRICH et al. 1985; EBINA et al. 1985; GOLDSTEIN et al. 1987). These differences could, however, be restricted to the nontranslated 5' and 3' regions of the RNA and not affect the receptor sequence.

The insulin receptor has been most extensively characterized in humans and rats. The sequence of the cytoplasmic domain of the rat receptor has been deduced from the sequence of rat genomic clones and found to be 95%–98% identical to the homologous region of the human receptor (see Fig. 2; LEWIS et al. 1986). However, based on antibody data, the extracellular domain may be less highly conserved than the cytoplasmic domain (MORGAN and ROTH 1986).

The only other species of insulin receptor for which amino acid sequence is available is the *Drosophila* insulin receptor. This insect receptor has an overall structure (an insulin-binding subunit of $M_r \sim 110$ and a subunit with kinase activity of $M_r \sim 95$) and specificity for insulin that is remarkably similar to the human receptor (FERNANDEZ-ALMONACID and ROSEN 1987). It binds insulin with higher affinity than IGF-I and II and does not recognize the insect insulin-like hormone, 4K-PTTH-II. Thus, it is a true insulin receptor. The portion of the receptor that has been sequenced that exhibits the highest homology to the human receptor is the kinase domain, sharing 64% sequence identity (see Fig. 2; PETRUZZELLI et al. 1986; NISHIDA et al. 1986). Within this region, a 90% sequence identity is found in the region that includes the three regulatory autophosphorylation sites of the receptor. Unfortunately, the sequence of the NH_2 terminus of the α subunit and the cysteine-rich region has not yet been determined. Since the *Drosophila* receptor binds insulin with high affinity and specificity, a comparison of the sequences of these regions of the human and insect receptors may help in localizing the insulin-binding region. The *Drosophila* receptor differs from the human receptor in having some nonprocessed prorecep-tor ($M_r \sim 170$) associated with the receptor complex in the plasma membrane. Although insulin could not be cross-linked to this 170 kdalton subunit, it was phosphorylated on tyrosine residues in response to insulin and was im-munoprecipitated with anti-peptide antibodies to the receptor (FERNANDEZ-ALMONACID and ROSEN 1987).

Another species whose insulin receptor does not appear to be processed is the stingray (STUART 1988). Although this receptor specifically binds insulin with high affinity and has an insulin-stimulated kinase activity, it has been reported to have a M_r of 210. This receptor was also reported to cross-react with antibodies to the human receptor, to exhibit a curvilinear Scatchard plot, and to exist as a dimer. Thus, the properties of this receptor are remarkably similar to the mammalian receptor, with the exception that the proreceptor is not cleaved to yield the α and β subunits.

E. Concluding Remarks

The insulin receptor has progressed from an insulin-binding activity to defined bands on polyacrylamide SDS gels and, most recently, to a specific sequence of amino acids. Based on the sequence of the receptor, a number of predictions could be made as to the role of different amino acids in particular functions of the receptor. These predictions have in part been tested by expression of receptors mutated at these particular residues. Via this approach, the critical role of Lys-1030 in the kinase activity of the receptor has been established. In addition, the role of Tyr-1162 and -1163 as regulatory autophosphorylation sites has been demonstrated. Moreover, Tyr-960 and the carboxy 43 amino acids of the β subunit were found not to play a role in regulating the receptor kinase, but to still play a critical function in mediating biological responses (WHITE et al. 1988; MAEGAWA et al. 1988). Additional studies should allow us to define other residues of the receptor which are involved in signal transduction as well as hormone binding. Moreover, the characterization of defective receptors from insulin-resistant patients should also help to define residues in the receptor critical for its proper function. Finally, with the overproduction of the receptor in various expression systems, it should eventually be possible to determine the three-dimensional structure of the receptor by X-ray crystallography.

References

Blackshear PJ, Nemenoff RA, Avruch J (1983) Insulin binds to and promotes the phosphorylation of a M_r 210000 component of its receptor in detergent extracts of rat liver microsomes. FEBS Lett 158:243–246

Bodsch W, Wedekind F, Sommer M-T, Brandenburg D (1988) On the insulin binding domain of the human insulin receptor. In: International symposium on insulin and the cell membrane, Slovak Academy of Sciences, Bratislava, Czechoslovakia, 27–30 June 1988

Böni-Schnetzler M, Scott W, Waugh SM, DiBella E, Pilch PF (1987) The insulin receptor: structural basis for high affinity ligand binding. J Biol Chem 262:8395–8401

Böni-Schnetzler M, Kaligian A, DelVecchio R, Pilch PF (1988) Ligand-dependent inter-subunit association within the insulin receptor complex activates its intrinsic kinase activity. J Biol Chem 263:6822–6828

Burant CF, Treutelaar MK, Allen KD, Sens DA, Buse MG (1987) Comparison of insulin and insulin-like growth factor I receptors from rat skeletal muscle and L-6 myocytes. Biochem Biophys Res Commun 147:100–107

Caro JF, Raju SM, Sinha MK, Goldfine ID, Dohm GL (1988) Heterogeneity of human liver, muscle, and adipose tissue insulin receptor. Biochem Biophys Res Commun 15:123–129

Chou CK, Dull TJ, Russell DS, Gherzi R, Lebwohl D, Ullrich A, Rosen OM (1987) Human insulin receptors mutated at the ATP-binding site lack protein tyrosine kinase activity and fail to mediate postreceptor effects of insulin. J Biol Chem 262:1842–1847

Czech MP, Massague J (1982) Subunit structure and dynamics of the insulin receptor. Fed Proc 41:2719–2723

Deger A, Krämer H, Rapp R, Koch R, Weber U (1986) The nonclassical insulin binding of insulin receptors from rat liver is due to the presence of two interacting α-subunits in the receptor complex. Biochem Biophys Res Commun 135:458–464

Donner DB, Yonkers K (1983) Hormone-induced conformational changes in the hepatic insulin receptor. J Biol Chem 258:9413–9418

Ebina Y, Ellis L, Jarnagin K, Edery M, Graf L, Clauser E, Ou J-h, Masiarz F, Kan YW, Goldfine ID, Roth RA, Rutter WJ (1985) The human insulin receptor cDNA: the structural basis for hormone-activated transmembrane signalling. Cell 40:747–758

Ebina Y, Araki E, Taira M, Shimada F, Mori M, Craik CS, Siddle K, Pierce SB, Roth RA, Rutter WJ (1987) Replacement of lysine residue 1030 in the putative ATP-binding region of the insulin receptor abolishes insulin- and antibody-stimulated glucose uptake and receptor kinase activity. Proc Natl Acad Sci USA 84:704–708

Ellis L, Clauser E, Morgan DO, Edery M, Roth RA, Rutter WJ (1986a) Replacement of insulin receptor tyrosine residues 1162 and 1163 compromises insulin-stimulated kinase activity and uptake of 2-deoxyglucose. Cell 45:721–732

Ellis L, Morgan DO, Clauser E, Edery M, Jong S-M, Wang L-H, Roth RA, Rutter WJ (1986b) Mechanisms of receptor-mediated transmembrane communication. In: Cold Spring Harbor Symp Quant Biol 51:773

Ellis L, Morgan DO, Clauser E, Roth RA, Rutter WJ (1987a) A membrane-anchored cytoplasmic domain of the human insulin receptor mediates a constitutively elevated insulin-independent uptake of 2-deoxyglucose. Mol Endocrinol 1:15–24

Ellis L, Morgan DO, Jong S-M, Wang L-H, Roth RA, Rutter WJ (1987b) Heterologous transmembrane signaling by a human insulin receptor-v-*ros* hybrid in Chinese hamster ovary cells. Proc Natl Acad Sci USA 84:5101–5105

Ellis L, Levitan A, Cobb MH, Ramos P (1988a) Efficient expression in insect cells of a soluble, active human insulin receptor protein-tyrosine kinase domain by use of a baculovirus vector. J Virol 62:1634–1639

Ellis L, Sissom J, Levitan A (1988b) Truncation of the ectodomain of the human insulin receptor results in secretion of a soluble insulin binding protein from transfected CHO cells. J Mol Recognition 1:25–31

Fernandez-Almonacid R, Rosen OM (1987) Structure and ligand specificity of the *Drosophila melanogaster* insulin receptor. Mol Cell Biol 7:2718–2727

Gammeltoft S (1984) Insulin receptors: binding kinetics and structure-function relationship of insulin. Physiol Rev 64:1321–1378

Goldstein BJ, Muller-Wieland D, Kahn CR (1987) Variation in insulin receptor messenger ribonucleic acid expression in human and rodent tissues. Mol Endocrinol 1:759–766

Grunfeld C, Shigenaga JK, Ramachandran J (1985) Urea treatment allows dithiothreitol to release the binding subunit of the insulin receptor from the cell membrane: implications for the structural organization of the insulin receptor. Biochem Biophys Res Commun 133:389–396

Gu J-L, Goldfine ID, Forsayeth JR, De Meyts P (1988) Reversal of insulin-induced negative cooperativity by monoclonal antibodies that stabilize the slowly dissociating state of the insulin receptor. Biochem Biophys Res Commun 150:694–701

Hedo JA, Simpson IA (1984) Internalization of insulin receptors in the isolated rat adipose cell. J Biol Chem 259:11083–11089

Hedo JA, Kahn CR, Hayashi M, Yamada KM, Kasuga M (1983) Biosynthesis and glycosylation of the insulin receptor. J Biol Chem 258:10020–10026

Heidenreich KA, Zahniser NR, Berhanu P, Brandenburg D, Olefsky JM (1983) Structural differences between insulin receptors in the brain and peripheral target tissues. J Biol Chem 258:8527–8530

Hendricks SA, Agardh C-D, Taylor SI, Roth J (1984) Unique features of the insulin receptor in rat brain. J Neurochem 43:1302–1309

Herrera R, Rosen OM (1986) Autophosphorylation of the insulin receptor in vitro. J Biol Chem 261:11980–11985

Herrera R, Lebwohl D, de Herreros AG, Kallen RG, Rosen OM (1988) Synthesis, purification, and characterization of the cytoplasmic domain of the human insulin receptor using a baculovirus expression system. J Biol Chem 263:5560–5568

Herzberg VL, Grigorescu F, Edge ASB, Spiro RG, Kahn CR (1985) Characterization of insulin receptor carbohydrate by comparison of chemical and enzymatic deglycosylation. Biochem Biophys Res Commun 129:789–796

Jacobs S, Cuatrecasas P (1980) Disulfide reduction converts the insulin receptor of human placenta to a low affinity form. J Clin Invest 66:1424–1427

Jacobs S, Cuatrecasas P (1981) Insulin receptor: structure and function. Endocr Rev 2:251–263

Jacobs S, Hazum E, Schechter Y, Cuatrecasas P (1979) Insulin receptor: covalent labeling and identification of subunits. Proc Natl Acad Sci USA 76:4918–4921

Jacobs S, Kull FC Jr, Cuatrecasas P (1983) Monensin blocks the maturation of receptors for insulin and somatomedin C: identification of receptor precursors. Proc Natl Acad Sci USA 80:1228–1231

Johnson JD, Wong ML, Rutter WJ (1989) Properties of the insulin receptor ectodomain. Proc Natl Acad Sci (USA) 85:7516–7520

Jonas HA, Newman JD, Harrison LC (1986) An atypical insulin receptor with high affinity for insulin-like growth factors copurified with placental insulin receptors. Proc Natl Acad Sci USA 83:4124–4128

Kadowaki T, Koyasu S, Nishida E, Tobe K, Izumi T, Takaku F, Sakai H, Yahara I, Kasuga M (1987) Tyrosine phosphorylation of common and specific sets of cellular proteins rapidly induced by insulin, insulin-like growth factor I, and epidermal growth factor in an intact cell. J Biol Chem 262:7342–7350

Kadowaki T, Bevins CL, Cama A, Ojamaa K, Marcus-Samuels B, Kadowaki H, Beitz L, McKeon C, Taylor SI (1988) Two mutant alleles of the insulin receptor gene in a patient with extreme insulin resistance. Science 240:787–790

Kahn CR (1976) Membrane receptors for hormones and neurotransmitters. J Cell Biol 70:261–286

Kahn CR, Maron R (1984) Immunology of the insulin receptor. In: Andreani D, Di Mario U, Federlin KF, Heding LG (eds) Immunology in diabetes. Kimpton Medical, London, p 209

Kasuga M, Karlsson FA, Kahn CR (1982a) Insulin stimulates the phosphorylation of the 95000-dalton subunit of its own receptor. Science 215:185–186

Kasuga M, Zick Y, Blithe DL, Crettaz M, Kahn CR (1982b) Insulin stimulates tyrosine phosphorylation of the insulin receptor in a cell-free system. Nature 298:667–669

Lewis RE, Tepper MA, Czech MP (1986) Characterization of a genomic clone encoding the rat insulin receptor cytoplasmic domain. In: Program and abstracts, 68th annual Meeting of the Endocrine Society, Anaheim, California, 25–27 June 1986

Maegawa H, McClain DA, Freidenberg G, Olefsky JM, Napier M, Lipari T, Dull TJ, Lee J, Ullrich A (1988) Properties of a human insulin receptor with a COOH-terminal truncation. J Biol Chem 263:8912–8917

Matsushime H, Wang L-H, Shibuya M (1986) Human c-ros-1 gene homologous to the v-ros sequence of UR2 sarcoma virus encodes for a transmembrane receptor-like molecule. Mol Cell Biol 6:3000–3004

McClain DA, Maegawa H, Lee J, Dull TJ, Ulrich A, Olefsky JM (1987) A mutant insulin receptor with defective tyrosine kinase displays no biologic activity and does not undergo endocytosis. J Biol Chem 262:14663–14671

McElduff A, Grunberger G, Gorden P (1985) An alteration in apparent molecular weight of the insulin receptor from the human monocyte cell line U-937. Diabetes 34:686–690

Morgan DO, Roth RA (1986) Mapping surface structures of the human insulin receptor with monoclonal antibodies: localization of main immunogenic regions to the receptor kinase domain. Biochem 25:1364–1371

Nishida Y, Hata M, Nishizuka Y, Rutter WJ, Ebina Y (1986) Cloning of a *Drosophila* cDNA encoding a polypeptide similar to the human insulin receptor precursor. Biochem Biophys Res Commun 141:474–481

Olson TS, Bamberger MJ, Lane MD (1986) Post-translational changes in tertiary and quaternary structure of the insulin proreceptor. J Biol Chem 263:7342–7351

Petruzzelli L, Herrera R, Arenas-Garcia R, Fernandez R, Birnbaum MJ, Rosen OM (1986) Isolation of a *Drosophila* genomic sequence homologous to the kinase domain of the human insulin receptor and detection of the phosphorylated *Drosophila* receptor with an anti-peptide antibody. Proc Natl Acad Sci USA 83:4710–4714

Pilch PF, Czech MP (1979) Interaction of cross-linking agents with the insulin effector system of isolated fat cells. J Biol Chem 254:3375–3380

Pilch PF, Czech MP (1980) Hormone binding alters the conformation of the insulin receptor. Science 210:1152–1153

Pottick LA, Moxley RT III, Livingston JN (1981) Tissue differences in insulin receptors: acute changes in insulin binding characteristics induced by wheat germ agglutinin. Diabetes 30:196–202

Rees-Jones RW, Hedo JA, Zick Y, Roth J (1983) Insulin-stimulated phosphorylation of the insulin receptor precursor. Biochem Biophys Res Commun 116:417–422

Rosen OM (1987) After insulin binds. Science 237:1452–1457

Roth RA, Cassell DJ (1983) Insulin receptor: evidence that it is a protein kinase. Science 219:299–301

Roth J, Grunfeld C (1981) Endocrine systems: mechanisms of disease, target cells, and receptors. In: Williams RH (ed) Textbook of endocrinology. Saunders, Philadelphia, p 15, chap 2

Roth RA, Mesirow ML, Cassell DJ (1983) Preferential degradation of the β subunit of purified insulin receptor. J Biol Chem 258:14456–14460

Roth RA, Morgan DO, Beaudoin J, Sara V (1986) Purification and characterization of the human brain insulin receptor. J Biol Chem 261:3753–3757

Schweitzer JB, Smith RM, Jarett L (1980) Differences in organizational structure of insulin receptor on rat adipocyte and liver plasma membranes: role of disulfide bonds. Proc Natl Acad Sci USA 77:4692–4696

Shia MA, Pilch PF (1983) The β subunit of the insulin receptor is an insulin-activated protein kinase. Biochem 22:717–721

Shia MA, Rubin JB, Pilch PF (1983) The insulin receptor protein kinase. J Biol Chem 258:14450–14455

Shoelson SE, White MF, Kahn CR (1988) Tryptic activation of the insulin receptor. J Biol Chem 263:4852–4860

Steele-Perkins G, Turner J, Edman JC, Hari J, Pierce SB, Stover C, Rutter WJ, Roth RA (1988) Expression and characterization of a functional human insulin-like growth factor I receptor. J Biol Chem 263:11486–11492

Stuart CA (1988) Characterization of a novel insulin receptor from stingray liver. J Biol Chem 263:7881–7886

Sweet LJ, Morrison BD, Pessin JE (1987) Isolation of functional $\alpha\beta$ heterodimers from the purified human placental $\alpha_2\beta_2$ heterotetrameric insulin receptor complex. J Biol Chem 262:6939–6942

Tornqvist HE, Gunsalus JR, Nemenoff RA, Frackelton AR, Pierce MW, Avruch J (1988) Identification of the insulin receptor tyrosine residues undergoing insulin-stimulated phosphorylation in intact rat hepatoma cells. J Biol Chem 263:350–359

Ullrich A, Bell JR, Chen EY, Herrera R, Petruzzelli LM, Dull TJ, Gray A, Coussens L, Liao Y-C, Tsubokawa M, Mason A, Seeburg PH, Grundfeld C, Rosen OM, Ramachandran J (1985) Human insulin receptor and its relationship to the tyrosine kinase family of oncogenes. Nature 313:756–761

Ullrich A, Gray A, Tam AW, Yang-Feng T, Tsubokawa M, Collins C, Henzel W, Le Bon T, Kathuria S, Chen E, Jacobs S, Francke U, Ramachandran J, Fujita-Yamaguchi Y (1986) Insulin-like growth factor I receptor primary structure: comparison with insulin receptor suggests structural determinants that define functional specificity. EMBO J 5:2503–2512

van Obberghen EB, Rossi A, Kowalski A, Gazzano H (1983) Receptor-mediated phosphorylation of the hepatic insulin receptor; evidence that the M_r 95000 receptor subunit is its own kinase. Proc Natl Acad Sci USA 80:945–949

Whitaker J, Okamoto A (1988) Secretion of soluble functional insulin receptors by transfected NIH3T3 cells. J Biol Chem 263:3063–3066

White MF, Shoelson SE, Keutmann H, Kahn CR (1988) A cascade of tyrosine autophosphorylation in the β-subunit activates the phosphotransferase of the insulin receptor. J Biol Chem 263:2969–2980

Yarden Y, Ullrich A (1988) Growth factor receptor tyrosine kinases. Ann Rev Biochem 57:443–478

Yip CC, Moule ML, Yeung CWT (1980) Characterization of insulin receptor subunits in brain and other tissues by photoaffinity labeling. Biochem Biophys Res Commun 96:1671–1678

Yoshimasa Y, Seino S, Whittaker J, Kakehi T, Kosaki A, Kuzuya H, Imura H, Bell GI, Steiner DF (1988) Insulin-resistant diabetes due to a point mutation that prevents insulin proreceptor processing. Science 240:784–787

CHAPTER 10

Insulin Receptor-Mediated Transmembrane Signalling

M. D. HOLLENBERG

A. Introduction

At least in part because of the intense interest generated by the dramatic therapeutic effects of insulin, when it was first used for patients like Leonard Thompson in the early 1920s, no single hormone has been more thoroughly investigated than has insulin. Yet, at this moment in time, over 60 years since the discovery of insulin, the precise series of reactions whereby insulin triggers a cellular response is not known. Nonetheless, as summarized by other chapters in this volume, an enormous amount of detailed information has been acquired about the structural requirements for the interaction of insulin with its receptor, about the oligomeric structure of the receptor complex which associates with other proteins in the plasma membrane, about the receptor's autophosphorylation and tyrosine kinase activity and about the precise amino acid sequences and biosynthesis of the receptor α and β subunits. There is no question in anyone's mind that all of the many actions of insulin stem from the initial interaction of insulin with its plasma membrane receptor. But, the main question yet to be answered is: how does receptor triggering cause such a diverse spectrum of responses in the cells with which insulin interacts? The diversity of response to insulin (Table 1) is seen not only in terms of the wide variety of processes regulated (ranging from changes in membrane polarization to the stimulation of DNA synthesis) but also in terms of the time frame of the responses, ranging from mil-

Table 1. Cellular actions of insulin

Response	Time course
Regulation of ion flux (Na^+, K^+, Ca^{2+})	Milliseconds to seconds
Stimulation of receptor tyrosine kinase	Seconds to minutes
Stimulation of receptor-substrate phosphorylation	Minutes
Stimulation of glucose transport	Minutes
Stimulation of enzyme activity	
Glycogen synthase	Minutes
Pyruvate dehydrogenase	Minutes
Acetyl-CoA-carboxylase	Minutes
Stimulation of amino acid transport	Tens of minutes
Stimulation of protein synthesis	Tens of minutes to hours
Regulation of gene transcription	Tens of minutes to hours
Stimulation of cell division	Tens of hours

liseconds (membrane polarization) to minutes (stimulation of glucose transport) to tens of hours (stimulation of DNA synthesis).

It is the goal of this chapter: first, to outline in general the basic mechanisms whereby ligands like insulin transmit a transmembrane signal to the cell interior; and second, to discuss how the insulin receptor per se might use one or more of these mechanisms to generate an amplified transmembrane signal.

B. Mechanisms of Transmembrane Signalling

I. Signal Generation

The basic mechanisms whereby a membrane receptor triggers an initial intracellular signal may turn out to be few in number. Three main processes can be singled out for comment:
1. Ligand-regulated ion channel activity (R_{CH}, Fig. 1);

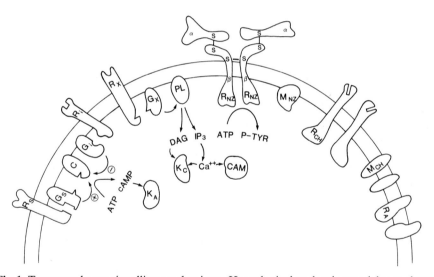

Fig. 1. Transmembrane signalling mechanisms. Hypothetical molecular models are shown for the various transmembrane signalling reactions discussed in the text. Receptors that either stimulate (R_s) or inhibit (R_i) adenylate cyclase (C) via the stimulatory (G_s) or inhibitory (G_i) guanine nucleotide regulatory proteins are shown on the left, as is a receptor (R_x) that acts via a G protein (G_x) to regulate membrane phospholipase (PL). Receptors that possess intrinsic enzymatic activity (R_{NZ}) are represented by the cartoon of the insulin receptor, shown to catalyse tyrosine phosphorylation (P-TYR, middle); M_{NZ} represents other membrane-associated enzymes that can be subject to receptor-mediated regulation. Ion channel receptors, like the nicotinic acetylcholine receptor (R_{CH}), are shown on the right along with other ion (M_{CH}) or metabolite/acceptor constituents (R_A) that may participate in transmembrane signalling. Potential messengers of the transmembrane signals like cyclic AMP (cAMP), calcium (Ca^{++}), diacylglycerol (DAG) or inositol trisphosphate (IP_3) are also shown along with their protein targets (K_A, cyclic AMP-dependent protein kinase; K_C, kinase C; CAM, calmodulin)

2. Ligand-regulated enzymatic activity, whereby the binding of a ligand to the extracellular portion of the receptor triggers a catalytic receptor domain that is located at the inner face of the cell membrane. It is now clear that the insulin receptor (shown as R_{NZ} in Fig. 1) fits into this category, along with the epidermal growth factor-urogastrone receptor and other tyrosine kinase receptors;
3. Ligand-modulated interaction of receptors with so-called guanine nucleotide regulatory G or N protein oligomers (receptors of the type R_s, R_i or R_x, interacting with G_s, G_i and G_x, Fig. 1).

The receptor–G protein interaction liberates the cryptic membrane-localized polypeptide mediators that form the G protein oligomer ($\alpha\beta\gamma$; see below). The process of receptor internalization, whereby the receptor and its bound ligand can find their way to the intracellular milieu (Fig. 2), provides a novel mechanism to translocate one of the three basic reactions triggered by receptors to an intracellular target. The key property of receptor function, irrespective of which of the three mechanisms is used, lies in the ability of the ligand (or its surrogate) to act as an allosteric regulator of the receptor's activity. In terms of insulin action, this property is expressed by the ability of insulin specifically to increase the receptors tyrosine kinase activity and to modulate receptor conformation and cell surface mobility (see below).

II. Signal Amplification

Once a ligand has triggered any of the three mechanisms outlined above, the initial signal must be greatly amplified to generate a transmembrane message that affects the entire cell. Here again, as with the triggering mechanisms, it is possible that a comparatively small subset of cellular reactions may provide the key elements of signal amplification. Two quite distinct amplification mechanisms that are of major importance comprise:

1. A change in membrane potential, resulting from a ligand-induced change in ion flux (e.g. via R_{CH}, or possibly via a receptor-mediated regulation of M_{CH}, in Fig. 1);
2. The initiation of a phosphorylation–dephosphorylation cascade, akin to the one that regulates the breakdown of glycogen to glucose.

As pointed out in Chap. 19 by ZIERLER, a small change in membrane polarization would have an immediate and profound effect on the orientation (and presumably, therefore, on the function) of many membrane proteins. Thus, the hyperpolarization caused by insulin in its target cells (Chap. 18) might readily affect a number of voltage-regulated ion channels, like the ones for sodium and calcium, thereby amplifying the initial insulin signal. The importance of calcium ion as a messenger of transmembrane signals via interactions with proteins like calmodulin (CAM, Fig. 1) and kinase C (K_C, Fig. 1) will be discussed below. In addition, the ability of a change in membrane potential to regulate the activity of an enzyme like phospholipase A_2 (THUREN et al. 1987) provides a means of generating a variety of cellular messengers that could be involved in the action of insulin.

It is relatively easy to envision how the triggering of a phospho-rylation–dephosphorylation cascade can yield an amplified receptor signal. Clas-sically, this mechanism is best understood in terms of the regulation of glycogenolysis whereby a few hundred molecules of cyclic AMP, acting via a kinase cascade, can trigger a very large breakdown of glycogen. In terms of in-sulin action, both phosphorylation (e.g. on ribosomal S6 protein) and dephosphorylation (e.g. pyruvate dehydrogenase activation) reactions are known to be important, thereby pointing to a key role for phosphorylation–dephosphorylation cascade reactions in amplifying the transmembrane signal generated by insulin.

III. Role of G Proteins

The intermediary action of the G or N proteins (so-called because of the key role of guanine nucleotides in the function of these proteins: RODBELL 1980; GILMAN 1987) is of utmost importance in triggering and amplifying a receptor-mediated signal. Those receptors that regulate the production of the second messenger, cyclic AMP, represent the best understood examples of G protein activity. It is now known that the activity of the adenylate cyclase enzyme (C, Fig. 1) is con-trolled not by a direct interaction with the receptor itself, but rather by an in-direct process, whereby the receptor triggers the release of a cryptic cyclase-regulatory polypeptide (α_s) from the oligomeric ($\alpha_s\beta\gamma$) G protein complex. Com-plex equilibria between the various constituents (receptor, G protein and cyclase) are possible (e.g. GILMAN 1987; LEVITZKI 1987; NORTHUP 1985). The β subunit of the G protein complex is in common to a variety of G proteins, whereas the α subunits (and γ subunits) are distinct for each G protein oligomer. In contrast with the G_s oligomer, the G_i oligomer responsible for mediating the inhibition of adenylate cyclase (e.g. via the α_2-adrenergic receptor) liberates an α_i subunit that does not appear (as does α_s) to interact directly with adenylate cyclase. Rather, as outlined elsewhere (NORTHUP 1985), it is the $\beta\gamma_i$ complex, liberated by the inter-action of the receptor (e.g. R_i, Fig. 1) with G_i, that combines with whatever α_s may be free in the membrane environment, so as to attenuate the level of cyclase activity. Much remains to be learned about the function of α_i and about the potential regulation by the $\beta\gamma$ complex of the activities of other α subunits (e.g. those that may regulate phospholipase activity ($G_x \rightarrow \alpha_x + \beta\gamma_x$, Fig. 1). To date, the activity of G proteins has been considered in terms of the regulation of adenylate cyclase, phospholipase and ion channel activity. In principle, additional mem-brane enzymes or other constituents (e.g. M_{NZ}, Fig. 1) may also be targets for the regulatory action of receptor-modulated G proteins. The possible involvement of a G protein in some of the actions of insulin will be discussed below.

IV. Messengers

For some time, cyclic AMP has been thought of as a prototype "second mes-senger" for the actions of agents like epinephrine or glucagon. Now, however, it can be appreciated that the receptor per se may be thought of as one of the mes-sengers and that the so-called α_s subunit of the G_s oligomer can also be considered as a "messenger" in the course of epinephrine (or glucagon) action. Further, as will be outlined below, it is quite likely that a single ligand-receptor interaction may trigger the release of not just one, but perhaps several messengers. Thus, it would appear to be inappropriate to focus on any one particular diffusible low molecular weight primary messenger for the multiple effects of a hormone like insulin. Rather, it may be more fruitful to think in terms of a matrix of diffusible messengers generated simultaneously by the combination of an individual ligand like insulin with its receptor. Substances that have been singled out for particular attention in terms of transmembrane signalling processes are: (1) sodium ion (nicotinic cholinergic receptor); (2) potassium ion (muscarinic cholinergic and somatostatin receptors acting via a G_K protein); (3) chloride ion (γ-aminobutyric acid/benzodiazepine and glycine receptors); (4) cyclic AMP (cyclase-coupled receptors like those for glucagon and ACTH); (5) cyclic GMP (atrial natriuretic factor receptor); (6) hydrogen ion (the triggering of growth factor receptors can increase intracellular pH); (7) calcium ion (for muscarinic cholinergic and α_1-adrenergic receptors); (8) diacylglycerol (many neurotransmitter receptors); along with the concomitantly released messenger, (9) inositol trisphosphate and other inositol polyphosphates (many neurotransmitter and peptide receptors); and (10) the glycan-containing messengers to be discussed in Chap. 14 in connection with insulin action. It is important to point out that in the case of many agents, not one, but three of these "messengers", namely calcium, diacylglycerol and inositol 1,4,5-trisphosphate may be involved in a complex bifurcating pathway triggered by the hydrolysis of membrane phosphoinositides (BERRIGDE 1987). Furthermore, the action of phospholipases to liberate arachidonate (either from membrane phospholipid or diacylglycerol) can result in the formation of prostanoids and/or leukotrienes that in turn can serve in a "cascade" manner as "messengers" triggering distinct receptor systems. Thus, the potential simultaneous release of multiple messengers during the course of the action of a single agonist like insulin renders difficult, if not impossible, the interpretation of overall cell response in terms of the generation of a *single* primordial messenger.

V. Receptor Dynamics and Transmembrane Signalling

1. The Mobile Receptor Paradigm

The concept of a receptor as a "mobile" or "floating" membrane constituent has evolved with the development of understanding of the general properties of cell surface proteins. Along with studies of immunoglobulin receptors (e.g. for IGE, METZGER and ISHIZAKA 1982), studies of the insulin receptor have contributed in a major way to the concept that receptor mobility and cross-linking are key factors in generating a transmembrane signal. The "mobile" or "floating" receptor model, described in more detail elsewhere (CUATRECASAS and HOLLENBERG 1976; HOLLENBERG 1985) permits the receptor to interact with effector moieties within the plane of the membrane. The key tenet of the model lies in the putative ability of the ligand, upon binding to the receptor, to change the ability of the receptor to interact with other membrane components. Thus, the entity LR, resulting from the combination of a ligand like insulin to its receptor

$$L + R \rightarrow LR \rightarrow LR\star$$

results in a conformationally active form of the receptor $R\star$, that can go on to form effector complexes of the kind

$$LR\star + E \rightarrow LR\star E.$$

Wherein E represents an effector molecule involved in the process of cell activation. A number of variations of this model have been developed (e.g. BOEYNAEMS and DUMONT 1977, 1980; DEHAEN 1976; LEVITZKI 1974). For instance, although the above equations illustrate as association model, wherein ligand binding promotes receptor-effector coupling, an alternative possibility is a "dissociation" model, wherein a precoupled inactive effector–receptor complex is dissociated to yield an active effector $E\star$ when the ligand binds to the receptor

$$L + RE \rightarrow LR + E\star.$$

In principle, the mobile receptor model does not restrict the number of distinct effector moieties with which the ligand–receptor complex might interact. This property could readily permit a single ligand-receptor complex to trigger concurrently a variety of transmembrane signals.

2. Receptor Microclustering, Patching, Internalization, and Cell Activation

Observations with a number of ligands, including insulin, epidermal growth factor-urogastrone (EGF–URO), low density lipoprotein (LDL), immunoglobulins and transferrin have revealed that, subsequent to ligand binding, many receptors (or acceptors, like the ones for LDL and transferrin) follow a common sequence of mobile reactions as outlined in Fig. 2 and summarized elsewhere (KING and CUATRECASAS 1981). In the absence of their specific ligands, receptors can be diffusely distributed over the cell surface. However, as illustrated in Fig. 2, at physiological temperatures, the binding of a ligand can lead to a rapid microclustering (receptor microclusters, containing perhaps two to ten receptors) and a reduction in receptor mobility, accompanied by the progressive aggregation of ligand receptor complexes into immobile patches (aggregates containing tens to hundreds of receptors) that can be visualized by fluorescence photomicrography (SCHLESSINGER et al. 1978). In cultured fibroblasts, the microclustering event is thought to precede the formation of patches that can be seen in the fluorescence microscope. Subsequent to the formation of the comparatively large receptor aggregates, the ligand–receptor complexes can be either shed into the medium or taken into the cell (internalized). Receptor internalization (PASTAN and WILLINGHAM 1981) appears to be an ongoing process that is accelerated when a ligand such as insulin binds to its receptor. It is not clear whether or not receptor occupation is a prerequisite for forming small receptor clusters in all cell types. For instance, in adipocytes, there are data to indicate

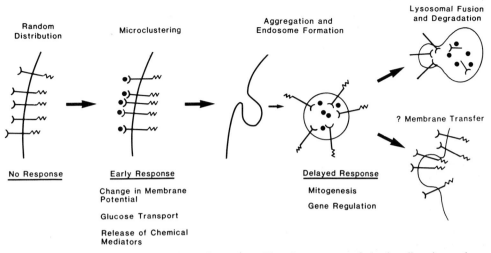

Fig. 2. Receptor dynamics and insulin action. The time course of the insulin-triggered clustering and internalization of the receptor is depicted along with the cellular responses that, as discussed in the text and summarized in Table 1, can take place over quite distinct time frames. The binding of insulin (*full circles*) is shown to activate both the receptor kinase activity (*wavy lines*) and the separate but concomitant clustering/internalization process

that insulin receptors exist as small clusters prior to the addition of insulin (SMITH and JARETT 1987). The mechanism(s) that lead to microclustering, aggregation and internalization of receptors (or acceptors) are poorly understood. In many cells, such as fibroblasts, internalization appears to occur at specific sites on the cell surface – the so-called bristle-coated pit. In some cell types (e.g. adipocytes or hepatocytes) receptors may be localized and internalized at sites other than the coated pit regions. Subsequent to aggregation, the receptor can be internalized via an endocytic process into a cellular compartment that appears to be distinct from the lysosome (Fig. 2). The intracellular receptor-bearing vesicles, which in contrast with lysosomes are not phase-dense in the electron microscope and are acid phosphatase negative, have been termed "endosomes" or "receptosomes". The latter term emphasizes the role of these specialized endocytic vesicles in the process of receptor-mediated endocytosis (PASTAN and WILLINGHAM 1981). One possible fate of such receptor-bearing endosomes is fusion with lysosomes, followed by the lysosomal degradation of the receptor (so-called receptor processing) and of the bound ligand. Several studies suggest that limited receptor processing may also occur at a prelysosomal site (possibly, endosome-associated). An alternative route that the receptosome may follow leads back to the cell surface via a recycling process that reintegrates the receptor into the plasma membrane. A possible fusion of the receptosome with other intracellular organelles (e.g. nuclear envelope) cannot be ruled out, but has yet to be documented. At present, little is known about the factors that control either the internalization process or the trafficking process that may lead on the one hand to lysosomal receptor processing or, on the other hand, to a recycling of the receptor back to the cell surface. The intracellular receptor domains very likely play an important role in this trafficking process. There is also little known about the possible role(s) for the degradation products (ligand or receptor fragments) that may be released into the cytoplasm as a result of the endosomal and lysosomal degradation (processing) events. In view of the peripatetic nature of the hormone–receptor complex, migrating from the cell surface to the cytoplasmic space, a key question to answer is: what role (if any) do these receptor migratory pathways play in the process of transmembrane signalling? The following paragraphs will deal with this question.

It has been appreciated for some time that cell surface antigens can be triggered to patch and cap. Now, it is apparent that the cross-linking of cell surface receptors for immunoglobulins (like IgE; METZGER and ISHIZAKA 1982; KANNER and METZGER 1983), or for polypeptides like insulin, is a key event for cell activation. In terms of polypeptide hormone action, essential observations underlining the importance of receptor microclustering came from work with anti-insulin receptor antibodies obtained either from insulin-resistant patients (KAHN et al. 1978) or from rabbits immunized with purified insulin receptor (JACOBS et al. 1978). In summary, intact antibodies or bivalent (Fab')$_2$ antibody preparations were able to mimic most of the actions of insulin (e.g. stimulation of glucose and amino acid transport) in a variety of target cells (hepatocytes, adipocytes, muscle, fibroblasts) except for the stimulation of DNA synthesis (KAHN et al. 1981). The intact antibodies, like insulin, were also able to cap insulin receptors on intact lymphocytes (SCHLESSINGER et al. 1980). However, monovalent Fab' anti-receptor antibody preparations which could compete effectively for insulin binding were not only unable to mimic insulin action, but were also effective competitive antagonists of insulin in an adipocyte glucose oxidation assay; the biological activity of the monovalent Fab' fragments could be restored by the addition of a second anti-Fab' antibody (KAHN et al. 1981). Results supporting the importance of receptor microaggregation for cell activation have come from work with antibodies directed against LHRH antagonists, wherein the antagonists could be turned into agonists in the presence of cross-linking antibodies (CONN et al. 1982; GREGORY et al. 1982; HOPKINS et al. 1981); the antibody-induced aggregation of the LHRH receptors could be observed at the electron microscopic level (HOPKINS et al. 1981). Although receptor microaggregation is associated with cell activation, as mentioned above, the microclustering of adipocyte insulin receptors can be observed in the absence of insulin (JARETT and SMITH 1977). Thus, receptor clustering may prove to be necessary, but not necessarily sufficient to initiate a cell response for insulin and other agents.

As outlined in Fig. 2, subsequent to receptor microclustering, both the ligand and the receptor can be internalized. An unresolved question is: does the internalized ligand-receptor complex play a role in cell activation? In terms of the rapid actions of hormones like insulin (e.g. stimulation of glucose transport), internalization per se would take place at too slow a rate to play a role in cell triggering. The "downregulation" process might, nonetheless, function in terms of regulating overall cell sensitivity. However, in relation to some of the delayed effects of agents like nerve growth factor (e.g. neurite outgrowth) or insulin (mitogenesis, gene regulation) a role for the internalized receptor–ligand complex has been hypothesized. The detection of the retrograde transport and nuclear binding of nerve growth factor in nerve cells (JOHNSON et al. 1978; YANKER and SHOOTER 1979) and observations of the nuclear binding of EGF–URO (JOHNSON et al. 1980; SAVION et al. 1981) and insulin (GOLDFINE et al. 1977; PODLECKI et al. 1987; SMITH and JARETT 1987) are in keeping with this hypothesis. The demonstrated action of intracellularly administered insulin in frog oocytes also argues strongly in favour of this possibility (MILLER 1988). Overall, the process of cell activation, related to receptor dynamics is pictured according to the outline in Fig. 2. Early responses (changes in membrane potential, metabolite transport) are thought to be triggered in concert with the microclustering of receptors. Delayed responses (mitogenesis, gene regulation) are pictured as possibly involving internalized receptor that is relocated to a specific cellular compartment. Thus, the temporally distinct actions of certain ligands like insulin may relate directly to the topographically distinct dynamic events (microclustering, followed by internalization) that occur over quite different time frames subsequent to ligand binding. In this context, the continued internalization of a receptor may be required to sustain a delayed cellular response to an agent like insulin or nerve growth factor. Thus, transmembrane signalling would have two cytoanatomic tiers and two associated time frames. More data on this issue relating directly to the action of insulin will be presented below.

VI. Feedback Regulation

Irrespective of the pathway that triggers a cell response (either microclustering or internalization) it is now evident that the entire activation process can be subject to feedback regulation. Thus, the same receptor-triggered reactions that initiate a cell response (e.g. activation of a phosphorylation cascade) can feed back on the receptor itself to turn off a receptor-driven process. For instance, phosphorylation of the β-adrenergic receptor, when it is occupied by an agonist, is believed to play a role in the desensitization of cells to adrenergic agents (LEFKOWITZ and CARON 1988); and kinase C-mediated phosphorylation is believed to play a role in receptor internalization and recycling (HUNTER et al. 1984; LIN et al. 1986). Further, β-adrenergic stimulation, presumably via a cyclic AMP-regulated kinase, causes a downregulation of insulin receptors (PESSIN et al. 1983); and elevation of cellular cyclic AMP causes a reduction in insulin receptor kinase activity (STADTMAUER and ROSEN 1986). Thus, in principle, each step of the mechanisms of signal generation and amplification outlined in this chapter, including the processes triggered by the insulin receptor can be a target for feedback regulation.

C. Insulin-Mediated Transmembrane Signalling

From the above discussion it is evident that there is no reason to assume that insulin causes all of its actions via a single mechanism and via a single soluble mediator. Rather, depending on the cell type, it appears likely that insulin may trigger two or possibly more of the amplification mechanisms discussed in

Sect. B.II; and several mediators of the type listed in Sect. B.II may, depending on the cell type and time frame of response, be involved in insulin action. In the sections that follow, I will attempt to single out those mechanisms and mediators which have been implicated in insulin action, so as to provide a specific model for insulin receptor function. For information complementary to the areas dealt with in this chapter, the reader is encouraged to consult two very useful recent reviews (GOLDFINE 1987; ROSEN 1987).

There are a number of factors that must be accounted for in any model of insulin receptor function. First, it is clear that the receptor per se contains all of the information necessary for a cellular response; insulin acts only as an allosteric regulator of the receptor that can otherwise be triggered by a variety of agents (monoclonal and polyclonal antibodies; plant lectins). Second, as discussed elsewhere in this volume (Chaps. 11 and 15), the receptor is a protein tyrosine kinase that can be stimulated by insulin or by anti-receptor antibodies; the autophosphorylation of the receptor appears to increase its activity towards a variety of substrates. Except for insulin-mediated changes in muscle membrane potential, occurring within 1 s (see Chap. 19), an insulin-mediated increase in receptor autophosphorylation (Chap. 11) represents one of the earliest observable cellular responses to insulin, preceding the stimulation of glucose transport (BERNIER et al. 1988). Third, receptor microclustering and mobility appear to be key properties involved in cell activation. In terms of the three cell triggering mechanisms discussed above in Sect. B.I, the one most likely to apply to insulin relates to the regulation by insulin of the receptors' cytoplasmic β subunit tyrosine kinase activity. Nonetheless, as will be discussed below, the participation of a G-related protein has also been implicated in insulin action. Thus, the main question to be considered is: is receptor tyrosine kinase activity obligatorily required for all of insulin's actions? The next section deals with this question.

I. Insulin Receptor Tyrosine Kinase Activity, Insulin Action and Receptor Internalization

Stemming from the discovery that the receptor for EGF–URO is a protein tyrosine kinase (COHEN et al. 1980) it has become apparent that a variety of other receptors, including the one for insulin, possess intrinsic ligand-regulated tyrosine kinase activity. The properties of the insulin receptor kinase activity, observed independently in a number of laboratories (KASUGA et al. 1982; VAN OB-BERGHEN and KOWALSKI 1982; ROTH and CASSELL 1983; ROSEN et al. 1983) are summarized in Chap. 11. Several approaches have been used to determine the relationship between the insulin receptor kinase activity and the ability of insulin to trigger a biological response. The most direct approach has come from site-directed mutagenesis studies, wherein the two groups that successfully cloned the insulin receptor gene have altered either the lysine residue responsible for ATP binding (CHOU et al. 1987; EBINA et al. 1987) or tyrosine residues present in one of the autophosphorylation domains of the receptor (ELLIS et al. 1986; RUTTER et al. 1988). When expressed in Chinese hamster ovary (CHO) cells, the intact human insulin receptor (which can be distinguished immunologically from the CHO cell insulin receptor) causes a leftward shift in the insulin concentration–response curve for a variety of responses, including deoxyglucose uptake, glycogen synthesis, ribosomal protein S6 kinase activity and thymidine incorporation (CHOU et al. 1987; EBINA et al. 1987). In contrast, when the mutant insulin receptors lacking or deficient in tyrosine kinase activity were expressed in CHO cells, there was either no effect on the insulin concentration–response curve (thymidine incorporation) or alternatively a *rightward* shift in the concentration–response curve (deoxyglucose uptake, glycogen synthesis, S6 kinase activity), pointing to a decrease, rather than the increase in cell sensitivity that had been observed upon transfection of the CHO cells with the intact human insulin receptor. The data were taken to indicate that the kinase-deficient mutant receptors were not biologically active in terms of triggering an insulin response and were, if anything, in some way competitive inhibitors of the function of the endogenous CHO insulin receptors present in the CHO cell line.

Unfortunately, all of these data have been obtained in the context of a cell that already possesses functional receptors for both IGF-I and insulin; these receptors alone are sufficient to generate a maximum cellular response, such that all conclusions are based on a shift of the concentration–response curve, rather than an "all-or-none" response in cells that have been transfected with the kinase-deficient receptor.

A second approach evaluating the relationship between receptor kinase activity and insulin action has employed monoclonal antibodies that can recognize specific domains on the β subunit of the insulin receptor and that can inhibit receptor kinase activity. When introduced into *Xenopus* oocytes, CHO cells or rat adipocytes, the kinase-inhibitory domain-specific monoclonals (recognizing the same autophosphorylation region for which Tyr→Phe substitutions were generated to do the receptor transfection experiments) were able to abrogate the action of insulin (oocyte maturation, deoxyglucose uptake). In contrast, β subunit-directed monoclonals that did not affect receptor tyrosine kinase activity, when introduced into the target cells, did not affect insulin action (MORGAN and ROTH 1987; MORGAN et al. 1986). These data are strongly supportive of the requirement of receptor kinase activity for insulin action.

The third approach to the question of the function of receptor kinase activity has employed anti-receptor antibodies (both polyclonal and monoclonal) that recognize extracellular receptor domains and that can cause receptor downregulation either with or without triggering a biological response in target cells. The most intriguing data have come from observations with antibodies which can trigger glucose transport or lipogenesis in adipocytes, but which do not appear to trigger receptor kinase activity either in intact adipocytes (SIMPSON and HEDO 1984) or in partially purified receptor preparations (ZICK et al. 1984; FORSAYETH et al. 1987a, b). Even though one of the polyclonal patient-derived antibodies (B-10) used in some of these studies has been found to trigger insulin receptor phosphorylation in another context (GHERZI et al. 1987), overall, the studies with the biologically active anti-receptor antibodies demonstrate a clear discordance between the ability of antibodies to trigger a biological response and the ability of the antibodies to stimulate receptor kinase activity. To complement the studies with the biologically active anti-receptor antibodies, work has been done with monoclonal anti-receptor antibodies that do not trigger either a biological response or receptor kinase activity, but which do cause receptor downregulation (FORSAYETH et al. 1987a, b). Thus, it would appear that under certain conditions, receptor kinase activity may not be required for stimulating lipogenesis and glucose transport in rat adipocytes or for triggering receptor downregulation in IM-9 lymphocytes.

Taken together, the above data are consistent with the hypothesis that the insulin receptor signal for some of insulin's actions can be triggered by a change in the receptor's conformational state and that this change can be brought about either by a site-specific monoclonal antibody or by receptor autophosphorylation. In this respect, it is of interest that autophosphorylation of the receptor for platelet-derived growth factor can cause a COOH terminal domain conformational change that may be related to the mitogenic response, but not to other "early responses" triggered by the receptor (KEATING et al. 1988; ESCOBEDO and WILLIAMS 1988). From this point of view, the receptor's kinase activity would be directed primarily towards the conformational change caused by receptor autophosphorylation, rather than towards the phosphorylation of other cellular substrates. This situation would be entirely in keeping with the mobile receptor paradigm outlined above, wherein insulin-triggered receptor autophosphorylation (or an equivalent antibody-directed conformational change, in the absence of phosphorylation) would permit the receptor to interact with an increased affinity with other membrane constituents. However, since the biologically active anti-receptor antibodies (especially the monoclonals) did not trigger receptor kinase activity per se (see FORSAYETH et al. 1987a, b), it is quite possible that some (but perhaps not all) of the responses to insulin do not involve a receptor-mediated phosphorylation of substrates other than the receptor β subunit.

II. Receptor-Triggered Phosphorylation and Signalling

Despite the possibility outlined above, that some of the actions of insulin may not require the receptor-mediated phosphorylation of cellular substrates, it is amply clear, as outlined in Chaps. 15–17, that phosphorylation–dephosphorylation reactions play a very important role in many of the metabolic processes triggered by insulin (DENTON et al. 1981; LARNER et al. 1982; KREBS and BEAVO 1979; AVRUCH et al. 1982). For instance, on the one hand, a dephosphorylation event can activate an enzyme like pyruvate dehydrogenase, whereas on the other hand the activation of a serine/threonine kinase, subsequent to insulin receptor triggering, can modulate the phosphorylation state (and presumably function) of ribosomal protein S6. Additionally, in view of the tyrosine kinase activity of the receptor, there has been a concerted search for cellular receptor substrates that can become phosphorylated on tyrosine residues in the course of insulin action. Such substrates might play an initiator role in a phosphorylation–dephosphorylation cascade triggered by insulin. Substrates phosphorylated on tyrosine in response to insulin, either in intact cells or in isolated membrane preparations, have been identified in a variety of cell types (rat adipocytes, mouse 3T3-L1 adipocytes, rat liver membranes, hepatoma cells). These substrates have apparent molecular weights ranging from 15000 (BERNIER et al. 1987, 1988) to 185000 (WHITE et al. 1985, 1987). Two key features pointing to the importance of these substrates are:

(a) the rapidity with which the substrates become phosphorylated, upon the addition of insulin (seconds to minutes); and (b) the dependence on the concentration of insulin, which at nanomolar or subnanomolar concentrations can trigger phosphorylation in a reversible manner. Of particular importance with respect to the relationship of these substrates to insulin action is the ability to monitor the degree of substrate phosphorylation within the context of a time course of a biological response triggered by insulin. For instance, the increase in insulin-stimulated phosphorylation of the 15 kdalton cytosolic protein detected at 2–3 min in 3T3-L1 adipocytes (the phosphoprotein accumulates only in the presence of phenylarsine oxide) can be seen to occur after receptor autophosphorylation (maximal at 1 min), but at a time earlier than the increase (at 4–6 min) in 2-deoxyglucose uptake (BERNIER et al. 1987, 1988). Similarly, insulin-stimulated tyrosine phosphorylation of the 185 kdalton substrate identified in hepatoma cells (within 30–60 s; WHITE et al. 1985, 1987), the 46 kdalton substrate in intact rat adipocytes (30–150 s; HARING et al. 1987) and a 160 kdalton cytosolic protein in 3T3-L1 cells (maximal at 60 s; MADOFF et al. 1988) occurs at a rate that preceeds many if not most of the actions of insulin in these target cells. Perhaps for technical reasons, the change in muscle cell polarization (Chap. 19) is the only insulin-mediated response that can be observed to occur more rapidly (maximal at 0.5–1 s) than it has been possible to monitor receptor and/or substrate tyrosine phosphorylation. Other proteins phosphorylated relatively rapidly on tyrosine in response to insulin include a 120 kdalton glycoprotein detected in rat liver membranes and hepatoma cells (PHILLIPS et al. 1987; PERROTTI et al. 1987) and a number of proteins (116, 62 kdalton and three proteins in the range 45–50 kdalton) in rat adipocytes (HARING et al. 1987). It is also of considerable interest, in terms of the signal amplification mechanisms discussed in Sect. B.II, that insulin can rapidly trigger the phosphorylation of calmodulin (17 kdalton; COLCA et al. 1987) as well as a 62 kdalton protein immunologically related to the R_{II} regulatory subunit of cyclic AMP-dependent protein kinase (LAWRENCE et al. 1986). The essence of the above discussion is that upon binding to its receptor, insulin can very rapidly and concurrently stimulate tyrosine phosphorylation in a variety of cellular proteins. Whether or not these phosphorylations are all catalysed by the receptor per se, it is quite likely that many of these interesting phosphoproteins will be found to play a role in the generation and/or amplification of the transmembrane signals triggered by the insulin receptor. The rapid phosphorylation on tyrosine of so many different proteins would suggest that many of these transmembrane phosphoprotein signals are generated early on in the action of insulin.

III. G Proteins and Insulin Action

Although many of the data summarized above highlight the importance of protein kinase reactions in generating an insulin-mediated transmembrane signal, there is information to indicate that novel G proteins may also play a role in some of the actions of insulin. Early on, it was appreciated that insulin could abrogate the rise in cyclic AMP caused by other agonists (summarized by HOL-LENBERG and CUATRECASAS 1975), thereby implicating an influence of insulin

either on adenylate cyclase, via a G protein, or on cyclic AMP phosphodiesterase. More recently, a protein possibly related to a G protein α subunit has
been identified in connection with the ability of insulin to prevent the cholera
toxin-catalysed ribosylation of a 25 kdalton protein in rat liver membranes
(HEYWORTH et al. 1985). Other evidence implicating a G protein in the action of
insulin relates to the effect of pertussis toxin which can block insulin's ability to
inhibit adenylate cyclase and to activate a "dense vesicle" cyclic AMP
phosphodiesterase in both liver and adipocytes (HEYWORTH et al. 1983, 1986).
Purified insulin receptor has also been shown to phosphorylate the G_i and G_0
oligomeric G proteins isolated from bovine brain (O'BRIEN et al. 1987). In other
studies, it has been shown that in intact rat adipocytes, pertussis toxin, known to
ADP-ribosylate and inactivate G_0, G_i and possibly other G proteins, can
abrogate the antilipolytic activity of insulin, but cannot abolish the ability of insulin to stimulate glucose oxidation (GOREN et al. 1985). Quite apart from the
studies in mammalian cells, it has been shown in frog oocytes that intracellular
injection of an antibody directed against the GTP-binding protein associated
with cell transformation, *ras* p21, can neutralize insulin-mediated germinal
vesicle breakdown (ROTH 1988); this study showed that purified *ras* p21 can also
serve as a substrate (albeit a poor one) of the insulin receptor kinase. Altogether,
the above results indicate that the interaction of the receptor with a G-related
protein could well be involved in some of the actions of insulin. Given the broad
impact that a G protein mechanism could have in amplifying a cellular signal, as
outlined above in Sect. B.III, this potential interaction of the receptor would
provide an added dimension to the multiple pathways that may be involved in
amplifying the initial insulin receptor signal.

IV. Interactions with Other Membrane Proteins

In the context of the mobile receptor paradigm outlined above, it is expected that
the insulin receptor will interact with a variety of membrane-associated proteins
during the course of insulin action. Although primary attention has been focused
on phosphorylated proteins that can serve as targets for the receptor kinase,
other approaches have been used to obtain evidence for the interaction of the
receptor with other membrane-associated proteins. In our work, it was observed
that the receptor binding properties and the receptor's hydrodynamic properties
could be altered by a membrane-localized nonreceptor glycoprotein fraction
(MATURO and HOLLENBERG 1978). Using radiation inactivation analysis,
HARMON and colleagues (1980, 1981, 1983) also found evidence for a membrane-
associated constituent with a molecular weight of about 300 000 that affected the
affinity of the receptor for insulin. By an independent approach using chemical
cross-linking or photoaffinity probes, receptor-associated proteins have been
identified having molecular weights in the 40–45 000 range (YIP et al. 1982;
GOREN et al. 1983; BROSSETTE et al. 1984; PHILLIPS et al. 1986), and in the region
of 65–85 000 (HEIDENREICH et al. 1983; YIP and MOULE 1983; BARON and
SONKSEN 1983). These proteins, while distinct from the insulin receptor α and β
subunits, may be able to form covalent attachments to the receptor via a disulphide–sulphydryl exchange reaction that may be triggered by the binding of

insulin (MATURO et al. 1983; HELMERHORST et al. 1986). Although the exact nature of the 40–45 kdalton receptor-associated proteins has not yet been clearly established, they are similar in size to the heavy chain of the class I major histocompatibility complex (MHC). Early work by EDIDIN and co-workers (LAFUSE and EDIDIN 1980; MERUELO and EDIDIN 1979) had pointed to a relationship between MHC antigens and cell surface receptors for insulin and EGF–URO; subsequent studies have shown that the photoaffinity-labeled insulin receptor can be coprecipitated by antibodies directed against class I, but not class II histocompatibility antigens (CHVATCHKO et al. 1983; FEHLMANN et al. 1985; DUE et al. 1986; PHILLIPS et al. 1986). Thus, the transmembrane signal generated by the insulin receptor may involve, *pari passu*, a specific interaction with class I MHC antigens. Presumably, the complex processes of insulin receptor clustering and internalization, represented schematically in Fig. 2, also involve the interaction of the receptor with a variety of membrane proteins. Mechanistically, receptor sulphydryls (AGLIO et al. 1985; HELMERHORST et al. 1986) and receptor sialic acid residues (CUATRECASAS and ILLIANO 1971; HAYES and LOCKWOOD 1986) may turn out to be of primary importance for the receptor–protein interactions that generate an overall transmembrane signal.

V. Low Molecular Weight Mediators of Insulin Action

As outlined above (Sect. B.IV) a variety of diffusible low molecular weight messengers, ranging from monovalent and divalent cations (Na^+, K^+, Ca^{2+}) to cyclic nucleotides (cAMP, cGMP) are thought to play a role in amplifying a receptor-triggered transmembrane signal. Chapter 19 points out the possible role of changes in cation flux and concomitant changes in membrane potential in generating an insulin signal. Some time ago, attention was also focused on the ability of insulin to modulate cellular levels of cyclic AMP and cyclic GMP (HOLLENBERG and CUATRECASAS 1975). Thus, in some respects, alterations in the cellular content of these two "second messengers" could play some overall role in insulin action. In a similar vein some data point to the involvement of either phosphoinositides or calcium in the action of insulin in selected cell types. For instance, in adipocytes, insulin can increase the incorporation of inositol into phosphatidylinositol and phosphatidylinositol 4,5-bisphosphate so as to enhance the ability of agonists like vasopressin and phenylephrine to stimulate phosphoinositide breakdown (PENNINGTON and MARTIN 1985); an insulin-mediated increase in phosphatidylinositol bisphosphate has also been observed in adipose tissue and cultured BCH3-I monocytes (FARESE et al. 1984, 1985). Nonetheless, although insulin can trigger and/or mimic phospholipase C activity in H-35 hepatoma cells (MATO et al. 1987), bovine liver membranes (SALTIEL and CUATRECASAS 1986) adipose tissue and BCH3-I myocytes (FARESE et al. 1985, 1986; SALTIEL et al. 1986), there does not appear to be a concomitant breakdown of phosphoinositides triggered in target cells by insulin (PENNINGTON and MARTIN 1985; TAYLOR et al. 1985). Thus, although insulin may play a synergistic role with respect to the actions of agents like vasopressin which stimulate phosphoinositide breakdown, the phosphoinositides per se would not appear to play a direct role in acting as messengers of insulin action. In this regard it is difficult to foresee a physiological role for the ability of the insulin receptor kinase in vitro to phosphorylate phosphatidylinositol (MACHICAO and WILLAND 1984). Nonetheless, the phospholipase C activity now associated with insulin action (see below) that may liberate compounds other than phosphoinositides, could also result in the formation of diacylglycerol. Thus, some participation of kinase C in modulating the cellular signal triggered by the insulin receptor can be hypothesized.

Although the ability of insulin to stimulate glucose oxidation and inhibit lipolysis is calcium-independent, insulin effects on protein synthesis, 2-deoxyglucose transport and lipoprotein lipase appear to require extracellular calcium (BONNE et al. 1978; FAIN 1980). Other data also implicate calcium as an important factor for certain of the actions of insulin (summarized by MCDONALD and PERSHADSINGH 1985). The main evidence pointing to a role for calcium as a messenger for insulin action comprises an insulin-mediated increase in adipocyte membrane calcium binding and an inhibition by insulin of adipocyte membrane Ca^{2+},Mg^{2+}-ATPase; the interaction of the insulin receptor with the calcium-binding cell regulator, calmodulin, is of considerable interest in this regard (COLCA et al. 1987; GRAVES et al. 1985). The ability of insulin to regulate Ca^{2+} channels in rat myoballs has also recently been documented (ZIERLER and WU 1988). Nonetheless, overall, extracellular/intracellular calcium levels would appear to play a permissive role for some actions on insulin rather than acting as a primary specific insulin receptor messenger.

By far the most interesting putative mediators of insulin action, as dealt with in detail in Chap. 14, are those low molecular weight compounds extracted from skeletal muscle (LARNER et al. 1979; JARRET and SEALS 1979) and from a variety of other sources, including liver membranes, hepatoma cells, IM-9 lymphocytes and BCH3-I cultured myocytes (see Chap. 14) and reviews by KIECHLE and JARETT 1985; GOTTSCHALK and JARETT 1985). A key feature of the mediator substances is their ability to mimic the action of insulin in a variety of cellular and subcellular assay systems (GOTTSCHALK and JARETT 1985). Most important is the evidence pointing to the existence of a family of multiple mediators, no single one of which may explain all of the actions of insulin (GOTTSCHALK and JARETT 1985; GOTTSCHALK et al. 1986). Intriguingly, the glycan-containing mediators (see Chap. 14) described by SALTIEL and co-workers and by others (SALTIEL et al. 1986, 1987; MATO et al. 1987) would appear to result from the activation of a phospholipase C that simultaneously releases a myristyl-containing diacylglycerol (ROMERO et al. 1988). As mentioned above, in view of these findings, and given the overall signalling mechanisms discussed in Sect. B, a potential role for a G_x-regulated phospholipase and for the participation of diacylglycerol-regulated kinase C cannot be ruled out in terms of the insulin-mediated transmembrane signalling process. In this context, kinase C and the glycan-containing mediators could function as amplifiers of the insulin-triggered signal as does cyclic AMP for the signal triggered by epinephrine. It may be of particular significance that the mediators so far described can mimic many of the actions of insulin, like antilipolysis, in intact adipocytes, but fail to stimulate glucose transport (KELLY et al. 1987). The mechanisms whereby insulin regulates glucose transport may thus turn out to be quite different from the mechanisms whereby insulin activates other metabolic processes.

VI. Receptor Internalization and Insulin Action

The ability of receptors to cluster and internalize during the course of cell activation, discussed in general in Sect. B.V.2, points to a possible role for internalized receptor in terms of the delayed actions of insulin. As summarized in more detail elsewhere (GOLDFINE 1987), the ability of insulin to trigger receptor internalization via at least two distinct routes (SMITH and JARETT 1985) and to bind to intracellular structures, including isolated nuclei, has been well documented. Further, upon internalization in 3T3 fibroblasts, insulin (and presumably its receptor) can be localized near the nucleus (SMITH and JARETT 1987; PODLECKI et al. 1987). Thus, there are substantial data that are consistent with the model outlined in Fig. 2, wherein the internalized receptor is viewed as migrating to different intracellular sites. What is more difficult to establish is whether or not the receptor, situated at an intracellular site, can generate a cellular signal essential for one of the delayed actions of insulin (e.g. stimulation of RNA synthesis). In support of a role for internalized receptor are data demonstrating effects of either insulin or anti-insulin receptor antibodies on the functions of isolated nuclei or nuclear membranes that might be expected to be involved in RNA processing or transport (GOLDFINE et al. 1985). The intracellular action of insulin in frog oocytes, mentioned above, also points to the existence of active intracellular receptors (MILLER 1988). On the other hand, receptor internalization per se cannot be viewed as sufficient for a delayed insulin action, since at least one of the biologically inactive monoclonal anti-receptor antibodies discussed above (MA-10, FORSAYETH et al. 1987a) has been observed to trigger receptor downregulation in IM-9 lymphocytes (FORSAYETH et al. 1987b). Thus, the receptor signal that triggers internalization (antibody MA-10 does not stimulate receptor autophosphorylation) would appear to be distinct from the receptor-triggered signal that causes cell activation. Nonetheless, a recent study of the tyrosine kinase properties of the internalized receptor, that can be recovered in endosomal vesicles from cell homogenates subsequent to insulin binding (KHAN et al. 1986) has revealed that there is an increase in the endosomal receptor enzymatic activity (compared with activity associated with the plasma membrane) that would appear to occur during the process of receptor internalization (POSNER et al. 1988). Thus, although internalization, on its own, would not appear to activate the receptor, the process of internalization that occurs once the receptor is activated by insulin could enhance the receptor's ability to phosphorylate intracellular substrates. The experiments done in vivo, wherein hepatic endosomal fractions were recovered after the administration of insulin, indicated a very rapid entry of activated receptor kinase (peak at about 5–6 min) into the cytoplasmic compartment (POSNER et al. 1988); at earlier time points after insulin administration (maximum activity at 30 s) receptor with a lower degree of activation (with respect to both autophosphorylation and substrate phosphorylation) was recovered in the plasma membrane fraction (POSNER et al. 1988). Taken together, the data summarized above point to an early activation of the receptor at the plasma membrane, followed by a concentration of a highly activated form of the receptor kinase in the endosomal compartment. The data are consistent with the model outlined in Fig. 2, wherein a role for the endosomal receptor kinase is suggested in terms of the delayed biological actions of insulin. Direct proof of this model may require the synthesis and testing of a receptor that lacks those domains required for internalization, but that retains its insulin-regulated tyrosine kinase activity. One would predict that such a receptor would trigger only the rapid actions of insulin (e.g. glucose transport) in a transfected test cell system.

VII. Feedback Regulation and Insulin Action

Although it is well recognized that feedback regulation is built into the many metabolic pathways that insulin regulates, only recently is it becoming apparent that feedback regulation may also take place at the level of the receptor-triggered signalling reactions. For instance, the phosphorylation state of the 15 kdalton polypeptide, thought to participate in the signalling reaction that modulates glucose transport (BERNIER et al. 1987, 1988) turns over so rapidly that it can only be trapped in its phosphorylated state in the presence of phenylarsine oxide. One also expects a rapid turnover and regulation of the putative glycan-containing mediators of insulin action. Similarly, as discussed above, just as there is evidence that the activity of the β-adrenergic receptor may be controlled by agonist-facilitated phosphorylation, evidence is now accumulating that the insulin receptor as well can be a target for a variety of protein kinases, including kinase C (JACOBS et al. 1983) and cyclic AMP-dependent kinase (STADTMAUER and ROSEN 1986). Although data indicate that the receptor kinase activity can be reversibly stimulated and dactivated by tyrosine autophosphorylation and dephosphorylation and that the activity can be diminished by subsequent phosphorylation of the active receptor with other protein kinases (e.g. STADTMAUER and ROSEN 1986), the exact sites at which receptor phosphorylation regulates the receptor kinase activity have been difficult to determine (TORNQVIST et al. 1988; KAHN et al. 1988; see also Chaps. 11 and 15). Thus, overall, much remains to be learned about the feedback regulation mechanisms that have as their target the transmembrane signalling reactions set in motion by insulin. This area would appear to be a fruitful one for further study.

D. Summary and Conclusions

The working hypothesis for the process whereby insulin, via its receptor, generates a transmembrane signal is outlined in Fig. 2. Central to the generation of the signal is the autophosphorylation/tyrosine kinase activity of the receptor, triggered by insulin in step with changes in receptor mobility, microclustering and aggregation. It appears quite likely that the differing time courses over which insulin causes its effects can be rationalized by the time courses of microclustering and the subsequent formation of active endosomes. Given the many cellular processes that insulin regulates, it is perhaps no surprise that instead of a single reaction pathway, a diverse subset of the signalling reactions summarized in Sect. B, as well as a family of mediator molecules, are probably involved in insulin action. This situation is entirely in keeping with the mobile receptor paradigm outlined above in Sect. B.V.1. These signalling reactions, ranging from substrate phosphorylation to the possible participation of G proteins and phospholipase-generated soluble mediators need not all be triggered simultaneously in an individual cell type in response to insulin. Rather, it is more likely that the wide repertoire available to the insulin receptor trigger will be restricted by the differentiation state of the specific target cell, which may possess only a small subset of the possible signalling reactions that the receptor may set in motion. Thus, the challenge now is not to find *the* single mediator or reaction mechanism responsible for generating a transmembrane insulin signal, but rather to identify the matrix of reactions that, in a given tissue type under specified physiological conditions (e.g. state of differentiation or state of cell growth), cause the cell to respond in a meaningful way to insulin.

Acknowledgments. Work in the author's laboratory, relating directly to many aspects of this chapter, has been made possible by the long-term support of the Medical Research Council of Canada. The help of Dr. John R. McDonald with part of the literature survey is gratefully acknowledged. Many thanks are also due to Ms. Judy Gayford for help with the preparation of this manuscript.

References

Aglio LS, Maturo JM III, Hollenberg MD (1985) Receptors for insulin and epidermal growth factor: interaction with organomercurial agarose. J Cell Biochem 28:143–157

Avruch J, Alexander MC, Plamer JL, Pierce MW, Nemenoff RA, Blackshear PJ, Tipper JP, Witters LA (1982) Role of insulin-stimulated protein phosphorylation in insulin action. Fed Proc 41:2629–2633

Baron MD, Sonksen PH (1983) Elucidation of the quaternary structure of the insulin receptor. Biochem J 212:79–84

Bernier M, Laird DM, Lane MD (1987) Insulin-activated tyrosine phosphorylation of a 15-kilodalton protein in intact 3T3-L1 adipocytes. Proc Natl Acad Sci USA 84:1844–1848

Bernier M, Laird DM, Lane MD (1988) Identification and role of a 15-kilodalton cellular target of the insulin receptor tyrosine kinase. In: Goren HJ, Hollenberg MD, Roncari DAK (eds) Insulin action and diabetes. Raven, New York, pp 117–128

Berridge MJ (1987) Inositol trisphosphate and diacylglycerol: two interacting second messengers. Ann Rev Biochem 56:159–193

Boeynaems JM, Dumont JE (1977) The two-step model of ligand-receptor interaction. Mol Cell Endocrinol 7:33–47

Boeynaems JM, Dumont JE (1980) Outlines of receptor theory. Elsevier/North-Holland Biomedical, Amsterdam New York Oxford

Bonne D, Belhadj O, Cohen P (1978) Calcium as modulator of the hormonal-receptors-biological response coupling system. Effects of Ca^{2+} ions on the insulin activated 2-deoxyglucose transport in rat fat cells. Eur J Biochem 87:261–266

Brossette N, van Obberghen E, Fehlmann M (1984) Interaction between insulin receptors and major histocompatibility complex antigens in mouse liver membranes. Diabetologia 27:74–76

Chou CK, Dull TJ, Russell DS, Gherzi R, Lebwohl D, Ullrich A, Rosen OM (1987) Human insulin receptors mutated at the ATP-binding site lack protein tyrosine kinase activity and fail to mediate postreceptor effects of insulin. J Biol Chem 262:1842–1847

Chvatchko Y, van Obberghen E, Kiger N, Fehlmann M (1983) Immunoprecipitation of insulin receptors by antibodies against class 1 antigens of the murine H-2 major histocompatibility complex. FEBS Lett 163:207–211

Cohen S, Carpenter G, King L Jr (1980) Epidermal growth factor-receptor-protein kinase interactions: copurification of receptor and epidermal growth factor-enhanced phosphorylation activity. J Biol Chem 255:4834–4842

Colca JR, DeWald DB, Pearson JD, Palazuk BJ, Laurino JP, McDonald JM (1987) Insulin stimulates the phosphorylation of calmodulin in intact adipocytes. J Biol Chem 262:11399–11402

Conn PM, Rogers DC, Stewart JM, Neidel J, Sheffield T (1982) Conversion of a gonadotropin-releasing hormone antagonist to an agonist. Nature 296:653–655

Cuatrecasas P, Hollenberg MD (1976) Membrane receptors and hormone action. Adv Protein Chem 30:251–451

Cuatrecasas P, Illiano G (1971) Membrane sialic acid and the mechanism of insulin action in adipose tissue cells. J Biol Chem 246:4938–4946

De Haen C (1976) The non-stoichiometric floating receptor model for hormone-sensitive adenylate cyclase. J Theor Biol 58:383–400

Denton RM, Brownsey RW, Belsham GJ (1981) A partial view of the mechanism of insulin action. Diabetologia 21:347–362

Due C, Simonsen M, Olsson L (1986) The major histocompatibility complex class I heavy chain as a structural subunit of the human cell membrane insulin receptor: implications for the range of biological functions of histocompatibility antigens. Proc Natl Acad Sci USA 83:6007–6011

Ebina Y, Araki E, Taira M, Shimida F, Mori M, Craik CS, Siddle K, Pierce SB, Roth RA, Rutter WJ (1987) Replacement of lysine residue 1030 in the putative ATP-binding region of the insulin receptor abolishes insulin- and antibody-stimulated glucose uptake and receptor kinase activity. Proc Natl Acad Sci USA 84:704–708

Ellis L, Clauser E, Morgan DO, Edery M, Roth RA, Rutter WJ (1986) Replacement of insulin receptor tyrosine residues 1162 and 1163 compromises insulin-stimulated kinase activity and uptake of 2-deoxyglucose. Cell 45:721–732

Escobedo JA, Williams LT (1988) Intracellular structural domains of the platelet-derived growth factor receptor have distinct functions. Clin Res 36:540A

Fain JN (1980) Hormonal regulation of lipid mobilization from adipose tissue. In: Litwack G (ed) Biochemical actions of hormones, vol VII. Academic, New York, pp 119–204

Farese RV, Farese RV Jr, Sabir MA, Larson RE, Trudeau WL III (1984) The mechanism of action of insulin on phospholipid metabolism in rat adipose tissue. Requirement for protein synthesis and a carbohydrate source, and relationship to activation of pyruvate dehydrogenase. Diabetes 33:648–655

Farese RV, Davis JS, Barnes DE, Standaert ML, Babischkin JS, Hock R, Rosic NK, Pollet RJ (1985) The de novo phospholipid effect of insulin is associated with increases in diacylglycerol, but not inositol phosphates or cytosolic Ca^{2+}. Biochem J 231:269–278

Farese RV, Kuo J-Y, Babischkin JS, Davis JS (1986) Insulin provokes a transient activation of phospholipase C in the rat epididymal fat pad. J Biol Chem 261:8589–8592

Fehlmann M, Peyron J-F, Samson M, van Obberghen E, Brandenburg D, Brossette N (1985) Molecular association between major histocompatibility complex class I antigens and insulin receptors in mouse liver membranes. Proc Natl Acad Sci USA 82:8634–8637

Forsayeth JR, Caro JF, Sinha MK, Maddux BA, Goldfine ID (1987a) Monoclonal antibodies to the human insulin receptor that activate glucose transport but not insulin receptor kinase activity. Proc Natl Acad Sci USA 84:3448–3451

Forsayeth JR, Montemurro A, Maddux BA, DePirro R, Goldfine ID (1987b) Effect of monocloncal antibodies on human insulin receptor autophosphorylation, negative cooperativity, and down regulation. J Biol Chem 262:4134–4140

Gherzi R, Russell DS, Taylor SI, Rosen OM (1987) Reevaluation of the evidence that an antibody to the insulin receptor is insulinmimetic without activating the protein tyrosine kinase activity of the receptor. J Biol Chem 262:16900–16905

Gilman AG (1987) G proteins: transducers of receptor-generated signals. Ann Rev Biochem 56:615–649

Goldfine ID (1987) The insulin receptor: molecular biology and transmembrane signaling. Endocr Rev 8:235–255

Goldfine ID, Vigneri R, Cohen D, Pliam NB, Kahn CR (1977) Intracellular binding sites for insulin are immunologically distinct from those on the plasma membrane. Nature 269:698–700

Goldfine ID, Purrello F, Vigneri R, Clawson GA (1985) Insulin and the regulation of isolated nuclei and nuclear subfractions: potential relationship of mRNA metabolism. In: DeFronzo R (ed) Diabetes metabolism reviews, vol 1. Wiley, New York, pp 119–137

Goren HJ, Elliott C, Dudley RA (1983) Adipocyte insulin-binding species: the size and subunit composition of the larger binding species. J Cell Biochem 21:161–177

Goren HJ, Northup JK, Hollenberg MD (1985) Action of insulin modulated by pertussis toxin in rat adipocytes. Can J Physiol Pharmacol 63:1017–1022

Gottschalk WK, Jarett L (1985) Intracellular mediators of insulin action. Diabetes Metab Rev 1:229–259

Gottschalk WK, Macaulay SL, Macaulay JO, Kelly K, Smith JA, Jarett L (1986) Characterization of mediators of insulin action. Ann NY Acad Sci 488:385–405

Graves CB, Goewert RR, McDonald JM (1985) The insulin receptor contains a calmodulin-binding domain. Science 230:827–829

Gregory H, Taylor CL, Hopkins CR (1982) Luteinizing hormone release from dissociated pituitary cells by dimerization of occupied LHRH receptors. Nature 300:269–271

Haring HU, White MF, Machicao F, Ermel B, Schleicher E, Obermaier B (1987) Insulin rapidly stimulates phosphorylation of a 46-kDa membrane protein on tyrosine residues as well as phosphorylation of several soluble proteins in intact fat cells. Proc Natl Acad Sci USA 84:113–117

Harmon JT, Kahn CR, Kempner ES, Schlegel W (1980) Characterization of the insulin receptor in its membrane environment by radiation inactivation. J Biol Chem 255:3412–3419

Harmon JT, Kempner ES, Kahn CR (1981) Demonstration by radiation inactivation that insulin alters the structure of the insulin receptor in rat liver membranes. J Biol Chem 256:7719–7722

Harmon JT, Hedo JA, Kahn CR (1983) Characterization of a membrane regulator of insulin receptor affinity. J Biol Chem 258:6875–6881

Hayes GR, Lockwood DH (1986) The role of cell surface sialic acid in insulin receptor function and insulin action. J Biol Chem 261:2791–2798

Heidenreich KA, Zahniser NR, Berhanu P, Brandenburg D, Olefsky JM (1983) Structural differences between insulin receptors in the brain and peripheral target tissues. J Biol Chem 258:8527–8530

Helmerhorst E, Ng DS, Moule ML, Yip CC (1986) High molecular weight forms of the insulin receptor. Biochemistry 25:2060–2065

Heyworth CM, Wallace AV, Houslay MD (1983) Insulin and glucagon regulate the activation of two distinct membrane-bound cyclic AMP phosphodiesterases in hepatocytes. Biochem J 214:99–110

Heyworth CM, Whetton AD, Wong S, Martin BR, Houslay MD (1985) Insulin inhibits the cholera-toxin-catalysed ribosylation of a M_r-25000 protein in rat liver plasma membranes. Biochem J 228:593–603

Heyworth CM, Grey A-M, Wilson SR, Hanski E, Houslay MD (1986) The action of islet activating protein (pertussis toxin) on insulin's ability to inhibit adenylate cyclase and activate cyclic AMP phosphodiesterases in hepatocytes. Biochem J 235:145–149

Hollenberg MD (1985) Receptor models and the action of neurotransmitters and hormones: some new perspectives. In: Yamamura HI et al. (ed) Neurotransmitter receptor binding, 2nd edn. Raven, New York, pp 1–39

Hollenberg MD, Cuatrecasas P (1975) Insulin: interaction with membrane receptors and relationship to cyclic purine nucleotides and cell growth. Fed Proc 34:1556–1563

Hopkins CR, Semoff S, Gregory H (1981) Regulation of gonadotropin secretion of the anterior pituitary. Philos Trans R Soc Lond [Biol] 296:73–81

Hunter T, Ling N, Cooper JA (1984) Protein kinase C phosphorylation of the GF receptor at a threonine residue close to the cytoplasmic face of the plasma membrane. Nature 311:480–483

Jacobs S, Chang K-J, Cuatrecasas P (1978) Antibodies to purified insulin receptor have insulin-like activity. Science 200:1283–1284

Jacobs S, Sahyoun N, Saltiel A, Cuatrecasas P (1983) Phorbol esters stimulate the phosphorylation of receptors for insulin and somatomedin-C. Proc Natl Acad Sci USA 80:6211–6213

Jarett L, Seals JR (1979) Pyruvate dehydrogenase activation in adipocyte mitochondria by an insulin-generated mediator from muscle. Science 206:1407–1408

Jarett L, Smith RM (1977) The natural occurrence of insulin receptors in groups on adipocyte plasma membranes as demonstrated with monomeric ferritin-insulin. J Supramol Struct 6:45–59

Johnson EM Jr, Andres RY, Bradshaw RA (1978) Characterization of the retrograde transport of nerve growth factor (NGF) using high specific activity [125I]NGF. Brain Res 150:319–331

Johnson LK, Vlodavsky I, Baxter JD, Gospodarowicz D (1980) Nuclear accumulation of epidermal growth factor in cultured rat pituitary cells. Nature 287:340–343

Kahn CR, Baird KL, Jarrett DB, Flier JS (1978) Direct demonstration that receptor crosslinking or aggregation is important in insulin action. Proc Natl Acad Sci USA 75:4209–4213

Kahn CR, Baird KL, Flier JS, Grunfeld C, Harmon JT, Harrison LC, Karlsson FA, Kasuga M, King GL, Lang UC, Podskalny JM, van Obberghen E (1981) Insulin receptors, receptor antibodies, and the mechanism of insulin action. Recent Prog Horm Res 37:477–538

Kahn CR, Sethu S, Reddy K, Shoelson S, Goren HJ, White MF (1988) Regulation of the insulin receptor by multi-site phosphorylation. In: Goren HJ, Hollenberg MD, Roncari DAK (eds) Insulin action and diabetes. Raven, New York (in press)

Kanner BI, Metzger H (1983) Crosslinking of the receptors for immunoglobulin E depolarizes the plasma membrane of rat basophilic leukemia cells. Proc Natl Acad Sci USA 80:5744–5748

Kasuga M, Zick Y, Blithe DL, Crettaz M, Kahn CR (1982) Insulin stimulates tyrosine phosphorylation of the insulin receptor in a cell-free system. Nature 298:667–669

Keating MT, Escobedo JA, Fantl WJ (1988) Ligand activation causes a phosphorylation-dependent change in platelet-derived growth factor receptor conformation. Clin Res 36:603A

Kelly KL, Mato JM, Merida I, Jarett L (1987) Glucose transport and antilipolysis are differentially regulated by the polar head group of an insulin-sensitive glycophospholipid. Proc Natl Acad Sci USA 84:6404–6407

Khan MN, Savoie S, Bergeron JJM, Posner BI (1986) Characterization of rat liver endosomal fractions: in vivo activation of insulin-stimulable receptor kinase in these structures. J Biol Chem 261:8462–8472

Kiechle F, Jarett L (1985) The molecular basis of insulin action: membrane-associated reactions and intracellular mediators. In: Hollenberg MD (ed) Insulin its receptor and diabetes. Dekker, New York, pp 181–204

King AC, Cuatrecasas P (1981) Peptide hormone-induced receptor mobility, aggregation, and internalization. New Engl J Med 305:77–88

Krebs EG, Beavo JA (1979) Phosphorylation-dephosphorylation of enzymes. Annu Rev Biochem 48:923–959

Lafuse W, Edidin M (1980) Influence of the mouse major histocompatibility complex, H-2, on liver adenylate cyclase activity and on glucagon binding to liver cell membranes. Biochemistry 19:49–54

Larner J, Galasko G, Chenk K, DePaoli-Roach AA, Huang L, Daggy P, Kellogg J (1979) Generation by insulin of a chemical mediator that controls protein phosphorylation and dephosphorylation. Science 206:1408–1410

Larner J, Cheng K, Schwartz C, Kikuchi K, Tamura S, Creacy S, Dubler R, Galasko G, Pullin C, Katz M (1982) Insulin mediators and their control of metabolism through protein phosphorylation. Recent Prog Horm Res 38:511–556

Lawrence JC Jr, Hiken JF, Inkster M, Scott CW, Mumby MC (1986) Insulin stimulates the generation of an adipocyte phosphoprotein that is isolated with a monoclonal antibody against the regulatory subunit of bovine heart cAMP-dependent protein kinase. Proc Natl Acad Sci USA 83:3649–3653

Lefkowitz RJ, Caron MG (1988) Adrenergic receptors: models for the study of receptors coupled to guanine nucleotide regulatory proteins. J Biol Chem 263:4993–4996

Levitzki A (1974) Negative cooperativity in clustered receptors as a possible basis for membrane action. J Theor Biol 44:367–372

Levitzki A (1987) Regulation of adenylate cyclase by hormones and G-proteins. FEBS Lett 211:113–118

Lin CR, Chen WS, Lazar CS, Carpenter CD, Gill GN, Evans RM, Rosenfeld MG (1986) Protein kinase C phosphorylation at Thr 654 of the unoccupied EGF receptor and EGF binding regulate functional receptor loss by independent mechanisms. Cell 44:839–848

Machicao E, Wieland OH (1984) Evidence that the insulin receptor-associated protein kinase acts as a phosphatidylinositol kinase. FEBS Lett 175:113–116

Madoff DH, Martensen TM, Lane MD (1988) Insulin and insulin-like growth factor-1 stimulate the phosphorylation on tyrosine of a 160 kDa cytosolic protein in 3T3-L1 adipocytes. Biochem J 252:7–15

Mato JM, Kelly KL, Abler A, Jarett L (1987) Identification of a novel insulin-sensitive glycophospholipid from H35 hepatoma cells. J Biol Chem 262:2131–2137

Maturo JM III, Hollenberg MD (1978) Insulin receptor: interaction with nonreceptor glycoprotein from liver cell membranes. Proc Natl Acad Sci USA 75:3070–3074

Maturo JM III, Hollenberg MD, Aglio LS (1983) Insulin receptor: insulin-modulated interconversion between distinct molecular forms involving disulfide-sulfhydryl exchange. Biochemistry 22:2579–2586

McDonald JM, Pershadsingh HA (1985) The role of calcium in the transduction of insulin action. In: Czech MP (ed) Molecular basis of insulin action. Plenum, New York, pp 103–117

Meruelo D, Edidin M (1979) The biological function of the major histocompatibility complex: hypotheses. In: Marchalonis JJ, Cohen N (eds) Self/non-self discrimination. Plenum, New York, pp 231–253 (Contemporary topics in immunobiology, vol 9)

Metzger H, Ishizaka T (1982) Transmembrane signalling by receptor aggregation: the mast cell receptor for IgE as a case study. Fed Proc Fed Am Soc Exp Biol 41:7–34

Miller DS (1988) Stimulation of RNA and protein synthesis by intracellular insulin. Science 240:506–509

Morgan DO, Roth RA (1987) Acute insulin action requires insulin receptor kinase activity: introduction of an inhibitory monoclonal antibody into mammalian cells blocks the rapid effects of insulin. Proc Natl Acad Sci USA 84:41–45

Morgan DO, Ho L, Korn LJ, Roth RA (1986) Insulin action is blocked by monoclonal antibody that inhibits the insulin receptor kinase. Proc Natl Acad Sci USA 83:328–332

Northup JK (1985) Overview of the guanine nucleotide regulatory protein systems, N_s and N_i, which regulate adenylate cyclase activity in plasma membranes. In: Cohen P, Houslay MD (eds) Molecular mechanisms of transmembrane signalling. Elsevier, Amsterdam, pp 91–116

O'Brien RM, Houslay MD, Milligan G, Siddle K (1987) The insulin receptor tyrosyl kinase phosphorylates holomeric forms of the guanine nucleotide regulatory proteins G_i and G_0. FEBS Lett 212:281–288

Pastan IH, Willingham MC (1981) Receptor-mediated endocytosis of hormones in cultured cells. Annu Rev Physiol 43:239–250

Pennington SR, Martin BR (1985) Insulin-stimulated phosphoinositide metabolism in isolated fat cells. J Biol Chem 260:11039–11045

Perrotti N, Accili D, Marcus-Samuels B, Rees-Jones RW, Taylor SI (1987) Insulin stimulates phosphorylation of a 120-kDa glycoprotein substrate (pp120) for the receptor-associated protein kinase in intact H-35 hepatoma cells. Proc Natl Acad Sci USA 84:3137–3140

Pessin JE, Gitomer W, Oka Y, Oppenheimer CL, Czech MP (1983) β-adrenergic regulation of insulin and epidermal growth factor receptors in rat adipocytes. J Biol Chem 258:7386–7394

Phillips ML, Moule ML, Delovitch TL, Yip CC (1986) Class I histocompatibility antigens and insulin receptors: evidence for interactions. Proc Natl Acad Sci USA 83:3474–3478

Phillips SA, Perrotti N, Taylor SI (1987) Rat liver membranes contain a 120 kDa glycoprotein which serves as a substrate for the tyrosine kinases of the receptors for insulin and epidermal growth factor. FEBS Lett 212:141–144

Podlecki DA, Smith RM, Kao M, Tsai P, Huecksteadt T, Brandenburg D, Lasher RS, Jarett L, Olefsky JM (1987) Nuclear translocation of the insulin receptor: a possible mediator of insulin's long term effects. J Biol Chem 262:3362–3368

Posner BI, Khan MN, Bergeron JJM (1988) The role of endosomal kinase activity in insulin action. In: Goren HJ, Hollenberg MD, Roncari DAK (eds) Insulin action and diabetes. Raven, New York, pp 141–150

Rodbell M (1980) The role of hormone receptors and GTP-regulatory proteins in membrane transduction. Nature 284:17–22

Romero G, Luttrell L, Rogol A, Zeller K, Hewlett E, Larner J (1988) Phosphatidylinositol-glycan anchors of membrane proteins: potential precursors of insulin mediators. Science 240:509–511

Rosen OM (1987) After insulin binds. Science 237:1452–1458

Rosen OM, Herrera R, Olowe Y, Petruzzelli LM, Cobb MH (1983) Phosphorylation activates the insulin receptor tyrosine protein kinase. Proc Natl Acad Sci USA 80:3237–3240

Roth RA (1988) Monoclonal antibodies as probes of insulin action and degradation. In: Goren HJ, Hollenberg MD, Roncari DAK (eds) Insulin action and diabetes. Raven, New York, pp 129–140

Roth RA, Cassell DJ (1983) Evidence that the insulin receptor is a protein kinase. Science 219:299–301

Rutter WJ, Morgan D, Ebina Y, Wang L-H, Roth R, Ellis L (1988) Membrane linked insulin receptor tyrosine kinase stimulates the insulin specific response. In: Goren HJ, Hollenberg MD, Roncari DAK (eds) Insulin action and diabetes. Raven, New York, pp 1–12

Saltiel AR, Cuatrecasas P (1986) Insulin stimulates the generation from hepatic plasma membranes of modulators derived from an inositol glycolipid. Proc Natl Acad Sci USA 83:5793–5797

Saltiel AR, Fox JA, Sherline P, Cuatrecasas P (1986) Insulin-stimulated hydrolysis of a novel glycolipid generates modulators of cAMP phosphodiesterase. Science 233:967–972

Saltiel AR, Sherline P, Fox JA (1987) Insulin-stimulated diacylglycerol production results from the hydrolysis of a novel phosphatidylinositol glycan. J Biol Chem 262:1116–1121

Savion N, Vlodavsky I, Gospodarowicz D (1981) Nuclear accumulation of epidermal growth factor in cultured bovine corneal endothelial and granulosa cells. J Biol Chem 256:1149–1154

Schlessinger J, Shechter Y, Willingham MC, Pastan I (1978) Direct visualization of binding, aggregation, and internalization of insulin and epidermal growth factor on living fibroblastic cells. Proc Natl Acad Sci USA 75:2659–2663

Schlessinger J, van Obberghen E, Kahn CR (1980) Insulin and antibodies against insulin receptor cap on the membrane of cultured human lymphocytes. Nature 286:729–731

Simpson IA, Hedo JA (1984) Insulin receptor phosphorylation may not be a prerequisite for acute insulin action. Science 223:1301–1304

Smith RM, Jarett L (1987) Ultrastructural evidence for the accumulation of insulin in nuclei of intact 3T3-L1 adipocytes by an insulin-receptor mediated process. Proc Natl Acad Sci USA 84:459–463

Stadtmauer L, Rosen OM (1986) Increasing the cAMP content of IM-9 cells alters the phosphorylation state and protein kinase activity of the insulin receptor. J Biol Chem 261:3402–3407

Taylor D, Uhing RJ, Blackmore PF, Prpic V, Exton JH (1985) Insulin and epidermal growth factor do not affect phosphoinositide metabolism in rat liver plasma membranes and hepatocytes. J Biol Chem 260:2011–2014

Thuren T, Tulkki A-P, Virtanen JA, Kinnunen PKJ (1987) Triggering of the activity of phospholipase A_2 by an electric field. Biochemistry 26:4907–4910

Tornqvist HE, Gunsalus JR, Avruch J (1988) Identification of the insulin receptor tyrosine residues autophosphorylated in vitro and in vivo: relationship to receptor kinase activation. In: Goren HJ, Hollenberg MD, Roncari DAK (eds) Insulin action and diabetes. Raven, New York (in press)

van Obberghen E, Kowalski A (1982) Phosphorylation of the hepatic insulin receptor: stimulating effect of insulin cells and in a cell-free system. FEBS Lett 143:179–182

White MF, Maron R, Kahn CR (1985) Insulin rapidly stimulates tyrosine phosphorylation of a M_r-185,000 protein in intact cells. Nature 318:183–186

White MF, Stegmann EW, Dull TJ, Ullrich A, Kahn CR (1987) Characterization of an endogenous substrate of the insulin receptor in cultured cells. J Biol Chem 262:9769–9777

Yankner BA, Shooter EM (1979) Nerve growth factor in the nucleus: interaction with receptors in the nuclear membrane. Proc Natl Acad Sci USA 76:1269–1273

Yip CC, Moule ML, Yeung CWT (1982) Subunit structure of insulin receptor of rat adipocytes as demonstrated by photoaffinity labeling. Biochemistry 21:2940–2945

Yip CC, Moule ML (1983) Structure of the insulin receptor of rat adipocytes: the three interconvertible redoc forms. Diabetes 32:760–767

Zick Y, Rees-Jones RW, Taylor SI, Gorden P, Roth J (1984) The role of antireceptor antibodies in stimulating phosphorylation of the insulin receptor. J Biol Chem 259:4396–4400

Zierler K, Wu F-S (1988) Insulin acts on Na, K, and Ca currents. Clin Res 36:624A

CHAPTER 11

The Insulin Receptor Tyrosine Kinase

P. Rothenberg, M. F. White, and C. R. Kahn

A. Introduction

The insulin receptor is a tyrosine protein kinase. This enzymatic activity of the insulin receptor was first recognized in 1982, and is an initial, critical component of the mechanism by which insulin controls cell metabolism. Insulin binds to the extracellular domain of the receptor, activating the intracellular catalytic domain and triggering a unique sequence of tyrosine phosphorylations on the receptor and also on specific cellular protein substrates. While many of the precise molecular details and biological nuances of insulin receptor kinase function remain a challenging puzzle, significant new insights have been gleaned from the concentrated research efforts focused on this subject.

In this chapter, the enzymology of the insulin receptor tyrosine kinase is reviewed, including a discussion of tissue and species distribution, purification and assay, kinetics, the nature and role of receptor autophosphorylation, substrate specificity, and regulation of the insulin receptor kinase by cofactors and other protein kinases. Emphasis is given to more recent advances, as comprehensive reviews have previously appeared (WHITE and KAHN 1986; GAMMELTOFT and VAN OBBERGHEN 1986; ROSEN 1987). A much more extensive review of insulin receptor structure is presented in Chap. 9, this volume.

B. Background

The current working model of the insulin receptor is shown in Fig. 1. The insulin receptor is an oligomeric, integral membrane glycoprotein. It is a symmetrical heterotetramer ($M_r \sim 350\,000$) as extracted from adipocytes and placenta (MASSAGUE et al. 1981a, b, 1982), although additional oligomeric forms ($M_r \sim 210$, 270, and 520 000) have been identified in other tissues (KASUGA et al. 1982b; CRETTAZ et al. 1984; CHRATCHKO et al. 1984; CHEN et al. 1986). Each heterodimer contains an extracellular, glycosylated, and hydrophilic α subunit ($M_r \sim 135\,000$), which is disulfide linked to a membrane-spanning β subunit ($M_r \sim 95\,000$). The heterotetramer ($\alpha_2\beta_2$) structure is maintained by additional disulfide bridges between the two α subunits. Affinity labeling with ^{125}I-labeled insulin strongly tags the α subunit (YIP et al. 1978, 1980; PILCH et al. 1979, 1980), but barely labels the β subunit, suggesting that the insulin binding site may be restricted to the α subunit. The exact stoichiometry of binding remains debated. Several reports suggest 1.1–1.6 insulin molecules per $M_r = 350\,000$ oligomer, (FUJITA-YAMAGUCHI et al. 1983; PANG and SHAFER 1984), although more recent

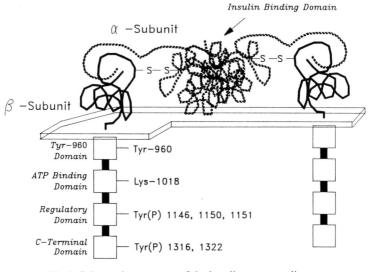

Fig. 1. Schematic structure of the insulin receptor dimer

studies indicate only 1 insulin molecule per $\alpha_2\beta_2$ heterotetramer (SWEET et al. 1987a).

The β subunit is a transmembrane protein. The NH_2 terminal, extracellular portion is anchored by a 23 amino acid transmembrane hydrophobic domain, while the cytoplasmic portion of the β subunit bears the catalytic domain of the insulin receptor kinase. The primary amino acid sequence deduced from the cloned human placental insulin receptor cDNA (EBINA et al. 1985; ULLRICH et al. 1985) indicates a highly conserved ATP binding site, with features homologous to other tyrosine kinases, including the EGF receptor and several membrane-associated viral transforming proteins. The β subunit is specifically labeled by ATP affinity reagents (ROTH et al. 1983; VAN OBBERGHEN et al. 1983; SHIA and PILCH 1983), and undergoes insulin-stimulated phosphorylation on multiple seryl, threonyl, and tyrosyl residues (KASUGA et al. 1982a) in the intact cell. In vitro, the purified insulin receptor undergoes insulin-stimulated β subunit autophosphorylation only on tyrosine residues and also catalyzes the tyrosine phosphorylation of other polypeptides (AVRUCH et al. 1982; PETRUZZELLI et al. 1982; VAN OBBERGHEN et al. 1982; ROTH et al. 1983; KASUGA et al. 1983a). Because protein phosphorylation dynamically regulates intermediary metabolism, cell growth, and differentiation (as exemplified by the oncogenic viral protein kinases and their cognate receptor kinases), the insulin receptor tyrosine kinase activity is clearly implicated as an essential feature of insulin action and has therefore received considerable attention.

I. Purification of the Insulin Receptor Kinase

The insulin receptor β subunit undergoes autophosphorylation upon insulin stimulation of the intact cell, but tyrosine kinase activity of the insulin receptor is generally poorly demonstrable in isolated membrane preparations (HAMMERMAN and GAVIN 1984). Detergent solubilization of membranes permits in vitro assay of insulin receptor kinase (KASUGA et al. 1982a; AVRUCH et al. 1982; BLACK-SHEAR et al. 1984; YU et al. 1987). The basis for low activity in intact membranes is unclear, since the kinase is clearly active in intact cells.

Purification of the insulin receptor kinase has employed wheat germ agglutinin (WGA) affinity chromatography, which extracts many sialated glycoproteins, such as the membrane receptors for EGF, PDGF, IGF-I, and achieves about 20-fold purification of the insulin receptor (HEDO et al. 1981). In WGA column eluates, the insulin receptor constitutes about 1% of all protein, depending on tissue source (WHITE et al. 1984a). Insulin receptor kinase activity is readily measured in such partially pure preparations, but copurifying proteases and protein phosphatases (KATHURIA et al. 1986; PANG et al. 1984; HARING et al. 1984a) can complicate enzymological analyses. Further insulin receptor kinase purification has been achieved by specific immunoprecipitation (ROTH and CASSEL 1983), or by insulin affinity chromatography (KASUGA et al. 1983a; ROSEN et al. 1983). Reducing SDS–PAGE analysis of such highly purified preparations reveals only a 135 kdalton α subunit, and a 95 kdalton β subunit, when sensitive silver stains are used (PIKE et al. 1986). Incubation of both the WGA- and insulin-affinity-purified preparations with insulin, [^{32}P]ATP, and Mn^{2+} generates only a single major phosphoprotein of 95 kdalton. However, the insulin-affinity-purified receptor is not suitable for all studies, because of an elevated basal autophosphorylation rate (ROSEN et al. 1983).

Insulin receptors seem almost ubiquitous in higher organisms (FERNANDEZ-ALMONACID and ROSEN 1987; HAVRANKOVA et al. 1978), often found on cell types not regarded as classic targets such as glia cells, erythrocytes, and oocytes, in addition to classically insulin-responsive target tissues such as liver, fat, and muscle (KAHN and CRETTAZ 1985). The insulin receptor kinase has been purified from diverse sources, including: rat hepatocytes (VAN OBBERGHEN et al. 1982; ZICK et al. 1983a), skeletal muscle (BURANT et al. 1984), adipocytes (HARING et al. 1982; TAMURA et al. 1984), hepatoma cells (BLACKSHEAR et al. 1984), 3T3-L1 adipocytes (PETRUZZELLI et al. 1982), melanoma (HARING et al. 1984b), human erythrocytes (GRIGORESCU et al. 1983; SUZUKI et al. 1987) and fibroblasts (GRIGORESCU et al. 1984), and placental membranes (PETRUZZELLI et al. 1984; FUJITA-YAMAGUCHI 1984; AVRUCH et al. 1982; KASUGA et al. 1983b). Recently, cultured Chinese hamster ovary cells or rat fibroblasts transfected with native or mutant human insulin receptor cDNA have been used as a rich source of insulin receptor (ELLIS et al. 1986; EBINA et al. 1987; CHOU et al. 1987; MCCLAIN et al. 1987; WHITE et al. 1988b). Overall, the structure and in vitro kinase properties of the insulin receptors isolated from all of these sources are quite similar, although subtle but physiologically significant differences in subunit size and/or activity may be present. For example, insulin receptor subunits isolated from brain have slightly smaller relative molecular masses than liver isolates, possibly due to differential glycosylation. This might also reflect functional specialization of the receptor (LOWE et al. 1986; HEIDENREICH et al. 1983).

Recent studies have indicated that all tissues express 2–5 distinct insulin receptor mRNA species (EBINA et al. 1985; GOLDSTEIN et al. 1987), suggesting the possibility of differential processing of a primary mRNA transcript (e.g. variable splicing, or alterations in 3' and 5' untranslated sequences). Putative insulin receptor isozymes have also been reported in placental tissue (Jonas et al. 1986).

II. Role of Autophosphorylation in Insulin Receptor Kinase Function

Self-catalyzed tyrosine phosphorylation of the insulin receptor β subunit stimulates or autoactivates the tyrosine kinase activity towards exogenous protein substrates (Kasuga et al. 1983b; Rosen et al. 1983; Yu and Czech 1984; Yu et al. 1986; Pang et al. 1985). When highly purified, insulin-Sepharose-immobilized receptors are incubated with ATP, Mn^{2+}, and histone 2B, the initial velocity of substrate tyrosine phosphorylation increases with time, rising during the course of the assay to a stable maximum, and thus producing curvilinear reaction velocity profiles. But when the same receptor is prephosphorylated by preincubation with ATP and Mn^{2+}, prior to adding histone, no such curvilinearity is observed (Kwok et al. 1986; Pike et al. 1986). Thus, autophosphorylation precedes activation of the phosphotransferase by insulin. The degree of kinase autoactivation can be varied with experimental conditions, and the extent of autoactivation correlates with $^{32}PO_4^{3}$ incorporation into specific β subunit domains (Kwok et al. 1986; Herrera and Rosen 1986; White et al. 1988b). After insulin-stimulated autophosphorylation occurs the insulin receptor kinase activity remains elevated, even after the liganded insulin is dissociated by mild acid elution (Rosen et al. 1983; Pang et al. 1985). This suggests that insulin is only transiently required to initiate insulin receptor kinase activation. Phosphorylation of the insulin receptor β subunit by an exogenous tyrosine kinase (src) similarly activates the insulin receptor kinase activity (White et al. 1984b; Yu et al. 1985). Dephosphorylation of the β subunit by exogenous alkaline phosphatase sharply decreases kinase activity (Yu and Czech 1984; Rosen et al. 1983). Autoactivation by autophosphorylation is not unusual, as known serine kinases (e.g., cAMP-dependent protein kinase; Flockhart and Corbin 1982) and some oncogenic protein tyrosine kinases (Lai et al. 1986; Schworer et al. 1986; Rangel-Aldao and Rose 1977) also demonstrate this property.

The effect of autophosphorylation on insulin receptor kinetic parameters has been examined, with conflicting results. In some instances substantial decreases in substrate K_m have been observed (Walker et al. 1987; Woo et al. 1986). However, most studies indicate that autophosphorylation does not alter the K_m for ATP or substrate, but increases the V_{max} (Kwok et al. 1986; Sweet et al. 1987c; Pike et al. 1986). Whether this discrepancy is due to differences in receptor preparation, incubation conditions, or use of different substrates in assays is unknown. Whether autophosphorylation modifies kinase substrate specificity is uncertain.

Autophosphorylation probably modifies receptor conformation. Polyclonal antisera elicited against limited peptide sequences containing the major β subunit autophosphorylation sites will immunoprecipitate receptors only after in vitro autophosphorylation, but do not bind the nonphosphorylated receptor (Herrera and Rosen 1986; Perlman et al. 1986). Whether such conformational changes are required for other aspects of insulin receptor function has not been explored. Autophosphorylation of the β subunit does not appear to affect insulin binding to the α subunit (JOOST et al. 1986; PIKE et al. 1986).

Is autophosphorylation solely a regulatory mechanism, or does it enhance insulin receptor kinase activity merely because the phosphotransferase mechanism obligatorily requires a transiently phosphorylated enzyme intermediate? These possibilities are not mutually exclusive, as β subunit is autophosphorylated at multiple sites. However, the autophosphorylated receptor does not transfer phosphate from itself to synthetic peptide substrates and kinetic analyses suggest that β subunit-catalyzed peptide tyrosine phosphorylation takes place through a random order, rapid equilibrium mechanism, wherein ATP binding and ADP release need not occur prior to peptide substrate binding, consistent with the lack of a transient phosphoenzyme intermediate (WALKER et al. 1987).

III. Autophosphorylation

1. Ligand Specificity

In vitro, the ED_{50} for insulin-stimulated autophosphorylation is 2–5 nM, corresponding closely to half-maximal insulin binding. Maximal stimulation usually occurs at 100 nM insulin, with a small ($\sim 15\%$) inhibition above this concentration (WHITE et al. 1984a). Insulin analogs promote autophosphorylation in proportion to their receptor binding affinities, and insulin-induced negative cooperativity of binding is not necessary for autophosphorylation (GRIGORESCU et al. 1983). IGF-I is about 5% as effective as insulin and IGF-II is even less effective, consistent with their low binding affinities.

2. Metal Ion Requirements

Divalent metal cations are necessary for insulin receptor kinase activity, serving as substrate cofactors and/or allosteric activators. At micromolecular ATP concentrations, Mn^{2+} maximally supports in vitro autophosphorylation at 2 mM, while Mg^{2+} is effective only at 15 mM. However, at ATP concentrations approaching physiological levels of 1–2 mM, Mn^{2+} and Mg^{2+} are equally effective (NEMENOFF et al. 1984). Mn^{2+} activates the kinase without affecting V_{max}, by decreasing the K_m for ATP, possibly binding allosterically to the receptor, rather than forming a metal ion–ATP complex (WHITE et al. 1984a). The K_m for Mn^{2+} decreases with increasing ATP concentrations; in vitro, high ATP concentrations therefore activate the receptor kinase, even at micromolecular Mn^{2+}, suggesting a possible physiological role of Mn^{2+} for in vivo receptor regulation.

3. Nucleotide Dependency

ATP is the only phosphoryl donor known to be accepted by the insulin receptor kinase, with a K_m highly dependent on reaction conditions and in the range 30–150 μM (PETRUZZELLI et al. 1984; NEMENOFF et al. 1984; WHITE et al. 1984a; PIKE et al. 1984).

4. Inhibitors

Certain divalent metal ions inhibit autophosphorylation; 1 µM Cu^{2+} inhibits autophosphorylation in vitro 50%, while 10 µM Zn^{2+} inhibits almost completely and irreversibly, without influencing insulin binding. Of the ions tested, the rank of inhibitory potency of ions is: Cu^{2+} > Zn^{2+}, Cd^{2+} > Co^{2+}, Ni^{2+}. Ca^{2+} is without effect (PANG and SHAFER 1985). Small organic inhibitors specific to the insulin receptor kinase are not known. Amiloride, a pyrazine diuretic, inhibits several tyrosine protein kinases, including the insulin receptor as well as other serine/threonine kinases, apparently acting as a competitive ATP analog (DAVIS and CZECH 1985). The isoflavone genistein specifically inhibits tyrosine protein kinases, but has not been tested on the insulin receptor (AKIYAMA et al. 1987).

Specific inhibition of the insulin receptor kinase has been attained through antibodies. A polyclonal antiserum elicited to a limited peptide sequence surrounding the Tyr-960 residue of the β subunit bound to the receptor, blocking autophosphorylation. Interestingly, once autophosphorylated the receptor was no longer susceptible to inhibition (HERRERA et al. 1985). Monoclonal antibodies to the β subunit kinase domain which bind to an antigenic region containing the Tyr-1150/1151 autophosphorylation sites also potently prevent kinase activation (MORGAN et al. 1986) and have been used in elegant in vivo cellular microinjection experiments to demonstrate the role of insulin receptor kinase activity in insulin action (MORGAN and ROTH 1987). Anti-phosphotyrosine antibodies also inhibit the autophosphorylation cascade in vitro (WHITE et al. 1988b; see also Sect. B.IV).

5. Activators

Mild trypsinization of intact isolated adipocytes is insulinomimetic in stimulating glucose oxidation and glycogen synthase (KONO and BARHAM 1971; KIKUCHI et al. 1981). Trypsinization of purified insulin receptors also stimulates in vitro autophosphorylation of the intact β subunit and a 72 kdalton degradation product (TAMURA et al. 1983; SHOELSON et al. 1988a). Moreover, trypsinization of intact ^{32}P-orthophosphate-labeled adipocytes causes in vivo β subunit autophosphorylation with retention of the enhanced kinase activity when receptors are subsequently purified under appropriate conditions (LEEF and LARNER 1987).

In vitro trypsinization of WGA-purified receptor cleaves the α and β subunits to 25 and 85 kdalton fragments, respectively, whereas trypsinization of whole cells cleaves only the α subunit to a 25 kdalton fragment. The truncated receptors have lost insulin binding capacity, but (both in vivo and in vitro forms) undergo spontaneous autophosphorylation in the absence of insulin, at rates comparable to insulin-stimulated rates of intact receptor autophosphorylation, and at the same sites, and without alterations in K_m for peptide substrate phosphorylation. These observations indicate that the unoccupied α subunit may tonically inhibit the β subunit catalytic activity, whereas insulin binding or removal of the α subunit relieves inhibition (SHOELSON et al. 1988). This is akin to the v-*erb*-B oncogene product, a homolog of the EGF receptor which lacks the extracellular EGF binding domain. The oncogene V-*erb*-B maintains a constitutively activated intracellular tyrosine protein kinase domain (DECKER 1985; GILMORE et al. 1985). Similarity of the insulin and EGF receptors transmembrane signaling mechanism between the extracellular ligand binding domain and the intracellular kinase domain is supported by experiments using fusion protein constructs. When the extracellular portion of the insulin receptor is fused to the intracellular part of the EGF receptor and expressed in cells, the new construct is processed and normally inserted into the plasma membrane, and remarkably, insulin binding promotes autophosphorylation of the EGF receptor kinase portion of the hybrid molecule (RIEDEL et al. 1986).

Hydrogen peroxide (H_2O_2) is another insulinomimetic agent for adipocytes, and fosters insulin receptor β subunit autophosphorylation and kinase activation in intact ^{32}P-orthophosphate-labeled adipocytes. However, in vitro treatment of purified insulin receptors with H_2O_2 does not activate the kinase, suggesting an indirect pathway for the in vivo effect (HAYES and LOCKWOOD 1987; KADOTA et al. 1987b). Activation of insulin receptor tyrosine kinase also occurs upon sodium vanadate treatment of whole cells; again the mechanism is unclear (KADOTA et al. 1987a; TAMURA et al. 1984), but probably involves inhibition of phosphotyrosyl protein phosphatases.

6. Receptor Activation by Anti-Receptor Antibodies

Some anti-receptor antibodies are biologically insulinomimetic and also activate the insulin receptor kinase in intact cells. Other insulinomimetic antibodies had not been seen to cause kinase activation (ZICK et al. 1984a; FORSAYETH et al. 1987). This dissociation had argued against necessary linkage of insulin kinase activation to insulin bioactivity. When the antiserum which was insulinomimetic on normal cells, but which did not activate the insulin receptor kinase (ZICK et al. 1984b; SIMPSON and HEDO 1984) is applied to transfected cells expressing mutant, *kinase-defective* insulin receptors, the antiserum was *not* insulinomimetic. It *was* insulinomimetic on control cells transfected with wild-type insulin receptors (EBINA et al. 1987; GHERZI et al. 1987). Moreover, when insulin receptors are purified from antibody-stimulated cells under conditions which preserve the in vivo phosphorylation state of the receptor, *all* insulinomimetic antisera significantly stimulated insulin receptor tyrosyl phosphorylation. Therefore, tyrosine kinase activation is probably essential for insulinomimetic activity of anti-receptor antibodies.

Antibody activation of insulin receptor kinase may require receptor cross-linking (HEFFETZ and ZICK 1986; O'BRIEN et al. 1987a). Bivalent polyclonal antibodies were agonists for the solubilized receptor kinase, whereas monovalent Fab' fragments were not. Cross-linking of Fab' fragments with a secondary antibody activated the receptor. Whether this cross-linking occurs between two $\alpha\beta$ heterodimers in a single heterotetramer, or between distinct $\alpha_2\beta_2$ heterotetramers is unknown. Monoclonal anti-receptor antibodies are also effective insulin receptor kinase agonists, but also only in bivalent form and in a narrow concentration range. When monoclonal antibody–receptor complexes are separated by size in sucrose gradients, both inter- and intramolecular cross-linked species are resolved. But activated receptors were predominantly associated with the intramolecularly cross-linked forms (O'BRIEN et al. 1987a).

Receptor cross-linking is necessary for antibody mimicry of insulin action on intact cells (KAHN et al. 1978) and the aggregation state of the receptor may change upon insulin binding (SHLESSINGER et al. 1980; HARMON et al. 1981; CHEN et al. 1986). But whether cross-linking is always required for insulin receptor kinase activation is unclear. Certain antibodies which compete for insulin binding can only partially activate the full kinase activity of the receptor, and in these complexes insulin cannot further increase insulin receptor kinase activity (HEFFETZ and ZICK 1986). In contrast, other antisera which do not compete for insulin binding can fully activate the receptor kinase. These observations indicate that antibodies and insulin may activate the receptor kinase through separate, distinct mechanisms (PONZIO et al. 1987).

IV. Sites of β Subunit Autophosphorylation

Insulin binding initiates rapid, multisite autophosphorylation in the intracellular portion of the β subunit, which proceeds in a sequential fashion, associated with incremental increases in tyrosine kinase activity. Direct microsequencing of partially purified rat (WHITE et al. 1988b) and human (TORNQVIST 1987, 1988) receptors autophosphorylated in vivo and in vitro reveals tyrosine phosphorylation on at least 6 of 13 candidate residues, mostly restricted to two distinct domains. Upon activation, two phosphotyrosines first appear within the tyrosine kinase domain at residues 1146 and either 1150 or 1151 (numbering of ULLRICH et al. 1985). Autophosphorylation in this domain had been detected in earlier studies (HERRERA and ROSEN 1986), and correlated well with kinase activation. Antiphosphotyrosine antibodies bind and trap this doubly phosphorylated receptor intermediate (2Tyr(P) form), preventing further tyrosine phosphorylation (WHITE et al. 1988b). The 2Tyr(P) form is not a maximally active protein kinase. Full activation requires phosphorylation of a third remaining tyrosyl residue in this domain, at either residue 1150 or 1151 (3Tyr(P) form). The exact position of this third phosphate ester is variable, and may in fact, not be critical (TORNQVIST et al. 1987). Tyr-1150 is homologous with the major autophosphorylation site in pp60*src* (HUNTER and COOPER 1986), and is important for kinase activation. When site-specific mutagenesis is employed to substitute phenylalanine at this position, the mutant insulin receptor has impaired exogenous substrate phosphorylating ability (ELLIS 1986), and diminished biological potency in transfected cells.

Concomitant with the appearance of the third phosphotyrosine at 1150/1151, phosphotyrosine also appears in residues 1316 and 1322, contained within the COOH terminal "tail" of the receptor. Removing these phosphotyrosines en bloc by mild trypsinolysis does not affect receptor kinase activity (GOREN et al. 1987), suggesting this tail region is less directly involved in the autoactivation process (HERRERA and ROSEN 1986; KATHURIA et al. 1986). The function of COOH terminal phosphotyrosines remains to be clarified.

Autophosphorylation of insulin receptor in vitro is not entirely identical to the in vivo process. In vitro, receptor autophosphorylation proceeds to the

3Tyr(P) form and maximal kinase activation, while in vivo, the 3Tyr(P) form is barely evident, even after maximal insulin stimulation. While this may be due to in vitro reaction conditions differing from cellular conditions (high Mn^{2+}, low ATP in vitro versus high Mg^{2+}, high ATP in vivo), it could also be due to inhibitory substrate concentrations and/or the action of phosphatases. In vitro, at sufficiently high saturating synthetic peptide concentrations, receptor autophosphorylation can be completely inhibited. The exogenous phosphate acceptor protein competitively prevents full autophosphorylation (WHITE et al. 1988b; KWOK et al. 1986; MORRISON and PESSIN 1987). Whether a similar competition is maintained in vivo (with potentially significant regulatory implications) will require quantitation of cellular substrates.

Additional autophosphorylation sites exist, but have not been directly localized. Tyr-960 had been implicated, since insulin stimulation of kinase activity was inhibited by an antibody against its domain (HERRERA et al. 1985), but mutant receptors substituting phenylalanine for Tyr-960 do not have altered β subunit autophosphorylation, although in vivo substrate phosphorylation is impaired (WHITE et al. 1988a).

V. Structural Requirements for Receptor Autophosphorylation

1. Quaternary Structure

In nonionic detergent solution, the rate of insulin-stimulated, receptor autophosphorylation is independent of receptor concentration (SHIA et al. 1983; PETRUZZELLI et al. 1984; WHITE et al. 1984a), implying that receptor autophosphorylation is an *intra*molecular reaction. However, an *inter*molecular phosphorylation reaction might seem concentration independent if receptors aggregate within detergent micelles. However, most evidence suggests the insulin receptor is monomeric in detergent solution (CUATRECASAS 1972; POLLET et al. 1981; AIYER 1983), and also β subunit autophosphorylation occurs in phospholipid vesicles containing, on average, only one receptor (SWEET et al. 1985).

The insulin receptor kinase phosphorylates exogenous peptide substrates synthesized to contain sequences of the receptor itself (STADTMAUER et al. 1986b) or homologous kinases (PIKE et al. 1986). Thus, autophosphorylation within the intact $\alpha_2\beta_2$ heterotetrameric receptor may proceed by "*trans*" phosphorylation: one β subunit acting on its neighboring counterpart. Alternatively, or perhaps coordinately, one β subunit may autophosphorylate itself ("*cis*" phosphorylation). When heterotetrameric $\alpha_2\beta_2$ receptors are cleaved into insulin-binding $\alpha\beta$ heterodimers (by a mild alkaline dithiothreitol treatment) and then separated by size, only the $\alpha_2\beta_2$ forms undergo insulin-stimulated autophosphorylation. The $\alpha\beta$ form does not – unless oxidized to re-form $\alpha_2\beta_2$ (BONI-SCHNETZLER et al. 1986). Also, once autophosphorylated, the $\alpha_2\beta_2$ form can be then cleaved into two $\alpha\beta$ forms, with retention of kinase activation in each half receptor. It thus appears that interaction between $\alpha\beta$ heterodimers is required for full insulin-stimulated β subunit autophosphorylation, but is not required for maintenance of the activated state. Moreover, insulin apparently promotes interactions between isolated heterodimers, such that the $\alpha\beta$ concentration dependence of β subunit autophosphorylation shows evidence for reaction between $\alpha\beta$ heterodimers, i.e., faster β subunit autophosphorylation at higher $\alpha\beta$ heterodimer concentrations (SWEET et al. 1987c). However, these results do not indicate whether *cis* or *trans* autophosphorylation mechanisms pertain, i.e., the $\alpha_2\beta_2$ oligomer may be the preferred structure for β subunit autophosphorylation because proximity between two β subunits permits *trans* autophosphorylation, or oligomerization may be required to induce a conformation with a single β subunit favorable to *cis* autophosphorylation. Preliminary data relevant to this as yet unresolved issue suggest that a trypsin-generated, truncated $\alpha\beta$ heterodimer autophosphorylates intramolecularly (SHOELSON et al. 1988b, submitted). Thus, it may be that insulin stimulation requires $\alpha\beta$–$\alpha\beta$ interactions, but that once activated, each β subunit catalyzes phosphorylation of its own tyrosines.

2. Role of Disulfide Bonds and Sulfhydryl Groups

Disulfide bridges of two types maintain insulin receptor quaternary structure: class I disulfides link $\alpha\beta$ heterodimers to form the $\alpha_2\beta_2$ heterotetrameric complex and are readily reduced; class II disulfides bridge α and β subunits, and are relatively resistant to reduction. Mild reduction with dithiothreitol enhances receptor kinase autophosphorylation (SHIA et al. 1983; PETRUZZELLI et al. 1984; FUJITA-YAMAGUCHI et al. 1983) under conditions where class I bonds are maintained such that the $\alpha_2\beta_2$ oligomer remains intact (SWEET et al. 1986). Dithiothreitol (DTT) increases the rate, but not extent of β subunit autophosphorylation, and increases the V_{max} of exogenous substrate phosphorylation, without altering substrate K_m (WILDEN and PESSIN 1987). Treatment of insulin-affinity-purified receptors with DTT under reaction conditions permissive of autophosphorylation enhances β subunit phosphorylation. In contrast, DTT treatment under conditions not permissive of autophosphorylation causes inhibition of subsequently assayed β subunit kinase activity. And in all conditions, DTT reduces insulin binding by $\sim 50\%$ (PIKE et al. 1986). The potential role of receptor disulfides in the physiological regulation of receptor kinase activity requires further definition.

Free sulfhydryl groups in the β subunit influence kinase function. Purified receptor autophosphorylation and exogenous substrate kinase activity are inhibited by –SH reagents such as NEM, and DTNB (SHIA et al. 1983; ZICK et al. 1983c; PIKE et al. 1984; WILDEN et al. 1986) and sensitivity to inhibition is increased by insulin (WILDEN and PESSIN 1987). [³H]NEM labels only the β subunit, and NEM does not inhibit insulin binding. Thus, there may be a critical –SH group(s) in the β subunit required for autophosphorylation. In isolated adipocyte plasma membranes, NEM treatment (prior to insulin treatment) enhances about twofold autophosphorylation of the β subunit, however, under these conditions NEM effects are likely to be complex (CHEN et al. 1986) and less readily interpretable.

3. Posttranslational Modifications

a) Glycosylation

Asparagine-linked, complex-type oligosaccharides are added to the α and β subunits of the insulin receptor during processing of the prorceptor prior to insertion into the plasma membrane. Inhibition of initial N-linked glycosylation with tunicamycin prevents insulin binding and receptor insertion into the plasma membrane (RONNETT et al. 1984). Swainsonine inhibits subsequent terminal glycosylation, generating incompletely processed receptors which, although inserted into the plasma membrane, are of greater than normal mobility on SDS–PAGE. These incompletely processed receptors have no detectable alterations in insulin binding or in vitro insulin-stimulated autophosphorylation (DURONIO et al. 1986; FORSAYETH et al. 1986). Similar conclusions were also obtained when intermediate glycosylation processing steps were specifically inhibited (DURONIO et al. 1988). Therefore, although glycosylation is required for synthesis of functional receptors, terminal glycosylation may not be essential.

Removing cell surface sialic acid with neuraminidase blocks cellular responses to insulin. But a very mild neuraminidase treatment of adipocytes preserves the sialic acid content of the insulin receptor, though still blocking insulin action. Insulin rceptors purified from such mildly treated adipocytes had no evident defects in kinase activity (HAYES et al. 1986). Thus, while sialic acid residues of other cell surface constituents may be vital in insulin action, sialic moieties on the receptor itself may not be directly involved in its tyrosine kinase activity.

b) Fatty Acid Acylation

Both α and β subunits of the insulin receptor contain covalently bound myristic and palmitic acid (HEDO et al. 1987; MAGEE and SIDDLE 1986). The role of this posttranslational modification in the tyrosine kinase activity of the receptor or insulin action is unkown.

4. Role of Tyrosines in the Kinase Domain

Structure–function relationships in the β subunit have been illuminated by site-specific mutagenesis of the cloned human insulin receptor. Such studies confirm the pivotal role of the twin tyrosines (1150, 1151; numbering of ULLRICH et al. 1985) in the kinase domain in the regulation of kinase activity. ELLIS et al. (1986) generated mutants where either Tyr-1150 or Tyr-1150 and –1151 were replaced by phenylalanine, and expressed the mutant receptors at high density in CHO cells. In vitro, WGA-purified mutant receptors had a threefold elevation of basal β subunit autophosphorylation, and autophosphorylation was largely not stimulated by insulin, i.e., the mutants were almost fully autophosphorylated in the absence of insulin. Since Tyr-1150/1151 are major autophosphorylation sites in vitro (see Sect. B.IV), this implies that in the mutant receptors some other tyrosine residues become more accessible to in vitro phosphorylation. When exogenous substrate phosphorylation was assayed, the single Tyr-1150 mutant had only an approximately 2-fold elevated basal rate, with weak insulin stimulation (about 1.3-fold compared to 12-fold for control receptors), and the double Tyr-1150/1151 mutant had a normal basal rate and no insulin stimulation. Thus, despite almost maximal autophosphorylation in these mutant receptors, exogenous substrate phosphorylation was markedly impaired and insulin-insensitive. This might indicate dissociability between autophosphorylation and the activation of the exogenous substrate protein kinase activity of the receptor (with the important caveat that autophosphorylation sites in the mutant receptors may not be equivalent to sites in the wild-type receptor).

In vivo, both mutant receptors evidenced low basal autophosphorylation and in contrast to the in vitro condition, were insulin-sensitive. This significant disparity between in vivo β subunit properties when studied in vivo may relate to the different sites of autophosphorylation observed in purified receptors as compared to receptors phosphorylated in vivo (WHITE et al. 1985a), and may indicate that the physiological function of the receptor kinase is not always accurately reflected during in vitro analyses.

5. Role of Lysine in the Kinase Domain

The tyrosine kinase domain of the β subunit begins at about residue 991, about 50 residues from the transmembrane domain, and contains a Gly-X-Gly-XX-Gly consensus ATP binding site (HUNTER and COOPER 1986). At position 1018, a charged lysine residue is believed to attract the ATP γ-phosphate, orienting the nucleotide for catalysis. Site-specific mutagenesis substituting alanine for Lys-1018 (CHOU et al. 1987), or substituting isosteric methionine or isostatic arginine at this position (EBINA et al. 1987) yields a receptor devoid of all kinase activity both in vitro and in vivo (McCLAIN et al. 1987). Such mutant receptors bind insulin normally, and no major β subunit structural perturbations are detected by conformation-sensitive, anti-β subunit monoclonal antibodies. The severely dysfunctional insulin responses of cells expressing the lysine-substituted, kinase-defective receptors supports the essential role of tyrosine kinase activity in insulin action.

C. Substrates of the Receptor Kinase

I. In Vitro Substrates

The in vitro substrate specificity of the insulin receptor kinase is slowly being defined. Comparisons with other tyrosine protein kinases, e.g., oncogene products, reveal important similarities and differences. The phosphate acceptor substrate specificity of the highly purified insulin receptor kinase demonstrates absolute preference for L-tyrosine; D-tyrosine, serine, and threonine are not phosphorylated (WALKER et al. 1987). Tyramine and acetyltyramine are low affinity tyrosine kinase substrates (BROWN et al. 1983) and an Arg-Tyr dipeptide is not phosphorylated by the insulin receptor (STADTMAUER and ROSEN 1983). Thus, steric and charge effects from neighboring groups influence substrate preference. This is true for many tyrosine kinases, e.g., the EGF and PDGF receptors and pp60^{src}, which best phosphorylate tyrosine residues COOH terminal to glutamate (HUNTER and COOPER 1986), e.g., as in gastrin, or a synthetic random copolymer poly(Glu,Tyr). However, this requirement is not absolute, e.g., [Val^5]-angiotensin II lacks anionic residues, but is a good insulin receptor substrate (STADTMAUER and ROSEN 1983). Vicinal cationic residues inhibit tyrosine phosphorylation (HOUSE et al. 1984) and protein secondary and tertiary structure also influence substrate K_m, e.g., enolase has a K_m of 2 μM while a synthetic polypeptide fragment containing the enolase phosphorylation site has a K_m of 150 μM (COOPER et al. 1984). For most protein kinases, protein substrates have not been readily predicted from their primary sequence, but have been empirically determined. Because phosphotyrosine-containing proteins are rare in vivo (EKMAN et al. 1987) and are largely unidentified, many in vitro studies of the insulin receptor kinase have employed proteins known to be phosphorylated by other tyrosine kinases. Additionally, proteins which are possibly subject to physiological regulation by insulin have also been evaluated. Known substrates of the insulin receptor kinase are listed in Table 1, some are discussed below.

Several glycolytic enzymes are tyrosine phosphorylated by oncogenic tyrosine protein kinases (COOPER et al. 1983, 1984). The insulin receptor phosphorylates phosphofructokinase (PFK), the major rate-limiting enzyme in glycolysis, in vitro with $K_m \sim 0.1$ μM which is near intracellular PFK concentrations. Insulin stimulates the phosphorylation 2- to 6-fold, but the stoichiometry is low, about 0.01 mol phosphate per mol enzyme tetramer (SALE et al. 1987). Proteins phosphorylated in vitro are not necessarily phosphorylated in vivo and whether tyrosine phosphorylation of PFK or other glycogenic enzymes (SALE et al. 1987) mediates insulin-stimulated glycolytic flux is unknown.

Transformed cells often have altered morphology, and some cytoskeletal proteins such as vinculin and myosin regulatory light chain are substrates of oncogenic tyrosine kinases (SEFTON et al. 1981). The insulin receptor phosphorylates tubulin and microtubule-associated proteins in vitro (KADOWAKI et al. 1985), inhibiting tubulin polymerization (WANDOSELL et al. 1987). This could relate to insulin effects on cell structure and vesicular transport between cytosol and plasma membrane, e.g., in regulation of glucose transporters.

Many cellular enzymes in the pathways of insulin action are known to be serine or threonine phosphorylated. Insulin stimulation increases cellular phosphoserine and phosphothreonine content, suggesting linkage between the insulin receptor kinase and other protein kinases and/or protein phosphatases (e.g., MALLER et al. 1986; BALLOTTI et al. 1986). In vitro insulin receptors phosphorylate calmodulin-dependent kinase, but not phosphorylase kinase, glycogen synthase, or casein kinase I and II (HARING et al. 1985). The insulin receptor apparently binds calmodulin (GRAVES et al. 1985) and the insulin receptor phosphorylates calmodulin in vitro (HARING et al. 1985; GRAVES et al. 1986) and also in intact adipocytes (COLCA et al. 1987) similar to *src* kinase in transformed fibroblasts (FUKAMI et al. 1986). Tyrosine phosphorylation could alter the calmodulin–Ca^{2+} binding affinity, with diverse effects on calcium-regulated processes. Purified hen oviduct progesterone receptor is an excellent in vitro sub-

Table 1. Substrates of insulin receptor kinase

Substrate	Reference
A. In vitro substrates	
Enzymes	
Phosphofructokinase	SALE et al. 1987
Lactate dehydrogenase	
Enolase	
Phosphoglyceromutase	
Malate dehydrogenase	
Structural proteins	
Tubulin	WANDOSELL et al. 1987
Microtubule-associated proteins	KADOWAKI et al. 1985
Actin	
Regulatory proteins	
Calmodulin	HARING et al. 1986b; GRAVES et al. 1986
Angiotensin II	
Myosin regulatory light chain	STADTMAUER and ROSEN 1983
Progesterone receptor	WOO et al. 1986
G proteins (G_i, G_0, T_a)	ZICK et al. 1987
Polypeptides	
poly (Glu:Tyr)	BROWN et al. 1983; REES-JONES et al. 1984
Histone 2B	
src-like peptide	KASUGA et al. 1983b
anti-*src* Ab	
Casein	
B. In vivo substrates	
pp240 (carcinoma)	KADOWAKI et al. 1987
pp185 (liver, fat, skeleton muscle, etc.)	WHITE et al. 1985a, 1987
pp160 (adipocyte)	YU et al. 1987
pp120 (liver)	REES-JONES and TAYLOR 1985
pp110 (liver, brown fat)	SADOUL et al. 1985
pp105 (lizard brain)	SHEMER et al. 1987
pp15 (adipocyte)	BERNIER et al. 1987
calmodulin (adipocytes)	COLCA et al. 1987

strate of several growth-factor-regulated tyrosine kinases, including the insulin receptor (WOO et al. 1986). Whether phosphorylation alters the function of this steroid hormone receptor is unknown.

Insulin modifies cellular responses to some hormones which utilize guanine-nucleotide-binding proteins (G proteins) in the mechanism by which they regulate intracellular second messenger (such as cAMP, or phosphoinositide) synthesis (THOMAS et al. 1985; HEYWORTH and HOULAY 1983). Under some in vitro conditions, the insulin receptor will phosphorylate the α subunit of three distinct G proteins: G_i, G_0, and transducin, but this has not been documented in vivo (O'BRIEN et al. 1987b; ZICK et al. 1986).

II. In Vivo Substrates

In vitro substrates of the insulin receptor may not be physiologically relevant targets of the receptor in the intact cells. To identify such proteins, alternative strategies have been employed. When polyclonal anti-phosphotyrosine antibodies are used to immunoprecipitate phosphotyrosine-containing proteins that appear in intact cells during insulin stimulation, a protein of $M_r \sim 185\,000$, designated pp185, appears in extracts of hepatoma cells (WHITE et al. 1985a, 1987; TORNQVIST et al. 1988), hepatocytes (OKAMOTO et al. 1988), kidney (IZUMI et al. 1987), 3T3-L1 adipocytes (GIBBS et al. 1986), L6 myocytes (BEGUINOT et al. 1988), and KB carcinoma cells (KADOWAKI et al. 1987). Although the identity and function of this protein is currently unknown, its phosphorylation is maximal within seconds after cell exposure to insulin and has a dose–response curve similar to insulin receptor autophosphorylation; it is biochemically distinct from the insulin or EGF receptor, suggesting it may be a physiological substrate which initiates insulin action at intracellular sites. Mutation of the insulin receptor at Tyr-960 does not block receptor autophosphorylation, but does impair the in vivo phosphorylation of pp185, and this correlates with impaired insulin action in cells expressing the mutant receptor, again supporting an important functional role of pp185 (WHITE et al. 1988a).

Common, as well as unique, substrates for several growth factor receptor kinases have been observed. Stimulation of KB carcinoma cells (KADOWAKI et al. 1987) or neuroblastoma cells (SHEMER et al. 1987) with insulin or IGF-I, elicits rapid tyrosine phosphorylation of an $M_r = 185\,000$ soluble protein possibly identical to pp185. EGF uniquely causes the appearance of a distinct 190 kdalton phosphoprotein. All three hormones slowly stimulated phosphorylation of a 240 kdalton protein (KADOWAKI et al. 1987).

A 120 kdalton insulin receptor substrate was first identified in WGA-purified insulin preparations from liver (REES-JONES and TAYLOR 1985). This glycoprotein is also present in purified rat (PHILLIPS et al. 1987) and human (CARO et al. 1987) liver plasma membranes. Insulin evokes its tyrosine phosphorylation in intact hepatoma cells (PERROTTI et al. 1987), and it may be relevant to insulin action in the liver, as it is absent from other tissues (ACCILI et al. 1986). Incubation of plasma membranes from insulin-treated adipocytes with $[\gamma\text{-}^{32}P]$-ATP increases phosphotyrosine in the insulin receptor β subunit and simultaneously in an unrelated glycoprotein of 160 kdalton (YU et al. 1987). Its role and relationships to other putative physiological substrates is unknown.

A significant experimental difficulty in isolating insulin receptor substrates from insulin-treated, intact cells stems from the rapid dephosphorylation of tyrosine phosphate esters by ubiquitous phosphotyrosyl protein phosphatases, but also some substrates may be only transiently phosphorylated intermediates in a tightly controlled, postreceptor signal cascade. The use of potent phosphatase inhibitors during preparation and/or rapid whole-cell denaturation prior to isolation partly addresses the former problem, while a different approach has been taken to resolve the latter issue. Phenylarsine oxide (PAO), a dithiol reagent, interrupts signaling between the insulin receptor and the glucose transport system. With PAO present, insulin stimulates tyrosine phosphorylation of a 15 kdalton protein in 3T3-L1 cells. The kinetic relationship is consistent with a position of pp15 as an intermediary between the insulin receptor and glucose uptake (BERNIER et al. 1987). Recent studies have suggested that pp15 is an abundant adipocyte-specific protein known as AP-2 (or protein 422). This protein has homology to fatty acid binding proteins and the significance of its phosphorylation is uncertain (HRESKO et al. 1988).

D. Regulation of the Insulin Receptor Kinase

I. Possible Role of cAMP-Dependent Kinase

Catecholamines induce insulin resistance in vivo (DEIBERT and DE FRONZO 1980; RIZZA et al. 1980). The exact mechanism of this desensitization is uncertain. Reduction of cell surface insulin receptor density, e.g. following β-adrenergic agonist treatment of adipocytes (PESSIN et al. 1983), could contribute to this desensitization. Another mechanism is suggested by the observation that in vitro, very high concentrations of cAMP-dependent protein kinase phosphorylate the insulin receptor β subunit on serine residues (stoichiometry of 1.2 mol phosphate per mol receptor). This inhibits tyrosine kinase activity towards poly(Glu:Tyr) about 25%, while not affecting insulin binding (ROTH and BEAUDOIN 1987; but see JOOST et al. 1986 for dissimilar results). In IM-9 lymphocytes, application of either 8-bromo cAMP or the adenyl cyclase agonist forskolin increases insulin β subunit phosphorylation twofold on serine/threonine residues and diminishes insulin-stimulated β subunit autophosphorylation and exogenous peptide phosphorylation (STADTMAUER et al. 1986a). Similar effects occur in adipocytes rendered insulin-resistant by catecholamine exposure (HARING et al. 1986b). This was attributed to an increased receptor K_m for ATP; β subunit serine/threonine phosphorylation was not assessed.

   Catecholamine-induced insulin resistance generated in rats by acclimatization to cold was associated with reduced kinase activity of insulin receptors subsequently purified from their brown adipose tissue (TANTI et al. 1987). Coincubation of purified receptors with cAMP-dependent protein kinase completely inhibited insulin-stimulated, but not basal, β subunit autophosphorylation. In this latter study, the mechanism of the effect was uncertain, and direct serine/threonine phosphorylation of the insulin receptor was not evident. Thus, cAMP-dependent protein kinase regulation of insulin receptor kinase is an unsettled issue. Hormones regulating cAMP levels like catecholamines or adenosine in fat cells may modulate cellular insulin sensitivity through a relatively direct mechanism.

II. Role of Protein Kinase C

Protein kinase C is a major intracellular serine/threonine-specific kinase which is activated by hormones mobilizing diacylglycerol and Ca^{2+}, as well as phorbol esters (e.g., 12-O-tetradecanoylphorbol-13-acetate, or TPA; for review see NISHIZUKA 1984). TPA blocks some insulin effects and stimulates serine phosphorylation of the insulin receptor in intact cells (JACOBS et al. 1983; TAKAYAMA et al. 1984; VAN DE WERVE et al. 1985). In cultured hepatoma cells, 30 min exposure to TPA doubled phosphoserine content of the β subunit coincident with a 50% reduction in insulin-stimulated tyrosine autophosphorylation. Partially purified receptors from these cells exhibited decreased V_{max} for in vitro autophosphorylation and exogenous substrate phosphorylation, and this inhibition was reversed upon treatment with alkaline phosphatase (TAKAYAMA et al. 1988). Similarly, in rat adipocytes TPA inhibits insulin-stimulated glucose uptake and WGA-purified insulin receptors from TPA-treated cells had only 3-fold insulin-stimulated autophosphorylation, compared to 12-fold for control receptors (HARING et al. 1986a). Consistent with these observations is the direct in vitro demonstration of serine phosphorylation of the insulin receptor β subunit by purified protein kinase C (BOLLAG et al. 1986). Phosphorylation did not affect insulin binding, but reduced kinase activity 65%.

Following insulin stimulation in vivo, the β subunit of the insulin receptor contains increased levels of phosphoserine and phosphothreonine (as well as phosphotyrosine), but the responsible kinases are unidentified. Protein kinase C may not be involved here because the tryptic phosphopeptide maps of the β subunit are not the same after insulin versus TPA treatment (JACOBS and CUATRECASAS 1986).

III. Role of Phospholipids

The conformation of integral membrane proteins is influenced by their lipid microenvironment, and the activity and/or substrate specificity of the insulin receptor kinase in the native plasma membrane lipid bilayer may not be identical with that of detergent-solubilized receptors. Addition of a soybean phospholipid mixture to solubilized human placental insulin receptors enhanced about sixfold both insulin-stimulated β subunit autophosphorylation and histone 2B phosphorylation. Phosphatidylinositol was found to be the most effective component of the lipid mixture and alone had a biphasic effect, inhibiting kinase activity at higher concentrations (SWEET et al. 1987b). Phosphatidylinositol constitutes a minor but tightly regulated percentage of plasma membrane phospholipids. These initial observations may have implications for physiological regulation of insulin receptor kinase function in vivo. A phosphatidylinositol kinase activity copurifies with the insulin receptor (SALE et al. 1986; SWEET et al. 1987c), but seems distinct from the major cellular phosphatidylinositol kinase and is weakly regulated by insulin.

IV. Role of Interactions with Other Proteins

Many hormone receptors interact noncovalently with G proteins. In turn, the G protein activates an intracellular effector enzyme system or ion channel, stimulating intracellular "second messenger" synthesis or altering transmembrane ion flux. As well, G proteins can reciprocally modify the activity of hormone receptors, sometimes affecting ligand binding (for review see CASEY and GILMAN 1988).

Insulin action has not been shown to involve a G protein-coupled second messenger system. However, insulin does modify the activity of certain hormones which do couple to G proteins (HEYWORTH and HOUSLAY 1983; THOMAS et al. 1985). Also, G proteins may be involved in the insulin activation of a PI-specific phospholipase C, an enzyme involved in phosphoinositol–glycan mediator synthesis (LOW and SALTIEL 1988). G proteins have also been implicated in the insulin activation of a specific phosphodiesterase. Insulin receptor interactions with G proteins are suggested by some recent studies. Autophosphorylation of the human insulin receptor is inhibited by the *ras* oncogene product, p21, a GTP-binding protein of incompletely defined function (O'BRIEN et al. 1987a). This in vitro observation is of interest in light of the inhibition of the insulin-induced *Xenopus* oocyte maturation by microinjected anti-*ras* antibodies (KORN et al. 1987). Also insulin inhibits pertussis-toxin-catalyzed ADP ribosylation of G proteins in liver plasma membranes (ROTHENBERG and KAHN 1988), and pertussis toxin blocks certain insulin bioeffects in some cell types (GOREN et al. 1985; LUTTRELL et al. 1988). The role of G proteins in insulin receptor transmembrane signal transduction needs to be more fully evaluated.

V. Physiological Aspects

The activity of the insulin receptor kinase has been measured across a spectrum of physiological states in a variety of tissues to assess whether modulation of the receptor kinase activity underlies pathological cellular insulin resistance. The role of insulin receptor kinase defects in the etiology of human diabetes mellitus is the subject of several recent reviews, and is not discussed here (KAHN 1985; HARING et al. 1987). While the biochemical mechanisms are unclear, several reports have characterized "up-activation" and "down-activation" of the insulin receptor kinase.

Isolated adipocytes exposed in vitro to elevated insulin concentrations which mimic in vivo hyperinsulinemia associated with insulin resistance, had insulin receptors which when immunopurified showed a 40% reduction in the V_{max} for poly(Glu:Tyr) phosphorylation (ARSENIS and LIVINGSTON 1986). Conversely, insulin receptor kinase activity towards poly(Glu:Tyr) is elevated in livers of rats fed a high carbohydrate diet (a state of increased hepatic insulin sensitivity) and was subnormal in fasted rats (FREIDENBERG et al. 1985). In this latter study, there was no evident change in the rate or extent of β subunit autophosphorylation between the different diet groups. This important observation suggests that physiologic modulation of kinase activity may be manifest with only certain substrates. The effects of diet may also be restricted to certain tissues (SIMON et al. 1986). Skeletal muscle is a major insulin target, and rates of basal and insulin-stimulated β subunit autophosphorylation and exogenous substrate phosphorylation in partially purified receptors are twofold higher in red than in white fibers of the same muscle group (JAMES et al. 1986), suggesting further levels of physiological regulation. Not all regulation of tissue insulin sensitivity may occur through the insulin receptor kinase. Skeletal muscle becomes severely insulin resistant after denervation, but no change in receptor kinase activity can be demonstrated, either in vivo or in vitro (BURANT et al. 1986), indicating the likelihood of additional "postreceptor" regulation.

References

Accili D, Perrotti N, Rees-Jones R, Taylor SI (1986) Tissue distribution and subcellular localization of an endogenous substrate (pp120) for the insulin receptor-associated tyrosine kinase. Endocrinology 119:1274–1280

Aiyer RA (1983) Structural characterization of insulin receptors. J Biol Chem 258:14992–14999

Akiyama T, Ishida J, Nakagawa S, Ogawara H, Watanabe S, Itoh N, Shibuya M, Fukami Y (1987) Genistein, a specific inhibitor of tyrosine-specific protein kinases. J Biol Chem 262:5592–5595

Arsenis G, Livingston JN (1986) Alterations in the tyrosine kinase activity of the insulin receptor produced by in vitro hyperinsulinemia. J Biol Chem 261:147–153

Avruch J, Nemenoff RA, Blackshear PJ, Pierce MW, Osathanondh R (1982) Insulin-stimulated tyrosine phosphorylation of the insulin receptor in detergent extracts of placental membranes. J Biol Chem 251:15162–15168

Ballotti R, Kowalski A, LeMarchand-Brustel Y, van Obberghen E (1986) Presence of an insulin-stimulated serine kinase in cell extracts from IM-9 cells. Biochem Biophys Res Commun 139:179–185

Beguinot F, Smith RJ, Kahn CR, Maron R, Moses AC, White MF (1988) Phosphorylational insulin-like growth factor I receptor by insulin receptor tyrosine kinase in intact cultured skeletal muscle cells. Biochemistry 27:3222–3228

Bernier M, Laird DM, Lane MD (1987) Insulin-activated tyrosine phosphorylation of a 15-kilodalton protein in intact 3T3-L1 adipocytes. Proc Natl Acad Sci USA 84:1844–1848

Blackshear PJ, Nemenoff RA, Avruch J (1984) Characteristics of insulin and EGF stimulation of receptor autophosphorylation in detergent extracts of rat liver and transplantable rat hepatomas. Endocrinology 114:141–152

Bollag GE, Roth RA, Beaudoin J, Mochly-Rosen D, Koshland DE Jr (1986) Protein kinase C directly phosphorylates the insulin receptor in vitro and reduces its protein kinase activity. Proc Natl Acad Sci USA 83:5822–5824

Boni-Schnetzler M, Rubin JB, Pilch PF (1986) Structural requirements for the transmembrane activation of the insulin receptor kinase. J Biol Chem 261:15281–15287

Brown S, Ghang MA, Racker E (1983) A rapid assay for protein kinase phosphorylating small polypeptides and other substrates. Anal Biochem 135:369–378

Burant CF, Treutelaar MK, Landreth GE, Buse MG (1984) Phosphorylation of insulin receptors solubilized from rat skeletal muscle. Diabetes 33:704–708

Burant CF, Treutelaar MK, Buse MG (1986) In vitro and in vivo activation of the insulin receptor kinase in control and denervated skeletal muscle. J Biol Chem 261:8985–8993

Caro JF, Shafer JA, Taylor SI, Raju SM, Perrotti N, Sinha MK (1987) Insulin stimulated protein phosphorylation in human liver plasma membranes: detection of endogenous or plasma membrane associated substrates for insulin receptor kinase. Biochem Biophys Res Comm 149:1008–1016

Casey PJ, Gilman AG (1988) G-protein involvement in receptor-effector coupling. J Biol Chem 263:2577–2580

Chen JJ, Kosower NS, Petryshyn R, London IM (1986) The effects of N-ethylmaleimide on the phosphorylation and aggregation of insulin receptors in the isolated plasma membranes of 3T3-F442A adipocytes. J Biol Chem 162:902–908

Chou CK, Dull TJ, Russell D, Gherzi R, Lebwohl D, Ullrich A, Rosen OM (1987) Human insulin receptors mutated at the ATP-binding site lack protein tyrosine kinase activity and fail to mediate post-receptor effects of insulin. J Biol Chem 262:1842–1847

Chratchko Y, Gazzano H, van Obberghen E, Fehlmann M (1984) Subunit arrangement of insulin receptors in hepatoma cells. Mol Cell Endocrinol 36:59–65

Colca JR, DeWald DB, Pearson JD, Palazuk BJ, Laurino JP, McDonald JM (1987) Insulin stimulates the phosphorylation of calmodulin in intact adipocytes. J Biol Chem 262:11399–11402

Cooper JA, Reiss NA, Schwartz RJ, Hunter T (1983) Three glycolytic enzymes are phosphorylated at tyrosine in cells transformed by Rous sarcoma virus. Nature 302:219–223

Cooper JA, Esch FS, Taylor SS, Hunter T (1984) Phosphorylation sites in enolase and lactate dehydrogenase utilized by tyrosine protein kinases in vivo and in vitro. J Biol Chem 259:7835–7841

Crettaz M, Jialal I, Kasuga M, Kahn CR (1984) Insulin receptor regulation and desensitization in rat hepatoma cells: the loss of the oligomeric forms of the receptor correlates with the change in receptor affinity. J Biol Chem 259:11543–11549

Cuatrecasas P (1972) Properties of the insulin receptor isolated from liver and fat cell membranes. J Biol Chem 247:1980–1991

Davis RJ, Czech MP (1985) Amiloride directly inhibits growth factor receptor tyrosine kinase. J Biol Chem 260:2543–2551

Decker SJ (1985) Phosphorylation of the erbB gene product from an AEV-transformed chick fibroblast cell line. J Biol Chem 260:2003–2006

Deibert DC, DeFronzo RA (1980) Epinephrine-induced insulin resistance in man. J Clin Invest 65:717–721

Duronio V, Jacobs S, Cuatrecasas P (1986) Complete glycosylation of the insulin and IGF-I receptors is not necessary for their biosynthesis and function: use of swainsonine as an inhibitor in IM-9 cells. J Biol Chem 261:910–975

Duronio V, Jacobs S, Romero PA, Herscovics A (1988) Effects of inhibitors of N-linked oligosaccharide processing on the biosynthesis and function of insulin and IGF-I receptors. J Biol Chem 263:5436–5445

Ebina Y, Ellis L, Jarnagin K, Edery M, Graf L, Clauser E, Ou JH, Masiarz F, Kan YW, Goldfine ID, Roth RA, Rutter WJ (1985) The human insulin receptor cDNA: the structural basis for hormone activated transmembrane signalling. Cell 40:747–758

Ebina Y, Araki E, Taira M, Shimada F, Mori M, Craik CS, Siddle K, Pierce SB, Roth RA, Rutter WJ (1987) Replacement of lysine residue 1030 in the putative ATP-binding region of the insulin receptor abolishes insulin- and antibody-stimulated glucose uptake and receptor kinase activity. Proc Natl Acad Sci USA 84:704–708

Ekman P, Ek B, Engstrom L (1987) The quantity of protein-bound [^{32}P]phosphotyrosine in hepatocytes and fibroblasts. The effects of tyrosine protein kinase activating agents. J Biochem (Tokyo) 101:863–870

Ellis L, Clauser E, Morgan DO, Edery M, Roth RA, Rutter WJ (1986) Replacement of insulin receptor tyrosine residue 1102 and 1163 compromises insulin-stimulated kinase activity and uptake of 2-deoxyglucose. Cell 45:721–732

Fernandez-Almonacid R, Rosen OM (1987) Structure and ligand specificity of the Drosophila melanogaster insulin receptor. Mol Cell Biol 7:2718–2727

Flockhart DA, Corbin JD (1982) Regulatory mechanisms in the control of protein kinases. CRC Crit Rev Biochem 12:133–186

Forsayeth J, Maddux B, Goldfine ID (1986) Biosynthesis and processing of the human insulin receptor. Diabetes 35:837–846

Forsayeth J, Caro JF, Sinha MK, Maddux BA, Goldfine ID (1987) Monoclonal antibodies to the human insulin receptor that activate glucose transport but not insulin receptor kinase activity. Proc Natl Acad Sci USA 84:3448–3451

Freidenberg GR, Klein HH, Cordera R, Olefsky JM (1985) Insulin receptor kinase activity in rat liver regulation by fasting and high carbohydrate feeding. J Biol Chem 260:12444–12453

Fujita-Yamaguchi Y (1984) Characterization of purified insulin receptor subunits. J Biol Chem 259:1206–1211

Fujita-Yamaguchi Y, Kathuria S (1985) The monomeric $\alpha\beta$ form of the insulin receptor exhibits much higher insulin-dependent tyrosine-specific protein kinase activity than the intact $\alpha_2\beta_2$ form of the receptor. Proc Natl Acad Sci USA 82:6095–6099

Fujita-Yamaguchi Y, Choi S, Sakamoto Y, Itakura K (1983) Purification of insulin receptor with full binding activity. J Biol Chem 258:5045–5049

Fukami Y, Nakamura T, Nakayama A, Kanehisa T (1986) Phosphorylation of tyrosine residues of calmodulin in RSV-transformed cells. Proc Natl Acad Sci USA 83:4190–4193

Gammeltoft S, van Obberghen E (1986) Protein kinase activity of the insulin receptor. Biochem J 235:1–11

Gherzi R, Russell DS, Taylor SI, Rosen OM (1987) Reevaluation of the evidence that an antibody to the insulin receptor is insulinomimetic without activating the protein tyrosine kinase activity of the receptor. J Biol Chem 262:16900–16905

Gibbs EM, Allard WJ, Lienhard GE (1986) The glucose transporter in 3T3-L1 adipocytes is phosphorylated in response to phorbol ester but not insulin. J Biol Chem 261:16597–16603

Gilmore T, DeClue JE, Martin GS (1985) Protein phosphorylation at tyrosine is induced by the v-*erb*B gene product in vivo and in vitro. Cell 40:609–618

Goldstein BJ, Muller-Wieland D, Kahn CR (1987) Variation in insulin receptor mRENA expression in human and rodent tissues. Mol Endocrinol 1:759–766

Goren HJ, Northrup JK, Hollenberg MD (1985) Action of insulin modulated by pertussis toxin in rat adipocytes. Can J Physiol Pharmacol 63:1017–1022

Goren HJ, White MF, Kahn CR (1987) Separate domains of the insulin receptor contain sites of autophosphorylation and tyrosine kinase activity. Biochemistry 26:2374–2381

Graves CB, Goewert RB, McDonald JM (1985) The insulin receptor contains a calmodulin binding domain. Science 230:827–829

Grigorescu F, White MF, Kahn CR (1983) Insulin binding and insulin-dependent phosphorylation of the insulin receptor solubilized from human erythrocytes. J Biol Chem 258:13708–13716

Grigorescu F, Flier JS, Kahn CR (1984) Defect in insulin receptor phosphorylation in erythrocytes are fibroblasts associated with severe insulin resistance. J Biol Chem 259:15003–15006

Hammerman MR, Gavin JR III (1984) Insulin-stimulated phosphorylation and insulin binding in canine renal basolateral membranes. Am J Physiol 247:F408–417

Haring HU, Kasuga M, Kahn RC (1982) Insulin receptor phosphorylation in intact adipocytes and in a cell-free system. Biochem Biophys Acta 108:1538–1545

Haring H, Kasuga M, White MF, Crettaz M, Kahn CR (1984a) Phosphorylation and dephosphorylation of the insulin receptor: evidence against an intrinsic phosphatase activity. Biochemistry 23:3298–3306

Haring HU, White MF, Kahn CR, Kasuga M, Lauris V, Fleischmann P, Murray M, Pawelek J (1984b) Abnormality of insulin binding and receptor phosphorylation in an insulin-resistant melanoma cell line. J Cell Biol 99:900–908

Haring HU, White MF, Kahn CR, Ahmad AA, DePaoli-Roach AA, Roach PJ (1985) Interaction of the insulin receptor kinase with serine/threonine kinases in vitro. J Cell Biochem 28:171–182

Haring H, Kirsch D, Obermaier B, Ermel B, Machicao F (1986a) Tumor-promoting phorbol esters increase the K_m of the ATP-binding site of the insulin receptor kinase from rat adipocytes. J Biol Chem 261:3869–3875

Haring H, Kirsch D, Obermaier B, Ermel B, Machicao F (1986b) Decreased tyrosine kinase activity of insulin receptor isolated from rat adipocytes rendered insulin-resistant by catecholamine treatment in vitro. Biochem J 234:59–66

Haring H, Obermaier B, Ermel B, Su Z, Mushack J, Rattenhuber E, Holzl J, Kirsch D, Machicao F, Herberg L (1987) Insulin receptor kinase defects as a possible cause of cellular insulin resistance. Diabete Metab 13:284–293

Harmon JT, Kempner EJ, Kahn CR (1981) Demonstration by radiation inactivation that insulin alters the structure of the insulin receptor in rat liver membranes. J Biol Chem 256:7719–7722

Havrankova J, Roth J, Brownstern M (1978) Insulin receptors widely distributed in the central nervous system of the rat. Nature 272:827–829

Hayes G, Lockwood DH (1986) The role of the cell surface sialic acid in insulin receptor function and insulin action. J Biol Chem 261:2791–2798

Hayes GR, Lockwood DH (1987) Role of insulin receptor phosphorylation in the insulinomimetic effects of hydrogen peroxide. Proc Natl Acad Sci USA 84:8115–8119

Hedo J, Collier E, Watkinson A (1987) Myristyl and palmityl acylation of the insulin receptor. J Biol Chem 262:954–957

Hedo JA, Kasuga M, van Obberghen E, Roth J, Kahn CR (1981) Direct demonstration of glycosylation of insulin receptor subunits by biosynthetic and external labelling: evidence for heterogeneity. Proc Natl Acad Sci USA 78:4791–4795

Heffetz D, Zick Y (1986) Receptor aggregation is necessary for activation of the soluble insulin receptor kinase. J Biol Chem 261:889–894

Heidenreich KA, Zahniser NR, Berhanu P, Brandenburg D, Olefsky J (1983) Structural differences between insulin receptors in the brain and peripheral target tissues. J Biol Chem 258:8527–8530

Herrera R, Rosen OM (1986) Autophosphorylation of the insulin receptor in vitro: designation of phosphorylation sites and correlation with receptor kinase activation. J Biol Chem 261:11980–11985

Herrera R, Petruzzelli L, Thomas N, Branson HN, Kaiser ET, Rosen OM (1985) An anti-peptide antibody that specifically inhibits insulin receptor autophosphorylation and protein kinase activity. Proc Natl Acad Sci USA 82:7899–7903

Heyworth CM, Houslay MD (1983) Insulin exerts actions through a distinct species of guanine nucleotide regulatory protein: inhibition of adenylyl cyclase. Biochem J 214:547–552

House C, Baldwin GS, Kemp BE (1984) Synthetic peptide substrates for the membrane tyrosine protein kinase stimulated by EGF. Eur J Biochem 140:363–367

Hresko RC, Bernier M, Hoffman RD, Flores-Riveros JR, Liao K, Laird DM, Lane MD (1988) Identification of phosphorylated 422(aP2) protein as pp15, the 15-kilodalton target of the insulin receptor tyrosine kinase in 3T3-L1 adipocytes. Proc Natl Acad Sci USA 85:8835–8839

Hunter T, Cooper JA (1986) Viral oncogenes and tyrosine phosphorylation. In: Boyer PD, Krebs E (eds) The enzymes, vol 17. Academic, Orlando, pp 191–246

Izumi T, White MF, Kadowaki T, Takaku F, Akanuma Y, Kasuga M (1987) Insulin-like growth factor rapidly stimulates tyrosine phosphorylation of a M_r 185000 protein in intact cells. J Biol Chem 262:1282–1287

Jacobs S, Cuatrecasas P (1986) Phosphorylation of receptors for insulin and IGF-I. Effects of hormones and phorbol esters. J Biol Chem 261:934–939

Jacobs S, Sahyoun NE, Saltiel AR, Cuatrecasas P (1983) Phorbol esters stimulate the phosphorylation of receptors for insulin and somatostatin C. Proc Natl Acad Sci USA 80:6211–6213

James DE, Zorzano A, Boni-Schnetzler M, Nemenoff RA, Powers A, Pilch PF, Ruderman NB (1986) Intrinsic differences of insulin receptor kinase activity in red and white muscle. J Biol Chem 261:14939–14944

Jonas HA, Newman JD, Harrison LC (1986) An atypical insulin receptor with high affinity for insulin-like growth factors co-purified with placental insulin receptors. Proc Natl Acad Sci USA 83:4124–4128

Joost HG, Steinfelder HJ, Schmitz-Salve C (1986) Tyrosine kinase activity of insulin receptors from human placenta. Effects of autophosphorylation and cAMP-dependent protein kinase. Biochem J 233:677–681

Kadota S, Fantus IG, Dragon G, Guyda HJ, Hersh B, Posner BI (1987a) Peroxide(s) of vanadium: a novel and potent insulinomimetic agent which activates the insulin receptor kinase. Biochem Biophys Res Comm 147:259–266

Kadota SI, Fantus IG, Deragon G, Guyda HJ, Posner BJ (1987b) Stimulation of IGF-II receptor binding and insulin receptor kinase activity in rat adipocytes: effects of vandate and H_2O_2. J Biol Chem 262:8252–8256

Kadowaki T, Fujita-Yamaguchi Y, Nishida E, Takaku F, Akiyama T, Kathuria S, Akanuma Y, Kasuga M (1985) Phosphorylation of tubulin and microtubule-associated proteins by the purified insulin receptor kinase. J Biol Chem 260:4016–4020

Kadowaki T, Koyasu S, Nishida E, Tobe K, Izumi T, Takaku F, Sakai H, Yahara I, Kasuga M (1987) Tyrosine phosphorylation of common and specific sets of cellular proteins induced by insulin, IGF-I, and EGF in an intact cell. J Biol Chem 262:7342–7350

Kahn CR (1985) The molecular mechanism of insulin action. Ann Rev Med 36:429–451

Kahn CR, Crettaz M (1985) Insulin receptors and the molecular mechanism of insulin action. Diabetes Metab Rev 1:5–32

Kahn CR, Baird KL, Jarett DB, Flier JS (1978) Direct demonstration that receptor cross-linking or aggregation is important in insulin action. Proc Natl Acad Sci USA 75:4209–4213

Kahn MN, Savoie S, Bergeron JJM, Posner BI (1986) Characterization of rat liver endosomal fractions. J Biol Chem 261:8462–8472

Kasuga M, Zick Y, Blithe DL, Crettaz M, Kahn CR (1982a) Insulin stimulates tyrosine phosphorylation of the insulin receptor in a "cell-free" system. Nature 298:667–669

Kasuga M, Hedo J, Yamada KM, Kahn CR (1982b) The structure of the insulin receptor and its subunits. Evidence for multiple non-reduced forms and a 210000 possible pro-receptor. J Biol Chem 257:10392–10399

Kasuga M, Karlsson FA, Kahn CR (1982c) Insulin stimulates the phosphorylation of the 95000-dalton subunit of its own receptor. Science 215:185–187

Kasuga M, Fujita-Yamaguchi Y, Blithe DL, Kahn CR (1983a) Tyrosine-specific protein kinase is associated with the purified insulin receptor. Proc Natl Acad Sci USA 80:2137–2141

Kasuga M, Fujita-Yamaguchi Y, Blithe DL, White MF, Kahn CR (1983b) Characterization of the insulin receptor kinase purified from human placental membranes. J Biol Chem 258:10973–10980

Kathuria S, Hartman S, Grunfeld C, Ramachandran J, Fujita-Yamaguchi Y (1986) Differential sensitivity of two functions of the insulin receptor to the associated proteolysis: kinase activity and hormone binding. Proc Natl Acad Sci USA 83:8570–8574

Kikuchi K, Schwartz C, Creasy S, Larner J (1981) Independent control of selected insulin-sensitive cell membrane and intracellular functions. Mol Cell Biochem 37:125–130

Koch R, Deger A, Jack HM, Klotz KN, Schenzle D, Kramer H, Kelm S, Muller G, Rapp R, Weber U (1986) Characterization of solubilized insulin receptors from rat liver microsomes. Existence of two receptor species with different binding properties. Eur J Biochem 154:281–287

Kono T, Barham FW (1971) Insulin-like effects of trypsin on fat cells. J Biol Chem 246:6204–6209

Korn LJ, Siebel CW, McCormick F, Roth RA (1987) Ras p21 as a potential mediator of insulin action in Xenopus oocytes. Science 236:840–843

Kwok YC, Nemenoff RA, Powers A, Avruch J (1986) Kinetic properties of the insulin receptor tyrosine protein kinase: activation through an insulin-stimulated tyrosine-specific intramolecular autophosphorylation. Arch Biochem Biophys 244:102–113

Lai Y, Nairn AC, Greengard P (1986) Autophosphorylation reversibly regulates the Ca^{2+}/calmodulin-dependence of Ca^{2+}/calmodulin-dependent protein kinase. II. Proc Natl Acad Sci USA 83:4253–4257

Leef JW, Larner J (1987) Insulin-mimetic effect of trypsin on the insulin receptor tyrosine kinase in intact adipocytes. J Biol Chem 262:14837–14842

Low MG, Saltiel AR (1988) Structural and functional roles of glycosyl-phosphatidylinositol in membranes. Science 239:268–275

Lowe WL Jr, Boyd FT, Clarke DW, Raizada MK, Hart C, LeRoith D (1986) Development of brain insulin receptors: structural and functional studies of insulin receptors from whole brain and primary cell cultures. Endocrinology 119:25–35

Luttrell LM, Hewlett EL, Romero G, Rogol AD (1988) Pertussis toxin treatment attenuates some effects of insulin in BC3HI murine myocytes. J Biol Chem 263:6134–6141

Magee AI, Siddle K (1986) Human insulin receptor contains covalently bound palmitic acid. Biochem Soc Trans 14:1103–1104

Maller JL, Pike LJ, Freidenberg GR, Cordera R, Stith BJ, Olefsky JM, Krebs EG (1986) Increased phosphorylation of ribosomal protein S6 following microinjection of insulin receptor-kinase into Xenopus oocytes. Nature 320:459–461

Massague J, Czech MP (1982) Role of disulfides in the subunit structure of the insulin receptor. J Biol Chem 257:6729–6738

Massague J, Pilch PF, Czech MP (1981a) Electrophoretic resolution of three major insulin receptor structures with unique subunit stoichiometries. Proc Natl Acad Sci USA 77:7137–7141

Massague J, Pilch PF, Czech MP (1981b) A unique proteolytic cleavage site on the β-subunit of the insulin receptor. J Biol Chem 256:3182–3190

McClain DA, Macgawa H, Lee J, Dull TJ, Ullrich A, Olefsky JM (1987) A mutant insulin receptor with defective tyrosine kinase displays no biologic activity and does not undergo endocytosis. J Biol Chem 262:14663–14671

Morgan DO, Roth RA (1987) Acute insulin action requires insulin receptor kinase activity: introduction of an inhibitory monoclonal antibody into mammalian cells blocks the rapid effects of insulin. Proc Natl Acad Sci USA 84:41–45

Morgan DO, Ho L, Korn LJ, Roth RA (1986) Insulin action is blocked by a monoclonal antibody that inhibits the insulin receptor kinase. Proc Natl Acad Sci USA 83:328–332

Morrison BD, Pessin JE (1987) Insulin stimulation of the insulin receptor kinase can occur in the complete absence of β-subunit autophosphorylation. J Biol Chem 262:2861–2868

Nemenoff RA, Kwok YC, Shulman GI, Blackshear PJ, Osathanondh R, Avruch J (1984) Insulin-stimulated tyrosine protein kinase. J Biol Chem 259:5058–5065

Nishizuka Y (1984) The role of protein kinase C in cell surface signal transduction and tumor promotion. Nature 308:693–698

O'Brien RM, Soos MA, Siddle K (1987a) Monoclonal antibodies to the insulin receptor stimulate the intrinsic tyrosine kinase activity by cross-linking receptor molecules. EMBO J 6:4033–4010

O'Brien RM, Houslay MD, Milligan G, Siddle K (1987b) The insulin receptor tyrosyl kinase phosphorylates holomeric forms of the guanine nucleotide regulatory proteins G_i and G_o. FEBS Lett 212:281–288

O'Brien R, Siddle K, Houslay MD, Hall A (1987c) Interaction of the human insulin receptor with the *ras* oncogene product p21. FEBS Lett 217:253–259

Okamoto M, White MF, Kahn CR (1988) Phosphorylation of the insulin receptor and its endogenous substrates in rat hepatocytes: characterization using anti-phosphotyrosine antibodies (submitted)

Pang DT, Shafer JA (1984) Evidence that insulin receptor from human placenta has a high affinity for only one molecule of insulin. J Biol Chem 259:8589–8596

Pang DT, Shafer JA (1985) Inhibition of the activation and catalytic activity of insulin receptor kinase by zinc and other divalent metal ions. J Biol Chem 260:5126–5130

Pang DT, Lewis SD, Sharma BR, Shafer JA (1984) Relationship between the subunit structure of insulin receptor and its competence to bind insulin and undergo phosphorylation. Arch Biochem Biophys 234:629–638

Pang D, Sharma B, Shafer JA (1985) Purification of the catalytically active phosphorylated form of insulin receptor kinase by affinity chromatography with O-phosphotyrosyl-binding antibodies. Arch Biochem Biophys 242:176–186

Perlman R, White MF, Kahn CR (1989) Conformational changes in the α- and β-subunits of the insulin receptor identified by anti-peptide antibodies. J Biol Chem 264:8946–8950

Perrotti N, Accili D, Marcus-Samuels B, Rees-Jones RW, Taylor SI (1987) Insulin stimulates phosphorylation of a 120-kDa glycoprotein substrate (pp120) for the receptor-associated protein kinase in intact H-35 hepatoma cells. Proc Natl Acad Sci USA 84:3137–3140

Pessin JE, Gitomer W, Oka Y, Oppenheimer CL, Czech MP (1983) β-Adrenergic regulation of insulin and EGF receptors in rat adipocytes. J Biol Chem 258:7386–7394

Petruzzelli LM, Ganguly S, Smith CJ, Cobb MH, Rubin CS, Rosen OM (1982) Insulin activates a tyrosine-specific protein kinase in extracts of 3T3-L1 adipocytes and human placenta. Proc Natl Acad Sci USA 79:6792–6796

Petruzzelli L, Herrera R, Rosen OM (1984) Insulin receptor in an insulin-dependent tyrosine protein kinase. Proc Natl Acad Sci USA 81:3327–3331

Petruzzelli L, Herrera R, Arenas-Garcia R, Fernandez R, Birnbaum M, Rosen OM (1986) Isolation of a *Drosophila* genomic sequence homologous to the kinase domain of the human insulin receptor and detection of the phosphorylated *Drosophila* receptor with an anti-peptide antibody. Proc Natl Acad Sci USA 83:4710–4714

Phillips SA, Perrotti N, Taylor SI (1987) Rat liver plasma membranes contain a 120 kDa glycoprotein which serves as a substrate for the tyrosine kinases of the receptors for insulin and EGF. FEBS Lett 212:141–144

Pike LJ, Kuenzel EA, Casneillie JE, Krebs EG (1984) A comparison of the insulin- and EGF-stimulated protein kinases from human placenta. J Biol Chem 259:9913–9921

Pike LJ, Eakes AT, Krebs EG (1986) Characterization of affinity-purified insulin receptors/kinase. Effects of dithiothreitol on receptors/kinase function. J Biol Chem 261:3782–3789

Pilch PF, Czech MP (1979) Interaction of cross-linking agents with the insulin effector system of isolated fat cells. J Biol Chem 254:3375–3381

Pilch PF, Czech MP (1980) Purification of the adipocyte insulin receptor by immunoaffinity chromatography. J Biol Chem 255:1722–1731

Pollet RJ, Haase BA, Standaert ML (1981) Characterization of detergent-solubilized membrane proteins. J Biol Chem 256:12118–12126

Ponzio G, Dolais-Kitabgi J, Louvard D, Gautier N, Rossi B (1987) Insulin and rabbit anti-insulin receptor antibodies stimulate additively the intrinsic receptor kinase activity. EMBO J 6:333–340

Rangel-Aldao R, Rose OM (1977) Effect of cAMP and ATP on the reassociation of phosphorylated and non-phosphorylated subunits of the cAMP-dependent protein kinase for bovine cardiac muscle. J Biol Chem 252:7140–7145

Reddy S, Kahn CR (1988) Insulin resistance: a look at the role of the insulin receptor kinase. Diabetic Med 5:621–629

Rees-Jones RW, Taylor SI (1985) An endogenous substrate for the insulin receptor-associated tyrosine kinase. J Biol Chem 260:4461–4467

Rees-Jones RW, Hendricks SA, Quarum M, Roth J (1984) The insulin receptor by rat brain is coupled to tyrosine kinase activity. J Biol Chem 259:3470–3474

Riedel H, Dull TJ, Schlessinger J, Ullrich A (1986) A chimaeric receptor allows insulin to stimulate tyrosine kinase activity of the EGF receptors. Nature 324:68–70

Rizza RA, Cryer PE, Haymond MW, Gerich JE (1980) Adrenergic mechanisms for the effects of epinephrine on glucose production and clearance in man. J Clin Invest 65:682–689

Ronnett GV, Knutson VP, Kohanski RA, Simpson TL, Lane MD (1984) Role of glycosylation in the processing of newly translated insulin proreceptor in 3T3-L1 adipocytes. J Biol Chem 259:4566–4575

Rosen OM (1987) After insulin binds. Science 237:1452–1458

Rosen OM, Herrera R, Olowe Y, Petruzzelli LM, Cobb MH (1983) Phosphorylation activates the insulin receptor tyrosine protein kinase. Proc Natl Acad Sci USA 80:3237–3240

Roth RA, Beaudoin J (1987) Phosphorylation of purified insulin receptor by cAMP kinase. Diabetes 36:123–126

Roth RA, Cassell MP (1983) Insulin receptor: evidence that it is a protein kinase. Science 219:299–301

Roth RA, Cassell DJ, Maddox BA, Goldfine ID (1983) Regulation of insulin receptor kinase activity by insulin mimickers and an insulin antagonist. Biochem Biophys Res Comm 115:245–252

Rothenberg PL, Kahn CR (1988) Insulin inhibits pertussis toxin catalyzed ADP ribosylation of G-proteins. J Biol Chem 263:15546–15552

Sadaoul J-L, Peyron J-F, Ballotti R, Debant A, Fehlman M, van Obberghen E (1985) Identification of a cellular 110 kDa protein substrate for the insulin-receptor kinase. Biochem J 227:887–892

Sale EM, White MF, Kahn CR (1987) Phosphorylation of glycolytic and gluconeogenic enzymes by the insulin receptor kinase. J Cell Biochem 33:15–26

Sale GJ, Fujita-Yamaguchi Y, Kahn CR (1986) Characterization of phosphatidylinositol kinase activity associated with the insulin receptor. Eur J Biochem 155:345–351

Schlessinger J, van Obberghen E, Kahn CR (1980) Insulin and antibodies against the insulin receptor cap on the membrane of cultured lymphocyte. Nature 286:729–731

Schworer CM, Colbran RJ, Soderling TR (1986) Reversible generation of a Ca^{2+}-independent form of Ca^{2+} (calmodulin)-dependent protein kinase II by an autophosphorylation mechanism. J Biol Chem 261:8581–8584

Sefton BM, Hunter T, Ball EH, Singer SJ (1981) Vinculin: a cytoskeletal target of the transforming protein of Rous sarcoma virus. Cell 24:165–174

Shemer J, Perrotti N, Roth J, LeRoith D (1987a) Characterization of an endogenous substrate related to insulin and IGF-I receptor in lizard brain. J Biol Chem 262:3436–3439

Shemer J, Adamo M, Wilson GL, Heffetz D, Zick Y, LeRoith D (1987b) Insulin and IGF-I stimulate a common endogenous phosphoprotein substrate (pp185) in intact neuroblastoma cells. J Biol Chem 262:15476–15482

Shia MA, Pilch PF (1983) The β-subunit of the insulin receptor is an insulin-activated protein kinase. Biochemistry 22:717–721

Shia MA, Rubin JB, Pilch PF (1983) The insulin receptor protein kinase. Physicochemical requirements for activity. J Biol Chem 258:14450–14455

Shoelson S, White MF, Kahn CR (1988a) Tryptic activation of the insulin receptor: proteolytic truncation of the α-subunit releases the β-subunit from inhibitory control. J Biol Chem 263:4852–4860

Shoelson SE, Boni-Schnetzler M, Kahn CR, Pilch PF (1988b) Autophosphorylation occurs with individual β-subunits of the insulin receptor. J Biol Chem (submitted)

Simon J, Rosebrough RW, McMurtry JP, Steele NC, Roth J, Adamo M, LeRoith D (1986) Fasting and refeeding alter the insulin receptor tyrosine kinase in chicken liver but fail to affect brain insulin receptors. J Biol Chem 261:17081–17088

Simpson IA, Hedo JA (1984) Insulin receptor phosphorylation may not be a prerequisite for acute insulin action. Science 223:1301–1304

Stadtmauer LA, Rosen OM (1983) Phosphorylation of exogenous substrates by the insulin receptor-associated protein kinase. J Biol Chem 258:6682–6685

Stadtmauer L, Rosen OM (1986a) Increasing the cAMP content of IM-9 cells alters the phosphorylation state and protein kinase activity of the insulin receptors. J Biol Chem 261:3402–3407

Stadtmauer L, Rosen OM (1986b) Phosphorylation of synthetic insulin receptor peptides by the insulin receptor kinase and evidence that the preferred sequence containing Tyr-1150 is phosphorylated in vivo. J Biol Chem 261:10000–10005

Suzuki S, Toyota T, Goto Y (1987) Characterization of the insulin receptor kinase from human erythrocytes. Endocrinology 121:972–979

Sweet LJ, Wilden PA, Spector AA, Pessin JC (1985) Incorporation of the purified human placental insulin receptor into phospholipid vesicles. Biochemistry 24:6571–6580

Sweet LJ, Wilden PA, Pessin JE (1986) Dithiothreitol activation of the insulin receptor/kinase does not involve subunit dissociation of the native receptor/kinase does not involve subunit dissociation of the native $\alpha_2\beta_2$ insulin receptor subunit complex. Biochemistry 25:7068–7074

Sweet LJ, Morrison BD, Pessin JE (1987a) Isolation of functional $\alpha\beta$ heterodimers from purified human placental $\alpha_2\beta_2$ heterotetrameric insulin receptor complex. J Biol Chem 262:6939–6942

Sweet LJ, Dudley DT, Pessin JE, Spector AA (1987b) Phospholipid activation of the insulin receptor kinase: regulation by phosphatidylinositol. FASEB J 1:55–59

Sweet LJ, Morrison BD, Wilden PA, Pessin JE (1987c) Insulin-dependent intermolecular subunit communication between isolated $\alpha\beta$ heterodimeric insulin receptor complexes. J Biol Chem 262:16730–16738

Takayama S, White MF, Lauris V, Kahn CR (1984) Phorbol esters modulate insulin receptor phosphorylation and insulin action in cultured hepatoma cells. Proc Natl Acad Sci USA 81:7797–7801

Takayama S, White MF, Kahn CR (1988) Phorbol ester-induced serine phosphorylation of the insulin receptor decreases its tyrosine kinase activity. J Biol Chem 262:3440–3447

Tamura S, Fujita-Yamaguchi Y, Larner J (1983) Insulin-like effect of trypsin on the phosphorylation of rat adipocyte insulin receptor. J Biol Chem 258:14749–14752

Tamura S, Brown TA, Whipple JH, Fujita-Tamaguchi Y, Dubler RE, Cheng K, Larner J (1984) A novel mechanism for the insulin-like effect of vanadate on glycogen synthase in rat adipocytes. J Biol Chem 259:6650–6658

Tanti J-F, Gremeaux T, Rochet N, van Obberghen E, LeMarchand-Brustel Y (1987) Effect of cAMP-dependent protein kinase on insulin receptor tyrosine kinase activity. Biochem J 245:19–26

Thomas AP, Martin-Requero A, Williamson JR (1985) Interactions between insulin and α_1-adrenergic agents in the regulation of glycogen metabolism. J Biol Chem 260:5963–5973

Tornqvist HE, Pierce MW, Frackelton AR, Nemenoff RA, Avruch J (1987) Identification of insulin receptor tyrosine residues autophosphorylated in vitro. J Biol Chem 262:10212–10219

Tornqvist HE, Gunsalus JR, Nemenoff RA, Frackelton AR, Pierce MW, Avruch J (1988) Identification of the insulin receptor tyrosine residues undergoing insulin-stimulated phosphorylation in intact rat hepatoma cells. J Biol Chem 263:350–359

Ullrich A, Bell JR, Chen EY, Herrera R, Petruzzelli LM, Dull TJ, Gray A, Coussens L, Liao Y-C, Tsubokawa M, Mason A, Seeburg PH, Grunfeld C, Rosen OM, Ramachandran J (1985) The human insulin receptor and its relationship to the tyrosine kinase family of receptor and its relationship to the tyrosine kinase family of oncogenes. Nature (London) 313:756–761

Van de Werve G, Proietto J, Jeanrenaud B (1985) Tumor-promoting phorbol esters increase basal and inhibit insulin-stimulated lipogenesis in rat adipocytes without decreasing insulin binding. Biochem J 225:523–527

van Obberghen E, Kowalski A (1982) Phosphorylation of the hepatic insulin receptor. FEBS Lett 143:179–182

van Obberghen E, Rossi B, Kowalski A, Gazzano H, Ponzion G (1983) Receptor-mediated phosphorylation of the hepatic insulin receptor: evidence that the M_r 95000 receptor subunit is its own kinase. Proc Natl Acad Sci USA 80:945–949

Walker DH, Kuppuswamy D, Visvanathan A, Pike LJ (1987) Substrate specificity and kinetic mechanism of human placental insulin receptor/kinase. Biochemistry 26:1428–1433

Wandosell F, Serrano L, Avila J (1987) Phosphorylation of α-tubulin carboxyl-terminal tyrosine prevents its incorporation into microtubules. J Biol Chem 262:8268–8273

White MF, Kahn CR (1986) The insulin receptor and tyrosine phosphorylation. In: Boyer PD, Krebs E (eds) The enzymes, vol 17. Academic, Orlando, pp 247–310

White MF, Haring HU, Kasuga M, Kahn CR (1984a) Kinetic properties and sites of autophosphorylation of the partially purified insulin receptor from hepatoma cells. J Biol Chem 259:255–264

White MF, Werth DK, Pasten I, Kahn CR (1984b) Phosphorylation of the solubilized insulin receptor by the gene product of the Rous sarcoma virus, pp60[src]. J Cell Biochem 26:169–179

White MF, Takayama S, Kahn CR (1985a) Differences in the sites of phosphorylation of the insulin receptor in vivo and in vitro. J Biol Chem 260:9470–9478

White MF, Maron R, Kahn CR (1985b) Insulin rapidly stimulates tyrosine phosphorylation of a M_r-185000 protein in intact cells. Nature 318:183–186

White MF, Stegman EW, Dull TJ, Ullrich A, Kahn CR (1987) Characterization of an endogenous substrate of the insulin receptor in cultured cells. J Biol Chem 262:9769–9777

White MF, Livingston JN, Backer JM, Lauris V, Dull T, Ullrich A, Kahn CR (1988a) Mutation of the insulin receptor at tyrosine 960 inhibits signal transduction but does not affect its tyrosine kinase activity. Cell 54:641–649

White MF, Shoelson SE, Keutmann H, Kahn CR (1988b) A cascade of tyrosine autophosphorylation in the β-subunit activates the phosphotransferase of the insulin receptor. J Biol Chem 263:2969–2980

Wilden PA, Pessin JE (1987) Differential sensitivity of the insulin-receptor kinase to thiol and oxidizing agents in the absence and presence of insulin. Biochem J 245:325–331

Wilden PA, Boyle TR, Swanson M, Sweet LJ, Pessin JE (1986) Alteration of intramolecular disulfides in insulin receptor kinase by insulin and dithiothreitol. Biochemistry 25:4381–4388

Woo DDL, Fay SP, Griest R, Coty W, Goldfine I, Fox CF (1986) Differential phosphorylation of the progesterone receptor by insulin, EGF and PDGF receptor tyrosine protein kinases. J Biol Chem 261:460–467

Yip CC, Moule ML (1983) Structure of the insulin receptor of rat adipocytes. Diabetes 32:760–767

Yip CC, Yeung CWT, Moule ML (1978) Photoaffinity labeling of insulin receptor of adipocyte plasma membrane. J Biol Chem 253:1743–1745

Yip CC, Yeung CWT, Moule ML (1980) Photoaffinity labeling of insulin receptor proteins of liver plasma membrane preparations. Biochemistry 19:70–76

Yu K-T, Czech MP (1984) Tyrosine phosphorylation of the insulin receptor β-subunit activates the receptor-associated tyrosine kinase activity. J Biol Chem 259:5277–5286

Yu K-T, Czech MP (1986) Tyrosine phosphorylation of insulin receptor β-subunit activates the receptor tyrosine kinase in intact H-35 hepatoma cells. J Biol Chem 261:4715–4722

Yu K-T, Werth, DK, Pastan IH, Czech MP (1985) src kinase catalyzes the phosphorylation and activation of the insulin receptor kinase. J Biol Chem 260:5838–5846

Yu K-T, Khalaf N, Czech MP (1987) Insulin stimulates the tyrosine phosphorylation of a $M_r = 160000$ glycoprotein in rat adipocyte plasma membranes. J Biol Chem 262:7865–7873

Zick Y, Whittaker J, Roth J (1983a) Insulin stimulated phosphorylation of its own receptor. J Biol Chem 258:3431–3434

Zick Y, Grunberger G, Podskalny JM, Moncada V, Taylor S, Gordon P, Roth J (1983b) Insulin stimulates phosphorylation of serine residues in soluble insulin receptors. Biochem Biophys Res Comm 116:1129–1135

Zick Y, Kasuga M, Kahn CR, Roth J (1983c) Characterization of insulin-mediated phosphorylation of the insulin receptor in a cell-free system. J Biol Chem 258:75–80

Zick Y, Rees-Jones RW, Taylor SI, Gorden P, Roth J (1984a) The role of antireceptor antibodies in stimulating phosphorylation of the insulin receptor. J Biol Chem 259:4396–4400

Zick Y, Sasaki N, Rees-Jones RW, Grunberger G, Nissley SP, Rechler MM (1984b) IGF-I stimulates tyrosine kinase activity in purified receptors from a rat liver cell line. Biochem Biophys Res Commun 119:6–13

Zick Y, Sagi-Eisenberg R, Pines M, Gierschik P, Spiegel AM (1986) Multisite phosphorylation of the α subunit of transducin by the insulin receptors and protein kinase C. Proc Natl Acad Sci USA 83:9294–9297

CHAPTER 12

Receptor-Mediated Internalization and Turnover

J. R. LEVY and J. M. OLEFSKY

A. Introduction

Binding of insulin to its specific cell surface receptor initiates a complicated array of events which encompass a variety of biochemical and physical steps. These include the generation of insulin's pleiotropic biologic effects, as well as endocytotic uptake of the hormone–receptor complex. Insulin receptor internalization facilitates the process of hormone degradation and also initiates the steps of receptor uptake, intracellular traffic, and recycling; these latter events can be referred to as the insulin receptor itinerary. We will review some recent findings and the current state of knowledge concerning the insulin receptor itinerary and speculate about its function.

B. The Insulin Receptor Itinerary

A general model for the endocytotic–intracellular pathway of the insulin receptor can be proposed (Fig. 1). The initial location of insulin–receptor complexes on the cell surface appears to vary with the particular cell line being studied. Quantitative ultrastructural analyses using monomeric feritin–insulin have dem-

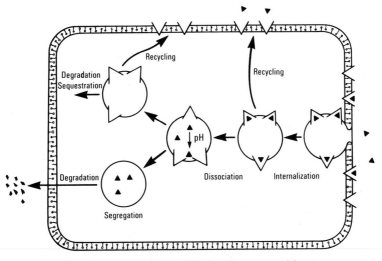

Fig. 1. Schematic model of the insulin receptor itinerary

onstrated that insulin receptors may be normally grouped together and randomly distributed over the cell surface such as in rat adipocytes (Jarett and Smith 1975; Smith and Jarett 1982, 1983) or are located primarily on microvillous projections of the cell and found as single or paired receptor sites such as in 3T3-L1 adipocytes (Smith et al. 1985). Once insulin binds to its receptor, the insulin–receptor complexes rapidly undergo endocytotic uptake. In some cell types this appears to involve a coated pit mechanism whereas in others it does not (Pilch et al. 1983; Posner et al. 1982; Carpentier et al. 1985; Goldberg et al. 1987). Following internalization the complexes are initially localized to vesicles that have been commonly referred to as "endosomes" (Bergeron et al. 1985). Within these endosomes, an ATP-dependent proton pump generates an acidic pH (<6.0) which facilitates rapid dissociation of insulin into the intravascular fluid compartment of the endosome, leaving the receptor in the endosomal membrane (Forgac et al. 1983; Galloway et al. 1983; Tycko et al. 1983; Levy and Olefsky 1988). After dissociation, the hormone and receptor physically segregate to follow separate pathways (Geuze et al. 1983; Levy and Olefsky 1988). The insulin remains dissolved in the fluid of the endosome, the bulk of which is transported forward to lysosomes and possibly other intracellular vesicular compartments (Bergeron et al. 1985). Once insulin dissociates and enters its separate pathway it is progressively degraded to intermediates and then to small molecular weight peptides. After ligand dissociation and segregation in discrete vesicles (CURL, compartment of uncoupling of receptor and ligand; Geuze et al. 1983), the receptor can recycle back to the plasma membrane (Brown et al. 1983; Heidenreich et al. 1984; Marshall et al. 1981), enter a degradative pathway (Heidenreich et al. 1984; Green and Olefsky 1982), or be sequestered within the cell (Deutsch et al. 1982; Krupp and Lane 1982). As indicated in Fig. 1, a "short loop" receptor recycling pathway may also exist, from which receptors can recycle early in this scheme prior to ligand dissociation. As suggested for the LDL receptor (Brown et al. 1983), this whole pathway can be termed the insulin receptor itinerary.

C. Insulin Processing and Dissociation

One important function of the endocytotic–intracellular pathway for insulin is to serve as a means to remove insulin from the circulation for subsequent hormone degradation. An early step in this process is internalization. Several laboratories have studied morphologically the binding and uptake of insulin. Like the non-hormonal ligands including low density lipoproteins, asialogylcoproteins, and α_2-macroglobulin (Goldstein et al. 1985; Goldberg et al. 1987), insulin was observed by autoradiography to undergo endocytosis by coated pits on rat hepatocytes, and IM-9-cultured human lymphocytes (Carpentier et al. 1981, 1985). In contrast, monomeric ferritin-labeled insulin and gold-labeled insulin (Goldberg et al. 1987) were observed to be excluded from coated pits on rat adipocytes and underwent endocytosis instead by micropinocytotic invaginations as shown in Fig. 2. Despite the various mechanisms in the different cell types, insulin, initially bound to the cell surface receptor, is taken into the cell via

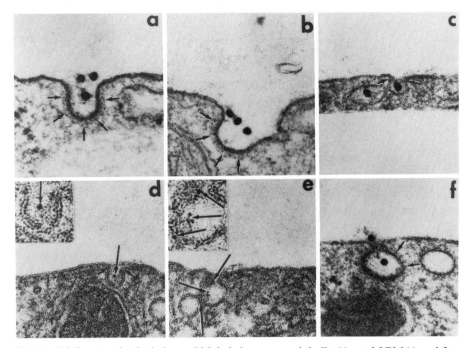

Fig. 2 a–f. Micrographs depicting gold-labeled α_2-macroglobulin (Au-α_2MGMA) and ferritin–insulin (Fm-I) on the cell surface at 4° C. Rat adipocytes were incubated with Au-α_2MGMA or FM-I for 2 h at 4° C (**a, b**) Au-α_2MGMA was concentrated within coated pits (*short arrows* indicate the clathrin coat). Although micrograph **b** is of lower contrast than micrograph **a**, the coat is still clearly defined. **c** Although Au-α_2MGMA was able to fit inside uncoated micropinocytotic invaginations, it was rarely found within these structures. Micropinocytotic invaginations were frequently connected to the cell surface by a narrow neck (*short arrows*). The innermost gold particle is contained within a membrane delimited structure that has no apparent connection to the cell surface; such particles were not classified as being on the cell surface. **d, e** Fm-I (*long arrows*) randomly occupied solitary uncoated micropinocytotic invaginations (**d**) and rosette-like complexes of these structures (**e**). The narrow neck of the invagination is indicated by *short arrows. Insets* show the Fm-I particles at higher magnification. **f** When cells incubated at 4° C were warmed to 37° C for 8 min, coated pits (*short arrows* indicate the clathrin coat) developed narrow necks, suggesting that they were in the process of pinching off to form coated vesicles. **a–e**, × 89 000; *inset* (Goldberg et al. 1987) × 184 000

an endocytotic process. Quantitative biochemical analyses have shown similar findings. This is demonstrated in Fig. 3, which summarizes experiments in which hepatocytes were incubated with ^{125}I-insulin at 37° C (Levy and Olefsky 1987). At the indicated times, the surface-bound radioactivity was removed by acid extraction; thus the nonextractable portion of the total cell-associated radioactivity represents internalized ^{125}I-insulin-derived material. Initially, almost all of the radiolabeled insulin is bound to surface receptors, as represented by acid-extractable radioactivity. After a short delay, insulin is internalized and represents between 40% and 50% of the total cell-associated radioactivity at the steady state. Internalization is higly temperature and energy sensitive; the frac-

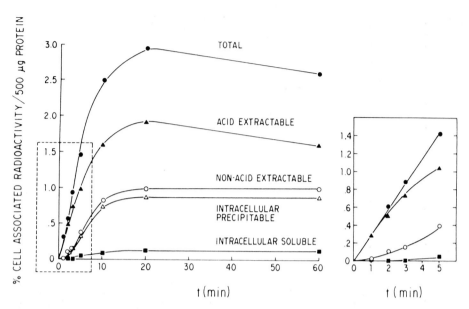

Fig. 3. Time course of binding and processing of $[^{125}I\text{-}A^{14}]$iodoinsulin in cultured hepatocytes. A time course of total specific cell-associated radioactivity (CAR) (*full circles*) was measured. At the above times, the cells were washed with ice-cold PBS, and binding buffer, titrated to pH 3.5, was added at 4° C to extract surface receptor-bound insulin. The time courses of acid-extractable radioactivity or surface-bound insulin (*full triangles*) and nonextractable CAR (*open circles*), representing internalized radioactivity, were calculated. Cells containing internalized radioactivity were solubilized and an equivolume of 15% TCA was added. The time course of intracellular TCA-precipitable radioactivity (*open triangles*) and soluble radioactivity (*full squares*) was measured. *Inset*, time course of the first 5 min of total binding (*full circles*), surface receptor binding (*full triangles*), internalization (*open circles*), and intracellular TCA-soluble radioactivity (FULL SQUARES). The data are identical, but the scale has been enlarged on the ordinate and abscissa to allow for a more detailed analysis

tion of the total cell-associated radioactivity which is non-acid-extractable in cells incubated at temperatures below 10° C or in the presence of the energy depletor, dinitrophenol, is negligible (OLEFSKY and KAO 1982). To estimate the elapsed time from insulin associated with surface receptors to intracellular ligand degradation to low molecular weight products, acid-washed cells were solubilized and the ability of the intracellular radioactivity to precipitate in trichloracetic acid (TCA) was then measured. TCA precipitates insulin and large insulin intermediates, but low molecular weight degradation products remain soluble. As shown in Fig. 3, low molecular weight degradation products appear by 5 min, and comprise about 10% of the total cell-associated radioactivity at the steady state. This finding has been confirmed by analyzing the intracellular radioactivity with Sephadex G-50 gel chromatography (LEVY and OLEFSKY 1987).

Since most of the internalized insulin receptors recycle back to the plasma membrane, as will be discussed in more detail below, the intracellular dissociation of the ligand–receptor complex must occur before the degradation of the

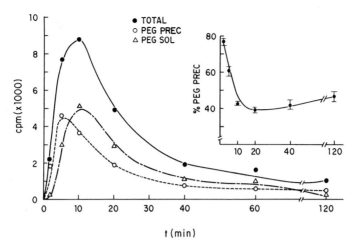

Fig. 4. Time course of intracellular, PEG-precipitable (insulin–receptor complexes) and PEG-soluble radioactivity in fibroblasts transfected with the cDNA for the human insulin receptor (HIRc). HIRc cells were incubated with 0.4 ng/ml ^{125}I-insulin for 2 h at 4° C. After washing the cells, fresh 37° C buffer was added at time 0 to initiate internalization in a relatively synchronous manner. At the indicated times the cells were cooled to 4° C, incubated with 4° C buffer (pH 3.5) to remove insulin bound to surface receptors, and solubilized. Total radioactive counts (*filled circles*), and the absolute amount of radioactive counts precipitable (*open circles*) and soluble (*open triangles*) in PEG were measured. *Inset*, time course of the percentage of intracellular radioactivity that is precipitable in PEG. The experiments were performed identically as above. The data are expressed as the absolute amount of radioactivity recovered in the precipitate divided by the total intracellular radioactivity × 100%

hormone, or before 5 min. Recent studies on insulin–receptor dissociation (LEVY and OLEFSKY 1988) have used a rat fibroblast cell line that has been transfected with a normal human insulin receptor gene; due to gene amplification, these cells express approximately 500 times the normal number of native rat insulin receptors. The insulin receptors of these cells display similar structure and binding characteristics as receptors of other insulin target tissues and internalize, recycle, and mediate insulin degradation in a normal manner (MCCLAIN et al. 1987). To quantitate the time course of insulin–receptor complex dissociation, radiolabeled insulin was incubated with cells for 2 h at 4° C. At time 0, the cells were reincubated with insulin-free 37° C buffer to initiate internalization in a relatively synchronous manner. After various times, the cells were cooled to 4° C, insulin bound to surface receptors was removed by incubation in acidic medium; the cells were then solubilized and the remaining radioactivity was analyzed for its ability to precipitate in polyethylene glycol (PEG). PEG selectively precipitates intact insulin–receptor complexes while free insulin and insulin degradation products remain soluble (CUATRECASAS 1972). As shown in Fig. 4, the total intracellular radioactivity increases rapidly as radiolabeled insulin is being internalized, reaches a peak by 10 min, and decreases thereafter as radioactivity is being released from the cell. By 2 min, almost all of the total intracellular radioactivity is composed of insulin–receptor complexes, confirming ultrastructural studies

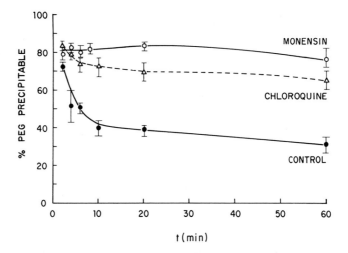

Fig. 5. Time course of the effect of monensin and chloroquine on intracellular dissociation of insulin from the insulin receptor. HIRc cells were incubated with ^{125}I-insulin at 37° C (in contrast to the "pulsed" experiments in Fig. 4) in the absence (*full circles*) or presence of 50 µ*M* monensin (*open circles*) or 100 µ*M* chloroquine (*open triangles*). At the indicated times, the cells were cooled, exposed to acid to remove surface-bound insulin, and then solubilized. The ability of the radioactivity to precipitate in 12.5% PEG was measured. The data are expressed as the absolute amount of radioactivity recovered in the precipitate divided by the total intracellular radioactivity × 100%

that insulin is internalized while associated with its receptor. However, by 5 min, a significant amount of intracellular dissociation has occurred, as evidenced by the appearance of PEG-soluble material. The rate of intracellular dissociation of insulin–receptor complexes can be estimated by calculating the percent of the total intracellular radioactivity which is PEG-precipitable with time. As seen in the inset in Fig. 4, the fractional rate of decline of insulin–receptor complexes is rapid, with a plateau by 10–20 min and a $t_{1/2}$ of approximately 7 min.

Studies in other ligand–receptor systems have demonstrated that dissociation may be linked to the acidification of specific endosomal vesicles, perhaps generated by proton pumps (HARFORD et al. 1983; ASCOLI 1984). It has been shown that α_2 macroglobulin (TYCKO and MAXFIELD 1982), and mannose–glycoprotein receptor complexes (WILEMAN et al. 1984) enter an acid environment very soon after internalization of the ligand–receptor complex has occurred, but before the ligand is transported to lysosomes. The carboxylic ionophore, monensin, reversibly raises the pH of these prelysosomal vesicles (TYCKO et al. 1983). As shown in Fig. 5, monensin completely inhibits the dissociation of insulin–receptor complexes (LEVY and OLEFSKY 1988); this most likely occurs by preventing acidification of prelysosomal vesicles. The weak base, chloroquine, is also very effective in inhibiting insulin–receptor dissociation.

Following intracellular dissociation, insulin and the insulin receptor physically segregate, enabling targeting of ligand to a degradative pathway and recycling of the receptor back to the cell surface. As shown in Fig. 5, monensin can prevent intracellular dissociation of insulin from its receptor and the ability of monensin

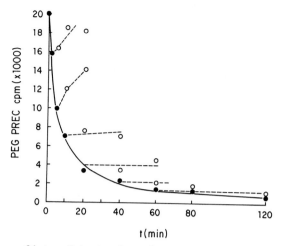

Fig. 6. Time course of intracellular insulin and receptor segregation. HIRc cells were incubated with 0.4 ng/ml ^{125}I-insulin and 50 µM monensin for 90 min at 37° C. The cells were washed in 4° C acidic binding buffer to remove surface receptor-bound insulin and reincubated at the indicated times in fresh, monensin-free, binding buffer at 37° C. At 2, 5, 10, 20, 40, and 60 min, monensin (50 µM) was re-added to parallel dishes of cells. Control cells (*full circles*) or cells retreated with monensin (*open circles*) were solubilized and the ability of the radioactivity to precipitate in 12.5% PEG was determined

to neutralize acidic vesicles can also be used to study segregation (WOLKOFF et al. 1984). The experiments are based on the theory that if the dissociated insulin and receptor remain in the same vesicle (i.e., not segregated) then monensin-induced vesicle neutralization should lead to reassociation or rebinding of insulin to the receptor. After physical segregation occurs, monensin will no longer promote reassociation, despite vesicle neutralization. Rat fibroblasts expressing high levels of insulin receptors due to the amplified human insulin receptor gene were incubated with radiolabeled insulin and monensin at 37° C for 2 h to permit the accumulation of intact intracellular insulin–receptor complexes (LEVY and OLEFSKY 1988). After removing surface–bound insulin with an acid wash, intracellular dissociation was initiated by reincubating the cells in fresh monensin–free buffer. As shown in Fig. 6, in the absence of monensin (*filled circles*) the amount of intact insulin–receptor complexes rapidly declines, due to intracellular dissociation. However, when monensin was re-added to cells before 5 min, there was a marked increase in the number of intact insulin–receptor complexes, reflecting reassociation of insulin with its receptor (*open circles*). However, after 5 min, the addition of monensin inhibited further insulin–receptor dissociation, but did not lead to any reassociation, suggesting that the dissociated insulin and receptor had now become physically segregated.

D. Insulin Degradation

The degradation of the insulin molecule has been extensively investigated and although it is possible that proteolytic cleavage begins at the cell surface, the

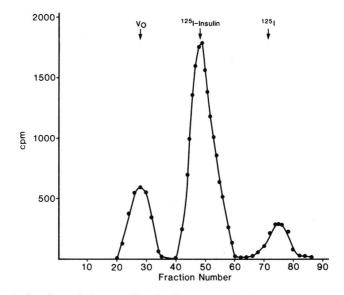

Fig. 7. Sephadex G-50 elution profile of cell-associated radioactivity from adipocytes incubated with $10^{-10}\,M$ ^{125}I-insulin. Adipocytes were incubated at 37° C for 1 h with $10^{-10}\,M$ ^{125}I-insulin. Following this, the cells were washed to remove all extracellular insulin and were then solubilized in a solution containing 0.1% Triton X-100 and 6 M urea, pH 2.5. The solubilized cell-associated radioactivity was then applied to a Sephadex G-50 column (50 × 1 cm) and eluted into 0.6 ml fractions. Data are from a representative experiment of seven separate studies

majority of the evidence shows that it occurs following internalization. Recent reports have shown that during the initial steps of insulin degradation by cells, products with molecular weight intermediate between intact insulin and individual amino acids or small peptides are generated (Duckworth et al. 1979; Misbin et al. 1980). These intermediate insulin degradative products are then ultimately metabolized to small peptides or amino acids. To better understand the intracellular processing of insulin, the ability of various insulin target tissues to produce such intermediates and the biologic activity and mechanisms of generation of these products has been examined.

Figure 7 demonstrates the gel filtration (Sephadex G-50) elution profile of cell-associated radioactivity derived from adipocytes incubated with ^{125}I-insulin at 37° C for 1 h. In these experiments, the cells were washed and extracellular radioactivity was removed prior to solubilization of cells and application of the solubilized radioactivity to the column. Thus, the data represent the elution profile of cell-associated (surface–bound plus intracellular) material. Three peaks can be defined: peak I contains large molecular weight material eluting with the void volume V_0; peak II represents the major peak eluting in the position of insulin; and peak III consists of small molecular weight degradation products (probably iodotyrosine). When the ability of 10% TCA to precipitate the radioactivity in these peaks was examined, it was found that these three peaks were 92%, 95%, and 12% TCA-precipitable respectively (Olefsky et al. 1979;

Table 1. Binding and biologic activity of "insulin" eluting in peak II of Sephadex G-50 column (% of control)[a]

	Fractions (Sephadex G-50)			
	40–45	46–50	51–55	56–60
Control cells				
Insulin binding	97	103	81	61
Biologic activity	94	107	78	57
Chloroquine-treated cells				
Insulin binding	95	106	98	90
Biologic activity	97	99	95	88

[a] Data represent the average of two studies in which the ability of the radioactive material eluting in the indicated fractions to bind to IM-9-cultured lymphocytes and stimulate ^{14}I-C-glucose oxidation by rat adipocytes was compared with an identical concentration of control ^{125}I-insulin. Results are expressed as % of control ^{125}I-insulin activity.

HAMMONS and JARETT 1980). However, although peak II comigrates with native insulin, Sephadex G-50 chromatography is relatively insensitive to subtle changes in molecular weight of the hormone. Consequently, the nature of the material in peak II was analyzed more precisely.

As seen in Fig. 7, peak II eluted in 20 fractions and the entire peak was divided into 4 parts (fractions 40–45, 46–50, 51–55, and 56–60). Next, the ability of the material in these fractions to bind to IM-9 lymphocyte insulin receptors or stimulate adipocyte glucose oxidation was measured. Table 1 shows the relative binding potency and biologic activity of these fractions, as compared to control ^{125}I-insulin. Not all the material in peak II is fully potent and the reduction in binding and biologic activity was most marked in the last quarter of peak II (fractions 56–60). This indicates that, in addition to insulin, peak II contains intermediate degradation products derived from insulin which have a similar molecular weight, but reduced biologic activity.

To further evaluate the possibility of intermediate degradation products in the cell-associated material which elutes in peak II, numerous investigators have pooled the fractions of peak II and further analyzed them by various techniques, including DEAE–Sephadex (OLEFSKY et al. 1979) and high performance liquid chromatography (HPLC) (STENTZ et al. 1983; PEAVY et al. 1985). Figure 8 shows the HPLC elution profile for insulin-sized material extracted from isolated rat hepatocytes (DUCKWORTH et al. 1988). Most of the labeled material eluted in the position of intact B26-[^{125}I]iodoinsulin, but predominent intermediate peaks were recovered at earlier elution times. Although the size of these insulin intermediates is approximately that of insulin, based on Sephadex G-50 column elution, they are markedly less hydrophobic than the intact molecule. Other studies have demonstrated that insulin intermediates were greater than 90% TCA-precipitable, but had impaired biological activity in regard to membrane binding, and decreased immunoprecipitability (STENTZ et al. 1985). These intermediates have been detected as early as 1 min after the addition of radiolabeled ligand to cells, and their formation has been decreased with various metabolic inhibitors of

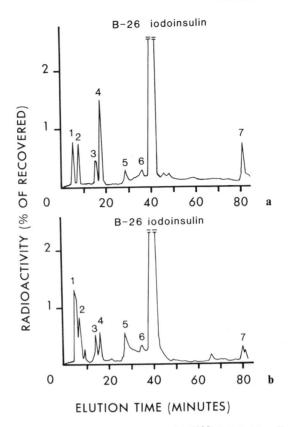

Fig. 8 a, b. Elution pattern from reverse-phase HPLC of ^{125}I-labeled insulin-sized material after incubation of [^{125}I]iodo (B26)insulin with insulin protease (**a**) and with isolated hepatocytes (**b**). Specifically labeled [^{125}I]iodo (B26)insulin was incubated either with purified insulin protease or with isolated hepatocytes, and the incubation medium was chromatographed on Sephadex G-50. The labeled material eluting in the position of [^{125}I]iodoinsulin was pooled, lyophilized, resuspended, and injection on a Zorbax C_8 reverse-phase column. Elution was with a series of gradient and isocratic steps with $(NH_4)_2HPO_4$/acetonitrile. The profile of radioactivity shown is typical of over ten experiments with enzyme and cells. Distinct and reproducible peaks are numbered 1–7, and the peak of intact [^{125}I]iodo (B26)insulin is identified. (DUCKWORTH et al. 1988)

insulin degradation such as chloroquine, dansylcadaverine, and bacitracin (PEAVY et al. 1985; KITABCHI and STENTZ 1985; HAMEL et al. 1987). To further examine their chemical nature, the insulin intermediate peaks detected by HPLC were pooled, lyophilized, and sulfitolized and the sulfitolized material was reanalyzed on HPLC. The products eluted in the position of authentic A chain showing that the intermediates contained intact A chain and that the metabolism of insulin occurs with initial alterations in the B chain (HAMEL et al. 1987). These results are consistent with the isolation of a protease which has a high affinity for insulin and thus was named insulin protease. This enzyme appears to cleave the B chain of insulin in regions between the disulfide bonds (DUCKWORTH et al. 1988), resulting in insulin intermediates which are more susceptible to nonspecific proteases that do not degrade the intact insulin molecule.

The subcellular compartment from which intermediate products are generated is currently unknown. It is possible that the initial cleavage steps resulting in intermediate insulin products occur while insulin is still bound to the receptor (i.e., prior to dissociation and segregation). In this event, one might expect release of some intermediate products into the extracellular space as a consequence of receptor recycling (see Fig. 1). Alternatively, intermediates might be generated within vesicles following dissociation prior to, or after physical segregation of insulin from its receptor. In this event, it is likely that intermediate products are committed to eventual terminal degradation, probably within lysosome-like structures, with ultimate release as small molecular weight products.

E. Retroendocytosis

In most insulin-sensitive target tissues, the majority of the internalized insulin is ultimately degraded and released as low molecular weight degradation products. However, a significant amount of the internalized hormone can be realesed intact, in proportions ranging from 20% in isolated rat adipocytes (MARSHALL 1985b); LEVY and OLEFSKY 1986) to 90% in endothelial cells (KING and JOHNSON 1985). The time course of release of intracellular radioactivity from cultured hepatocytes is shown in Fig. 9a. In this experiment, the cells were incubated with radiolabeled insulin for 20 min at 37° C, surface-bound insulin was extracted with a 4° C acidic buffer, and the cells were reincubated in fresh insulin-free medium at 37° C (LEVY and OLEFSKY 1987). Intracellular radioactivity was released from the cells in a time- and temperature-dependent manner. The appearance of radioactivity in the medium and its ability to precipitate with 7.5% TCA is shown in Fig. 9b. TCA-soluble material (i.e., degraded insulin) was quickly released over the first 10 min and more gradually thereafter. TCA-precipitable material was also released; after 45 min, approximately 40% of the total radioactivity released into the medium was TCA-precipitable. Further analysis with Sephadex G-50 chromatography and HPLC of radioactivity in the medium confirmed that intact insulin was released from the intracellular compartment of cells. This implies that a certain portion of the internalized insulin does not enter the degradative pathway. Most likely, this intact material is released from cells during the process of receptor recycling. Thus, as indicated in Fig. 1, a short loop, or rapid pathway for receptor recycling may exist and internalized insulin-receptor complexes may be recycled to the plasma membrane via this pathway prior to ligand dissociation. In this event, insulin, bound to receptors which enter this possible short loop recycling pathway, would be protected from intracellular degradation and would be released into the extracellular space following reinsertion of the receptor into the membrane. A similar process, termed retroendocytosis has been described for a variety of other ligand receptors, including the LDL receptor in smooth muscle cells (AULINSKAS et al. 1981). As a variation on this theme, some of the receptors recycled through the major

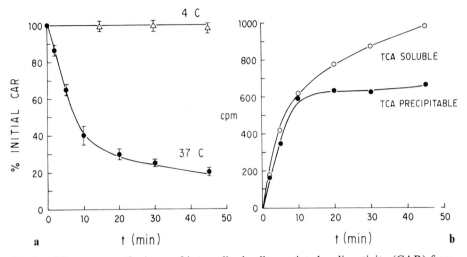

Fig. 9. a Time course of release of internalized cell-associated radioactivity (CAR) from hepatocytes. Hepatocytes were incubated for 20 min at 37° C with 0.4 µg/ml [$^{125}I = A^{14}$]iodoinsulin, and surface-bound insulin was removed with a 4° C acidic (pH 3.5) buffer. After washing five times with ice-cold PBS, insulin-free binding buffer at 4° or 37° C was added, and at the indicated times, CAR was measured. Each data point represents the mean ± SEM of three experiments, each done in triplicate. **b** Time course of appearance of TCA-precipitable (*full circles*) and TCA-soluble (*open circles*) material from the internal compartment of hepatocytes. The experiments were performed at 37° C using the protocol described above. The radioactivity in medium was analyzed for its ability to precipitate in 7.5% TCA at the given times after fresh insulin-free medium was added

pathway could still retain bound intact insulin, either through incomplete acidification-induced dissociation or incomplete segregation of insulin from the vesicles containing receptors targeted for recycling. This too would lead to release of undegraded insulin.

The process of insulin internalization and intracellular metabolism serves to clear bound hormone from the cell surface and deliver the ligand to intracellular sites of degradation. The structural signals which allow cells to target insulin receptors for sorting through the endocytotic recycling pathway are unknown.

F. Insulin Receptor Processing

After internalization, insulin receptors can be recycled, degraded, or retained within the cells, and these pathways have been the subject of considerable investigation. To study these events, the technique of photoaffinity labeling has been utilized. With this method, a photoreactive derivative of insulin is covalently attached to the binding site of the insulin receptor as the result of a photochemical reaction. The insulin derivative contains a functional group such as an arylazide moiety which upon irradiation yields a highly reactive intermediate (in the case of arylazide derivatives a nitrene free radical) capable of inserting into C–H or C–C bonds within the immediate vicinity of the binding site. The use of photoreactive analogs to study insulin receptors was first introduced

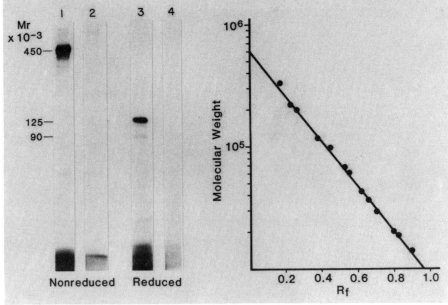

Fig. 10 a, b. Identification and molecular weight determination of photoaffinity-labeled insulin receptors on rat adipocytes. **a** Autoradiogram showing proteins on rat adipocytes covalently labeled with ^{125}I-NAPA-DP-insulin. Isolated adipocytes were incubated with ^{125}I-NAPA-DP-insulin (40 ng/ml) at 16° C for 30 min in the absence (lanes 1 and 3) or in the presence (lanes 2 and 4) of excess unlabeled insulin. After photolysis, adipocytes were solubilized and subjected to electrophoresis on a 5%–15% porous acrylamide gel as described in the text. The apparent molecular weights of the specifically labeled bands are indicated at the left of the figure, **b** semilog plot of molecular weight vs electrophoretic mobility of standard proteins subjected to electrophoresis as described above. (HEIDENREICH et al. 1983)

by Yip et al. who modified the NH_2 terminal of the B chain of insulin by addition of a phenylazide moiety (YIP et al. 1978, 1980). Other photoreactive insulin derivatives have been synthesized and characterized by Brandenburg et al. (WISHER et al. 1980; THAMM et al. 1980). The photoreactive derivatives of insulin are receptor agonists and are highly selective in labeling insulin receptors. The insulin analog used in the studies from our laboratory is iodinated (2-nitro, 4-azidophenylacetyl)-des-Phe^{B1}-insulin (^{125}I-NAPA-DP-insulin). This derivative has 80%–90% of insulin's biological activity (BRANDENBURG et al. 1980). Another attractive feature of this molecule is that it can be activated at a wavelength of light (360 nm) which does not damage cellular proteins. Thus, photoaffinity labeling offers advantages over other affinity cross-linking techniques for functional studies of viable cells because covalent attachment of insulin to its receptor occurs specifically at the ligand binding site without cross-linking other cellular proteins and without loss of cell viability. Thus, this technique allows one to selectively follow the metabolism of occupied receptors.

In the studies depicted in Fig. 10, adipocytes were covalently labeled with ^{125}I-NAPA-DP-insulin at 16° C (HEIDENREICH et al. 1984). The proteins

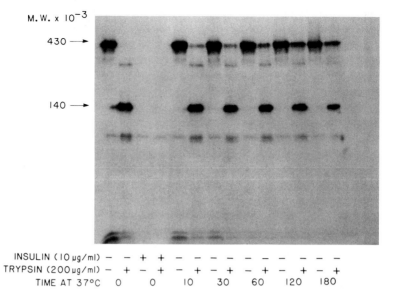

M. W. x 10⁻³

INSULIN (10 µg/ml)	−	−	+	+	−	−	−	−	−	−	−	−	−
TRYPSIN (200 µg/ml)	−	+	−	+	−	+	−	+	−	+	−	+	−	+
TIME AT 37°C	0		0		10		30		60		120		180	

Fig. 11. Autoradiogram of surface and internalized insulin receptors following incubation of adipocytes at 37° C. Adipocytes were photoaffinity labeled at 16° C. Following incubation at 37° C, half of the cells were solubilized immediately and the remainder were treated with trypsin (200 µg/ml, 37° C, 10 min) prior to solubilization. The solubilized samples were subjected to electrophoresis on 5%–15% acrylamide gels. (HEIDENREICH et al. 1984)

covalently labeled with ¹²⁵I-NAPA-DP-insulin were then identified by electrophoresis and autoradiography. When electrophoresis was performed under nonreducing conditions, only one protein band with an apparent molecular weight of 430000 was labeled (Fig. 10, lane 1). The labeling of this band was specific since labeling was abolished in the presence of excess unlabeled porcine insulin (Fig. 10, lane 2). When photolabeled receptors were reduced with dithiothreitol prior to electrophoresis, the major labeled protein was 125 kdalton, consistent with the fact that the 125 kdalton α subunit contains the binding site of the insulin receptor.

Having demonstrated that insulin receptors could be specifically photoaffinity labeled at 16° C, the cells were warmed to 37° C to determine if the labeled receptors were internalized and processed. Results from a typical experiment are seen in the autoradiogram of Fig. 11 (HEIDENREICH et al. 1984). At 16° C, the specifically labeled insulin receptor with an apparent molecular weight of 430000 (Fig. 11, lane 1) is readily seen. If cells were treated with trypsin prior to solubilization, all of the labeled insulin receptors were converted into smaller tryptic fragments (Fig. 11, lane 2) and the predominant tryptic fragment had an apparent molecular weight of 140000. This indicated that all of the photoaffinity-labeled receptors were on the cell surface at 16° C. When cells were photoaffinity labeled at 16° C and then incubated at 37° C, the amount of radiolabeled insulin receptors decreased as a function of time (Fig. 11, lanes 5, 7, 9, 11, 13). To determine the distribution of receptors between the cell surface and the intracellular compartment over these time intervals, a separate aliquot of cells was exposed to trypsin prior to solubilization at each time point. These experiments demonstrated that an increasing proportion of the labeled insulin receptors became insensitive to enzymatic degradation by trypsin (Fig. 11, lanes 6, 8, 10, 12, 14), indicating internalization of photolabeled insulin receptors. Thus, the autoradiogram of Fig. 11 shows that: (a) there was a decrease in the total amount of radioactivity in the cellular insulin receptor band, indicating degradation of hormone receptor complexes; and (b) there was an increase in the trypsin-insensitive (intracellular) pool of insulin receptors.

In order to quantitatively measure the time course of internalization and degradation of insulin receptors at 37° C, adipocytes were photoaffinity labeled at 16° C and further incubated at 37° C as described in Fig. 11. At appropriate times, the cells were solubilized, subjected to electrophoresis, sliced, and counted in a gamma counter. Degradation was assessed by counting the radioactivity in the insulin receptor peak (430 kdalton) as a function of time at 37° C. Susceptibility to tryptic digestion was used to distinguish between insulin receptors on the cell surface and those inside the cell. The results from these experiments are seen in Fig. 12 (HEIDENREICH et al. 1984). The amount of labeled insulin receptors recovered from the cell (indicated by the *full circles*) decreased rapidly during the first 60 min at 37° C. After this time, there was little net loss of radiolabeled insulin receptors from the cell. If plotted on a first-order rate plot, the degradation curve was clearly biphasic. An estimated $t_{1/2}$ of the initial phase was 1.5 h. The internalization of insulin receptors (indicated by the *open circles*) occurred at a significantly faster rate than the loss of receptors from the cell. This resulted in an accumulation of receptors inside the cell as seen in the inset of Fig. 12. The proportion of intracellular insulin receptors reached an apparent steady state after 30 min, representing about 20% of the labeled receptors originally on the cell surface. The faster rate of internalization relative to the net loss of receptors from the cell and the steady state distribution of receptors between the intracellular and surface pools indicates that internalization is not the rate-limiting step in the overall loss of insulin receptors from the cell.

Receptor occupancy appears to be an important factor in modulating the rate and, perhaps, the mechanism of receptor degradation. When cells are incubated in the absence of hormone, insulin receptors have a half-life of approximately 10 h. This rate of degradation has been measured in 3T3-L1 preadipocytes (RONNETT et al. 1982; REED et al. 1981) and chick liver cells (KRUPP and LANE 1981, 1982) using heavy isotope density shift procedures and IM-9 lymphocytes with biosynthetic labeling techniques (THAMM et al. 1980). When receptors are occupied with insulin they undergo more rapid endocytosis and some of them are degraded via a lysosomal pathway. Data from our studies indicate that degradation is relatively rapid ($t_{1/2} = 2$–3 h) and sensitive to cycloheximide and chloroquine (HEIDENREICH et al. 1983). Results from other laboratories also support the concept that occupancy regulates degradation of receptors. For example, KASUGA et al. (1981) reported that the half-life of insulin receptors decreased from 9–12 h to 3 h when cultured cells were exposed to insulin. In addition, insulin enhanced the degradation rate of its receptors in 3T3-L1 adipocytes as assessed by heavy isotope density shift procedures (REED et al. 1981).

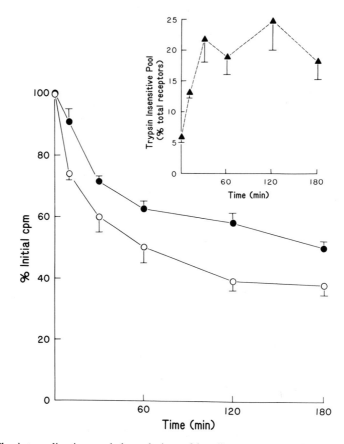

Fig. 12. The internalization and degradation of insulin receptors at 37° C. Experiments were carried out as described in Fig. 6. Instead of autoradiography, the gels were sliced and counted on a gamma counter. Degradation (*full circles*) was assessed by quantitating the amount of radioactivity in the insulin receptor band as a function of time at 37° C. Susceptibility of tryptic digestion was used to distinguish between insulin receptors on the cell surface and those inside the cell. A decrease in the number of receptors susceptible to trypsin represents loss of receptors from the cell surface (*open circles*). *Inset,* the accumulation of insulin receptors inside the cell. (Heidenreich et al. 1984)

Initial evidence of recycling of insulin receptors was produced with studies using the weak base, chloroquine (Heidenreich et al. 1984). As shown in Fig. 13, chloroquine markedly inhibited the net loss of insulin receptors from the cell (Fig. 13a), but had no effect on the internalization of surface receptors (Fig. 13b). As a result, the intracellular pool of insulin receptors increased by more than twofold compared to control cells (Fig. 13c). The effect of chloroquine on receptor degradation raises the possibility that lysosomes are involved in the degradation of internalized receptors. On the other hand, chloroquine can also interfere with the processing of internalized cell surface receptors by inhibiting acidification of endosomal vesicles (Levy and Olefsky 1988) or fusion of endocytotic vesicles with other intracellular membrane structures (McKanna et al. 1979).

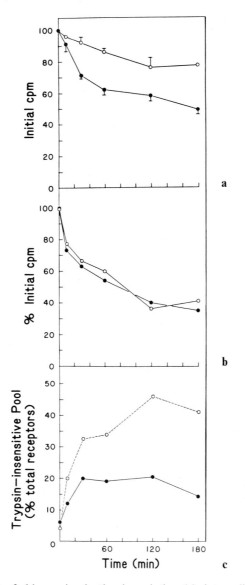

Fig. 13 a–c. The effect of chloroquine in the degradation (**a**), internalization (**b**) and accumulation (**c**) of insulin receptors within the cell. Experiments were performed as described in Fig. 12 in the absence (*open circles*) and presence (*full circles*) of 0.2 µ*M* chloroquine. (HEIDENREICH et al. 1984)

Regardless of the precise mechanism of action of this agent, the results clearly indicate that insulin receptors are degraded by a chloroquine-sensitive pathway.

As shown in Fig. 12, after 1 h at 37° C, the net disappearance of insulin receptors from the cell surface slowed markedly and the overall loss of insulin receptors from the cell was minimal. The apparent plateau in the time course of inter-

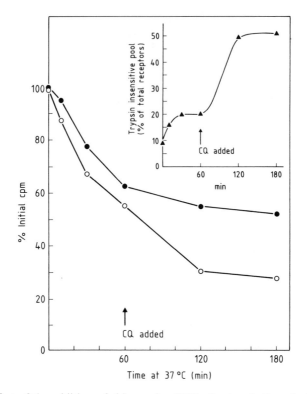

Fig. 14. Effect of the addition of chloroquine (CQ) after incubating photoaffinity-labeled cells at 37° C for 1 h. Experiments were performed as described in Fig. 12 with the exception that chloroquine (0.2 m*M*) was added at 1 h (Heidenreich et al. 1984)

nalization and overall degradation could be due to cessation of these processes over time. Alternatively, this plateau (approximately 20–60 min at 37° C) could reflect a new dynamic steady state in which the process of receptor internalization is nearly balanced by receptor recycling with only minimal net degradation of the accumulated intracellular receptors. To distinguish between these possibilities, the photolabeled cells were treated with chloroquine at 60 min when net loss of receptors was minimal. One would predict that if both internalization and intracellular degradation of insulin receptors had ceased at 60 min, then chloroquine should have little or no effect when added at this time. However, if the apparent plateau reflects a new equilibrium reached between the processes of internalization and recycling, then chloroquine should cause a perturbation in the plateau. The experiments were carried out by photoaffinity labeling surface receptors and allowing internalization and degradation to proceed for 60 min. At this time, 0.2 m*M* chloroquine was added and the cells were incubated for an additional 2 h at 37° C. The results are shown in Fig. 14 (Heidenreich et al. 1984). As can be seen, chloroquine had no significant effect on the overall loss of total cellular receptors. However, a marked change in the distribution of cellular receptors between the intracellular and surface pools was observed. Thus, follow-

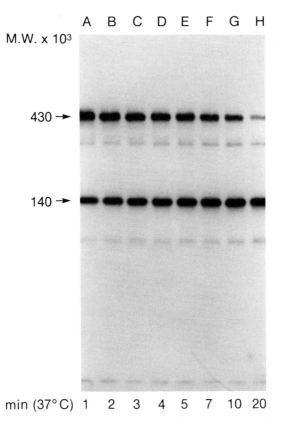

Fig. 15. Recycling of photoaffinity-labeled insulin receptors. Following photoaffinity labeling at 16° C, adipocytes were incubated at 37° C for 20 min. Trypsin (100 µg/ml) was then added to the cells, and the samples were further incubated at 37° C for the indicated times (*lanes A–H*)

ing the addition of chloroquine, there was a marked increase in net loss of receptors from the cell surface and this was accompanied by a corresponding twofold increase in the intracellular pool of surface-derived insulin receptors. This redistribution was clearly evident 1 h after the addition of chloroquine and a new apparent steady state was reached. These results strongly suggest that the plateau of net receptor loss from the cell surface observed in control cells represents a new steady state between the loss of receptors from the cell surface (i.e., internalization) and recycling of receptors back to the plasma membrane.

Recently, experiments using photoaffinity-labeled insulin receptors were performed that provide direct evidence that receptors are recycled in rat adipocytes (HUECKSTEADT et al. 1986). Receptor recycling was examined using trypsin to distinguish receptor location. Following photoaffinity labeling at 16° C, adipocytes were incubated at 37° C for 20 min. Trypsin was then added to the cells, and the samples were further incubated at 37° C for various times. At 37° C, trypsin digested all of the surface insulin receptors by 1 min; therefore, any remaining intact receptors represent those inside the cell (Fig. 15, lane A). As seen in Fig. 15,

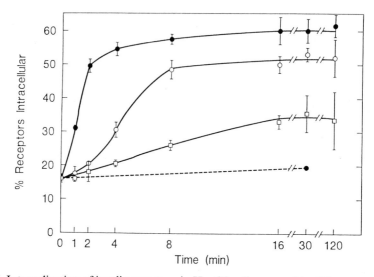

Fig. 16. Internalization of insulin receptors in HepG2 cells exposed to different concentrations of insulin. Insulin at 100 ng/ml (*full circles*), 10 ng/ml (*open circles*), or 1 ng/ml (*open squares*) was added to dishes of HepG2 cells at 37° C. Control cultures (*dashed line*) were exposed to insulin at 4° C. At various times (abscissa), cultures were removed, washed in ice-cold buffer at pH 4, and either trypsinized, or not, in duplicate as described. Total insulin receptors in nontrypsinized cells and intracellular receptors in trypsinized cells were determined as described and plotted as the percentage of intracellular receptors at each time. Results are the means ($\pm$ SEM) of four experiments, each done in duplicate

with increasing time of exposure to trypsin, the size of the intracellular pool of receptors decreased, and the tryptic fragments of the insulin receptors increased. This indicated that internalized receptors (that were once trypsin-resistant) became trypsin-sensitive by returning to the cell surface. Thus, the loss of receptors from the intracellular pool represented insulin receptor recycling. Internalized photoaffinity-labeled insulin receptors recycled rapidly to the cell surface at 37° C with a $t_{1/2}$ of recycling of 6 min and nearly all of the internalized receptors returned to the cell surface by 40 min. Insulin receptor recycling was both temperature- and energy-dependent. Thus, at 12° C or in the presence of dinitrophenol, the loss of radioactivity in the 430 kdalton band was negligible.

Other studies performed by measuring radiolabeled ligand binding to total, intracellular, and cell surface receptors have also convincingly demonstrated in isolated rat adipocytes (Marshall 1985a) and cultured hepatocytes (McClain and Olefsky 1988) that internalized receptors are recycled back to the plasma membrane in an energy- and temperature-dependent manner. Furthermore, exposure to native insulin results in internalization and recycling of receptors, the rate and extent of which is dependent on the insulin concentration (McClain and Olefsky 1988). The results of an experiment of insulin receptor processing in hepatocytes is shown in Fig. 16. Cells were incubated with various concentrations of insulin at 37° C, and at different times the numbers of total and intracellular (trypsin-resistant) receptors were quantified. In resting cells, 16% of the insulin

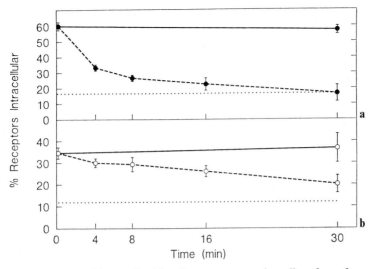

Fig. 17 a, b. Recycling of internalized insulin receptors to the cell surface after removal of insulin from HepG2 cells. Cells were treated with either 100 ng/ml (*filled circles,* **a**) or 1 ng/ml (*open circles,* **b**) of insulin for 30 min at 37° C. After this treatment, parallel cultures were acid washed (*dashed lines*) to remove bound insulin and returned to 37° C in insulin-free medium at 0 min. At various times duplicate washed and unwashed cultures were removed and the percentage of intracellular receptors was determined as described in the text. The *dotted line* represents the level of intracellular receptors in cells that had not been exposed to insulin. Results are the means (± SEM) of five independent experiments

receptors were intracellular. When exposed to 100 ng/ml insulin at 37° C, receptors rapidly internalize and reach a steady state by 8 min with approximately 60% of the receptors intracellular. Cells exposed to lower concentrations of insulin internalize fewer receptors and require longer to reach a new equilibrium. When extracellular and surface receptor-bound insulin is removed by an acid wash, the percentage of intracellular insulin receptors returns to that of resting cells. However, the rate of decrease of intracellular receptors (i.e., the rate of recycling) is much greater in cells incubated with the higher concentrations of insulin as shown in Fig. 17. These results imply qualitatively different fates for receptors internalized after exposure to different concentrations of insulin. Drugs or treatments known to affect receptor metabolism support the concept of distinct intracellular fates of receptor processing. For instance, hypotonic shock and hypokalemia, which arrest coated pit formation, blocked internalization of insulin and insulin receptors at low concentrations of insulin, but allowed internalization in response to high concentrations of insulin. Monensin and chloroquine caused intracellular accumulation of receptors internalized at low concentrations of insulin, but had a relatively smaller effect on receptors internalized at higher concentrations of insulin. These experiments raise the possibility that the high dose pathway, unlike the low dose pathway, may proceed independently of coated pits and endosomal acidification. This high dose pathway would be consistent with the short loop, recycling, pathway as shown in the schematic model in Fig. 1.

The trafficking of insulin receptors may differ between various cell types. For instance, treatment of chick liver cells with insulin for 18 h resulted in a 60% loss of cell surface receptors with no change in the number of total receptors (KRUPP and LANE 1981). Since insulin affected neither receptor synthesis nor degradation, it was concluded that insulin-induced receptor downregulation occurs through translocation of receptors from the plasma membrane to the cell interior where they are sequestered. In summary, once inside the cell there are at least three possible fates for the internalized insulin receptors. Receptors can be recycled back to the cell surface, degraded, or sequestered undegraded within the cell. Although the initial internalization event is clearly accelerated when receptors are occupied with insulin, the balance between recycling, degradation, and intracellular sequestration appears to differ among cell types. The relative magnitudes of recycling, degradation, and sequestration of internalized receptors modulate the final concentration of cellular receptors and the proportion of these receptors found at the cell surface or in the cell interior. It seems likely that these events determine the acute effect of insulin to downregulate its receptor, and thus regulate the sensitivity of the cell to insulin.

The mechanisms underlying the initial internalization event are still unknown. However, recent advances using molecular biologic techniques have provided important information regarding the initial cellular signal for receptor internalization and the subsequent degradation of the ligand. The β subunit of the insulin receptor is a tyrosine kinase which is activated after insulin binds to the extracellular α subunit (see Chap. 10). Several recent reports have indicated

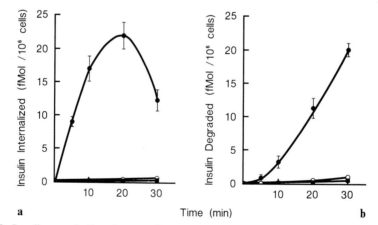

Fig. 18. Insulin metabolism by transfected cell lines. The cDNAs encoding the normal human insulin receptor (HIRc) and a receptor that had lysine residue 1018 replaced by alanine (A/K 1018) were used to transfect Rat 1 fibroblasts. HIRc-B (*full circles*), A/K 1018-B (*open circles*), or Rat 1 (*filled squares*) cells ($2-5 \times 10^5$ cells/35-mm dish) were exposed to 0.07 nM ^{125}I-insulin at 37° C as described. At various times after insulin addition, the medium was removed to assay for insulin degradation, and the cells were rinsed with ice-cold PBS. Surface-bound insulin counts/min internalized at 0.07 nM and 40% of the total counts/min degraded at 0.07 nM, was subtracted to yield specific internalization (**a**) and degradation (**b**) of insulin. Results of four experiments $\pm$ SE were normalized to fmol insulin/10^6 cells

that this tyrosine kinase activity may be central to insulin action and may mediate the internalization of insulin-receptor complexes (ROSEN 1987), MCCLAIN et al. 1987; HARI and ROTH 1987). The latter evidence is based on experiments where rat fibroblasts were transfected with either the normal cDNA for the human insulin receptor, or with a cDNA that lacks tyrosine kinase activity due to mutagenesis of lysine 1018 which disrupts the ATP binding sites of the α subunit. It is unlikely that this single amino acid substitution (alanine for lysine) alters the tertiary structure of the receptor in other domains. Since the α subunit remains uncharged, the transfected cells, invariably demonstrate normal insulin binding (MCCLAIN et al. 1987).

The transfected cells were then exposed to tracer concentrations of radiolabeled insulin to investigate whether the transfected normal or mutated receptors (A/K 1018) mediated insulin internalization and degradation. As can be seen in Fig. 18, transfected normal receptors (1.2×10^6 receptors/cell) in rat fibroblasts mediate insulin metabolism. Significant amounts of insulin were internalized and degraded by these cells; by 30 min insulin degradation is so extensive that the extracellular insulin concentration falls, leading to a decrease in the amount of intracellular insulin. In contrast, internalization and degradation of insulin is markedly inhibited in cells transfected with the mutated insulin

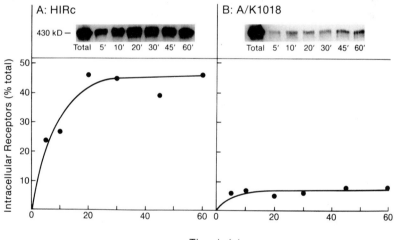

Time (min)

Fig. 19 A, B. Internalization of photoaffinity-labeled HIRc and A/K 1018 receptors. Receptors on HIRc-B (**A**) or A/K 1018-B (**B**) cells at 0° C were labeled with ^{125}I-NAPA-DP-insulin. Cells were warmed to 37° C to allow internalization to proceed, and at various times cultures were treated with trypsin in order to quantitate trypsin-resistant (intracellular) receptors. Control cells not warmed and not treated with trypsin ("total," *insets*) were used to quantitate total receptors. In cells treated with trypsin without warming (not shown), 95%–98% of the receptors were degraded by trypsin. Solubilized cells were fractionated by nonreducing SDS-polyacrylamide gel electrophoresis and subjected to autoradiography. The insulin receptor bands at 430-kdalton (*inset*) from three independent experiments were excised and counted in a gamma counter. The trypsin-resistant (intracellular) counts/min are plotted as a percentage of total counts/min from nontrypsinized cells as a function of time

receptor cDNA. When internalization of insulin receptors, rather than of insulin itself, was measured, a similar pattern of results emerged. Figure 19 demonstrates time courses of intracellular (trypsin-resistant) photoaffinity-labeled normal and A/K 1018 receptors. The kinetics of internalization of the normal receptors are very similar to those seen in other mammalian cells (see above) with a steady state reached at approximately 20 min. In A/K 1018 cells, very little receptor internalizes with time; even after 60 min, only 5% of the A/K 1018 receptors have escaped trypsin sensitivity. Finally, the insulin-induced loss of insulin receptors (downregulation) was examined. In fibroblasts transfected with normal receptors, insulin causes a time- and dose-dependent loss of total and surface receptors such that after 24 h of treatment, 72% of the total cellular insulin receptors have been lost. Similar treatment led to no loss of A/K 1018 receptors. Similar findings have been reported by other investigators (Russel et al. 1987) and demonstrate that autophosphorylation of the β subunit is a necessary step in insulin receptor endocytosis and emphasize the central role of receptor-mediated internalization on hormone degradation and receptor downregulation.

Insulin association with the α subunit clearly induces internalization of the insulin receptor through stimulation of β subunit tyrosine kinase. However, there is, presently, convincing evidence that receptors also internalize and recycle independent of ligand binding (Hari and Roth 1987). In a study utilizing Chinese hamster ovary (CHO) cells transfected with the cDNA for wild-type human insulin receptors or mutated human receptors lacking tyrosine kinase activity, all plasma membrane proteins were labeled nonspecifically with ^{125}I. Human insulin receptors were detected by immunoprecipitation with an anti-human insulin receptor antibody that did not cross-react with CHO insulin receptors. After initiating internalization by warming cells to 37° C, a time course of appearance of trypsin-resistant radiolabeled receptors was measured. As shown in Fig. 20, in the absence of insulin, normal receptors and receptors lacking tyrosine kinase constitutively internalize at a similar rate. However, insulin markedly increases the rate and magnitude of internalization of normal receptors, but has no effect on the mutant receptors.

In summary, insulin receptors are in a dynamic steady state and are actively being processed even in resting, nonstimulated cells. The constitutive internalization and degradation of receptors must be balanced by the cellular processes of receptor recycling and synthesis of new receptors since the concentration of surface receptors remains constant. However, when insulin binds to the α subunit of the insulin receptor, this balance is perturbed. In the short term, internalization of receptors, mediated through the stimulation of tyrosine kinase activity of the β subunit, is markedly enhanced. After prolonged exposure to insulin, receptors downregulate. This is due to an insulin-induced increase in receptor turnover as a result of enhanced receptor degradation. Interestingly, the insulin-induced receptor downregulation in the HIRc cells provides supporting evidence for this concept. Thus, the cDNA encoding HIRc receptor does not contain genomic, regulatory sequences. Therefore, the downregulation observed in these cells does not involve changes at the level of gene transcription. Consistent with this is the recent finding that downregulation of insulin receptors in hepatoma cells is not accompanied by changes in levels of insulin receptor on RNA (Hatada et al. 1988).

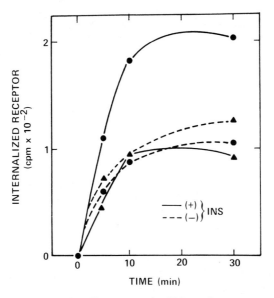

Fig. 20. Internalization of the insulin receptor in Chinese hamster ovary cells transfected with a normal human insulin receptor cDNA (*full circles*) and a mutant cDNA coding for an insulin receptor that lacks tyrosine kinase activity (*full triangles*). Cells were incubated in the presence or absence of 1 μM insulin (*INS*) and the amount of internalized receptor was determined after the indicated period of time, as described in the text. (HARI and ROTH 1987)

In addition to the biochemical techniques now available to distinguish between surface and intracellular receptors, and the rate of flux of receptors through the various pathways, subcellular fractionation and morphologic techniques have contributed to the localization of intracellular receptors and their potential physiologic function. Reports have shown localization of internalized insulin receptor complexes to Golgi-associated vesicles (BERGERON et al. 1985) which probably represent endocytotic vesicles traversing the Golgi area. Whether these internalized receptors associate with lysosomes, or lysosomal-related structures is currently unresolved. Overall, the published evidence is consistent with the idea that somewhere after formation of the endosomal vesicles, insulin–receptor complexes traverse the Golgi region in endocytotic vesicles where dissociation and segregation occurs. The initial steps of insulin degradation may occur in these compartments, but the peptide components targeted for degradation are probably delivered to lysosomal structures (which may themselves reside in the Golgi region, at least initially) after passing through the Golgi region in endosomal-derived vesicles. Consistent with this, Posner et al. (KHAN et al. 1981) have found that internalized ^{125}I-insulin radioactivity can be sequentially localized to heavy, intermediate, then light Golgi-associated fractions before the major association with acid phosphatase-containing vesicles occurs. Furthermore, the Golgi-associated vesicles which contain the internalized insulin are endosomal in nature and do not contain galactosyl transferase.

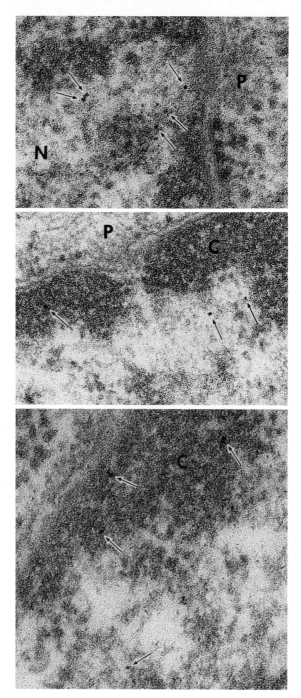

Fig. 21. Nuclear localization of ferritin–insulin (Fm-I). 3T3-L1 adipocytes were incubated for 60 min at 37° C with 20 ng/ml Fm-I. These three electron micrographs demonstrate Fm-I particles (*arrow*) inside the nucleus (*N*). The particles were frequently found near a nuclear pore (*P*), and there usually were multiple Fm-I particles in the same general area. Most of the nuclear-associated Fm-I particles were found in or at the periphery of the condensed chromatin (*C*). (SMITH and JARETT 1987) (× 158 000)

Internalized receptors may exert biologic effects at their subcellular localization by serving as a transducing mechanism conveying information from the surface to the cell interior. Recent reports have localized ferritin–insulin receptors (SMITH and JARETT 1987) and photoaffinity-labeled insulin receptors (PODLECKI et al. 1987) to the nuclei of 3T3-L1 adipocytes and isolated hepatocytes, respectively. As shown in Fig. 21, electron micrographs demonstrate ferritin–insulin particles inside the nucleus after incubating 3T3-L1 adipocytes with the ligand for 60 min at 37° C. The particles were frequently found near a nuclear pore. However, most of the nuclear-associated ferritin-insulin particles were found in or at the periphery of the condensed chromatin. These findings are supported by biochemical studies showing that radiolabeled insulin or photoaffinity-labeled insulin receptors (PODLECKI et al. 1987) accumulate in isolated nuclei after subfractionation. Although the receptors that translocate to the nucleus involve only a small percentage of the total receptors internalized, this may be a potential mechanism for mediating some of insulin's long-term cellular actions such as control of gene transcription.

The insulin receptor itinerary has many potential physiologic functions. Ligand internalization and intracellular metabolism may serve as one homeostatic mechanism for removing ligand from the circulation. This mechanism is particularly plausible in the liver where the endocytotic uptake and degradation of insulin is responsible for removing approximately 40%–50% of the insulin that is secreted from the pancreas into the portal circulation (KADEN et al. 1973). In addition, the itinerary of ligand–receptor complexes may modulate the biologic responsiveness of insulin-sensitive target tissues. Activation of a biologic response depends on the association of insulin with its receptor and the subsequent stimulation of the β subunit tyrosine kinase. Little is known about the termination or deactivation of a cellular response. However, it seems possible that the dissociation of intracellular insulin–receptor complexes, mediated through the acidification of intracellular endosomes, may be important in terminating the signalling events for insulin action. The balance between internalization, recycling, degradation, synthesis, and intracellular sequestration of receptors after short and prolonged exposure to insulin regulates the number of cellular receptors, and thus the sensitivity of the cell to insulin. Finally, internalized receptors may be specifically directed to intracellular sites, such as the nucleus, where they elicit some of insulin's pleiotropic biologic effects. Further investigations with the use of biochemical, subfractionation, and morphologic techniques may further clarify the role of the insulin receptor itinerary.

References

Ascoli M (1984) Lysosomal accumulation of the hormone-receptor complex during receptor-mediated endocytosis of human choriogonadotropin. J Cell Biol 99:1242–1250

Aulinskas TH, van der Westhuyzen DR, Bierman EL, Gevers W, Coetzze GA (1981) Retroendocytosis of low density lipoproteins by cultured bovine aortic smooth muscle cells. Biochem Biophys Acta 664:255–261

Bergeron JJM, Cruz J, Khan MN, Posner BI (1985) Uptake of insulin and other ligands into receptor-rich endocytic components of target cells: the endosomal apparatus. Ann Rev Physiol 47:383–403

Brandenburg D, Diaconescu C, Saunders D, Thamm P (1980) Covalent linking of photoreactive insulin to adipocytes produces a prolonged signal. Nature 286:821–822

Brown MS, Anderson RGW, Goldstein JL (1983) Recycling receptors: the round-trip itinerary of migrant membrane proteins. Cell 32:633–667

Carpentier J-L, van Obberghen EV, Gorden P, Orci L (1981) Surface redistribution of ^{125}I-insulin in cultured human lymphocytes. J Cell Biol 91:17–25

Carpentier J-L, Fehlman M, van Obberghen E, Gordon P, Orci L (1985) Redistribution of ^{125}I-insulin on the surface of rat hepatocytes as a function of dissociation time. Diabetes 34:1002–1007

Cuatrecasas P (1972) Isolation of the insulin receptor of liver and fat-cell membranes. Proc Natl Acad Sci USA 69:318–322

Deutsch PJ, Rosen OM, Rubin CS (1982) Identification and characterization of a latent pool of insulin receptors in 3T3-L1 adipocytes. J Biol Chem 257:5350–5358

Duckworth WC, Stentz F, Heinemann M, Kitabchi AE (1979) Initial site of cleavage of insulin by insulin protease. Proc Natl Acad Sci USA 76:635–642

Duckworth WC, Hamel FG, Peavy DE, Liepnicks JJ, Ryan MP, Hemodson MA, Frank BH (1988) Degradation products of insulin generated by hepatocytes and by insulin protease. J Biol Chem 263:1826–1833

Forgac M, Cantley L, Wiedenmann B, Altstiel L, Branton D (1983) Clathrin-coated vesicles contain an ATP-dependent protein pump. Proc Natl Acad Sci USA 80:1300–1303

Galloway CJ, Dean GE, Marsh M, Rudnick G, Mellman I (1983) Acidification of macrophage and fibroblast endocytic vesicles in vitro. Proc Natl Acad Sci USA 80:3334–3338

Geuze HJ, Slot JW, Strous GJAM, Lodish HF, Schwartz AL (1983) Intracellular site of asialoglycoprotein receptor-ligand uncoupling: double-label immunoelectron microscopy during receptor-mediated endocytosis. Cell 32:277–287

Goldberg RI, Smith RM, Jarett L (1987) Insulin and α_2-macroglobulin-methylamine undergo endocytosis by different mechanisms in rat adipocytes. J Cell Physiol 133:203–212

Goldstein JL, Brown MS, Anderson RGW, Russell DW, Schneider WJ (1985) Receptor-mediated endocytosis: concepts emerging from the LDL receptor system. In: Annual review of cell biology, vol 1. Annual Reviews, Palo Alto

Green A, Olefsky JKM (1982) Evidence for insulin-induced internalization and degradation of insulin receptors in rat adipocytes. Proc Natl Acad Sci USA 79:427–431

Hamel FG, Peavy DE, Ryan MP, Duckworth WC (1987) HPLC analysis of insulin degradation products from isolated hepatocytes. Diabetes 36:702–708

Hammons GT, Jarett L (1980) Lysosomal degradation of receptor-bound ^{125}I-labeled insulin by rat adipocytes: its characterization and dissociation from the short-term biologic effects of insulin. Diabetes 29:475–486

Harford J, Bridges K, Ashwell G, Klausner RD (1983) Intracellular dissociation of receptor bound asialoglycoproteins in cultured hepatocytes. J Biol Chem 258:3191–3197

Hari J, Roth RA (1987) Defective internalization of insulin and its receptor in cells expressing mutated insulin receptors lacking kinase activity. J Biol Chem 262:15341–15344

Hatada EN, McClain DA, Potter E, Ullrich A, Olefsky JM (1989) Effects of growth and insulin treatment on the levels of insulin receptors and their mRNA in HepG2 cells. J Biol Chem 264:6741–6747

Heidenreich KA, Berhanu P, Brandenburg D, Olefsky JM (1983) Degradation of insulin receptors in rat adipocytes. Diabetes 32:1001–1009

Heidenreich KA, Brandenburg D, Berhanu P, Olefsky JM (1984) Metabolism of photoaffinity-labeled insulin receptors by adipocytes. Role of internalization, degradation, and recycling. J Biol Chem 259:6511–6515

Huecksteadt T, Olefsky JM, Brandenburg D, Heidenreich KA (1986) Recycling of photoaffinity-labeled insulin receptors in rat adipocytes. J Biol Chem 261:8655–8659

Jarett L, Smith RM (1975) Ultrastructural localization of insulin receptors on adipocytes. Proc Natl Acad Sci USA 72:3526–3530

Kaden M, Harding P, Field JB (1973) Effect of intraduodenal glucose administration on hepatic extraction of insulin in the anesthetized dog. J Clin Invest 52:2016–2022

Kasuga M, Kahn CR, Hedo JA, van Obberghen E, Yamada KM (1981) Insulin-induced receptor loss in cultured human lymphocytes is due to accelerated receptor degradation. Proc Natl Acad Sci USA 78:6917–6921

Khan MN, Posner BI, Verma AK, Khan JR, Bergeron JJM (1981) Intracellular hormone receptors: evidence for insulin and lactogen receptors in a unique vesicle sedimenting in lysosome fraction of rat liver. Proc Natl Acad Sci USA 78:4980–4984

King GL, Johnson SM (1985) Receptor mediated transport of insulin across endothelial cells. Science 227:1583–1585

Kitabchi AE, Stentz FB (1985) The effect of inhibitors of insulin processing on generating of insulin intermediate products from human fibroblast as detected by high performance liquid chromatography. Biochem Biophys Res Commun 128:163–170

Krupp MN, Lane MD (1981) On the mechanism of ligand-induced down regulation of insulin receptor level in the liver cell. J Biol Chem 256:1689–1694

Krupp MN, Lane MD (1982) Evidence for different pathways for the degradation of insulin and insulin receptor in the chick liver cell. J Biol Chem 257:1372–1377

Levy JR, Olefsky JM (1986) Retroendocytosis of insulin in rat adipocytes. Endocrinology 119:572–579

Levy JR, Olefsky JM (1987) The trafficking and processing of insulin and insulin receptors in cultured rat hepatocytes. Endocrinology 121:2075–2086

Levy JR, Olefsky JM (1988) Intracellular insulin-receptor dissociation and segregation in a rat fibroblast cell line transfected with a normal human insulin receptor gene. J Biol Chem 263:6101–6108

Marshall S (1985a) Kinetics of insulin receptor internalization and recycling in adipocytes: shunting of receptors to a degradative pathway by inhibitors of recycling. J Biol Chem 260:4136–4142

Marshall S (1985b) Dual pathways for the intracellular processing of insulin. J Biol Chem 260:13524–13530

Marshall S, Green A, Olefsky JM (1981) Evidence for recycling of insulin receptors in isolated rat adipocytes. J Biol Chem 256:11464–11470

McClain DA, Olefsky JM (1988) Evidence for two independent pathways of insulin receptor internalization in hepatocytes and hepatoma cells. Diabetes 37:806–815

McClain DA, Maegawa H, Lee J, Dull TJ, Ulrich A, Olefsky JM (1987) A mutant insulin receptor with defective tyrosine kinase displays no biologic activity and does not undergo endocytosis. J Biol Chem 262:14663–14671

McKanna JA, Haigler HT, Cohen S (1979) Hormone receptor topology and dynamics: morphological analysis using ferritin-labeled epidermal growth factor. Proc Natl Acad Sci USA 76:5689–5693

Misbin RI, Davies JG, Offord RE, Halban PA, Mehl TD (1980) Binding and degradation of semisynthetic tritiated insulin by IM-9 cultured human lymphocytes. Diabetes 29:730–735

Olefsky JM, Kao M (1982) Surface binding and rates of internalization of ^{125}I-insulin in adipocytes and IM-9 lymphocytes. J Biol Chem 257:8667–8673

Olefsky JM, Kobayashi M, Chang H (1979) Interactions between insulin and its receptors after the initial binding event. Functional heterogeneity and relationships to insulin degradation. Diabetes 28:460–471

Peavy DE, Hamel FG, Kincke VL, Duckworth WC (1985) Evidence that bacitracin alters intracellular insulin metabolism in isolated rat hepatocytes. Diabetes 34:217–225

Pilch PF, Shia MA, Benson RJJ, Fine RE (1983) Coated vesicles participate in the receptor-mediated endocytosis of insulin. J Cell Biol 93:133–138

Podlecki DA, Smith RM, Kao M, Tsai P, Huecksteadt T, Brandenburg D, Lasher RS, Jarett L, Olefsky JM (1987) Nuclear translocation of the insulin receptor. J Biol Chem 262:3362–3368

Posner BI, Khan MN, Bergeron JJM (1982) Endocytosis of peptide hormones and other ligands. Endocrine Rev 3:280–298

Reed BC, Ronnett GV, Clements PR, Lane MD (1981) Regulation of insulin receptor synthesis and degradation. J Biol Chem 256:3917–3925

Ronnett GV, Knutson VP, Lane MD (1982) Insulin-induced downregulation of insulin receptors in 3T3-L1 adipocytes. Altered rate of receptor inactivation. J Biol Chem 257:4285–4291

Rosen O (1987) After insulin binds. Science 237:1452–1458

Russell DS, Gherzi R, Johnson EL, Chou OM, Rosen OM (1987) The protein-tyrosine kinase activity of the insulin receptor is necessary for insulin-mediated receptor downregulation. J Biol Chem 262:11833–11840

Smith RM, Jarett L (1982) A simplified method of producing biologically ultrastructural marker for occupied insulin receptors. J Histochem Cytochem 30:650–656

Smith RM, Jarett L (1983) Quantitative ultrastructural analysis of receptor-mediated insulin uptake into adipocytes. J Cell Physiol 115:199–207

Smith RM, Jarett L (1987) Ultrastructural evidence for the accumulation of insulin in nuclei of intact 3T3-L1 adipocytes by an insulin-receptor mediated process. Proc Natl Acad Sci USA 84:459–463

Smith RM, Cobb MH, Rosen OM, Jarett L (1985) Ultrastructural analysis of the organization and distribution of insulin receptors on the surface of 3T3-L1 adipocytes: rapid microaggregation and migration of occupied receptors. J Cell Physiol 123:167–179

Stentz FB, Harris HC, Kitabchi AE (1983) Early detection of degraded A_{14}-^{125}I-insulin in human fibroblasts by the use of high performance liquid chromatography. Diabetes 32:474–480

Stentz FB, Harris HL, Kitabchi AE (1985) Characterization of insulin-degrading activity of intact and subcellular components of human fibroblasts. Endocrinology 116:926–934

Thamm P, Saunders D, Brandenburg D (1980) Photoreactive insulin derivatives: preparation and characterization. In: Brandenburg D, Wollmer A (eds) Insulin, chemistry, structure, and function of insulin and related hormones: proceedings of the second international insulin symposium, 4–7 September 1979, Aachen. Gruyter, Berlin, p 309

Tycko B, Maxfield FR (1982) Rapid acidification of endocytic vesicles containing α_2-macroglobulin. Cell 28:643–651

Tycko B, Keith CH, Maxfield FR (1983) Rapid acidification of endocytic vesicles containing asialoglycoprotein in cells of a human hepatoma line. J Cell Biol 97:1762–1776

Wileman T, Boshans RL, Schlesinger P, Stahl P (1984) Monensin inhibits recycling of macrophage mannose-glycoprotein receptors and ligand delivery to lysosomes. Biochem J 220:665–675

Wisher MH, Baron MD, Jones RH, Sonksen PH, Saunders DJ, Thamm P, Brandenburg D (1980) Photoreactive insulin analogues used to characterize the insulin receptor. Biochem Biophys Res Commun 92:492–498

Wolkoff AW, Klausner RD, Ashwell G, Harford J (1984) Intracellular segregation of asialoglycoproteins and their receptor: a prelysosomal event subsequent to dissociation of the ligand-receptor complex. J Cell Biol 98:375–381

Yip CC, Yeung CW, Moule ML (1978) Photoaffinity labeling of insulin receptors of rat adipocyte plasma membrane. J Biol Chem 253:1743–1745

Yip CC, Yeung CW, Moule ML (1980) Photoaffinity labeling of insulin receptor proteins of liver plasma membrane preparations. Biochemistry 19:70–76

CHAPTER 13

Insulin-like Growth Factor I Receptors

S. JACOBS

A. Introduction

The somatomedins, insulin-like growth factor-I or somatomedin-C, and insulin-like growth factor-II or multiplication-stimulating activity, are polypeptides that play an important, but not completely defined role in regulating growth and differentiation (FROESCH et al. 1985). IGF-I is synthesized and secreted by a variety of tissues in response to growth hormone and mediates many of the classical effects of growth hormone (SALMON and DAUGHADAY 1957; FROESCH et al. 1985; RECHLER and NISSLEY 1985). This is the most firmly established physiological role of IGF-I. However, it is becoming increasingly clear that several other factors control IGF-I production and that it probably has a far wider role than just a growth hormone mediator. IGF-II is also synthesized by a large variety of tissues. Its physiological role is less well defined. Its production is not regulated by growth hormone to the same extent as that of IGF-I and it is probably not a true somatomedin (ROTH 1988).

There are two types of high affinity receptors for the insulin-like growth factors (MASSAGUE and CZECH 1982; KASUGA et al. 1982). The type II receptor binds IGF-II with a higher affinity than IGF-I and does not bind insulin at all. It is a 260000 molecular weight glycoprotein, composed of a single polypeptide chain, which has no structural relationship to the insulin receptor (ROTH 1988). Careful analysis of the relative potency of IGF-I and IGF-II to produce various known biological responses, and the ability of specific anti-receptor antibodies to block these responses, has indicated that they are probably not mediated by the type II receptor (MOTTOLA and CZECH 1984; VERSPOHL et al. 1984; CONOVER et al. 1986; SHIMIZU et al. 1986). It has been demonstrated recently that this receptor is identical to the cation-independent mannose 6-phosphate receptor, a receptor whose main function is to transport and target lysosomal enzymes to the lysosomes (MORGAN et al. 1987; MACDONALD et al. 1988; KEISS et al. 1988). What role if any the type II receptor has in transmembrane signaling is currently very much of a mystery.

The type I receptor, which is thought to mediate most or all of the well-recognized effects of both insulin-like growth factors, is the receptor that will be considered in this chapter. It binds IGF-I with high affinity, IGF-II with slightly lower affinity, and insulin with still lower affinity (RECHLER and NISSLEY 1985). Although it probably is the receptor that mediates the effects of both IGF-I and IGF-II, because it has a higher affinity for IGF-I, it has commonly been called the IGF-I receptor. Like the insulin receptor, it is a disulfide-linked heterote-

```
INSR  mgtggrrgaaaaplivanaalllgaaaghLYP--GEVC-PGMDIRNNLTRLHELENCSVIEGHLQILLMFKTRPEDFRDLSFPKLIMITDYLLLFRVGLE   70
IGFR  mksgsgggsptslwglllsaalslwptsgEICGPGIDIRNDYQQLKRLENCTVIEGYLHILLISK--AEDYRSYRFPKLTVITEYLLFRVAGLE          64

INSR  SLKDLFPNLTVIRGSRLFFNYALVIFEMVHLKELGLYNLMNITRGSVRIEKNNELCYLATIDWSRILDSVEDNYIVLNKDDNEECGDICPGTAKGKTNCP   170
IGFR  SLGDLFPNLTVIRGWKLFYNYALVIFEMTNLKDIGLYNLRNITRGAIRIEKNADLCYLSTVDWSLILDAVSNNYIVGNKPPK-ECGDLCPGTMEEKPMCE   163

INSR  ATVINGQFVERCWTHSHCQKVCPTICKSHGCTAEGLCCHSECLGNCSQPDDPTKCVACRNFYLDGRCVETCPPPYYHFQDWRCVNFSFCQDLHHKCKNSR   270
IGFR  KTTINNEYNYRCWTTNRCQKMCPSTCGKRACTENNECCHPECLGSCSAPDNDTACVACRHYYAGVCVPACPPNTYRFEGWRCVDRDFCANILSAESSDS    263

INSR  RQGCHQYVIHNNKCIPECPSGYTMNSSNLL-CTPCLGPCPKVCHLLEGEKTIDSVTSAQELRGCTVINGSLIINIRGGNNLAAELEANLGLIEEISGYLK   369
IGFR  E----GFVIHDGECMQECPSGFIRNGSQSMYCIPCEGPCPKVCEEEKKTKIDSVTSAQMLQGCTIFKGNLLINIRRGNNIASELENFMGLIEVVTGYVK    359

INSR  IRRSYALVSLSFFRKLRLIRGETLEIGNYSFYALDNQNLRQLWDWSKHNLTITQGKLFFHYNPKLCLSEIHKMEEVSGTKGRQERNDIALKTNGDQASCE   469
IGFR  IRHSHALVSLSFLKNLRLLGEEQLEGNYSFYVLDNQNLQQLWDWDHRNLTIKAGKMYFAFNPKLCVSEIYRMEEVTGTKGRQSKGDINTRNNGERASCE    459

INSR  NELLKFSYIRTSFDKILLRWEPYWPPDFRDLLGFMLFYKEAPYQNVTEFDGQDACGSNSWTVVDIDPPLRSNDPKSQNHPGWLMRGLKPWTQYAIFVKTL   569
IGFR  SDVLHFTSTTTSKNRIIITWHRYRPPDYRDLISFTVYYKEAPFKNVTEYDGQDACGSNSWNMVDVD--L---PPNKDVEPGILLHGLKPWTQYAVYVKAV   554

INSR  -VTFSDERRTYGAKSDIIYVQTDATNPSVPLDPISVSNSSSQIILKWKPSDPNGNITHYLVFWERQAEDSELFELDYCLKGLKLPSRTWS--PPFESEDS   667
IGFR  TLTMVENDHIRGAKSEILYIRTNASVPSIPLDVLSASNSSSQLIVKWNPSLPNGNLSYYIVRWQRQPQDGYLYRHNYCSKD-KIPIRKYADGTIDIEEV    653

INSR  QKHNQSEY-EDSAGECCSCPKTDSQILKELEESSFRKTFEDYLHNVVFVPRPSRKRRSLGDVGNVTVAVPT-VAAFPNTSSTSVPTSPEEHRP-FEK-VV   763
IGFR  TENPKTEVCGGEKGPCCACPKTEAEKQAEKEEAEYRKVFENFLHNSIFVPRPERKRRDVMQVANTTMSSRSRNTTAADTYNITDPEELETEYPFFESRVD   753

INSR  NKESLVISGLRHFTGYRIELQACNQDTPEERCSVAAYVSARTMPEAKADDIVGPVTHEIFENNVVHLMWQEPKEPNGLIVLYEVSYRRYGDEELHLCDTR   863
IGFR  NKERTVISNLRPFTLYRIDIHSCNHEAEKLGCSASNFVFARTMPAEGADDIPGPVTWEPRPENSIFLKWPEPENPNGLILMYEIKYGSQVEDQRE-CVSR   852

INSR  KHFALERGCRLRGLSPGNYSVRIRATSLAGNGSWTEPTYFYVTDYLDVPSNIAKiiiigpliifvvvigsiylfiRKRQPDGPLG--PLYASSNPEYL    961
IGFR  QEYRKYGGAKLNRLNPGNYTARIQATSLSGNGSWTDPVFFYVQAKTGYENFIHliialpvavllivgglvimluvfhRKRNNSR-LGNGVLYASVNPEYF   951

INSR  SASDVFPCSVYVPDEWEVSREKITLLRELGQGSFGMVYEGNARDIIKGEAETRVAVKTVNESASLRERIEFLNEASVMKGFTCHHVVRLLGVVSKGQPTL  1061
IGFR  SAAD-----VVVPDEWEVAREKITMSRELGQGSFGMVYEGVAKGVVKDEPETRVAIKTVNEAASMRERIEFLNEASVMKEFNCHHVVRLLGVVSQGQPTL  1046

INSR  VVMELMAHGDLKSYLRSLRPEAENNPGRPPPTLQEMIQMAAEIADGMAYLNAKKFVHRDLAARNCMVAHDFTVKIGDFGMTRDIYETDYYRKGGKGLLPV  1161
IGFR  VIMELMTRGDLKSYLRSLRPEMENNPVLAPPSLSKMIQMAGEIADGMAYLNANKFVHRDLAARNCMVAEDFTVKIGDFGMTRDIYETDYYRKGGKGLLPV  1146

INSR  RWMAPESLKDGVFTTSSDMWSFGVVLWEITSLAEQPYQGLSNEQVLKFVMDGGYLDQPDNCPERVTDLMRMCWQFNPMRPTFLEIVNLLKDDLHPSFPE   1261
IGFR  RWMSPESLKDGVFTTYSDVWSFGVVLWEIATLAEQPYQGLSNEQVLRFVMEGGLLDKPDNCPDMLFELMRMCWQYNPKMRPSFLEIISSIKEEMEPGFRE  1246

INSR  VSFFHSEENKAPESELEMEFEDMENVPLDRSSHCQREEAGGRDGG----------SSLGFKRSYEEHIPYTHMNGGKKNGRILTLPRSNPS           1343
IGFR  VSFYYSEENKLPEPEELDLEPENMESVPLDPSASSSSLPLDRHSGHKAENGPGPGVVLRASFDERQPYAHMNGGRKNERALPLPQSSTC              1343
```

Fig. 1a

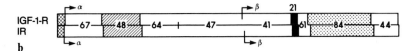

Fig. 1 a, b. Comparison of insulin and IGF-I receptors. **a** Aligned sequence–areas of significance designated as follows: *lower case*, signal sequence; *arrow*, start of β subunit; *lowercase underlined*, membrane-spanning region; *bold face*, tyrosine kinase domain; *bold italics*, ATP binding residues; *bold italics underlined*, major site of autophosphorylation; *underlined*, site of insulin receptor threonine phosphorylation. **b** Schematic representation of receptor domains. *Numbers* indicate percent identical residues in each region. *Crosshatched*, Signal sequence; *diagonal shading*, cysteine-rich region; *black bar*, membrane-spanning region; *stippled*, tyrosine kinase domain

tramer composed of two extracellular α subunits and two transmembrane β subunits (CHERNAUSEK et al. 1981; KULL et al. 1983; LEBON et al. 1985; MORGAN et al. 1986; ULLRICH et al. 1986; STEELE-PERKINS et al. 1988). Like the insulin receptor, the type I IGF receptor is synthesized as a single-chain polypeptide precursor, which dimerizes and subsequently is proteolytically cleaved and glycosylated to form the mature tetrameric receptor (JACOBS et al. 1983a; DURONIO et al. 1986, 1988). The sequence of this precursor (ULLRICH et al. 1986) is compared with that of the insulin receptor in Fig. 1. The receptors share considerable sequence homology throughout their entire extent. Regions of particularly high homology are the tyrosine kinase domains (84%), the NH₂ terminal region of the α subunit (67%), and the mid-region of the α subunit (64%). Although both receptors have a cysteine-rich region, which has been thought to participate in hormone binding, the hormology in this region is relatively low (48%).

Because of their extensive similarities, comparing the IGFs and the type I receptor with insulin and its receptor provides an excellent opportunity for understanding how these related signaling systems work. While they are certainly worthy of study in their own right, it is this aspect of the IGFs and the IGF-I receptor that will be the point of view of this chapter.

B. Hormonal Determinants of Binding and Specificity

The affinities of insulin and the IGFs for binding to the insulin receptor and the IGF-I receptor have been studied extensively in many different tissues. Both absolute and relative values vary somewhat from tissue to tissue and even from study to study in the same tissue. Typically, for the insulin receptor the affinity of insulin is 40 times higher than IGF-II and 150–500 times higher than IGF-I. For the IGF-I receptor, the affinity of IGF-I is 3–10 times higher than IGF-II and 50–500 times higher than insulin (RECHLER and NISSLEY 1985).

A comparison of the amino acid sequence of IGF-I (RINDERKNECHT and HUMBEL 1978a) and IGFF-II (RINDERKNECHT and HUMBEL 1978b) with insulin (Fig. 2) provides some insight into the basis of this cross-reactivity. Approximately 45% of the residues in the B chain of insulin and 60% in the A chain are identical in the IGFs. These include six cysteines which form essential disulfide bonds. In contrast to insulin, the IGFs have a short C peptide which remains intact in

A Chain

	1	2	3	4	5	6	7	8	9	10	11	12	13	14	15	16	17	18	19	20	21
Ins.	Gly	Ile	Val	Glu	Gln	Cys	Cys	Thr	Ser	Val	Cys	Ser	Leu	Tyr	Gln	Leu	Glu	Asn	Tyr	Cys	Asn
IGFI 42-62	Gly	Ile	Val	Asp	Glu	Cyc	Cyc	Phe	Arg	Ser	Cys	Asp	Leu	Arg	Arg	Leu	Glu	Met	Tyr	Cys	Ala
IGFII 41-61	Gly	Ile	Val	Glu	Glu	Cys	Cys	Phe	Arg	Ser	Cys	Asp	Leu	Ala	Leu	Leu	Glu	Thr	Tyr	Cys	Ala

B Chain

	-2	-1	1	2	3	4	5	6	7	8	9	10	11	12	13	14	15	16	17	18	19	20	21	22	23	24	25	26	27	28	29	30
Ins.			Phe	Val	Asn	Gln	His	Leu	Cys	Gly	Ser	His	Leu	Val	Glu	Ala	Leu	Tyr	Leu	Val	Cys	Gly	Glu	Arg	Gly	Phe	Phe	Tyr	Thr	Pro	Lys	Thr
IGFI 1-29			Gly	Pro	Glu	Thr	Leu	Cys	Gly	Ala	Glu	Leu	Val	Asp	Ala	Leu	Gln	Phe	Val	Cys	Gly	Asp	Arg	Gly	Phe	Tyr	Phe	Asn	Lys	Pro	Thr	
IGFII 1-32	Ala	Tyr	Arg	Pro	Ser	Glu	Thr	Leu	Cys	Gly	Gly	Glu	Leu	Val	Asp	Thr	Leu	Gln	Phe	Val	Cys	Gly	Asp	Arg	Gly	Phe	Tyr	Phe	Ser	Arg	Pro	Ala

C Peptide

IGFI 30-41	Gly	Tyr	Gly	Ser	Ser	Ser	Arg	Ala	Pro	Gln	Thr
IGFII 33-40	Ser	Arg	Val	Ser	Arg	Arg	Ser	Arg			

D Peptide

IGFI 63-70	Pro	Leu	Lys	Pro	Ala	Lys	Ser	Ala
IGFII 62-67	Thr		Pro	Ala	Lys	Ser	Glu	

Fig. 2. Sequence of human insulin, IGF-I, and IGF-II. Numbered at the top to correspond to the A, B, and C chain of insulin. Numbered consecutively on left from the amino terminus of the mature peptide

the mature molecule and a D peptide carboxy terminal extension of the B chain. X-ray coordinates for neither IGF-I nor IGF-II have been determined directly. However, extensive similarity with insulin allowed BLUNDELL et al. (1983) to construct a reliable model for these peptides based upon the structure of insulin. Since those residues which comprise the hydrophobic core of insulin are identical in the IGFs, the main chain of the IGFs corresponding to the A and B chains of insulin were assumed to have the same conformation. The side chains of the invariant residues of the hydrophobic core were assigned identical positions to those of insulin and held constant in subsequent adjustments. Residues in the IGFs which are identical or highly conserved to those on the surface of insulin were initially given similar positions, but were moved when remaining residues were inserted to be consistent with charge and steric constraints and to minimize energy calculations.

The surface of insulin that is believed to contact its receptor is composed of residues B22-B25 and surrounding residues from the NH_2 and COOH terminus of the A chain. Assuming the similar main chain conformation and side chain positions proposed by Blundell's model and the likelihood that similar surfaces are involved in receptor contact, specificity of binding could result from amino acid substitutions in the contact region or from the presence of the D peptide extension or the intact C peptide of the IGFs, which are both in the vicinity of the presumed contact surface and could either interfere with binding or contribute to it.

I. D Peptide

Several hormone analogs have been constructed to evaluate the importance of the D peptide. These clearly indicate that the D peptide has an adverse effect on binding to the insulin receptor. Deletion of the D octapeptide of IGF-I increases its affinity for the insulin receptor fourfold (CASCIERI et al. 1988a). Similarly, appending the D octapeptide of IGF-I or the D hexapeptide of IGF-II onto the A chain of insulin or various insulin IGF-I hybrids results in a four- to tenfold loss of affinity for the insulin receptor (KING et al. 1982; DEVROEDE et al. 1985; JOSHI et al. 1985a, b).

The importance of the D peptide for enhancing binding to the IGF-I receptor is less clear. While deleting the D peptide of IGF-I has no effect on its affinity for the IGF-I receptor, deleting the D peptide of an IGF-I mutant in which B25 tyrosine is replaced with leucine decreases its already low affinity for the IGF-I receptor an additional 2.5-fold (CASCIERI et al. 1988a). Extending the A chain of insulin with the D octapeptide of IGF-I increases its affinity for the IGF-I receptor threefold (KING et al. 1982), but extending the A chain of insulin with the D hexapeptide of IGF-II has no significant effect on binding to the IGF-I receptor (DEVROEDE et al. 1986). The affinity for the IGF-I receptor of a two-chain insulin–IGF-I hybrid in which the B chain of insulin is replaced by the B peptide of IGF-I is not affected by appending the D hexapeptide of IGF-II (DEVROEDE et al. 1985; TSENG et al. 1987). However, the affinity of a hybrid in which the A chain of insulin was replaced by the A peptide of IGF-I was increased threefold, but the potency for stimulating thymidine incorporation was actually decreased somewhat (TSENG et al. 1987). While it is difficult to interpret these results coherently, it would seem that the D peptide if it has any effect on binding to the

IGF-I receptor, may enhance the affinity of the IGFs somewhat, and that the octapeptide of IGF-I may be more effective than the hexapeptide of IGF-II perhaps accounting in part for the lower affinity of IGF-II for that receptor.

II. C Peptide

The low affinity of proinsulin for the insulin receptor illustrates the adverse effect of the C peptide on binding to the insulin receptor. The C peptide of the IGFs, on the other hand enhances binding to the IGF-I receptor. Preliminary studies indicate that a two-chain IGF-I analog in which the C peptide is deleted binds to the IGF-I receptor with a tenfold lower affinity than native IGF-I (TSENG et al. 1987). The C peptide could enhance binding by masking charge at the NH_2 terminus of the A chain or COOH terminus of the B chain, by altering the conformation of the A or B chains, or by providing residues which could directly participate in binding to the receptor. A series of C peptide mutants of IGF-I support the latter alternative (CASCIERI et al. 1988 b). Replacement of the C peptide with four glycines results in a hormone analog with 30-fold lower affinity for the IGF-I receptor, but with the same affinity for the insulin receptor, suggesting that its A and B peptide conformations are not greatly altered. A single mutation in the C peptide, C2 tyrosine to alanine, results in an 18-fold loss of affinity for the IGF-I receptor with no change in affinity for the insulin receptor. Furthermore, proinsulin has more than a 10-fold lower affinity than insulin for the IGF-I receptor in spite of the presence of the intact C peptide. The relatively major differences between IGF-I and IGF-II in their C peptide may be important determinants for their specificity for the IGF-I receptor.

III. B Peptide

Residues B22-B25 of the insulin B chain have been implicated in insulin receptor binding. This region is highly conserved in the IGFs. While it also appears to be intimately involved in binding to the IGF-I receptor, it contributes little to specificity for this receptor relative to the insulin receptor. Thus, mutating this region of IGF-I produces roughly parallel changes in its affinity for both the IGF-I receptor and the insulin receptor (CASCIERI et al. 1988 a). Mutating B24-B26 of IGF-I to PhePheTyr, the sequence found in insulin, decreases its affinity for the IGF-I receptor by only 20% and doubles its affinity for the insulin receptor, changes which were not felt to be significant. Mutating B25 Tyr to Leu or Ser, changes which drastically reduce the affinity of insulin for its receptor (see Chap. 3), decrease the affinity of IGF-I for the IGF-I receptor 32-fold and 17-fold respectively, and decrease its affinity for the insulin receptor 10-fold and 2-fold. Similarly, altering residues of insulin in this region produce roughly parallel changes in its affinity for both receptors, although here too minor quantitative differences occur (CARA et al. 1988). Removing the last five residues of the B chain of insulin (despentapeptide insulin) has no significant effect on insulin receptor binding, but actually increased its affinity for the IGF-I receptor about twofold, suggesting that these residues are not only not important for binding to the IGF-I receptor, as is the case for the insulin receptor, but that they actually sterically interfere with binding to that receptor. Removing additional residues from the B chain of insulin (deshexapeptide insulin, and desheptapeptide insulin) produces drastic reductions in its affinity for the IGF-I receptor as is the case for the insulin receptor. Replacing B25 Phe of despentapeptide insulin with Tyr, the homologous residue in IGF-I, increases its affinity for both receptors: threefold for the insulin receptor, fivefold for the IGF-I receptor. Replacing B25 with non-aromatic residues decreased its affinity for both receptors moderately.

IV. A Peptide

The A chain is surprisingly important in determining selectivity for the insulin versus the IGF-I receptor. A two-chain hybrid, consisting of the A chain of IGF-I and the B chain of insulin is 13-fold more potent than insulin (about 1.5% as potent as IGF-I) in competing with IGF-I for binding to the IGF-I receptor of human placenta (TSENG et al. 1987), but is only 41% as potent as insulin in competing for binding to the insulin receptor (CHEN et al. 1988). Of 21 residues, 10 differ in this peptide. Several can be excluded as being responsible for this selectivity. Perhaps the most obvious differences are the replacement of $Ala^8Ser^9Val^{10}$ of insulin with PheArgSer and $Tyr^{14}Gln^{15}$ with ArgArg. These, however, are probably not important determinants of specificity because an IGF-I mutant containing residues found in insulin at these positions has the same af-finity and specificity as IGF-I for both receptors (CASCIERI et al. 1988b). The difference at position 18 is not likely to be important, since replacing methionine with threonine at this position has no effect on the affinity of IGF-I for either receptor. The difference at position 4, Glu versus Asp, is also not likely to be important; it is a very conservative change, and rat insulin, which has high affinity for the insulin receptor actually has Asp at this position while IGF-II has Glu. This leaves positions A5, 12, and 21.

Both IGF-I and IGF-II have Asp at position A12, while all the high affinity insulins have serine at this position. Although several fish insulin's have either Asp or Asn at position A12, these insulins have a reduced affinity for the insulin receptor and reduced specificity for the insulin receptor relative to the IGF-I receptor (KING and KAHN 1981). Therefore, A12 deserves serious consideration as contributing to the receptor specificity of IGF-I. However, this residue is not part of the surface that is thought to make contact with the receptor. If it does contribute to determining receptor specificity, it probably does so by a small conformation change which is reflected over a distance at the binding site.

A5 and A21 are both thought to be near the receptor binding region of insulin and probably also of IGF-I. They are very highly conserved. All known insulins contain Gln at A5 and Asn at A21, while IGF-I and IGF-II both contain Glu at A5 and Ala at A21. Therefore, these positions are likely to contribute to the specificity of insulin and IGFs for their respective receptors. Analogs in which these residues are individually exchanged will be very helpful in evaluating this possibility.

In summary, homologous surfaces of insulin and the IGFs are probably involved in binding to their respective receptors. Residues homologous to those of insulin that are thought to participate in insulin receptor binding, B21-B26, A1,

A5, A19, and A21, probably participate in IGF binding to its receptor. Most differences between insulin and the IGFs at these positions are conservative and with possible exceptions of A5 or A21 probably contribute little to determining receptor specificity. Receptor specificity is probably determined largely by the C and D peptides of the IGFs. The C peptide provides residues which probably contribute to the IGF-I receptor binding site and therefore enhance IGF-I receptor binding, but interfere, probably because of steric reasons, with binding to the insulin receptor. The D peptide although it probably contributes little in a positive fashion to binding to the IGF-I receptor, is accommodated by that receptor, but probably for steric reasons interferes with binding to the insulin receptor.

C. The Receptor Binding Site

Less is known about sites on the receptors which determine binding affinity and specificity. In affinity labeling studies, the α subunits of both receptors are labeled predominantly (YIP et al. 1980; MASSAGUE et al. 1980; JACOBS et al. 1979; MASSAGUE and CZECH 1982; KASUGA et al. 1982; CHERNAUSEK et al. 1981; BHAUMIC et al. 1981). This suggests that the α subunit binds the hormone. However, part of the β subunit, which is also labeled, although considerably less intensely, is probably near the hormone binding site and may contribute to it (YIP et al. 1980; MASSAGUE et al. 1980). Tryptic digestion of the affinity-labeled insulin receptor generates an affinity-labeled fragment which has been identified as the NH_2 terminal 55000 residues of the α subunit (WAUGH et al. 1989). It has been pointed out that within this region, there is a sequence (residues 84–91) which is similar to B20–B26 of insulin, and that since this region of insulin is involved in insulin–insulin interaction in dimer formation, it could also be involved in insulin–receptor interaction (DE MEYTS et al. 1988). Mutation of several residues within this region of the receptor markedly decreases its affinity for insulin (WHITTAKER et al. 1988; BODSH 1988). This is not definitive proof that this region is directly involved in insulin binding; mutation could produce conformational changes which are reflected at distant sites. However, it is interesting to note that a similar sequence is found in the IGF-I receptor (residues 78–85). Since desoctapeptide insulin, which is missing this region, retains considerable activity (about 1% of native insulin) it is likely that other portions of the receptor are also involved.

Scatchard plots of hormone binding to both insulin and IGF-I receptors have been reported to be curvilinear by some investigators, but not by others. One possible explanation for curvilinearity is an interaction between multiple binding sites within the receptor tetramer. Studies in which tetrameric receptors have been dissociated into $\alpha\beta$ dimers by mild reduction of intersubunit disulfide bonds support such an explanation (BONI-SCHNETZLER et al. 1986, 1987; SWEET et al. 1987a; TOLLEFSEN and THOMPSON 1988). These studies have shown that each dissociated $\alpha\beta$ dimer can bind one molecule of hormone. The properties of the binding site on the dissociated $\alpha\beta$ dimer are different than on the intact tetramer. BONI-SCHNETZLER et al. (1987) found that whereas the intact insulin receptor exhibited high affinity curvilinear Scatchard plots, the dissociated dimers exhibited only a low affinity linear Scatchard plot. SWEET et al. (1987a) obtained slightly different results. They also found that dissociated insulin receptor dimers had a lower affinity than the intact tetramer. However, they found that dissociation resulted in a twofold increase in binding capacity. In the tetramer, a pair of dimers together bound only one molecule of insulin, whereas the dissociated dimers each bound one molecule of insulin. TOLLEFSEN and THOMPSON (1988) obtained similar results for the IGF-I receptor. Dissociation of the receptor into $\alpha\beta$ dimers decreased its affinity for IGF-I binding, but increased its binding capacity. FELTZ et al. (1988) however, found that dissociation of IGF-I receptors into dimers had no effect on the number of binding sites or on its affinity for IGF-I. However, there was a tenfold decrease in the potency of insulin to inhibit IGF-I binding.

There are discrepancies between the findings of these groups. They may be partially related to the arbitrary nature of deciding what is nonspecific binding and what is low affinity specific binding, therefore either including or discounting the low affinity portion of the Scatchard plot. In addition differences between the results of TOLLEFSEN and THOMPSON (1988) and FELTZ et al. (1988) may be due to the fact that the latter group used a less purified receptor preparation which may have contained hybrid receptors (see Sect. F). However, all these findings suggest that for both receptors there is an interaction between $\alpha\beta$ dimers when they are assembled into tetramers. This interaction leads to an asymmetry whereby one dimer acquires a higher affinity for hormone binding, while its neighbor acquires a lower or negligible affinity. Whether this asymmetry exists in the absence of hormone binding or whether it is induced by a negatively cooperative interaction between dimers that occurs when hormone binds to one of the dimer pairs is not clear from these data. However, the extensive kinetic studies of DE MEYTS et al. (1976) in the insulin receptor would suggest negative cooperativity.

D. Biological Responses

Physiologically, IGF-I acts mainly as a growth factor while insulin acts mainly as a metabolic regulator. However, this may be due more to the cellular distribution of their receptors than to a fundamental difference in their mode of action (VERSPOHL et al. 1988). Thus, although many cell types express both receptors, classical metabolic targets for insulin, liver, and fat, have insulin receptors, but few if any IGF-I receptors, while fibroblasts have a preponderance of IGF-I receptors.

Insulin, acting through its own receptor can stimulate mitogenesis. This has been most clearly demonstrated in H-35 cells. In these cells, which have insulin and type II IGF receptors, but lack IGF-I receptors (MASSAGUE et al. 1982), insulin at concentrations as low as 1 ng/ml is mitogenic (KOONTZ and IWAHASHI 1981; KOONTZ 1984; MASSAGUE et al. 1982). Although IGF-I and IGF-II also produce a comparable mitogenic response, concentrations greater than 500-fold higher are required. Furthermore, antibodies to the insulin receptor that act as agonists are also mitogenic while an antibody to the type II IGF receptor is without effect (KOONTZ 1984; MOTTOLA and CZECH 1984). Similar results, indicating that insulin acting through its own receptor is mitogenic, have also been obtained in F9 embryocarcinoma cells (NAGARAJAN and ANDERSON 1982). Another approach to demonstrating that the insulin receptor can produce a mitogenic response is by overexpressing cloned insulin receptor cDNA in cells which have few insulin receptors. This results in a dramatic increase in the sensitivity of the resulting cell line to the mitogenic effects of insulin (CHOU et al. 1987).

Human fibroblasts have a preponderance of IGF-I receptors compared with insulin receptors. In these cells the major component of the mitogenic response to high concentrations of insulin is mediated through the IGF-I receptor, since it can be blocked by α IR-3, a monoclonal antibody specific for the IGF-I receptor (KING et al. 1980; FLIER et al. 1986; VAN WYK et al. 1985). However, even in fibroblasts, provided the cell line has a sufficient number of insulin receptors, a small component of the mitogenic response to insulin, which occurs at low, physiologic concentrations of insulin and can not be blocked by α IR-3, is mediated through the insulin receptor (FLIER et al. 1986).

Similarly, IGF-I acting through its own receptor can produce metabolic responses which are usually attributed to insulin. It has been long recognized that in cells, such as fibroblasts, that are classical targets for the growth-promoting effects of IGF-I, IGF-I also stimulates glucose and amino acid uptake. This has been generally considered to be part of a pleiotropic growth-promoting response. However, it is becoming increasingly clear, that even in cells which serve as models of classical metabolic targets for insulin, the IGF-I receptor, when expressed at sufficient levels, can mediate these metabolic responses. Differentiated 3T3-F442A cells are an excellent model of metabolically active adipocytes. However, unlike adipocytes, they express IGF-I receptors (MASSAGUE and CZECH 1982). Even though cells are terminally differentiated and do not have a growth response to IGF-I, IGF-I acting through the IGF-I receptor actively stimulates glucose transport in a very similar manner to insulin (SCHWARTZ et al. 1985). HepG2 cells are a well-differentiated human hepatoma cell line, which maintain many of the normal metabolic responses of liver to insulin, but unlike normal liver contain levels of IGF-I receptors which are comparable to those of insulin receptors. These cells respond to physiological concentrations of IGF-I with an increase in glucose incorporation into glycogen and an increase in aminoisobutyric acid uptake, which can be blocked by a monoclonal antibody specific for the IGF-I receptor, indicating that it is mediated through that receptor (VERSPOHL et al. 1984, 1988). Unlike most classical targets for the metabolic effects of insulin, skeletal muscle and muscle cell lines have significant numbers of IGF-I receptors, and IGF-I, acting at its own receptor, effectively stimulates glucose transport and other metabolic effects (YU and CZECH 1984; POGGI et al. 1979; MEULI and FROESCH 1977; DEVROEDE et al. 1984; BEGUINOT et al. 1985; SHIMIZU et al. 1986).

Although when insulin receptors and IGF-I receptors are expressed at similar levels in the same cell, they are able to mediate similar responses, they appear to have distinct quantitative differences in their abilities to produce specific responses. For example, in HepG2 cells, insulin and IGF-I receptors produce similar maximal effects on glucose incorporation into glycogen and on aminoisobutyric acid uptake. The effect of each hormone is blocked only by an antibody specific for its own receptor, indicating that each hormone is acting through its own receptor. However, even though both hormones have similar receptor binding affinities and similar numbers of receptors per cell, the effects of insulin start as low as 10^{-11} M and occur over a three log concentration range while the effects of IGF-I begin at 10^{-9} M and occur over only a single log concentration range (VERSPOHL et al. 1988). Thus, insulin receptor does appear to be more effective at mediating these metabolic responses than IGF-I receptors. On the other hand, when a human insulin receptor is transfected into Rat 1 cells and expressed at levels similar to endogenous IGF-I receptors, IGF-I is more effective than insulin in stimulating thymidine incorporation, even though the effect of insulin is mediated through the transfected insulin receptor (McCLAIN et al. 1988). The ability of insulin and IGF-I receptors to mediate similar but perhaps quantitatively different responses should be reflected in their tyrosine kinase activities and ultimately in their structure.

E. Tyrosine Kinase Activity

Like the insulin receptor, the IGF-I receptor is a hormonally stimulated, tyrosine-specific protein kinase (JACOBS et al. 1983b; RUBIN et al. 1983; ZICK et al. 1984). Many of its enzymatic properties are remarkably similar to those of the insulin receptor (SASAKI et al. 1985). It has a strict requirement for ATP as a substrate and will not use GTP. It has a strong preference for Mn over Mg as a cation cofactor. It is autophosphorylated, and autophosphorylation results in activation of its kinase activity (SASAKI et al. 1985; YU et al. 1986). Autophosphorylation is an intramolecular reaction since its rate is independent of receptor concentration (SAZUKI et al. 1985) and it occurs when the receptor is immobilized on IGF-I agarose beads (YU et al. 1986). Hormonal stimulation of IGF-I receptor tyrosine kinase activity requires intact tetramers (FELTZ et al. 1988; TOLLEFSEN and THOMPSON 1988), as is the case with the insulin receptor (BONI-SCHNETZLER et al. 1986; SWEET et al. 1987b). Hormone binds to dissociated dimers, but does not stimulate autophosphorylation, suggesting that an interaction between dimers is required. It is not known if autophosphorylation is an intrasubunit reaction as has been demonstrated for the insulin receptor (SHOELSON et al. 1988).

The specificity of the IGF-I receptor for a variety of synthetic and in vitro substrates has been compared to that of insulin receptor (SASAKI et al. 1985). Overall they are quite similar. There is however a clear discrepancy in that poly(Glu, Tyr) 1:1 is a much better substrate for the IGF-I receptor than for the insulin receptor. Perhaps more significantly, in KB cells, treatment with either insulin or IGF-I results in the tyrosine phosphorylation of pp185, the major endogenous substrate for insulin receptor (see Chaps. 11 and 15), and a 240000 M_r protein (KADOWAKI et al. 1987). Two pieces of evidence indicate that each hormone is stimulating phosphorylation of these proteins by acting through its own receptor. First, the dose–response curve for stimulation of pp180 and pp240 phosphorylation by insulin and IGF-I is similar to that for each hormone to stimulate autophosphorylation of its own receptors. Second, phosphorylation of pp180 and pp240 by IGF-I, but not by insulin is blocked by α-IR-3, an IGF-I receptor-specific monoclonal antibody. Phosphorylation of pp180 by IGF-I has also been reported to occur in neuroblastoma cells (SHEMER et al. 1987). Thus, the substrate specificities of the two receptor tyrosine kinases are quite similar, although subtle differences exist which could explain their overlapping but perhaps quantitatively different biological activities.

Possible structural bases for both these similarities and differences are suggested by examining the sequences (ULLRICH et al. 1986) of the intracellular portions of both receptors, which can be divided into three main regions: a juxtamembrane region, a main central region, which is clearly recognized as the tyrosine kinase domain based upon its similarity with other tyrosine kinases, and a COOH terminal region. The highest degree of homology between the two receptors exists in their tyrosine kinase domains; 84% of the residues are identical. The NH_2 terminal part of the tyrosine kinase domain is involved in ATP binding. It contains conserved glycines (976, 978, and 981 of the IGF-I receptor) and a conserved lysine (1003 of the IGF-I receptor) which fit the consensus sequence for the ATP binding site of kinases. Indeed, mutation of this lysine in the insulin receptor completely inactivates its tyrosine kinase activity and makes it ineffective in transmembrane signaling (CHOU et al. 1987; EBINA et al. 1987). Tyrosines 1131, 1135, and 1136 are homologous to the major sites of autophosphorylation of the insulin receptor. Residues in this region are perfectly conserved in the two receptors. Based on similar phosphopeptide maps, this region also appears to be the major site of autophosphorylation of the IGF-I receptor (JACOBS and CUATRECASAS 1986), and appears to be a conserved regulatory region of kinase activity.

Most of the sequence differences between the two receptors within their tyrosine kinase domains are clustered within three small regions that follow residues 986, 1072, and 1208 of the IGF-I receptor. The region following 1072 is particularly interesting as possibly determining selectivity of growth-promoting activity. The PDGF receptor has a 100-residue insertion in a similar position that is not found in other tyrosine kinases (YARDEN et al. 1986). Deletion of this insertion results in loss of the ability of PDGF to stimulate a mitogenic response, but does not interfere with the other biological actions of PDGF (ESCOBEDO and WILLIAMS 1988).

The COOH terminal regions of the insulin receptor and the IGF-I receptor also differ considerably. Only 44% of the residues are identical. This region too, has received consideration as contributing to specificity of their biological effects and to their substrate specificity. An insulin receptor mutant with a deleted carboxy terminus loses its ability to initiate the metabolic responses to insulin, such as stimulation of 2-deoxyglucose uptake and glycogen synthesis, but is more effective in initiating mitogenic responses (McCLAIN et al. 1988; MAEGAWA et al. 1988). This shift in specificity is associated with a subtle change in substrate specificity of its tyrosine kinase activity for both in vitro substrates and in vivo cellular substrates.

Recently, it has been shown that the IGF-I receptor is a substrate of the insulin receptor (BEQUINOT et al. 1988). Receptor cross-phosphorylation is another possible mechanism by which activation of one receptor could produce effects normally attributed to the other.

F. Receptor Heterogeneity

Early affinity labeling studies that defined the type I and type II IGF receptors also suggested that there were significant differences in their relative affinities for IGF-I, IGF-II, and insulin in different tissues (MASSAGUE and CZECH 1982; RECHLER et al. 1980). Since then, a considerable body of experimental data has accumulated, which is difficult to explain exclusively on the basis of the three distinct, well-characterized receptors for IGF-I, IGF-II, and insulin. Because of this, investigators have postulated the existence of receptor subtypes or atypical receptors (see JONAS 1988 for a recent review). While it is difficult to precisely characterize the binding properties of a receptor when it is present in a mixture of cross-reacting receptors, the evidence, which is based upon hormone binding and immunological techniques, and methods to physically separate receptors, probably cannot be explained by this alone.

IM-9 lymphocytes contain an IGF-II binding site from which labeled IGF-II can be displaced by unlabeled IGF-II with $IC_{50} = 2.5$ nM (HINTZ et al. 1984). It also has a high affinity for both insulin ($IC_{50} = 1.5$ nM) and IGF-I ($IC_{50} = 7$ nM). Affinity labeling shows that this binding is due to a receptor with a subunit structure indistinguishable from classical insulin or IGF-I receptors (MISRA et al. 1986), although its affinity profile is clearly different from either. A variety of receptor antibodies have similar effects on insulin and IGF-II binding in these cells, but diverge in their effects on IGF-I binding. For example, serum B10 (1 : 500) inhibits insulin, IGF-II, and IGF-I binding by 51%, 41%, and 18% respectively while α-IR-3 inhibits IGF-I binding by 45%, but has no effect on insulin or IGF-II binding. A component of IGF-I binding with similar properties, which can be resolved from the majority of IGF-I binding activity by chromatography on insulin–agarose, has been identified in human placenta (JONAS et al. 1986; TOLLEFSEN et al. 1987). Like the atypical IM-9 receptor, this receptor binds insulin, IGF-II, and IGF-I with relative potencies of 1 : 10 : 50. It has a subunit structure similar to classical insulin and IGF-I receptors and is recognized by several antibodies specific for the insulin receptor, but not by α-IR-3.

A second type of atypical receptor appears to exist in human placenta (JONAS and HARRISON 1985). B2 antiserum immunoprecipitates approximately 30% of the [^{125}I]IGF-I binding activity of human placenta. The immunoprecipitable component has lower affinity for IGF-I than the classical receptor and labeled IGF-I is displaced by IGF-I, IGF-II, and insulin with relative potencies of 1 : 40 : 120. Analysis of the receptor responsible for this binding by affinity labeling indicates that it has a disulfide-linked tetrameric structure indistinguishable from classical insulin or IGF-I receptors. Interestingly, disulfide reduction with dithiothreitol converts this low affinity IGF-I binding, B2-immunoprecipitable component to a higher affinity form, not precipitable by B2 (JONAS and HARRISON 1986).

A receptor with similar properties was purified from human placenta with an α-IR-3 affinity column (CASELLA et al. 1986). It binds IGF-I and IGF-II with similar affinities and also binds insulin with about tenfold lower affinity. This receptor appears to have two high affinity hormone binding sites with distinct properties. One site has a higher affinity for IGF-I and is blocked by α-IR-3. The other has a higher affinity for IGF-II and is not blocked by α-IR-3.

ALEXANDRIDES and SMITH (1989) identified a developmentally regulated species of muscle IGF-I receptor that was present in fetal muscle, but disappeared shortly after birth. The fetal form had higher basal and higher hormonally stimulated activity and was stimulated by low concentrations of both insulin and IGF-I. It had a higher molecular weight β subunit than the adult receptor.

Different species of IGF-I receptors have also been identified based on their reactivity with 5D9, an insulin receptor monoclonal antibody (MORGAN and ROTH 1986). 5D9 inhibits IGF-I receptor binding in IM-9 cells and human placenta, but has minimal effect on IGF-I binding in HepG2 cells, human muscle cells, and human fibroblasts. In contrast, in all these cells α-IR-3 inhibits IGF-I receptor binding and 5D9 inhibits insulin binding.

G. A Hybrid IGF-I Receptor

In most tissues in which it has been studied, the β subunit of the IGF-I receptor migrates as a doublet on SDS polyacrylamide gel electrophoresis. When compared in the same tissue, the lower component of the doublet has a mobility similar to that of the insulin receptor β subunit. It is possible that this structural heterogeneity could partially account for the different receptor subtypes described above.

The basis for this structural heterogeneity is becoming increasingly clear. Based on the ability of inhibitors of glycosylation or endoglycosidases to obliterate the differences in mobility between the two subunits, it is partially due to differences in glycosylation. However, there are, as well, more profound structural and immunological differences. In many structural respects the lower component of the IGF-I receptor β subunit doublet resembles the insulin receptor β subunit and is different from the upper component. In intact HepG2 cells, the upper component is susceptible to trypsin cleavage, but the lower component, like the insulin receptor β subunit, is not (DURONIO et al. 1988). While it is conceivable that differences in glycosylation could result in differing trypsin sensitivity, it would be surprising if more extensive glycosylation would result in increased sensitivity. Furthermore, when analyzed by tryptic phosphopeptide maps the two components of the IGF-I receptor β subunit doublet are clearly different from each other (C.P. MOXAM et al. in preparation). The lower component, but not the upper component, contains a phosphothreonine-containing phosphopeptide, which is one of the most prominent phosphopeptides in the basal state and is also a major site of phosphorylation by protein kinase C. It also contains several phosphoserine-containing tryptic peptides which are not present in the upper component. On the other hand, the phosphothreonine and the phosphoserine-containing peptides that are unique to the lower component are also present in the insulin receptor β subunit. In fact the insulin receptor phosphothreonine has been determined to be threonine 1336 (LEWIS et al. 1988). There is no homologous threonine in the IGF-I receptor sequence reported by ULLRICH et al. (1986).

The upper and lower components of the IGF-I β subunit are covalently associated in the same disulfide-linked heterotetramer. Both α-IR-3, which recognizes the α subunit of the IGF-I receptor (KULL et al. 1983), and P5, an antibody produced against a synthetic peptide corresponding to the carboxy terminal residues (1328–1343) of the insulin receptor (HERRERA et al. 1986), immunoprecipitate both components of the β subunit doublet of unreduced IGF-I receptor. However, if the receptor is dissociated into αβ dimers with 2 mM dithiothreitol, α-IR-3 will only immunoprecipitate the upper component while P5 will only immunoprecipitate the lower component (C. P. MOXAM et al. in preparation). Thus, P5 does not directly recognize the upper component, but immunoprecipitates it because of its covalent association with the lower component. Similarly, α-IR-3 does not recognize the α subunit that is associated with the lower component in the αβ dimer, only the α subunit associated with the upper component. It immunoprecipitates both components because of their covalent association in the intact tetramer.

In many structural and immunological aspects, the lower component of the β subunit is similar to the insulin receptor β subunit. The most likely explanation for the data described above is that the lower component of the β subunit is the insulin receptor β subunit, and that the species of IGF-I receptor described above is a hybrid tetramer composed of an αβ dimer which is a product of the IGF-I receptor gene and an αβ dimer that is a product of the insulin receptor gene. The existence of such a hybrid receptor is consistent with evidence indicating that a panel of monoclonal antibodies cross-react with both receptors (K. SIDDLE, personal communication). This hybrid receptor is an IGF-I receptor because autophosphorylation of both of its subunits is stimulated by equivalent low concentrations of IGF-I and higher concentrations of insulin.

If this model is correct, the insulin receptor α subunit must be much less effective in binding insulin or stimulating autophosphorylation when it is part of the hybrid receptor than when it is part of the insulin receptor. This is consistent with recent reports, discussed above, indicating that the binding properties of both insulin and IGF-I receptor αβ dimers are altered when they are incorporated into tetramers, and that the binding affinity of one dimer may be suppressed (BONI-SCHNETZLER et al. 1987; SWEET et al. 1987a; TOLLEFSEN and THOMPSON 1988; FELTZ et al. 1988). The high affinity binding of IGF-I to its α subunit in the hybrid tetramer could equivalently stimulate the phosphorylation of both subunits either through a symmetric allosteric activation of both subunits or because once one subunit becomes activated, it rapidly catalyzes the phosphorylation of its neighbor.

Such a hybrid receptor could have very unique and complicated properties. A negatively cooperative interaction between binding sites on the hybrid tetramer could result in a receptor with affinities quite different from either symmetric tetramer. This would be particularly apparent when affinities were expressed as IC_{50} values for inhibition of labeled hormone binding. For example, labeled insulin binding to the hybrid tetramer would be predominantly to the insulin receptor dimer. Unlabeled IGF-I could inhibit this labeled insulin binding by direct competition at the insulin binding site, which would be expected to have a low affinity. But it could also, through a negatively cooperative interaction, inhibit labeled insulin binding to the insulin site by binding to the IGF-I receptor site, for which it would be expected to have a higher affinity. It is possible that such a hybrid receptor could account for the atypical low affinity IGF-I receptor (JONAS and HARRISON 1986), which is immunoprecipable by B2, but which is converted into a high affinity IGF-I receptor by disulfide reduction, or for the IGF-I receptor having two distinct binding sites, only one of which is blocked by α-IR-3 (CASELLA et al. 1986), or for the finding that insulin is less effective in competing with IGF-I binding to the dimeric than to the tetrameric IGF-I receptor (FELTZ et al. 1988). Such a hybrid receptor could also help explain the overlapping biological effects mediated by these two receptors.

References

Alexandrides TK, Smith RJ (1989) A novel fetal insulin-like growth factor (IGF) I receptor. J Biol Chem 264:12922–12930

Beguinot F, Kahn CR, Moses AC, Smith RJ (1985) Distinct biologically active receptors for insulin, insulin-like growth factor I, and insulin-like growth factor II in cultured skeletal muscle cells. J Biol Chem 260:15892–15898

Bequinot F, Smith RJ, Kahn CR, Maron R, Moses AC, White MF (1988) Phosphorylation of insulin-like growth factor I receptors by insulin receptor tyrosine kinase in intact cultured skeletal muscle cells. Biochemistry 27:3222–3228

Bhaumic B, Bala RM, Hollenberg MD (1981) Somatomedin receptor of human placenta: solubilization, photolabeling, partial purification and comparison with insulin receptor. Proc Natl Acad Sci USA 78:4279–4283

Blundell TL, Bedarkar S, Humbel RE (1983) Tertiary structures, receptor binding, and antigenicity of insulin-like growth factors. Fed Proc 42:2592–2597

Bodsh W (1988) In vitro mutagenesis in the insulin binding domain of the human insulin receptor. In: Proceedings of The Action of Insulin and Related Growth Factors in Diabetes Mellitus. Joslin Clinic Boston, October 10

Boni-Schnetzler M, Rubin JB, Pilch PF (1986) Structural requirement for the transmembrane activation of the insulin receptor kinase. J Biol Chem 261:15281–15287

Boni-Schnetzler M, Scott W, Waugh SM, Di Bella E, Pilch PF (1987) The insulin receptor: structural basis for high affinity ligand binding. J Biol Chem 262:8395–8401

Cara JF, Satoe HN, Tager HS (1988) Structural determinants of ligand recognition by type I insulin-like growth factor receptors: use of semisynthetic insulin analog probes. Endocrinology 122:2881–2887

Cascieri MA, Chicchi GG, Applebaum J, Hayes NS, Green BG, Bayne ML (1988a) Mutants of human insulin-like growth factor I with reduced affinity for the type I insulin-like growth factor receptor. Biochemistry 27:3229–3233

Cascieri MA, Chicchi GG, Applebaum J, Hayes NS, Green BG, Bayne ML (1988b) Identification of the domains of IGF I responsible for high affinity binding to the types 1 and 2 receptors, insulin receptor, and binding proteins. FASEB J2:A1773

Casella SJ, Han VK, D'Ercole J, Svoboda ME, Van Wyk JJ (1986) Insulin-like growth factor II binding to the type I somatomedin receptor: evidence for two high affinity binding sites. J Biol Chem 261:9268–9273

Chen ZZ, Schwartz GP, Zong L, Burke GT, Chanley JD, Katsoyannis PG (1988) Determinants of growth-promoting activity reside in the A-domain of insulin-like growth factor-I. Biochemistry 27:6105–6111

Chernausek SK, Jacobs S, Van Wyk JJ (1981) Structural similarities between human receptors for somatomedin C and insulin: analysis by affinity labeling. Biochemistry 20:7345–7350

Chou CK, Dull TJ, Russel DS, Cheryi R, Bebwohl D, Ullrich A, Rosen OM (1987) Human insulin receptors mutated at the ATP-binding site lack protein tyrosine kinase activity and fail to mediate post receptor effects of insulin. J Biol Chem 262:1842–1847

Conover CA, Misra P, Hintz RL, Rosenfeld RG (1986) Effect of an anti-insulin-like growth factor antibody on insulin-like growth factor II stimulation of DNA synthesis in human fibroblasts. Biochem Biophys Res Commun 139:501–508

De Meyts P, Bianco AR, Roth J (1976) Site-site interactions among insulin receptors. Characterization of the negative cooperativity. J Biol Chem 251:1877–1888

De Meyts P, Gu J-L, Shymko R, Kaplan B, Bell G, Whittaker J (1988) Insulin receptor alpha subunit sequence 83–103 contains a binding domain complementary to the cooperative site of insulin. In: Proceedings of The Action of Insulin and Related Growth Factors in Diabetes Mellitus. Joslin Clinic Boston, October 10

De Vroede MA, Romananus JA, Standaert ML, Pollet RJ, Nissley SP, Rechler MM (1984) Interaction of insulin-like growth factors with a nonfusing mouse muscle cell line: binding, action, and receptor down-regulation. Endocrinology 114: 1917–1929

De Vroede MA, Rechler MW, Nissley SP, Joshi S, Burke GT, Katsoyannis PG (1985) Hybrid molecules containing the B-domain of insulin-like growth factor I are recognized by carrier proteins of the growth factor. Proc Natl Acad Sci USA 82:3010–3014

De Vroede MA, Rechler MM, Nissley SP, Ogawa H, Joshi S, Burke GT, Katsoyannis PG (1986) Receptor reactivity of hybrid molecules containing portions of the insulin-like growth factor-I and insulin molecules. Diabetes 35:355–361

Duronio V, Jacobs S, Cuatrecasas P (1986) Complete glycosylation of the insulin and insulin-growth factor I receptors is not necessary for their biosynthesis and function: use of swainsonine as an inhibitor in IM-9 cells. J Biol Chem 261:970–975

Duronio V, Jacobs S, Romero PA, Herscovics A (1988) Effects of inhibitors of N-linked oligosaccharide processing on the biosynthesis and function of insulin and insulin-like growth factor-I receptors. J Biol Chem 263:5436–5445

Ebina Y, Araki E, Taira M, Shimada R, Mari M, Craik C, Siddle K, Pierce SB, Roth RA, Rutter WJ (1987) Replacement of lysine residue 1030 in the putative ATP-binding region of the insulin receptor abolishes insulin- and antibody-stimulated glucose uptake and receptor kinase activity. Proc Natl Acad Sci USA 84:704–708

Escobedo JA, Williams LT (1988) A PDGF receptor domain essential for mitogenesis but not for many other responses to PDGF. Nature 335:85–87

Feltz SM, Swanson ML, Wemmie JA, Pessin JE (1988) Functional properties of an isolated alpha-beta heterodimeric human placenta insulin-like growth factor I receptor complex. Biochemistry 27:3234–3242

Flier JS, Usher P, Mosses AC (1986) Monoclonal antibody to the type I insulin-like growth factor (IGF-I) receptor blocks IGF-I receptor-mediated DNA synthesis: clarification of the mitogenic mechanisms of IGF-I and insulin in human skin fibroblasts. Proc Natl Acad Sci USA 83:664–668

Froesch ER, Schmid C, Schwander J, Zapf J (1985) Actions of insulin-growth factors. Annu Rev Physiol 47:443–467

Herrera R, Petruzzelli LM, Rosen OM (1986) Antibodies to deduced sequences of the insulin receptor distinguish conserved and nonconserved regions in the IGF-I receptor. J Biol Chem 261:2489–2491

Hintz RL, Thorsson AV, Enberg G, Hall K (1984) IGF II binding on human lymphoid cells: demonstration of a common high affinity receptor for insulin like peptides. Biochem Biophys Res Commun 127:929–936

Jacobs S, Cuatrecasas P (1986) Phosphorylation of receptors for insulin and insulin-like growth factor I: effects of hormons and phorbol esters. J Biol Chem 261:934–939

Jacobs S, Hazum E, Shechter Y, Cuatrecasas P (1979) Insulin receptor: covalent labeling and identification of subunits. Proc Natl Acad Sci USA 76:4918–4921

Jacobs S, Kull FC Jr, Cuatrecasas P (1983a) Monensin blocks the maturation of receptors for insulin and somatomedin C: identification of receptor precursors. Proc Natl Acad Sci USA 80:1228–1231

Jacobs S, Kull FC Jr, Earp HS, Svoboda ME, Van Wyk JJ, Cuatrecasas P (1983b) Somatomedin-C stimulates the phosphorylation of the B subunit of its own receptor. J Biol Chem 258:9581–9584

Jonas HA (1988) Heterogeneity of receptors for insulin and IGF-I. In: Kahn CR, Harrison LC (eds) Evidence for receptor subtypes, receptor biochemistry and methodology, vol 12B. Liss, New York, pp 19–36

Jonas HA, Harrison LC (1985) The human placenta contains two distinct binding and immunoreactive species of insulin-like growth factor I receptors. J Biol Chem 260:2288–2294

Jonas HA, Harrison LC (1986) Disulfide reduction alters the immunoreactivity and increases the affinity of insulin-like growth factor I receptors in human placenta. Biochem J 236:417–423

Jonas HA, Newman JD, Harrison LC (1986) An atypical insulin receptor with high affinity for insulin-like growth factors copurified with placenta insulin receptors. Proc Natl Acad Sci USA 83:4124–4128

Joshi S, Burke GT, Katsoyannis PG (1985a) Synthesis of an insulin-like compound consisting of the A chain of insulin and a B chain corresponding to the B domain of human insulin-like growth factor I. Biochemistry 24:4208–4214

Joshi S, Ogawa H, Burke GT, Tseng L Y-H, Rechler MM, Katsoyannis PG (1985b) Structural features involved in the biological activity of insulin and the insulin-like growth factors: A^{27} insulin/BIGF I. Biochem Biophys Res Commun 133:423–429

Kadawaki T, Koyashu S, Nishida E, Tobe K, Izumi T, Takaku T, Sakai H, Yahara I, Kasuga M (1987) Tyrosine phosphorylation of common and specific sets of cellular proteins rapidly induced by insulin, insulin-like growth factor I, and epidermal growth factor in an intact cell. J Biol Chem 262:7342–7350

Kasuga M, Van Obberghen E, Nissley SP, Rechler MM (1982) Structure of the insulin-like growth factor receptor in chicken embryo fibroblasts. Proc Natl Acad Sci USA 79:1864–1868

Kiess W, Blickenstaff GD, Sklar MM, Thomas CL, Nissley SP, Sahagian GG (1988) Biochemical evidence that the type II insulin-like growth factor receptor is identical to the cation-independent mannose-6-phosphate receptor. J Biol Chem 263:9339–9344

King GL, Kahn CR (1981) Non-parallel evolution of metabolic and growth promoting function of insulin. Nature 292:644–646

King GL, Kahn CR, Rechler MM, Nissley SP (1980) Direct demonstration of separate receptors for growth and metabolic activities of insulin and multiplication-stimulating activity (an insulin-like growth factor) using antibodies to the insulin receptor. J Clin Invest 66:130–140

King GL, Kahn CR, Samuels B, Danho W, Bullesbach EE, Gattner HG (1982) Synthesis and characterization of molecular hybrids of insulin and insulin-like growth factor I. J Biol Chem 257:10869–10873

Koontz JW (1984) The role of the insulin receptor in mediating the insulin-stimulated growth response in Reuber H-35 cells. Mol Cell Biochem 58:139–146

Koontz JW, Iwahashi M (1981) Insulin as a potent specific growth factor in a rat hepatoma cell line. Science 211:947–949

Kull FC Jr, Jacobs S, Su Y-F, Svoboda ME, Van Wyk JJ, Cuatrecasas P (1983) Monoclonal antibodies to receptors for insulin and somatomedin-C. J Biol Chem 258:6561–6566

LeBon TR, Jacobs S, Cuatrecasas P, Kathuria S, Fujita-Yamaguchi Y (1986) Purification of insulin-like growth factor-I receptor from human placenta membranes. J Biol Chem 261:7685–7689

Lewis RE, Cao L, Perregaux DG, Czech MP (1988) Identification of a major threonine phosphorylation site on the insulin receptor induced by phorbol ester. In: Proceedings of The Action of Insulin and Related Growth Factors in Diabetes Mellitus. Joslin Clinic Boston, October 10

MacDonald RG, Pfeffer SR, Coussens L, Tepper MA, Brocklebank CM, Mole JE, Anderson JK, Chen E, Czech MP, Ullrich A (1988) A single receptor binds both insulin-like growth factor II and mannose-6-phosphate. Science 239:1134–1137

Maegawa H, McClain DA, Freudenberg G, Olefsky JM, Napier M, Lipari T, Pull TJ, Lee J, Ullrich A (1988) Properties of a human insulin receptor with a COOH-terminal truncation. II. Truncated receptors have normal kinase activity but are defective in signaling metabolic effects. J Biol Chem 263:8912–8917

Massague J, Blinderman LA, Czech MP (1982) The high affinity insulin receptor mediates growth stimulation in rat hepatoma cells. J Biol Chem 257:13958–13963

Massague J, Czech MP (1982) The subunit structure of two distinct receptors for insulin-like growth factors I and II and their relationships to insulin receptor. J Biol Chem 257:5038–5045

Massague J, Pilch PF, Czech MP (1980) Electrophoretic resolution of three major insulin receptor structures with unique subunit stoichiometries. Proc Natl Acad Sci USA 77:7137–7141

McClain DA, Maegawa H, Levy J, Hueckstaadt T, Dull TJ, Lee J, Ullrich A (1988) Properties of a human insulin receptor with a COOH-terminal truncation. I. Insulin binding, autophosphorylation, and endocytosis. J Biol Chem 263:8904–8911

Meuli C, Froesch ER (1977) Insulin and nonsuppressible insulin-like activity (NSILA-S) stimulate the same glucose transport system via two separate receptors in rat heart. Biochem Biophys Res Commun 75:689–695

Misra P, Hintz RL, Rosenfeld R (1986) Structure and immunological characterization of insulin-like growth factor II binding to IM-9 cells. J Clin Endocrinol Metab 63:1400–1405

Morgan DO, Roth RA (1986) Identification of a monoclonal antibody which can distinguish between two distinct species of the type I receptor for insulin-like growth factor. Biochem Biophys Res Commun 138:1341–1347

Morgan DO, Jarnagin K, Roth RA (1986) Purification and characterization of the receptor for insulin-like growth factor I. Biochemistry 25:5560–5564

Morgan DO, Edman JC, Standrin DM, Fried VA, Smith MC, Roth RA, Rutter WJ (1987) Insulin-like growth factor II receptor as a multifunctional binding protein. Nature 3296:300–307

Mottola C, Czech MP (1984) The type II IGF receptor does not mediate increased DNA synthesis in H-35 hepatoma cells. J Biol Chem 259:12705–12713

Nagarajan L, Anderson NB (1982) Insulin promotes the growth of F9 embryonal carcinoma cells apparently by acting through its own receptor. Biochem Biophys Res Commun 106:974–980

Poggi C, Le Marchand-Brustel Y, Zapf J, Froesch ER, Freychet P (1979) Effects and binding of insulin-like growth factor I (IGF I) in the isolated soleus muscle of lean and obese mice: comparison with insulin. Endocrinology 105:723–729

Rechler MM, Nissley SP (1985) The nature and regulation of the receptors for insulin-like growth factors. Annu Rev Physiol 47:425–442

Rechler MM, Zapf J, Nissley SP, Froesch ER, Mosses AC, Podskalny JM, Schilling EE, Humbel RE (1980) Interactions of insulin-like growth factors I and II and multiplication-stimulating activity with receptors and serum carrier proteins. Endocrinology 107:1451–1459

Rinderknecht E, Humbel RE (1978a) The amino acid sequence of insulin-like growth factor I and its structural homology with proinsulin. J Biol Chem 253:2769–2776

Rinderknecht E, Humbel RE (1978b) The amino acid sequence of human insulin-like growth factor II. FEBS Lett 89:283–289

Roth RA (1988) Structure of the receptor for insulin-like growth factor II: the puzzle amplified. Science 239:1269–1271

Rubin JB, Shia MA, Pilch PF (1983) Stimulation of tyrosine-specific phosphorylation in vitro by insulin-like growth factor I. Nature 305:438–450

Salmon WD, Daughaday WH (1957) A hormonally controlled serum factor which stimulates sulfate incorporation by cartilage in vivo. J Lab Clin Med 49:825–836

Sasaki N, Rees-Jones RW, Zick Y, Nissly SP, Rechler MM (1985) Characterization of insulin-like growth factor-stimulated tyrosine kinase activity associated with the β subunit of type I insulin-like growth factor receptors of rat liver cells. J Biol Chem 260:9793–9804

Schwartz J, Foster CM, Satin MS (1985) Growth hormone and insulin-like growth factors I and II produce distinct alterations in glucose metabolism in 3T3-F442A adipocytes. Proc Natl Acad Sci USA 82:8724–8728

Shemer J, Adamo M, Wilson GL, Heffez D, Zick Y, Le Roith D (1987) Insulin and insulin-like growth factor-I stimulate a common endogenous phosphoprotein substrate (pp185) in intact neuroblastoma cells. J Biol Chem 262:15476–15482

Shimizu M, Webster C, Morgan DO, Blau HM, Roth RA (1986) Insulin and insulin-like growth factor receptors and responses in cultured human muscle cells. Am J Physiol 251:E611–E615

Shoelson SE, Boni-Schnetzler M, Pilch PF (1988) Autophosphorylation occurs within individual subunits of the insulin receptor. Diabetes 37 [Suppl 1]:29A

Steele-Perkins G, Turner J, Edman JC, Hari J, Pierce SB, Stover C, Rutter WJ, Roth RA (1988) Expression and characterization of a functional human insulin-like growth factor I receptor. J Biol Chem 263:11486–11492

Sweet LJ, Morrison BD, Pessin JE (1987a) Isolation of functional alpha beta heterodimers from the purified human placental alpha2-beta2 heterotetrameric insulin receptor complex. J Biol Chem 262:6939–6942

Sweet LJ, Morrison BD, Wilden PA, Pessin JE (1987b) Insulin-dependent intermolecular subunit communication between isolated alpha beta hetrodimeric insulin receptor complex. J Biol Chem 262:16730–16738

Tollefsen SE, Thompson K (1988) The structural basis for insulin-like growth factor I receptor high affinity binding. J Biol Chem 263:16267–16273

Tollefsen SE, Thompson K, Petersen DJ (1987) Separation of the high affinity insulin-like growth factor I receptor from low affinity binding sites by affinity chromatography. J Biol Chem 262:16461–16469

Tseng L-Y, Schwartz GP, Sheikh M, Chen ZZ, Joshi S, Wang J-F, Nissley SP, Burke GT, Katsoyannis PG, Rechler MM (1987) Hybrid molecules containing the A-domain of insulin-like growth factor-I and the B-chain of insulin have increased mitogen activity relative to insulin. Biochem Biophys Res Commun 149:672–679

Ullrich A, Gray A, Tam AW, Yang-Feng T, Tsubokawa M, Collins C, Henzel W, Le Bon T, Kathuria S, Chen E, Jacobs S, Franke U, Ramachandran J, Fujita-Yamaguchi Y (1986) Insulin-like growth factor I receptor primary structure: comparison with insulin receptor suggests structural determinants that define functional specificity. EMBO J 5:2503–2512

Van Wyk JJ, Graves DC, Casella SJ, Jacobs S (1985) Evidence from monoclonal antibody studies that insulin stimulates deoxyribonucleic acid synthesis through the type I somatomedin receptor. J Clin Endocrinol Metab 61:639–643

Verspohl EJ, Roth RA, Vigneri R, Goldfine ID (1984) Dual regulation of glycogen metabolism by insulin and insulin-like growth factors in human hepatoma cells (Hep-G2): analysis with an anti-receptor monoclonal antibody. J Clin Invest 74:1436–1443

Verspohl EJ, Maddux BA, Goldfine ID (1988) Insulin and IGF I regulate the same biological functions in Hep G2 cells via their own specific receptors. J Clin Endocrinol Metab 67:169–174

Waugh SM, DiBella EE, Pilch (1989) Isolation of a proteolytically-derived domain of the insulin receptor containing the major site of cross-linking/binding. Biochemistry (in press)

Whittaker J, Seino S, Bell GJ, Gu JL, Shymko RM, DeMeyts P (1988) Identification of an insulin binding domain of the insulin receptor. In: Proceedings of The Actions of Insulin and Related Growth Factors in Diabetes Mellitus. Joslin Clinic Boston, October 10

Yarden Y, Escobedo JA, Kuang WJ, Yanzfing TL, Daniel TO, Tremble PW, Chen EY, Ando ME, Harkins RW, Franke U (1986) Structure of the receptor for platelet-derived growth factor helps define a family of closely related growth factor receptors. Nature 323:226–232

Yip CC, Yeung CWT, Moule ML (1980) Photoaffinity labeling of insulin receptor proteins of liver plasma membrane preparations. Biochemistry 19:70–76

Yu K-T, Czech MP (1984) The type I insulin-like growth factor receptor mediates the rapid effects of multiplication-stimulating activity on membrane transport systems in rat soleus muscle. J Biol Chem 259:3090–3094

Yu K-T, Peters MA, Czech MP (1986) Similar control mechanisms regulate the insulin and type I insulin-like growth factor tyrosine kinases. J Biol Chem 261:11341–11349

Zick Y, Sasaki N, Rees-Jones RW, Grunberger G, Nissley PS, Rechler MW (1984) Insulin-like growth factor-I (IGF-I) stimulates tyrosine kinase activity in purified receptors from a rat liver cell line. Biochem Biophys Res Commun 119:6–13

Part IV   Effect of Insulin on Cellular Metabolism

CHAPTER 14

Second Messengers of Insulin Action

A. R. Saltiel and P. Cuatrecasas

A. Overview

Despite significant advances in the past few years on the chemistry and biology of insulin and its receptor, the molecular events that couple the insulin–receptor interaction to the regulation of cellular metabolism remain uncertain. Progress in this area has been complicated by the pleiotropic nature of insulin's actions. These most likely involve a complex network of pathways resulting in the coordination of mechanistically distinct cellular effects. Since the well-recognized mechanism of signal transduction (i.e., cyclic nucleotides, ion channels) appear not to be central to insulin action, investigators have searched for a novel second messenger system. A low molecular weight substance has been identified that mimics certain actions of insulin on metabolic enzymes. This substance has an inositol glycan structure, and is produced by the insulin-sensitive hydrolysis of a glycosylphosphatidylinositol (glycosyl-PI) in the plasma membrane. This hydrolysis reaction, which is catalyzed by a specific phospholipase C, also results in the production of a structurally distinct diacylglycerol, that may selectively regulate one or more of the protein kinases C. The glycosyl-PI precursor for the inositol glycan enzyme modulator is structurally analogous to the recently described glycosyl-PI membrane protein anchor. Preliminary studies suggest that a subset of proteins anchored in this fashion might be released from cells by a similar insulin-sensitive, phospholipase-catalyzed reaction. Future efforts will focus on the precise role of the metabolism of glycosyl-PI in insulin action.

B. Introduction

Few polypeptide hormones have received the attention accorded to insulin, which was the first to be isolated, sequenced, and to have its gene cloned and its secondary and tertiary structure elucidated (Blundell et al. 1972; Sanger 1959; Ullrich 1977). The metabolic effects of the hormone, evaluated in the 1950s, revealed that insulin is the most potent physiological anabolic agent known (Cahill 1971; Levine 1982). Studies on the initial phase of insulin action, binding to its cell surface receptor (Cuatrecasas 1971; Freychet et al. 1971), served as a model of hormone–receptor interactions for other peptide hormones. The insulin receptor was one of the first to be solubilized with detergent and purified by affinity chromatography (Cuatrecasas 1972a). Elucidation of its subunit structure followed shortly thereafter (Jacobs et al. 1977; Pilch and Czech 1980). The cDNA for the human proreceptor has been cloned and its amino acid sequence deduced (EBINA et al. 1985; ULLRICH et al. 1985). Moreover, the regulation of receptor function by tyrosine phosphorylation has been extensively explored (GOLDFINE 1987; ROSEN 1987). Despite these past successes, however, our current understanding of the molecular events that link insulin–receptor interactions to the regulation of cellular metabolism lags far behind that of other peptide hormones. One possible explanation for this slow progress may lie in the complicated nature of insulin action. The cellular effects of the hormone are very broad, including modulation of: (a) transport of molecules across the plasma membrane; (b) levels of cyclic nucleotides; (c) activities of key enzymes in intermediary metabolism; (d) rates of protein synthesis; (e) rates of DNA and RNA synthesis, including specific gene expression; and (f) cellular growth and differentiation. The relative activation and coordination of these distinct cellular processes by insulin varies with cell type, state of differentiation of the cell, and presence of other hormones, as well as the insulin dose–response and time course. This suggests that insulin action involves a network of interrelated and independent pathways with differing levels of divergence regarding mechanisms of regulation (CZECH 1977; KAHN et al. 1981). Thus, it is important to recognize that the search for a single mechanism to explain all of the many effects of insulin is an oversimplistic expectation.

There have been numerous attempts over the past 20 years to elucidate the biochemical pathways involved in signal transduction for insulin. Among the most notable of these early events include changes in membrane potential, cyclic nucleotides, ion flux, polyphosphoinositide hydrolysis, generation of hydrogen peroxide, and internalization of insulin itself or its proteolytic fragments (AVRUCH et al. 1982; CZECH 1977, 1985; DENTON 1986; FAIN 1974; GOLDFINE 1987; JACOBS and CUATRECASAS 1983; KAHN et al. 1981; ROSEN 1987). One of the earliest documented cellular effects of insulin was its ability to decrease the accumulation of cAMP induced by catecholamines in fat cells (BUTCHER et al. 1966). Early suggestions that the glycogenic, lipogenic, and antilipolytic effects of insulin might be secondary to decreases in cAMP were complicated by later studies (FAIN and ROSENBERG 1972) in which changes in cAMP levels were dissociated from these metabolic activities. Attention was also focused early on the release of membrane-bound calcium ions by insulin (CLAUSEN 1975). A role for calcium is supported by more recent evidence demonstrating that certain actions of insulin in the adipocyte can be blocked by the calcium chelator quin 2 (PERSHADSINGH et al. 1987), and that calmodulin can regulate insulin receptor function (GRAVES et al. 1986). However, there are also numerous reports indicating that calcium does not have a direct role as a mediator of insulin action. For example, intracellular calcium is not elevated in response to insulin (KLIP et al. 1984), extracellular calcium is not required for insulin to act (VENKATESON and DAVIDSON 1983), and calcium ionophores or calcium mobilizing agents do not mimic the effect of insulin (KLIP et al. 1984; VENKATESON and DAVIDSON 1983). In fact, in most cases elevation of cellular calcium levels tends to promote catabolic activities, effects which are opposite to those of insulin. Although these early proposals of a role for calcium or cyclic nucleotides as primary mediators of insulin action did not survive later scrutiny, it is possible that changes in either or both of these intracellular components may contribute to some of the actions of the hormone.

As it became clear that the well-recognized mechanisms of signal transduction (i.e., cyclic nucleotides, ion channels) were not primarily responsible for explaining the actions of insulin, many investigators have more recently focused on the role of protein phosphorylation. Early studies suggested that one of the primary actions of the hormone was to promote dephosphorylation reactions, presumably by activation of protein phosphatases or inhibition of kinases (AVRUCH et al. 1982; DENTON 1986). The list of proteins that can undergo dephosphorylation in response to insulin includes glycogen synthetase (LARNER 1971; VILLAR-PALASI and LARNER 1960), pyruvate dehydrogenase (COORE et al. 1971; JUNGAS 1971), hormone-sensitive lipase (STRALFORS et al. 1984), pyruvate kinase (CLAUS et al. 1979), HMG CoA reductase (KENNELLY and RODWELL 1985), acetyl CoA carboxylase (JAMIL and MADSEN 1987; KRAKOWER and KIM 1981; THAMPY and WAKIL 1985), and others (DENTON 1986). In these cases, the insulin-induced decrease in phosphorylation appears to be at least partly responsible for the regulation of enzyme activity. Later studies showed that insulin stimulated the phosphorylation of certain proteins, including ribosomal S6 (ROSEN et al. 1981), ATP citrate lyase (ALEXANDER et al. 1979), acetyl CoA carboxylase (BROWNSEY and DENTON 1982; WITTERS 1981), and others with no known function (AVRUCH et al. 1982; DENTON 1986). In fact, in these studies protein phosphorylation was at least as prominent an action of insulin as was the induction of dephosphorylation.

Two major hypotheses have emerged to explain the coupling of the insulin receptor to intracellular changes in protein phosphorylation (Fig. 1): (a) the existence of a phosphorylation cascade, initiated by the tyrosine kinase activity of the receptor (KASUGA et al. 1982); and (b) the generation of a second messenger, which acts in a manner analogous to cyclic nucleotides or inositol phosphates (LARNER et al. 1979). These two pathways need not be mutually exclusive, and in fact they may operate synergistically to coordinate the series of cellular responses to insulin. Evaluation of the phosphorylation cascade hypothesis has centered on site-directed mutagenesis (CHOU et al. 1987; ELLIS et al. 1986) and anti-receptor antibody (MORGAN and ROTH 1987; MORGAN et al. 1986) experiments which suggest that the receptor tyrosine kinase is necessary for many of the actions of insulin. Although several proteins can be phosphorylated on tyrosine residues in response to insulin (DENTON 1986), the biologically relevant substrates for the receptor kinase that might be responsible for initiating the intracellular cascade have not been identified.

The search for an insulin-dependent "second messenger" has been under way since the early 1970s. LARNER et al. (1979) first reported the existence of an insulin-sensitive substance in skeletal muscle that could acutely modulate glycogen synthetase in vitro. Similar kinds of extracts or substances of elusive chemical identity were subsequently identified in a variety of cell types. These were reported capable of regulating the activities of several insulin-sensitive enzymes, such as pyruvate dehydrogenase (KIECHLE et al. 1980; SALTIEL et al. 1981; SEALS and CZECH 1980), low K_m cAMP phosphodiesterase (PARKER et al. 1982), adenylate cyclase (SALTIEL et al. 1982), glucose 6-phosphatase (SUZUKI et al. 1984), and acetyl CoA carboxylase (SALTIEL et al. 1983). The reader is referred to a review of these early studies (JARETT and KIECHLE 1984). Although these

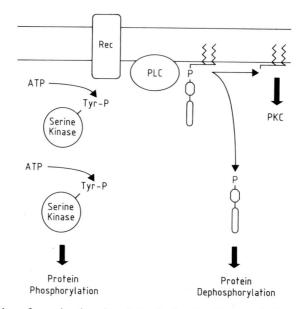

Fig. 1. Regulation of protein phosphorylation by insulin. The metabolic actions of insulin involve the acute regulation of the state of phosphorylation of several proteins. Certain of these are phosphorylated in response to insulin, including ribosomal S6, ATP citrate lyase (*ATP CL*), acetyl CoA carboxylase (*ACC*), and others. The dephosphorylation of other proteins is observed in response to the hormone, including glycogen synthetase (*GS*), pyruvate dehydrogenase (*PDH*), pyruvate kinase (*PK*), acetyl CoA carboxylase (*ACC*), and triglyceride lipase (*TL*). Two possible pathways that might link the insulin receptor to the modulation of the relevant protein kinases and phosphatases are illustrated: *left* a phosphorylation cascade, initiated by the tyrosine kinase activity of the receptor; *right* the generation of a second messenger produced by a receptor-coupled effector system

enzyme-modulating activities were detected in several laboratories, there were some inconsistencies in their reported properties and little progress was made in the elucidation of chemical composition or structure.

C. Biological Characterization of the Putative Insulin Second Messengers

I. Purification

The search for a defined chemical substance that could "mediate" some of the actions of insulin resulted initially in the isolation of two structurally similar substances that were released from hepatic plasma membranes in response to insulin (SALTIEL and CUATRECASAS 1986). The purification of these enzyme-modulating activities relied mainly on ion exchange, molecular sizing, and phase partitioning procedures. The basic protocol for the purification of these substances is:

Treatment of liver membranes with insulin
Centrifugation of membranes and extraction of supernatant
C-18 reverse-phase chromatography
DEAE–cellulose chromatography
QAE–Sephadex chromatography
Strong anion exchange HPLC
P-2 gel permeation chromatography

The two substances were resolved on ion exchange columns or by high voltage electrophoresis, specifically due to their distinct net negative charge. They were not soluble in organic solvents, nor were they adsorbed to reverse-phase columns, indicating a relatively high degree of polarity. The chemical properties of these substances, inferred from their chromatographic behavior and susceptibility to specific chemical modification, are summarized as follows:

Two activities, resolved by distinct net negative charges
Acid stable
Alkaline labile
Molecular weight 800–1000 on gel permeation chromatography
Hydrophilic
Inactivated by acetylation, methylation, periodate oxidation, nitrous acid deamination

The dissimilar net negative charge of the two substances was observed even at pH 2.0, suggesting that phosphate may be the major charged species. The substances were identical in all other chemical properties and enzyme-modulating activities. Their relative hydrophilicity, negative charge, apparent molecular weight of 800–1000 daltons, and sensitivity to periodate and nitrous acid suggested the existence of an oligosaccharide phosphate moiety.

II. Biological Activities

Initial studies on the biological activities of these substances focused on the modification of the activity of the low K_m cAMP phosphodiesterase in fat cell membranes (SALTIEL and STEIGERWALT 1986). This enzyme is cAMP-specific, insensitive to calmodulin, and stimulated in intact cells by physiological concentrations of insulin (SALTIEL and STEIGERWALT 1986). The activity of this cAMP phosphodiesterase was activated acutely by the modulator, as reflected by an increase in the V_{max} of the enzyme, with no appreciable effect on its K_m (SALTIEL 1987). Interestingly, the insulin-independent, low affinity form of the enzyme was not affected. In addition to altering this phosphodiesterase activity, the purified substances could also modify in vitro other insulin-sensitive enzymes assayed in subcellular fractions, including adenylate cyclase (SALTIEL 1987), pyruvate dehydrogenase (SALTIEL 1987), phospholipid methyltransferase (KELLY et al. 1986), and acetyl CoA carboxylase (SALTIEL, unpublished). The kinetic parameters for the modulation of these enzymes are summarized in Table 1. In some cases, the modulation of enzyme activity was biphasic with respect to the concentration of the modulator reminiscent of the paradoxical insulin dose dependencies observed for some of the metabolic activities of the hormone.

Table 1. Enzyme modulation by the inositol glycan in cell-free assays

Enzyme	Source	Effect	Kinetics
Pyruvate dehydrogenase	Liver mitochondria	Biphasic	V_{max}
Low Km cAMP phosphodiesterase	Adipocyte membranes	Biphasic	V_{max}
Adenylate cyclase	Adipocyte membranes	Inhibitory	V_{max}
Acetyl CoA carboxylase	Liver cytosol	Stimulatory	V_{max}
Phospholipid methyltransferase	Adipocyte membranes	Inhibitory	V_{max}

Although the precise biochemical mechanism(s) by which these substances elicit their effects on these enzymes is unclear, the regulation of the activity of each of these enzymes might be explained by alterations in the state of phosphorylation of the enzyme or closely related regulatory factors (SALTIEL 1987). The specific regulation of protein phosphatase activities by these substances was observed in lysates from brain and liver (SALTIEL, unpublished). Although the precise phosphatase subtype(s) that is apparently activated by the enzyme modulator has not been identified, preliminary characterization suggests that this activity is dependent on divalent cations. Acute modulation of these rate-limiting enzymes of intermediary metabolism suggests a possible role for these molecules in the regulation of carbohydrate and lipid synthesis and degradation. Moreover, the inhibition of adenylate cyclase and the stimulation of cAMP phosphodiesterase indicate that these substances may play a part in reducing intracellular cAMP levels, especially influential in the actions of insulin in liver cells.

D. Chemical Characterization
of the Enzyme Modulator as an Inositol Glycan

I. Similarity with the Glycosyl-PI Protein Anchor

Preliminary compositional analyses of the enzyme-modulating substances suggested the existence of inositol as a component. Several well-known inositol phosphate-containing compounds were evaluated, but none exhibited significant enzyme-modulating activity, and they did not share the chemical properties, chromatographic, or electrophoretic behavior of the enzyme modulators. Additionally, neither the composition nor concentration of the known inositol phosphates were affected by insulin. These results suggested that the enzyme modulators might be unusual derivatives containing inositol phosphate. A potential clue was identified when a novel glycosylated derivative of inositol was found in certain cell surface proteins (Fig. 2). This novel molecular species was shown to result from a covalent bond between certain proteins and PI (see LOW and SALTIEL 1988; LOW et al. 1986 for review). This unusual linkage at the carboxy terminus of these proteins serves as an anchor for attachment to the plasma membrane. The protein is coupled via an amide bond to ethanolamine which is then attached through a phosphodiester linkage to an oligosaccharide that ex-

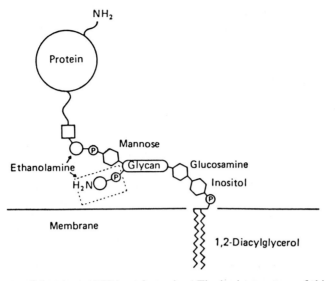

Fig. 2. Structure of the glycosyl-PI protein anchor. The basic structure of this membrane protein anchor is illustrated. The carboxy terminal amino acid is linked by an amide bond to ethanolamine, which is in turn connected through a phosphodiester linkage to an oligosaccharide of variable composition and structure. The terminal monosaccharide of this glycan is non-*N*-acetylated glucosamine, which is linked at the C-1 position to one of the hydroxyls of the inositol ring on phosphatidylinositol. The glycerol lipid moiety serves as the membrane-anchoring domain. (Adapted from Low et al. 1986)

hibits a terminal non-*N*-acetylated hexosamine glycosidically linked to the inositol ring of PI. The membrane-bound form of the protein can be converted to a water-soluble form that contains a carboxy terminal glycosylinositol phosphate by digestion with a bacterial PI-specific phospholipase C (PLC) (Low and Finean 1977), with the simultaneous liberation of diacylglycerol. Additionally, the membrane-bound form of the protein can be hydrolyzed with a specific phospholipase D, producing a water-soluble protein with terminal glycosylinositol and phosphatidic acid (Davitz et al. 1987; Low and Prasad 1988).

The unique ability of the bacterial PI-PLC to release proteins from intact membranes or cells has revealed that over 30 proteins are anchored in this manner to the plasma membrane (Low and Saltiel 1988). These proteins are both evolutionarily and functionally highly diverse, encompassing hydrolytic enzymes, complement regulatory proteins, adhesion molecules, parasitic coat proteins, a number of cell surface antigens with specific cellular distribution, and additional proteins of unknown significance. The only property common to all of these proteins is their location on the cell surface.

To evaluate the possibility that the insulin-dependent enzyme modulators might arise from the phosphodiesteratic hydrolysis of a structurally similar glycolipid, the PI-specific bacterial PLC was added to liver plasma membranes, and the release of the modulators into the medium was assayed. In this series of experiments, PI-PLC was found to reproduce the effect of insulin in facilitating the generation of the enzyme modulators (Saltiel and Cuatrecasas 1986; Saltiel 1987). The PI-PLC digestions generated substances which were chromatographically, electrophoretically, and chemically identical to those produced by insulin treatment, suggesting that the modulators may contain a basic inositol phosphate glycan structure. Moreover, a potential precursor of the PI-PLC-generated substance could be extracted with organic solvents from liver membranes and chromatographically resolved from other known phosphoinositides (Saltiel and Cuatrecasas 1986; Saltiel et al. 1986, 1987). These preliminary experiments raised the possibility that the enzyme modulators were produced as a result of a hormone-stimulated hydrolysis of this novel membrane-associated glycosyl-PI.

II. Metabolic Labeling of the Inositol Glycan

The generation of this inositol glycan enzyme modulator was also evaluated in a cultured myocyte cell line, BC$_3$H1, by following the incorporation of radioactive precursors. In preincubated cells, the enzyme-modulating substances were rapidly labeled with [^{3}H]myo-inositol in response to physiological concentrations of insulin. The presence of nonacetylated glucosamine in the molecule was suggested by the sensitivity of the enzyme-modulating activity to treatment with nitrous acid under conditions in which non-N-substituted hexosamine undergoes deamination with subsequent cleavage of adjacent glycosidic bonds. This specific reaction was used previously to probe the structure of the glycosyl-PI protein anchor (Low et al. 1986). Exposure of the cells to insulin also stimulated the incorporation of [^{3}H]glucosamine into the two peaks of enzyme-modulating activity in parallel to the incorporation of [^{3}H]inositol (Saltiel et al. 1986). The chromatographic, chemical, and electrophoretic behavior of the metabolically labeled compounds was identical to that of the biologically active enzyme modulators, suggesting a structure consisting of inositol phosphate glycosidically linked to the C-1 position of glucosamine, which in turn is glycosidically linked to additional monosaccharides, the precise composition and orientation of which have not been determined (Fig. 3).

The hormone-sensitive turnover of glycosyl-PI has been studied in a number of cell types. This or similar molecules have been metabolically labeled with inositol (Saltiel et al. 1986, 1987), glucosamine (Igareshi and Chambaz 1987; Mato et al. 1987a; Saltiel et al. 1986, 1987), phosphate (Igareshi and Chambaz 1987; Mato et al. 1987a), and saturated fatty acids (Igareshi and Chambaz 1987; Mato et al. 1987a; Saltiel et al. 1986, 1987) in different cultured cells, including BC$_3$H1 myocytes (Saltiel et al. 1986, 1987), H-4 hepatoma cells (Mato et al. 1987a), 3T3-L1 preadipocytes, T-lymphocytes (Gaulton et al. 1988), and primary cultures of adrenocortical cells (Igareshi and Chambaz 1987). The glycolipid is rapidly turned over in response to insulin in BC$_3$H1, 3T3-L1, T-lymphocytes, and hepatoma cells. A similar inositol-containing glycolipid apparently was metabolized in adrenocortical cells in response to serum (Igareshi and Chambaz 1987).

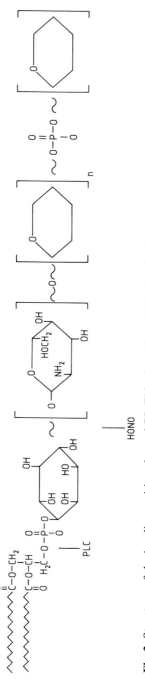

Fig. 3. Structure of the insulin-sensitive glycosyl-PI. This glycophospholipid molecule consists of 1,2-diacylglycerol (or alkylacyl-glycerol) linked by a phosphodiester bond to inositol. The inositol is glycosidically linked to the C-1 position of glucosamine. Glucosamine is then attached at the 4 or 6 position to additional monosaccharides. The terminal monosaccharide may be linked through a phosphodiester bond. The sites of hydrolysis for phospholipase C (*PLC*) and nitrous acid (*HONO*) are shown

E. Structure and Biosynthesis of Glycosyl-PI

I. Structural Studies

Considerable interest has been focused on the structure and biosynthesis of the glycosyl-PI precursor for the enzyme modulator, and on its possible relationship to the glycosyl-PI protein anchor. Some of the properties of this glycolipid precursor are summarized as follows:

Metabolically labeled with inositol, glucosamine, phosphate, saturated fatty acids
Hydrophobic domain consists of 1,2-diacylglycerol or 1,2-alkylacylglycerol
Phospholipase C digestion produces inositol phosphate glycan
Nitrous acid cleavage produces phosphatidylinositol

Hydrolysis of the T-lymphocyte or BC_3H1 cell-derived glycolipid with PI-PLC produced the inositol phosphate glycan and diacylglycerol (SALTIEL et al. 1987). MATO et al. (1987a) suggested that in hepatoma cells the glycerolipid moiety contained a 1,2-alkylacylglycerol structure. More recent studies (MATO et al. 1987b) have indicated another possible structural variation in the liver-derived glycosyl-PI, the presence of significant but variable amounts of *chiro*-inositol, which perhaps accounts for the apparent lack of [^3H]*myo*-inositol labeling in hepatoma cells.

II. Relationship to the Glycosyl-PI Protein Anchor

The insulin-sensitive glycosyl-PI appears to exhibit considerable similarity to the glycosyl-PI protein anchor. The two types of glycolipid contain similar glycerolipid domains, sensitivity to PI-PLC and nitrous acid, the presence of inositol, nonacetylated glucosamine, and a variable glycan region. However, the insulin-sensitive glycosyl-PI apparently lacks two of the features commonly observed in the protein anchor: ethanolamine and amino acids. Additionally, the molecular size of the insulin-sensitive glycosyl-PI is smaller than similar molecules bound to protein.

Although the topological distribution of the insulin-sensitive glycolipid in the plasma membrane is uncertain, preliminary studies suggest a cytoplasmic orientation, since treatment of cells with PI-PLC (presumably exhaustive) does not block the insulin-induced intracellular accumulation of the inositol glycans (SALTIEL et al. 1989). In contrast, ALVAREZ et al. (1988) have suggested an extracellular location for the lipid; 85% of the total cellular glycosyl-PI was apparently surface labeled with isethionyl acetimide. However, these investigators failed to consider spontaneous activation of the PLC that is induced by cell lysis, so that the total cellular lipid may have been significantly underestimated. Thus, this issue must remain open until definitive evidence is available. Although the free release from cells of the inositol glycan in response to insulin has not been reported, the exogenous addition of PI-PLC to intact cells produces an insulin-like activity, exogenous addition of purified inositol glycan to intact cells mimics many of the effects of insulin, and insulin may lead to the release from the cell surface of certain glycosyl-PI-anchored proteins (see below). The possibility that the protein-bound and free forms of glycosyl-PI are located on opposite sides of the plasma membrane leads to further uncertainty regarding their respective biosynthetic processing. One possibility to explain this apparent dilemma is that the early stages of glycosylation of PI occur on the cytoplasmic aspect of the endoplasmic reticulum. Upon attaining a certain level of glycosylation, a fraction of the glycolipid molecules ultimately destined for protein anchoring might be translocated from the cytoplasmic domain of the endoplasmic reticulum membrane to the luminal surface, in analogy to the translocation of the $(Man)_5$-$(GlcNAc)_2$-lipid utilized for N-linked glycosylation of proteins (HIRSCHBERG and SNIDER 1987). This translocation step may serve to segregate further biosynthetic modifications of the lipid molecules ultimately destined for protein attachment from those that will remain on the cytoplasmic face. Alternately, a final processing event, such as addition of a terminal sugar phosphate, may serve to segregate those molecules not destined for translocation. In either case, the subsequent membrane trafficking to the cell surface might then result in a cytoplasmically oriented free glycolipid and a cell surface-oriented protein-anchored glycolipid.

F. Glycosyl-PI Hydrolysis is Catalyzed by an Insulin-Sensitive Glycosyl-PI-Specific Phospholipase C

I. Characterization of a Specific PLC

While insulin is known to cause increased labeling of several phospholipids (DETORRONTEGUI and BERTHET 1966; FARESE et al. 1989; MANCHESTER 1963; PENINGTON and MARTIN 1985; STEIN and HALE 1974), it has not been found to immediately stimulate the hydrolysis of PI or the polyphosphoinositides, and it does not induce calcium mobilization through the generation of inositol trisphosphate (FARESE et al. 1985b; PENINGTON and MARTIN 1985). In contrast, insulin does stimulate the hydrolysis of the [^{3}H]inositol- or [^{3}H]myristate-labeled glycosyl-PI, with the simultaneous production of the [^{3}H]inositol glycan and [^{3}H]myristate diacylglycerol. The rapid production of this specifically labeled diacylglycerol is not observed with agonists known to stimulate the hydrolysis of the polyphosphoinositides (SALTIEL et al. 1987). Thus, the specifically labeled diacylglycerol and the inositol glycans probably arise from the specific, insulin-sensitive hydrolysis of the free glycosyl-PI. Furthermore, these data suggest that the relevant PLC might be highly selective for glycosyl-PI substrates.

These observations led to the search for a glycosyl-PI-specific PLC. Such an enzyme has been isolated from a plasma membrane fraction of liver, using as an assay the liberation of diacylglycerol from the glycosyl-PI-anchored variant surface glycoprotein from *Trypanosoma brucei* or the inositol glycan from the BC$_3$H1 cell-derived glycosyl-PI (Fox et al. 1987). The properties of this enzyme are summarized as follows:

Catalyzes the phosphodiesteratic hydrolysis of glycosyl-PI
Ineffective in hydrolysis of PI, PIP$_2$, or other phospholipids
No requirement for calcium ions
Purified on ion exchange, hydrophobic interaction chromatography
Major silver-stained band at 52 000 daltons
Active as a monomer in nonionic detergent

The catalytic activity appears to reside in a single polypeptide with an apparent molecular weight of about 52 000. The enzyme is calcium-independent, and it is specific for glycosyl-PI; no hydrolysis of PI, PIP_2, or other phospholipids is observed under a variety of conditions.

II. Coupling of the Insulin Receptor to the Glycosyl-PI-Specific PLC

How is the regulation of glycosyl-PI hydrolysis coupled to the activity of the insulin receptor? Although a PLC capable of catalyzing this reaction has been purified, it has thus far been difficult to demonstrate activation of the enzyme by insulin in isolated plasma membranes comparable to that observed in intact cells (SALTIEL et al. 1986). It is possible that the enzyme is activated upon homogenization, as is observed for the glycosyl-PI-specific enzyme in T. brucei. Such a constitutively activated phospholipase might also account for the large amount of enzyme-modulating activity which is spontaneously released from isolated membrane preparations (SALTIEL and CUATRECASAS 1986). At present, the exact mechanism(s) whereby the insulin receptor is coupled to the stimulation of the phospholipase activity remains a critical unknown.

Recent studies with anti-receptor antibodies (MORGAN et al. 1986; MORGAN and ROTH 1987) or site-directed mutagenesis (ELLIS et al. 1986) indicate that the tyrosine kinase activity of the receptor may be necessary for the expression of all of the biological actions of insulin. This suggests that the activation of the glycosyl-PI-specific PLC by the receptor might occur as a consequence of a tyrosine kinase-induced cascade, possibly leading to changes in the state of phosphorylation of the enzyme or perhaps a putative regulatory factor (Fig. 4). Alternatively, the autophosphorylation of the receptor on tyrosine residues could constitute the activation signal that allows the initiation of a membrane coupling event, involving noncovalent interactions with a regulatory factor. The latter possibility is supported by recent studies demonstrating that certain monoclonal antibodies which do not stimulate the receptor tyrosine kinase activity retain other insulin-mimetic properties regarding metabolic regulation (FORSAYETH et al. 1987). These findings suggested that perhaps the interaction of these antibodies with the insulin receptor induces a conformational change similar to that caused by autophosphorylation, resulting in some intramembrane coupling event. In either case, the regulatory factor for such a coupling event might be a specific GTP-binding protein, which in turn could activate the PLC, similar to the coupling of other hormone-sensitive phospholipases via specific G proteins. The involvement of a G protein in insulin action has been suggested in studies demonstrating that pertussis toxin (ELKS et al. 1987; GOREN et al. 1985) or antibodies to the GTP-binding ras p21 protein (DESHPANDE and KUNG 1987; KORN et al. 1987) can block certain actions of insulin. Along these lines, a recent report has suggested that the insulin-dependent production of myristate-labeled diacylglycerol and hydrolysis of glycosyl-PI can be blocked by pretreatment of BC_3H1 cells with pertussis toxin (LUTTRELL et al. 1988). Certain of the G proteins are relatively good substrates for the insulin receptor kinase in vitro (KAMATA et al. 1987; O'BRIEN et al. 1987; ZICK et al. 1986). Although the direct phosphorylation of a G protein on tyrosine residues in response to insulin has not been observed in vivo, these in vitro data suggest at least the possibility of a high affinity interaction between certain G proteins and the receptor.

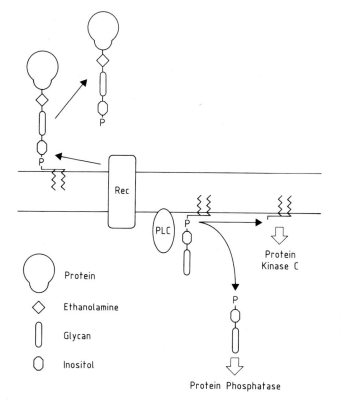

Fig. 4. The metabolism of glycosyl-PI in insulin action. A hypothetical model is presented illustrating the hydrolysis of glycosyl-PI in plasma membranes. The interaction of insulin with its receptor causes the activation of the receptor tyrosine kinase, probably the initial signal for receptor function. The activated receptor is then coupled by an unknown mechanism that may involve an intermediate G protein to the stimulation of one or more PLC specific for glycosyl-PI. This enzyme or group of enzymes then catalyzes the hydrolysis of a free glycosyl-PI that might be on the cytoplasmic side of the plasma membrane, resulting in the intracellular generation of the enzyme-modulating inositol phosphate glycan. Similarly, a glycosyl-PI-specific phospholipase might also cause the release of anchored proteins like heparan sulfate proteoglycan, alkaline phosphatase, or lipoprotein lipase. In both cases the hydrolysis also results in the production of diacyl-glycerol that may cause a selective activation of the protein kinases C

III. Release of Glycosyl-PI-Anchored Proteins

Another result of the insulin-induced activation of a glycosyl-PI-PLC might be the release of glycosyl-PI-anchored proteins. Insulin has been reported to cause reduced levels of cellular alkaline phosphatase in the rat osteogenic sarcoma line ROS 17/2.8 (LEVY et al. 1986). The acute release of this enzyme from BC$_3$H1 cells was observed in response to insulin (ROMERO et al. 1988), perhaps as a result of a phospholipase-catalyzed cleavage of the glycosyl-PI anchor for this enzyme. Like PI-PLC, insulin has been reported to acutely promote the release of the glycosyl-PI-anchored heparan sulfate proteoglycan from rat hepatocytes (ISHIHARA et al.

1987). Interestingly, the released proteoglycan with terminal glycosylinositol phosphate behaved as an autocrine growth regulator in these cells, due to its specific internalization at sites which recognize the inositol phosphate moiety.

Lipoprotein lipase is also acutely released from 3T3-LI adipocytes in response to insulin (SPOONER et al. 1979). Recent studies suggest that a form of this enzyme is anchored to the cell surface by glycosyl-PI, since enzyme labeled metabolically with [^{3}H]glucosamine or [^{32}P]orthophosphate can be specifically immunoprecipitated from the media of 3T3-L1 cells after treatment with PI-PLC (CHAN et al. 1988). Additionally, membrane-associated lipoprotein lipase can be labeled at the cell surface with biotin and subsequently solubilized with PI-PLC. The kinetics of release of lipoprotein lipase activity from 3T3-L1 cells by insulin and PI-PLC are identical, indicating that the acute-phase release by insulin may be due to activation of a glycosyl-PI-specific phospholipase (CHAN et al. 1988). The observation that lipoprotein lipase is a glycosyl-PI-anchored protein is of special significance, since this represents the first example of a protein anchored in this fashion that is known to be released from cells in response to hormones. Tissue or circulating levels of certain of the glycosyl-PI-anchored proteins are altered in diabetic states, including alkaline phosphatase, 5'-nucleotidase, lipoprotein lipase, and heparan sulfate proteoglycan. However, it is unclear whether insulin will lead to the release of all accessible glycosyl-PI-anchored proteins, or only a specific subset. The exploration of this issue may help to resolve whether there are distinct hormonally sensitive and insensitive "structural" pools of glycosyl-PI, similar to what has been proposed for metabolic pooling of the inositol lipids. Although it is not yet known whether the insulin-induced release of these glycosyl-PI-anchored proteins is due to a PLC, the hydrolysis of glycolipid molecules on opposite sides of the membrane in response to insulin also raises questions concerning the polarity of orientation of the relevant hydrolytic enzymes. If the free glycosyl-PI is located on the cytoplasmic face of the plasma membrane, it may be necessary to invoke at least two separate enzymes, located on different sides of the membrane. One possibility is that the hormone-sensitive, cell surface-oriented enzyme is a specific phospholipase D (DAVITZ et al. 1987; LOW and PRASAD 1988).

G. The Role of Diacylglycerol in Insulin Action

Another controversial issue in insulin has been the possible involvement of the calcium- and phospholipid-dependent protein kinase C. Some studies have demonstrated that in intact cells, phorbol esters (the tumor-promoting activators of protein kinase C which can substitute for diacylglycerol) can mimic the actions of insulin on glucose transport (FARESE et al. 1985a), lipogenesis (VAN DE WERVE et al. 1985), glucose oxidation (FARESE et al. 1985a), pyruvate dehydrogenase (FARESE et al. 1985a), mitogenesis (FARESE et al. 1985a), and the phosphorylation of certain proteins (GRAVES and McDONALD 1985; HONEYMAN et al. 1983; TREVILLYAN et al. 1985). On the other hand, phorbol esters inhibit insulin-stimulated lipogenesis (VAN DE WERVE et al. 1985) and antagonize insulin action on glycogen synthesis and glycogenolysis (AHMAD et al. 1984). Phosphorylation of the glucose transport protein was observed in response to phorbol esters, but not insulin (GIBBS et al. 1986; JOOST et al. 1987). Downregulation of kinase C by prolonged exposure to phorbol esters prevented the reactivation of glucose transport after a second challenge with phorbol esters, but did not alter the stimulation of transport by insulin in L6 muscle cells (KLIP et al. 1984). However, downregulation of kinase C did cause a 60% reduction in insulin-activated glucose transport in rat adipocytes (CHERGUI et al. 1987), and a 40% loss of insulin-activated lipogenesis (SMAL and DEMEYTS 1987). In addition, the potent kinase C inhibitor sphingosine completely blocked insulin-stimulated glucose transport in adipocytes (ROBERTSON et al., submitted). The picture has been further complicated by studies examining the hormonal activation of kinase C. Several investigators have reported that while phorbol esters induce the rapid loss of kinase C activity from cytosolic fractions, insulin had no effect (GLYNN et al. 1986; VAARTGES et al. 1986), indicating that the hormone was ineffective in promoting the redistribution of the enzyme from a cytosolic to a membrane fraction thought to reflect kinase C activation. However, insulin has been reported to stimulate both a cytosolic and membrane-associated kinase C in BC$_3$H1 cells (COOPER et al. 1987), and a membrane-associated kinase C in rat diaphragm (WALAAS et al. 1987), indicating a potential mechanism for the activation of the kinase that does not involve its translocation to the membrane.

It may be possible to accommodate these apparently conflicting observations by invoking the involvement of distinct chemical forms or metabolic pools of diacylglycerol produced in response to insulin. Most agonists that cause kinase C activation do so by stimulating the hydrolysis of the polyphosphoinositides, leading to the generation of inositol phosphates and diacylglycerol that contains arachidonate in the C-2 position. The absence of phosphoinositide turnover in response to insulin, as well as the scarcity of arachidonate in the insulin-generated diacylglycerol suggests that this diacylglycerol is indeed different and may arise from an alternate source. Three potential alternate routes of diacylglycerol synthesis can be considered: (a) de novo synthesis from phosphatidic acid (FARESE et al. 1987); (b) hydrolysis of phosphatidylcholine (FARESE et al. 1987); and (c) hydrolysis of glycosyl-PI (SALTIEL et al. 1986). These distinct mechanisms can all lead to the production of a structurally distinct species of diacylglycerol without inositol trisphosphate-induced calcium mobilization. Thus, it is possible that insulin can cause a selective activation of kinase C, depending upon the cell type, extent of activation, enzyme compartmentalization, substrate specificity, or susceptibility to proteolysis. Perhaps the most interesting possibility includes the selective activation of isoforms of kinase C by structurally distinct diacylglycerols. Multiple forms of the enzyme were predicted by the cloning of multiple cDNAs (COUSSENS et al. 1986; KNOPF et al. 1986), and several isozymes have now been chromatographically resolved (HUANG et al. 1986; PELOSIN et al. 1987). Some evidence suggests that these isoforms may exhibit distinct regulatory properties, especially with regard to calcium and diacylglycerol sensitivity as well as substrate specificity. Moreover, these isoforms may exhibit different tissue and/or subcellular distributions, or may be differentially susceptible to downregulation or proteolysis. Thus, the selective activation of protein kinase C or a fraction of isozymes may explain the apparent discrepancies between the biological actions of phorbol esters and insulin.

H. Inositol Glycans as Second Messengers of Insulin Action

Despite the progress made in identifying the structure and biogenesis of the inositol glycans, it is still premature to regard these compounds as second messengers for any of the actions of insulin. The apparent insulin dependency, rapidity, and extent of the generation of the inositol glycans are consistent with the properties expected for second messengers. However, many questions remain concerning their precise chemical structures as well as the nature of their insulin-mimetic properties. Thus far, the actions and properties of the inositol glycans have been explored mainly in subcellular assays, so that the extent to which these molecules reproduce the actions of insulin in intact cells remains unclear. In recent studies, these issues have been addressed by evaluating the effects of the inositol glycan in fat and liver cells, summarized in Table 2. Purified preparations of these compounds mimic the lipogenic (Saltiel and Sorbara-Cazan 1987) and antilipolytic (Kelly et al. 1987b) actions of insulin, as well as the regulation of phosphorylase a (Alvarez et al. 1987), pyruvate kinase (Alvarez et al. 1987), cAMP levels (Alvarez et al. 1987), specific protein phosphorylation (Alemany et al. 1987), and pyruvate dehydrogenase (Gottschalk and Jarett 1988) in intact fat or liver cells, but do not appear to modulate glucose transport (Kelly et al. 1987; Saltiel and Sorbara-Cazan 1987). An oligosaccharide with chemical and chromatographic properties similar to the inositol glycan has recently been isolated from conditioned media of Reuber hepatoma cells (Witters and Watts 1988). This substance stimulates both [^{3}H]thymidine uptake and activation of acetyl CoA carboxylase in this cell line in a manner kinetically indistinguishable from and not additive with insulin. In contrast, this glycan does not stimulate amino acid uptake or tyrosine aminotransferase induction in hepatoma cells nor glucose transport in 3T3-L1 cells. Thus, the selective ability of the inositol glycan

Table 2. Insulin-mimetic activities of the inositol glycan in intact cells

Activity	Cell type	References
Antilipolysis	Adipocyte	Kelly et al. 1987b
Lipogenesis	Adipocyte	Saltiel and Sorbara-Cazan 1987
Glucose oxidation	Adipocyte	Saltiel and Sorbara-Cazan 1987
Pyruvate oxidation	Adipocyte	Gottschalk and Jarett 1988
Protein phosphorylation	Adipocyte	Alemany et al. 1987
	Hepatoma	Witters et al. 1988
Glycogen phosphorylase	Hepatocyte	Alvarez et al. 1987
Pyruvate kinase	Hepatocyte	Alvarez et al. 1987
Phospholipid methyltransferase	Adipocyte	Kelly et al. 1987
Acetyl CoA carboxylase	Hepatoma	Witters et al. 1988
DNA synthesis	Hepatoma	Witters et al. 1987
cAMP levels	Hepatocyte	Alvarez et al. 1987

to mimic only a subset of the actions of insulin provides further evidence for diverse pathways of signal transduction in the actions of the hormone.

The insulin-mimetic actions of the inositol glycan on intact cells suggest that this substance might exert its metabolic effects after entering the cell. However, the chemical structure of the compound suggests that passive transport is unlikely. Interestingly, at millimolar concentrations inositol monophosphate blocks the lipogenic effect of the inositol glycan in intact adipocytes, but is ineffective in attenuating the stimulation of cAMP phosphodiesterase activity in adipocyte membranes (SALTIEL and SORBARA-CAZAN 1987). This observation suggests the possible existence of an active transport system that specifically recognizes the inositol phosphate portion of the glycan. However, there is no evidence that the inositol glycan is released freely from cells upon hormone stimulation, and the significance of a putative cellular uptake for PLC-released proteins or glycans requires further investigation.

Although the inositol glycan appears to be a promising candidate for a second messenger of insulin action, a number of issues remain to be resolved. The ultimate proof of a role for these compounds as second messengers will critically depend, among other things, on the determination of their precise structures. This may be complicated by the chromatographic resolution of multiple species of these molecules, perhaps reflecting the existence of distinct forms with different enzyme-modulating functions. It will also be important to produce these compounds in large quantity and homogeneous form, perhaps by organic synthesis, in order to reevaluate each of their biological activities in detail. Moreover, the biosynthetic route, mode of production and degradation, relationship to the insulin receptor, and the precise biochemical actions of these molecules need further exploration. More detailed molecular characterization of the glycosyl-PI-specific PLC will be necessary. Development of inhibitors or neutralizing antisera to this enzyme, and eventually site-directed mutagenesis studies, should help to define the functional role of this reaction in the pleiotropic actions of insulin.

While there is no reason to expect that these molecules will answer all of the questions regarding the mechanisms of signal transduction in insulin action, the evaluation of the metabolism of glycosyl-PI in pathological states of insulin resistance may lead to the identification of possible critical postreceptor defects in diabetes. Thus, the ultimate goal of these investigations includes the potential development of novel therapeutic strategies for diabetes with the emphasis on bypassing insulin itself to directly intervene in cellular metabolism.

References

Ahmad Z, Lee F-T, DePaoli-Roach A, Roach PJ (1984) Phosphorylation of glycogen synthase by the Ca^{2+} and phospholipid-activated protein kinase (protein kinase C). J Biol Chem 259:8743–8747

Alemany S, Mato JM, Stralfors P (1987) Phosphodephospho control by insulin is mimicked by a phospho-oligosaccharide in adipocytes. Nature 330:77–79

Alexander MC, Kowaloff EM, Witters LA, Dennihy DT, Avruch J (1979) Purification of a hepatic 120 000-dalton hormone stimulated ^{32}P peptide and its identification as ATP-citrate lyase. J Biol Chem 254:8052–8056

Alvarez JF, Cabello MA, Feliu JE, Mato JM (1987) A phospho-oligosaccharide mimics insulin action on glycogen phosphorylase and pyruvate kinase activities in isolated rat hepatocytes. Biochem Biophys Res Commun 147:765–771

Alvarez JF, Varela I, Ruiz-Albusac JM, Mato JM (1988) Localisation of the insulin-sensitive phosphatidylinositol glycan at the outer surface of the cell membrane. Biochem Biophys Res Commun 152:1455–1462

Avruch J, Alexander MC, Palmer JL, Pierce MW, Nemenoff RA, Blackshear PJ, Tipper JP, Witters LA (1982) Role of insulin-stimulated protein phosphorylation in insulin action. Fed Proc 41:2629–2633

Blundell TL, Dodson GG, Hodgkin DC, Mercola DA (1972) Insulin: the structure in the crystal and its reflection in chemistry. Adv Protein Chem 26:279–402

Brownsey RW, Denton RM (1982) Evidence that insulin activates fat cell acetyl CoA carboxylase by increased phosphorylation at a specific site. Biochem J 202:77–86

Butcher RW, Sneyd JGT, Park CR, Sutherland EW (1966) Effect of insulin on adenosine 3′,5′-monophosphate in rat epididymal fat pad. J Biol Chem 241:1651–1653

Cahill GF Jr (1971) Physiology of insulin in man. Diabetes 20:785–797

Chan BL, Lisanti MR, Rodriguez-Boulan E, Saltiel AR (1988) Insulin-stimulated release of lipoprotein lipase by metabolism of its PI anchor. Science 241:1670–1672

Chergui G, Caron M, Wicek D, Lascols O, Capeau J, Picard J (1987) Decreased insulin responsiveness in fat cells rendered protein kinase C-deficient by a treatment with a phorbol ester. Endocrinology 120:2192–2194

Chou CD, Cull TJ, Russel OS, Cherzi R, Lebwohl D, Ullrich A, Rosen OM (1987) Human insulin receptors mutated at the ATP binding site lack protein kinase activity and fail to mediate postreceptor effects of insulin. J Biol Chem 262:1842–1847

Claus TH, El-Maghrabi MR, Pilkis SJ (1979) Modulation of the phosphorylation state of rat liver pyruvate kinase by allosteric effectors and insulin. J Biol Chem 254:7855–7862

Clausen T (1975) Effect of insulin on glucose transport in muscle cells. Curr Top Membr Transp 6:169–226

Cooper DR, Konda TS, Standaert ML, Davis JS, Pollet RJ, Farese RV (1987) Insulin increases membrane and cytosolic protein kinase C activity in BC3H-1 myocytes. J Biol Chem 262:3633–3639

Coore HG, Denton RM, Martin BR, Randle PJ (1971) Regulation of adipose tissue pyruvate dehydrogenase by insulin and other hormones. Biochemistry 125:115–127

Coussens L, Parker P, Chee L, Yang-Feng TL, Chem E, Waterfield MD, Francke V, Ullrich A (1986) Multiple, distinct forms of bovine and human protein kinase C suggest diversity in cellular signalling pathways. Science 233:859–866

Cuatrecasas P (1971) Insulin-receptor interactions in adipose tissue cells: direct measurement and properties. Proc Natl Acad Sci USA 68:1264–1268

Cuatrecasas P (1972a) Affinity chromatography and purification of the insulin receptor of liver cell membranes. Proc Natl Acad Sci USA 69:1277–1281

Cuatrecasas P (1972b) Isolation of the insulin receptor of liver and fat cell membranes. Proc Natl Acad Sci USA 69:318–322

Czech MP (1977) Molecular basis of insulin action. Annu Rev Biochem 46:359–384

Czech MP (1985) The nature and regulation of the insulin receptor: structure and function. Annu Rev Physiol 47:357–381

Davitz MA, Herald D, Shek S, Krakow J, Englund PT, Nussenzweig V (1987) A glycan-phosphatidylinositol-specific phospholipase D in human serum. Science 238:81–84

Denton RM (1986) Early events in insulin actions. Adv Cyclic Nucleotide Protein Phosphorylation Res 26:293–341

Deshpande AK, Kung H-F (1987) Insulin induction of Xenopus laevis oocyte maturation is inhibited by monoclonal antibody against p21 ras proteins. Mol Cell Biol 1:1285–1288

Detorrontegui G, Berthet J (1966) The action of insulin on the incorporation of [^{32}P] phosphate in the phospholipids of rat adipose tissue. Biochim Biophys Acta 116:477–481

Ebina Y, Ellis L, Jarnagin K, Edery M, Graf L, Clauser E, Ouj JH et al. (1985) The human insulin receptor cDNA: the structural basis for hormone activated transmembrane signalling. Cell 259:11 083–11 092

Elks ML, Jackson M, Manganiello CVC, Vaughan M (1987) Effect of NG-(2,2-phenylisopropyl)adenosine and insulin on cAMP metabolism in 3T3-L1 adipocytes. Am J Physiol 252:C342–C348

Ellis L, Clauser E, Morgan DO, Edery M, Roth RA, Rutter WJ (1986) Replacement of insulin receptor tyrosine residues 1162 and 1163 compromises insulin-stimulated kinase activity and uptake of 2-deoxyglucose. Cell 45:721–729

Fain JN (1974) Mode of insulin action. In: Richenberg HV (ed) Biochemistry of hormones, vol 8. MTP Press, Lancaster, pp 1–23 (MTP International review of science biochemistry, ser 1)

Fain JN, Rosenberg L (1972) Antilipolytic action of insulin on fat cells. Diabetes 21:414–425

Farese RV, Standaert ML, Barnes DE, David JS, Pollet RJ (1985a) Phorbol ester provokes insulin-like effects on glucose transport, amino acid uptake and pyruvate dehydrogenase activity in BC$_3$H1 cultured myocytes. Endocrinology 116:2650–2653

Farese RV, Davis JT, Barnes DE, Standaert ML, Bebishkin JS, Hock R, Rosie NK, Pollet RJ (1985b) The de novo phospholipid effect of insulin is associated with increases in diacylglycerol, but not inositol phosphates or cytosolic Ca. Biochem J 231:269–278

Farese RV, Konda TS, David JS, Standaert ML, Pollet RJ, Cooper DR (1987) Insulin rapidly increases diacylglycerol by activating de novo phosphatidic acid synthesis. Science 238:586–589

Farese RV, Cooper DR, Kanda TS, Nair GP, Standaert ML, Pollet RJ (1989) Insulin provokes coordinated increases in the synthesis of phosphatidylinositol, phosphatidylinositol-phosphates and the phosphatidylinositol-glycan in BC$_3$H1 myocytes. Biochem J (in press)

Forsayeth JR, Caro JF, Sinha MK, Maddux BA, Goldfine ID (1987) Monoclonal antibodies to the human insulin receptor that activate glucose transport but not insulin receptor kinase activity. Proc Natl Acad Sci USA 84:3448–3453

Fox JA, Soliz NM, Saltiel AR (1987) Purification of a phosphatidylinositol-glycan specific phospholipase C from liver plasma membranes: a possible target of insulin action. Proc Natl Acad Sci USA 84:2663–2667

Freychet P, Roth J, Neville DM Jr (1971) Insulin receptors in the liver: specific binding of [^{125}I]insulin into the plasma membrane and its relationship to insulin bioactivity. Proc Natl Acad Sci USA 68:1833–1837

Gaulton GN, Kelly KL, Pawlowski J, Mato JM, Jarett L (1988) Regulation and function of an insulin-sensitive glycosyl-phosphatidylinositol during T lymphocyte activation. Cell 538:963–970

Gibbs EM, Allard WJ, Lienhard GE (1986) The glucose transporter in 3T3-L1 adipocytes is phosphorylated in response to phorbol ester but not in response to insulin. J Biol Chem 261:16 597–16 603

Glynn BP, Calliton JW, McDermott JM, Witters LA (1986) Phorbol esters, but not insulin, promote depletion of cytosolic protein kinase C in rat adipocytes. Biochem Biophys Res Commun 135:1119–1125

Goldfine ID (1987) The insulin receptor: molecular biology and transmembrane signalling. Endocr Rev 8:235–255

Goren JJ, Northrup JK, Hollenberg MD (1985) Action of insulin modulated by pertussis toxin in rat adipocytes. Can J Physiol Pharmacol 63:1017–1022

Gottschalk WK, Jarett L (1988) The insulinomimetic effect of the polar head group of an insulin-sensitive glycophospholipid on pyruvate dehydrogenase in both subcellular and whole cell assays. Arch Biochem Biophys 261:175–185

Graves CB, McDonald JM (1985) Insulin and phorbol ester stimulated phosphorylation of a 40 kDa protein in adipocyte plasma membranes. J Biol Chem 260:11 286–11 292

Graves CB, Gale RD, Laurino JP, McDonald JM (1986) The insulin receptor and calmodulin. J Biol Chem 261:10 429–10 438

Hirschberg CB, Snider MD (1987) Topography of glycosylation in the rough endoplasmic reticulum and Golgi apparatus. Annu Rev Biochem 56:63–87

Honeyman TW, Strohsnitter W, Scheiel CR, Schimmel J (1983) Phosphatidic acid and phosphatidylinositol labeling in adipose tissue. Biochem J 212:489–498

Huang K-P, Nakabayashi H, Huang FL (1986) Isozymic forms of rat brain C^{2+}-activated and phospholipid-dependent protein kinase. Proc Natl Acad Sci USA 83:8535–8539

Igareshi Y, Chambaz EM (1987) A novel inositol glycophospholipid (IGPC) and the serum dependence on its metabolism in bovine adrenocortical cells. Biochem Biophys Res Commun 145:249–256

Ishihara M, Fedarko NS, Conrad HE (1987) Involvement of phosphatidylinositol and insulin in the coordinate regulation of proteoheparan sulfate metabolism and hepatocyte growth. J Biol Chem 262:4708–4716

Jacobs S, Cuatrecasas P (1983) Insulin receptors. Annu Rev Pharmacol Toxicol 23:461–479

Jacobs S, Shechter Y, Bissell K, Cuatrecasas P (1977) Purification and properties of insulin receptors from rat liver membranes. Biochem Biophys Res Commun 77:981–988

Jamil H, Madsen NB (1987) Phosphorylation state of acetyl CoA carboxylase. J Biol Chem 262:638–642

Jarett L, Kiechle PL (1984) Intracellular mediators of insulin action. Vitam Horm 41:51–78

Joost HG, Weber TM, Cushman SW, Simpson IA (1987) Activity and phosphorylation state of glucose transporters in plasma membranes from insulin-, isoproterenol- and phorbol ester-treated rat adipose cells. J Biol Chem 262:11 262–11 267

Jungas RL (1971) Hormonal regulation of pyruvate dehydrogenase. Metabolism 20:43–53

Kahn CR, Baird KL, Flier JS, Grunfeld C, Harmon JT, Harrison LC, Karlsson FA et al. (1981) Insulin receptors, receptor antibodies, and the mechanism of insulin action. Recent Prog Horm Action 37:477–537

Kamata T, Kathuria S, Fujita-Yamaguchi Y (1987) Insulin stimulates the phosphorylation level of v-Ha-*ras* protein in membrane fraction. Biochem Biophys Res Commun 144:19–25

Kasuga M, Karlsson FA, Kahn CR (1982) Insulin stimulates the phosphorylation of the 95 000 dalton subunit of its own receptor. Science 215:185–187

Kelly KL, Mato JM, Jarett L (1986) The polar head group of a novel insulin-sensitive glycophospholipid mimics insulin action on phospholipid methyltransferase. FEBS Lett 209:238–242

Kelly KL, Merida I, Wong EHA, DiCenzo D, Mato JM (1987a) A phospho-oligosaccharide mimics the effect of insulin to inhibit isoproterenol-dependent phosphorylation of phospholipid methyltransferase in isolated adipocytes. J Biol Chem 262:15 285–15 290

Kelly KL, Mato JM, Merida I, Jarett L (1987b) Glucose transport and antilipolysis are differentially regulated by the polar head group of an insulin sensitive glycophospholipid. Proc Natl Acad Sci USA 84:6404–6407

Kennelly PJ, Rodwell VW (1985) Regulation of 3-hydroxy-3-methylglutaryl coenzyme A reductase by reversible phosphorylation-dephosphorylation. J Lipid Res 26:903–914

Kiechle FL, Jarett L, Kotagal N, Popp DA (1980) Isolation from rat adipocytes of a chemical mediator for insulin activation of pyruvate dehydrogenase. Diabetes 29:852–855

Klip A, Ramlal T (1987) Protein kinase C is not required for insulin stimulation of hexose uptake in muscle cells in culture. Biochem J 242:131–136

Klip A, Li G, Logan WJ (1984) Role of calcium ions in insulin action on hexose transport in L6 muscle cells. Am J Physiol 247:E297–E304

Knopf JL, Lee M-H, Sultzman LA, Kriz RW, Loomis CR, Hewick RM, Bell RM (1986) Cloning and expression of multiple protein kinase C cDNAs. Cell 46:491–502

Korn LJ, Siebel CW, McCormick F, Roth RA (1987) *Ras* p21 as a potential mediator of insulin action in *Xenopus* oocytes. Science 236:840–843

Krakower GR, Kim K-H (1981) Purification and properties of acetyl CoA carboxylase phosphatase. J Biol Chem 256:2408–2413

Larner J (1971) Insulin and glycogen synthase. Diabetes 21:428–432

Larner J, Galasko G, Cheng K, DePaoli-Roach AA, Huang L, Daggy P, Kellog J (1979) Generation by insulin of a chemical mediator that controls protein phosphorylation and dephosphorylation. Science 206:1408–1410

Levine R (1982) Insulin: the effects and mode of action of the hormone. Vitam Horm 39:145–173

Levy JR, Murray E, Manolagas S, Olefsky JM (1986) Demonstration of insulin receptors and modulation of alkaline phosphatase activity by insulin in rat osteoblastic cells. Endocrinology 119:1786–1792

Low MG, Finean JB (1977) Release of alkaline phosphatase from membranes by a phosphatidylinositol-specific phospholipase C. Biochem J 167:281–284

Low MG, Prasad ARS (1988) A phospholipase D specific for the phosphatidylinositol anchor of cell surface proteins is abundant in plasma. Proc Natl Acad Sci USA 85:980–984

Low MG, Saltiel AR (1988) Structural and functional roles of glycosyl-phosphatidylinositol in membranes. Science 239:268–275

Low MG, Ferguson MAJ, Futterman AH, Silman I (1986) Covalently attached phosphatidylinositol as a hydrophobic anchor for membrane proteins. Trends Biochem Sci 11:212–214

Luttrell LM, Hewlett EL, Romero G, Rogol AD (1988) *Bordetella pertussis* toxin treatment attenuates some of the effects of insulin in BC$_3$H-1 murine myocytes. J Biol Chem 263:6134–6141

Manchester KL (1963) Stimulation by insulin of incorporation of [^{32}P] phosphate and [^{14}C] acetate into lipid and protein of isolated rat diaphragm. Biochim Biophys Acta 70:208–210

Mato JM, Kelly KL, Abler A, Jarett L (1987a) Identification of a novel insulin-sensitive glycophospholipid from H35 hepatoma cells. J Biol Chem 262:2131–2137

Mato JM, Kelly KL, Abler A, Jarett L, Corkey BE, Cashel JA, Zopf D (1987b) Partial structure of an insulin-sensitive glycophospholipid. Biochem Biophys Res Commun 146:764–772

Morgan DO, Roth RA (1987) Acute insulin action requires insulin receptor kinase activity: introduction of an inhibitory monoclonal antibody into mammalian cells blocks the rapid effects of insulin. Proc Natl Acad Sci USA 84:41–45

Morgan DO, Ho L, Korn LJ, Roth RA (1986) Insulin action is blocked by a monoclonal antibody that inhibits the insulin receptor kinase. Proc Natl Acad Sci USA 83:328–332

O'Brien RM, Siddle K, Houslay MD, Hall A (1987) Interaction of the human insulin receptor with the *ras* oncogene product p21. FEBS Lett 217:253–259

Parker JC, Kiechle FL, Jarett L (1982) Partial purification from hepatoma cells of an intracellular substance which mediates the effects of insulin on pyruvate dehydrogenase and low K_m cAMP phosphodiesterase. Arch Biochem Biophys 215:339–344

Pelosin J-M, Vilgrain I, Chambaz EM (1987) A single form of protein kinase C is expressed in bovine adrenocortical tissue, as compared to four chromatographically resolved isozymes in rat brain. Biochem Biophys Res Commun 147:382–391

Penington SR, Martin BR (1985) Insulin-stimulated phosphoinositide metabolism in isolated fat cells. J Biol Chem 260:11 309–11 045

Pershadsingh HA, Gale RD, McDonald JM (1987) Chelation of intracellular calcium prevents stimulation of glucose transport by insulin and insulinomimetic agents in the adipocyte. Evidence for a common mechanism. Endocrinology 121:1727–1732

Pilch PF, Czech MP (1980) The subunit structure of the high affinity insulin receptor. J Biol Chem 255:1722–1731

Romero G, Luttrell L, Rogol A, Zeller K, Hewlett E, Larner J (1988) Phosphatidylinositol-glycan anchors of membrane proteins: potential precursors of insulin mediators. Science 240:509–511

Rosen OM (1987) After insulin binds. Science 237:1452–1458

Rosen OM, Rubin CS, Cobb MH, Smith CJ (1981) Insulin stimulates the phosphorylation of ribosomal protein S6 in a cell free system derived from 3T3-L1 adipocytes. J Biol Chem 256:3630–3633

Saltiel AR (1987) Insulin generates an enzyme modulator from hepatic plasma membranes: regulation of cAMP phosphodiesterase, pyruvate dehydrogenase and ade nylate cyclase. Endocrinology 120:967–972

Saltiel AR, Cuatrecasas P (1986) Insulin stimulates the generation from hepatic plasma membranes of modulators derived from an inositol glycolipid. Proc Natl Acad Sci USA 83:5793–5797

Saltiel AR, Sorbara-Cazan LR (1987) Inositol glycan mimics the action of insulin on glucose utilization in rat adipocytes. Biochem Biophys Res Commun 149: 1084–1092

Saltiel AR, Steigerwalt RW (1986) Purification of a putative insulin-sensitive cAMP phosphodiesterase or its catalytic domain from rat adipocytes. Diabetes 35:698–704

Saltiel AR, Jacobs S, Siegel M, Cuatrecasas P (1981) Insulin stimulates the release from liver plasma membranes of a chemical modulator of pyruvate dehydrogenase. Biochem Biophys Res Commun 102:1041–1047

Saltiel AR, Siegel MI, Jacobs S, Cuatrecasas P (1982) Putative mediators of insulin action: regulation of pyruvate dehydrogenase and adenylate cyclase activities. Proc Natl Acad Sci USA 79:3513–3517

Saltiel AR, Doble A, Jacobs S, Cuatrecasas P (1983) Putative mediators of insulin action regulate hepatic acetyl CoA carboxylase activity. Biochem Biophys Res Commun 110:789–795

Saltiel AR, Fox JA, Sherline P, Cuatrecasas P (1986) Insulin stimulated hydrolysis of a novel glycolipid generates modulators of cAMP phosphodiesterase. Science 233:967–972

Saltiel AR, Sherline P, Fox JA (1987) Insulin-stimulated diacylglycerol production results from the hydrolysis of a novel phosphatidylinositol glycan. J Biol Chem 262: 1116–1121

Saltiel AR, Osterman DG, Darnell JC (1989) The role of glycosyl-phosphoinositides in insulin action. Cold Spring Harbor Symp Quant Biol 53:955–963

Sanger F (1959) Chemistry of insulin. Science 129:1340–1344

Seals JR, Czech MP (1980) Evidence that insulin activates an intrinsic plasma membrane protease in generating a secondary chemical mediator. J Biol Chem 255:6529–6531

Smal J, DeMeyts P (1987) Role of kinase C in the insulin-like effects of human growth hormone in rat adipocytes. Biochem Biophys Res Commun 147:1232–1240

Spooner PM, Chernick SS, Garrison MM, Scow RO (1979) Insulin regulation of lipoprotein lipase activity and release in 3T3-L1 adipocytes. J Biol Chem 254:10021–10029

Stein JM, Hale CN (1974) The effect of insulin on ^{32}P incorporation into rat-fat cell phospholipids. Biochim Biophys Acta 337:41–49

Stralfors P, Fredrickson G, Olsson H, Belfrage P (1984) Reversible phosphorylation of hormone-sensitive lipase in the hormonal control of adipose tissue lipolysis. Horm Cell Regul 8:153–162

Suzuki S, Toyoto T, Suzuki H, Goto Y (1984) A putative second messenger of insulin action regulates hepatic microsomal glucose-6-phosphatase. Biochem Biophys Res Commun 118:40–46

Thampy KG, Wakil SJ (1985) Activation of acetyl CoA carboxylase. J Biol Chem 260:6318–6323

Trevillyan JM, Perisic O, Traugh JA, Byus CV (1985) Insulin and phorbol ester-stimulated phosphorylation of ribosomal protein S6. J Biol Chem 260:3041–3044

Ullrich A (1977) Rat insulin genes: construction of plasmids containing the coding sequences. Science 196:1313–1319

Ullrich A, Bell JR, Chen EY, Herrera R, Petrozelli LM, Dull TJ, Gray A et al. (1985) Human insulin receptor and its relationship to the tyrosine kinase family of oncogenes. Nature 313:756–761

Vaartges WJ, deHaas CGM, van den Bergh SG (1986) Phorbol esters, but not epidermal growth factor or insulin, rapidly decrease soluble protein kinase C activity in rat hepatocytes. Biochem Biophys Res Commun 130:1328–1333

Van de Werve G, Proielto J, Jenrenaud B (1985) Tumor-promoting phorbol esters increase basal and inhibit insulin stimulated lipogenesis in rat adipocytes without decreasing insulin binding. Biochem J 225:523–527

Venkateson N, Davidson MB (1983) Hepatic insulin effects independent of changes in Ca metabolism. Life Sci 32:467–474

Villar-Palasi C, Larner J (1960) Insulin mediated effect on the activity of UDPG-glycogen transglucosylase in muscle. Biochem Biophys Acta 39:171–173

Walaas SI, Horn RS, Adler A, Albert KA, Walaas O (1987) Insulin increases membrane protein kinase C activity in rat diaphragm. FEBS Lett 220:311–318

Witters LA (1981) Insulin stimulates the phosphorylation of acetyl CoA carboxylase. Biochem Biophys Res Commun 100:872–878

Witters LA, Watts TD (1988) An autocrine factor from Reuber hepatoma cells that stimulates DNA synthesis and acetyl CoA carboxylase: characterization of biologic activity and evidence for a glycan structure. J Biol Chem 263:8027–8036

Witters LA, Watts TD, Gould GW, Lienhard GE, Gibbs EM (1987) Regulation of protein phosphorylation by insulin and an insulin-mimetic oligosaccharide in 3T3-L1 adipocytes and FaO hepatoma cells. Biochem Biophys Res Commun 153:992–998

Zick Y, Sagi-Eisenberg R, Pines M, Gierschik P, Spiegel AM (1986) Multisite phosphorylation of the α subunit of transducin by the insulin receptor kinase and protein kinase C. Proc Natl Acad Sci USA 83:9294–9297

CHAPTER 15

Insulin Regulation of Protein Phosphorylation

J. AVRUCH, H. E. TORNQVIST, J. R. GUNSALUS, E. J. YURKOW, J. M. KYRIAKIS, and D. J. PRICE

A. Introduction

This chapter will consider the role of regulatory protein phosphorylation in the sequence of reactions intermediate between insulin binding at the cell surface, and the distal biologic effects. The major biologic actions and in vivo physiologic roles of insulin (at least in postnatal life) have been understood for some time (CAHILL 1971): as the primary hormone of energy storage in adult mammals, insulin's dual physiologic roles are directed toward promoting nutrient entry and assimilation into macromolecules/storage forms (anabolism), as well as inhibiting the breakdown of these macromolecules/energy stores and their release into circulation for oxidation by tissues (anticatabolism). The changes in the cellular physiology of skeletal muscle, liver, and adipose tissue (CAHILL and STEINER 1972), which determine the integrated in vivo response have been well described. This situation is unique among hormones which act through receptor tyrosine kinases; the physiologic roles of these so-called growth factors in the intact organism are largely unknown, although plausible scenarios can be constructed for several, e.g., IGF-I and PDGF. It was anticipated that because the physiologically relevant molecular targets of insulin regulation could be clearly identified (e.g., glucose transport and the enzymes regulating glycogen, fatty acid, and triglyceride metabolism), the pertinent components of the intracellular signaling pathway, which couple the receptor tyrosine kinase to these targets, would also be most readily identified in the insulin system. Although this supposition may yet prove true, and progress has been steady (CZECH 1985; BELFRAGE et al. 1986b; ROSEN 1987, GOREN et al. 1988), it is still not possible to enumerate the complete series of steps which couple insulin receptor activation to any of the distal biologic responses.

The insulin–insulin receptor interaction initiates a complex cascade of intracellular regulatory reactions. It is now quite clear that the major catalytic events in this network involve alterations in tyrosine/serine/threonine phosphorylation. Although the wiring diagram can be drawn in several ways in light of present evidence, the phosphorylation events can be considered in three general categories:

1. The insulin receptor itself is a tyrosine-specific protein kinase, and inhibition of this activity leads to complete inhibition of the insulin-activated signaling function of the receptor (CHOU et al. 1987; EBINA et al. 1987; MCCLAIN et al. 1988). Thus, insulin-stimulated tyrosine-specific phosphorylation of cellular proteins clearly includes regulatory modifications critical for insulin action.

The identity of the relevant molecules, with the exception of the insulin receptor (IR), is not yet known: some candidate substrates will be reviewed briefly.

2. Insulin increases the serine/threonine phosphorylation of a variety of proteins (AVRUCH et al. 1985) and a growing list of insulin-activated serine/threonine-specific protein kinases has accumulated (CZECH et al. 1988). We have previously suggested that these insulin-responsive serine protein kinases serve as the dominant signaling intermediates in this pathway, coupling the activated insulin receptor to more distal biologic effects, including perhaps some of those ultimately mediated by enzyme dephosphorylation (AVRUCH et al. 1979a, b, 1982, 1985). This hypothesis has recently gained direct support from studies on the insulin regulation of an S6(Ser) protein kinase and a cAMP phosphodiesterase; this evidence and the substantial volume of descriptive information on this topic which has appeared in the past several years merit review at this time.

3. At the distal end of the reaction sequence, many of the best characterized metabolic responses to insulin in vivo and in intact cells are attributable to alterations in the activities of enzymes such as glycogen synthase, pyruvate dehydrogenase, and hormone-sensitive triglyceride lipase, wherein the proximate regulatory event is the dephosphorylation of one or more serine and/or threonine residues (CZECH 1985; BELFRAGE et al. 1986b). The regulation of each of these metabolic enzymes is complex, and has been reviewed in detail in this volume and elsewhere (BOYER and KREBS 1986). The present discussion will consider briefly selected aspects of the mechanisms by which insulin regulates the protein kinases and phosphatases mediating these important serine/threonine dephosphorylations.

B. Insulin-Stimulated Tyrosine Phosphorylation

In 1982, KASUGA showed that in intact cells, insulin stimulated the phosphorylation of serine, threonine, and tyrosine residues on the 95000 dalton (β) subunit of its own receptor. The stimulation of IR β subunit tyrosine phosphorylation immediately suggested a functional analogy between the IR and the EGF receptor. COHEN and co-workers (CARPENTER 1979; USHIRO and COHEN 1980; COHEN et al. 1980) had shown earlier that the EGF receptor was ligand-activated, tyrosine-specific protein kinase, which catalyzed an extensive tyrosine-specific autophosphorylation reaction in isolated membranes or after purification to homogeneity. Work carried out in a number of laboratories established that partially purified IR catalyzed a tyrosine-specific autophosphorylation reaction. The tyrosine kinase activity of the IR persisted in stoichiometric relation to the insulin binding function as the receptor was purified to homogeneity (NEMENOFF et al. 1984; PETRUZZELLI et al. 1984). In 1985, the primary structure of the human IR was established by molecular cloning of cDNA (ULLRICH et al. 1985; EBINA et al. 1985a). A region in the intracellular extension of the β subunit, exhibiting homology with other tyrosine kinases, was identified; expression of the molecule encoded by this cDNA yielded an insulin-activated, tyrosine-specific protein kinase

(EBINA et al. 1985 b). Attention was then directed to determining the role of the tyrosine kinase activity in insulin's numerous actions. Two lines of evidence indicate that the phosphotransferase activity is necessary for insulin activation of the signaling function: inactivation of the ATP binding site via mutagenesis (of Lys 1018) concomitantly abolishes phosphotransferase activity and all aspects of insulin-induced signaling (CHOU et al. 1987; EBINA et al. 1987; McCLAIN et al. 1987). Similarly, microinjection into intact cells of anti-receptor monoclonal antibodies which prevent insulin-induced receptor autophosphorylation and kinase activity blocks all aspects of insulin action; microinjection of monoclonal anti-receptor antibodies which bind to another epitope on the β subunit without blocking autophosphorylation/kinase does not modify insulin signaling (MORGAN et al. 1986a, MORGAN and ROTH 1987).

Such experiments establish that the tyrosine-specific phosphotransferase function of IR is absolutely required in order for insulin to initiate biologic responses. A detailed examination of the structural and enzymatic properties of the IR tyrosine kinase is presented elsewhere in this volume (Chap. 11). The following discussion will focus on consideration of the way the receptor kinase carries out its signaling function in situ, i.e., the identity of the functionally relevant Tyr(P)-containing protein targets. Much evidence supports the conclusion that the very first intracellular regulatory step involves an intramolecular, tyrosine-specific autophosphorylation of the receptor β subunit itself; the status of the other candidate substrates is more problematic.

I. Insulin Receptor β Subunit

Virtually all serine/threonine and tyrosine-specific protein kinases exhibit some degree of autophosphorylation, at least when examined in vitro after partial purification; the tyrosine-specific autophosphorylation of the IR β subunit, however, is one of a small number that have been firmly established as a crucial regulatory reaction. All reports agree that autophosphorylation of the partially purified receptor markedly augments the phosphotransferase activity toward exogenous protein substrates and renders the kinase activity ligand-independent (ROSEN et al. 1983; YU and CZECH 1984; PANG et al. 1985; KWOK 1986; KOHANSKI and LANE 1986; MORRISON and PESSIN 1987; HERRERA and ROSEN 1986; TORNQVIST and AVRUCH 1988); the results of KWOK et al. (1986) and TORNQVIST and AVRUCH (1988) indicate that autophosphorylation is absolutely required for the receptor to phosphorylate exogenous substrates at all. If autophosphorylation is blocked, receptor-catalyzed phosphorylation of other proteins will not occur despite full insulin binding and the presence of metal/ATP. Insulin capacitates the receptor to carry out the autophosphorylation rapidly and in a fashion most conducive to kinase activation. Conversely, once β subunit autophosphorylation has occurred, with concomitant activation of the tyrosine kinase, insulin may be removed from the binding site with no loss of kinase activity; dephosphorylation is required for deactivation. This property has been employed to capture the activated state of the IR kinase in intact cells; detergent solubilization of insulin-treated cells in the presence of tyrosine phosphatase inhibitors preserves an activated form of the IR kinase, which can be partially deactivated by treatment with alkaline phosphatase (YU and CZECH 1986; KLEIN et al. 1986). Thus, IR β subunit tyrosine autophosphorylation is an obligatory intermediate step in the insulin stimulation of receptor protein kinase, in situ as well as with the partially purified IR, and can now be viewed as the first intracellular reaction in insulin action.

The relationship of site-specific autophosphorylation to the acquisition of kinase activity has been clarified by a combination of protein chemistry and genetic analyses. The receptor tyrosine residues which participate in the autophosphorylation reaction in vitro (TORNQVIST et al. 1987) and in the intact rat hepatoma (H4) cells (TORNQVIST et al. 1988) were recently identified by direct amino acid sequence analysis: at least 6 of the 13 tyrosines on the β subunit intracellular extension are modified, and are distributed in clusters on three segments of the linear sequence. Three Tyr(P) are located in the heart of the "tyrosine kinase" domain at residues 1146, 1150, and 1151, and are found predominantly in double- (1146/1150 and 1146/1151) and triple-phosphorylated arrays. A second Tyr(P) cluster is near the carboxyl terminus, wherein tyrosine 1316 and 1322 are both phosphorylated. Finally, several lines of indirect evidence (TORNQVIST et al. 1987, 1988) indicate that one or more of the tyrosines at 953/960/972, just inside the membrane, also undergo insulin-stimulated phosphorylation. The progress of site-specific autophosphorylation, examined in vitro using purified human placental IR, was shown to occur in a very concerted, almost "all-or-none" fashion, especially in the presence of insulin (TORNQVIST and AVRUCH 1988); once the initial site (or sites) has been modified, the receptor proceeds abruptly to multisite autophosphorylation, with concomitant acquisition of kinase activity toward exogenous protein substrates.

Autophosphorylation of partially purified IR in the absence of insulin proceeds slowly, and although the extent of overall ^{32}P incorporation approaches 50% of that observed with insulin, the corresponding acquisition of kinase activity as a function of overall ^{32}P incorporation is disproportionately lower in the absence of insulin. This observation indicates that activation requires the phosphorylation of sites that are not efficiently phosphorylated in the absence of insulin. Insulin accelerates phosphorylation into all sites, but does not cause the appearance of "new" sites of autophosphorylation. Nevertheless, inasmuch as hormone binding must direct phosphorylation preferentially into those tyrosine residues whose modification is most productive of kinase activation, the identity of these crucial phosphorylation sites can be inferred by comparison of the extent of site-specific, β subunit Tyr phosphorylation to the concomitant extent of kinase activation in the presence and absence of insulin (TORNQVIST and AVRUCH 1988; WHITE et al. 1988a; HERRERA and ROSEN 1986); such analyses indicate that, in the presence or absence of insulin, multiple modification of the Tyr at 1146/1150/1151 correlates best with acquisition of kinase activity, whereas phosphorylation of the Tyr at 1316/1322 can be extensive (e.g., in the absence of insulin) in the face of little kinase activation. Consistent with the negligible role of Tyr 1316/1322 in kinase activation are the observations that receptor molecules which have been truncated by removal from the β subunit carboxyl terminus of a short segment which includes Tyr 1316/1322, either by partial proteolysis (GOREN et al. 1987) or genetic engineering

(McCLAIN et al. 1988; MAEGAWA et al. 1988a), exhibit a tyrosine kinase function which is largely intact, both in vitro and in the intact cell, and an ability to effectuate signal transmission which is only slightly impaired. By contrast, conversion of Tyr 1150 (or 1150 plus 1151) to Phe by mutagenesis (ELLIS et al. 1986) yields a receptor whose insulin-stimulated tyrosine kinase activity and ability to carry out signal transmission is severely impaired (or totally inhibited). Recently, DEBANT et al. (1988) have presented evidence that suggests that the 1150/1151 Tyr-Phe mutant receptor, although unable to signal activation of glucose transport and glycogen deposition, may be relatively unimpaired in its ability to signal augmented [³H]thymidine incorporation and S6 phosphorylation. Nevertheless, exclusive of DNA synthesis, it can be stated with considerable confidence that receptor autophosphorylation of the residues at 1146/1150/1151 is necessary for activation of the protein kinase activity toward exogenous substrates, and for activation of the cellular response. Specific functional roles, apart from kinase activation, for the other β subunit Tyr(P) residues seems likely, e.g., promoting interaction with native substrates, and recent experiments (see below) support this view.

II. Is Tyr Phosphorylation of Other Proteins the Next Step?

Two general models can be envisioned for the next step in signal transmission. In the first model, the autophosphorylated receptor is now able to interact with, and phosphorylate on tyrosine residues, one or more target proteins whose function is thereby altered; these proteins serve to transmit, amplify, and diversify the signal. Alternatively, the only relevant substrate for insulin-stimulated tyrosine phosphorylation may be the receptor β subunit itself. Autophosphorylation causes a very extensive alteration in the conformation of the receptor, as shown by altered reactivity of the autophosphorylated receptor with a variety of monoclonal and anti-peptide polyclonal antibodies (HERRERA et al. 1985; MORGAN and ROTH 1986b; HERRERA and ROSEN 1986). This new β subunit conformation may enable the receptor to interact effectively (but noncovalently) with one or more coupling elements, such as G proteins (KORN et al. 1987), which in turn serve to transmit the signal. Both formulations emphasize that receptor autophosphorylation is a critical step in the response to insulin, in that it causes the receptor to adopt a new conformation, one that allows productive interaction with other proteins. The essential and critical difference in the two hypotheses is that in one version, these coupling factors are substrates, whose tyrosine phosphorylation is obligatory to further signal transmission, whereas the second hypothesis views their (tyrosine) phosphorylation (if it occurs at all) as irrelevant. While distinct, these two models are obviously not mutually exclusive.

Several observations indicate that a functional, insulin-stimulated tyrosine phosphotransferase activity (at least as reflected by the ability to carry out β subunit autophosphorylation and catalyze the tyrosine phosphorylation of model substrates in broken cell preparations or after partial purification) while perhaps necessary, is not sufficient to ensure that a receptor, when situated in an intact cell, is capable of activating the cellular response upon binding insulin. ELLIS et al. (1987a) constructed a hybrid IR which contained the extracellular domain of the IR (except for 8 residues just prior to the β subunit transmembrane segment),

grafted onto p68$^{gag-ros}$, starting with about 10 residues of the *gag* segment fol
lowed by *ros* sequences which included 6 extracellular residues, the transmembrane (29 residues) segment and the entire cytoplasmic (367 residues)
domain. The cytoplasmic portion of v-*ros* is about 50% identical in amino acid
sequence to the homologous region of the IR. This chimeric receptor exhibited
considerable insulin-stimulated receptor autophosphorylation in intact cells, but
did not mediate insulin-stimulated glucose transport or [^{3}H]thymidine uptake.
The Tyr phosphorylation of putative intracellular substrates (e.g., p180) was not
described. Thus, the ability to "couple" requires more than simply an insulin-
activated transmembrane "tyrosine protein kinase." Even more striking are the
properties of a human IR whose Tyr at 960 has been converted to Phe through
site-directed mutagenesis (WHITE et al. 1988b). This mutant exhibits a relatively
normal autophosphorylation (in terms of overall ^{32}P incorporation) in vitro and
in intact cells, and a vigorous, essentially normal kinase activity, measured by
phosphorylation of a synthetic peptide substrate after receptor solubilization and
partial purification. When expressed in intact CHO cells, these 960 Tyr→Phe
mutant receptors catalyzed β subunit autophosphorylation, but failed to catalyze
the Tyr phosphorylation of the ubiquitous 180 kdalton polypeptide "substrate"
endogenous to many cell types, and were otherwise completely inactive biologically. It seems unlikely that this single Tyr→Phe alteration radically disturbs
overall β subunit conformation, given the fairly normal overall autophosphorylation and kinase activity after solubilization. A more likely conclusion is
that the mutation produces a moderate alteration in β subunit conformation, but
essentially eliminates the ability of the receptor to interact with its native
targets.

It thus appears that the structural requirements for productive coupling in
situ are rather stringent and are poorly reflected by the ability of the partially
purified receptor to phosphorylate model substrates in vitro. Analogous situations have been observed with other enzymes (e.g., dihydrofolate reductase;
BENKOVIC et al. 1988), wherein a mutation may severely impair interaction with
the native substrate (or cofactor), but a model substrate (or cofactor), whose
structure lacks some of the specificity determinants of the native substrate, can be
tolerated and will permit a productive reaction. Thus, the data of ELLIS et al.
(1987a) and WHITE et al. (1988b) argue neither for or against the importance of
the tyrosine kinase in signaling, nor permit a choice between the two hypotheses
posed earlier; rather they indicate that the structural requirements for productive
coupling, certainly for the receptor, and by inference, for the as yet unidentified
"coupling elements"/substrate(s) are specified within a rather narrow tolerance.
Little is known concerning structural features which constitute the specificity
determinants for phosphorylation of native substrates. As discussed below, "native" substrates have not yet been identified with certainty, and the amino acid
sequences surrounding the sites of phosphorylation of the candidate substrates
have not yet been analyzed (PIKE et al. 1984; KLEIN et al. 1985; ZICK et al. 1985;
SAHAL et al. 1988). Studies with synthetic peptides modeled initially on the
autophosphorylation site (Tyr 416) of v-*src* (SMART et al. 1981; HUNTER 1982), indicate that most tyrosine kinases exhibit some preference for peptides with acidic
residues amino terminal to the tyrosine (HUNTER and COPPER 1985). This amino

acid sequence motif, however, is unlikely to be a critical determinant, inasmuch as the K_m values for such peptide substrates are generally in the 0.1–1.0 mM range at best, and tyrosines lacking this context are phosphorylated in a variety of model proteins which can serve as in vitro substrates, e.g., enolase (COOPER et al. 1984) and myosin light chain (GALLIS et al. 1983). We (TORNQVIST and AVRUCH, unpublished observations) have synthesized peptides which correspond to the sequences surrounding each of the 13 tyrosine residues on the intracellular extension of the β subunit. Although the best substrates (highest V_{max}/K_m) for the IR kinase among these peptides are those that encompass the tyrosine residues which serve as native autophosphorylation sites, the peptides all exhibit $K_m \geq 1$ mM, except for the segment surrounding Tyr 1146/1150/1151, which exhibits $K_m \sim 0.2$ mM (see also STADTMAUER and ROSEN 1986). Thus, although primary sequence has some influence in specifying the autophosphorylation sites, it is unlikely to be a major determinant. Moreover, the primary sequence surrounding the β subunit autophosphorylation sites are probably not closely related to phosphorylation site sequences in native substrates, inasmuch as antipeptide antibodies raised against synthetic peptides corresponding to the IR autophosphorylation sites, while reactive with the native Tyr(P) IR, do not react with the Tyr(P) p180 candidate substrate (TORNQVIST et al. 1988).

One clear-cut requirement for "coupling" receptor to at least certain targets in situ is the topological constraint imposed by the necessity for the kinase to be anchored in the membrane in order to activate glucose transport. ELLIS et al. (1987b) have shown that a truncated form of the insulin receptor, which retains its extracellular β subunit and transmembrane domains (and is thus anchored in the membrane), although exhibiting a tyrosine kinase activity (assayed after solubilization) which is less than one-third of the basal (minus insulin) activity of the wild-type IR, confers a constitutive activation of glucose transport in the intact cell. By contrast, a mutant IR whose NH$_2$ terminus is the sixth intracellular residue of the β subunit, and which is an entirely cytoplasmic protein, exhibits very high tyrosine kinase activity, but fails to produce any activation of glucose transport when expressed in the intact cell. Whether the requirement for membrane localization extends to other actions, e.g., activation of glycogen synthase or S6 kinase, has not yet been described, and is of interest, given the report that microinjection of purified IR kinase into the cytoplasm of *Xenopus* oocytes is capable of activating the S6(Ser) protein kinase (MALLER et al. 1986).

Another set of provocative and unexplained phenomena are provided by the observation that certain monoclonal anti-receptor antibodies including several against the IR (FORSAYETH et al. 1987a, b; SIDDLE et al. 1987; TAYLOR et al. 1987) and one anti-IGF-I receptor antibody. α-IR-3 (STEELE-PERKINS et al. 1988), which bind to the receptor extracellular domain and exhibit (insulin-like) agonist activity, can elicit their biologic effects without causing detectable stimulation of β subunit tyrosine phosphorylation; in the case of α-IR-3, agonist activity on several distal functions occurs without detectable tyrosine phosphorylation of p180. Interestingly, however, mutagenic inactivation of the IR ATP binding site (via Lys 1018→Ala) abolishes the cellular response to at least one of these agonist anti-IR antibodies (antibody 25-49 of EBINA et al. 1987). The explanation of these data is unknown; the failure to visualize receptor autophosphorylation/p180 Tyr(P), under conditions wherein partial activation of more distal responses are detectable, may simply reflect the substantial amplification of the signal which can occur between receptor occupancy and the final biologic responses (e.g., see Fig. 1). It is also possible that the antibody-induced clustering or internalization of the wild-type receptor permits the "basal", i.e, unstimulated level of receptor kinase, to perform the signaling function more efficiently, such that a much lower level of autophosphorylation and/or substrate phosphorylation is able to initiate the subsequent step (e.g., see SIDDLE et al. 1985; KAHN et al. 1978). Alternatively, if "coupling" does not require substrate tyrosine phosphorylation, then perhaps the antibodies induce a β subunit conformation which mimics that conferred by autophosphorylation, and the receptor is thus rendered competent to "couple"; this transition is prevented by mutagenesis of Lys 1018. Although puzzling, these data do not compel the adoption of a model that discounts the role of substrate Tyr phosphorylation in signaling.

Given that IR autophosphorylation clearly activates the IR kinase and that, once activated, the IR displays a vigorous phosphotransferase activity, with a turnover number measured in vitro which is at least 10% that of the cAMP-dependent protein kinase, and comparable to that of the cGMP-dependent protein kinase (NEMENOFF et al. 1984; HERRERA et al. 1988), why is it necessary to entertain hypotheses that discount the role of protein substrate Tyr phosphorylation in signaling? The major reason that "noncovalent coupling" models cannot be overlooked is that no example has been identified in any system of a physiologically relevant protein substrate, whose Tyr phosphorylation in situ is necessary for some aspect of the program of hormone action (or oncogenic transformation), other than the tyrosine kinase molecules themselves. However, the absence of known substrates is not unique to the tyrosine kinases; several Ser/Thr kinases lack known physiologically relevant substrates in situ (EDELMAN et al. 1987). No native substrate has been identified for casein kinase 1, a ubiquitous Ser/Thr-specific protein kinase, and the proposed substrates for casein kinase 2 are inferred almost entirely from the unique specificity determinants of this enzyme. Very few "native" substrates have been identified for the cGMP-dependent protein kinase or kinase C; in fact, no substrate has been identified for either of these two well-studied kinases to which can be assigned a key role in mediating the major biologic response of cGMP (e.g., smooth muscle relaxation) or active phorbol esters (tumor "promotion"). Thus, although definitive proof is not yet at hand, receptor-catalyzed tyrosine phosphorylation of one or more intracellular proteins remains the most likely "next step" in insulin action.

III. Strategies for the Detection of Physiologically Relevant Receptor Kinase Substrates

Two general approaches have been taken toward the identification of the "relevant" Tyr(P) substrates of the IR and other tyrosine kinases. A large body of literature describes the phosphorylation in vitro of purified/unpurified, identified/unidentified proteins, by partially purified IR; this literature will not be reviewed. Although many proteins can be phosphorylated in such reactions, it is now clear from the results of ELLIS et al. (1987a) and WHITE et al. (1988b), described above, that this approach will give many "false-positive" responses, i.e., proteins which can be phosphorylated in vitro which do not undergo phosphorylation in the intact cell. Moreover, "false-negatives" may also be engendered, e.g., because it is clear that proper "coupling"/substrate recognition in situ requires that the receptor conformation not be disturbed by solubilization/purification (a property which is not accurately reflected by continued phosphorylation of "model" substrates), and, at least for certain substrates, that the receptor be appropriately inserted in the membrane (as is true for v-*src* and other nonreceptor tyrosine kinases). Few studies have reported the phosphorylation of potential substrates after reincorporating the IR into phospholipid vesicles; for example, KRUPINSKI et al. (1988) recently reported insulin-dependent tyrosine phosphorylation of $G_{o\alpha}$ and $G_{i\alpha}$ in such a reconstituted system, but not transducin and $G_{s\alpha}$. Notwithstanding, insulin-stimulated tyrosine phosphorylation of $G_{i\alpha}$ in insulin-treated rat hepatocytes is not detectable by immunoprecipitation of $G_{i\alpha}$ from ^{32}P-labeled cells (ROTHENBERG and KAHN 1988). Thus, once a candidate substrate has been detected by its in vitro phosphorylation by the IR, it is necessary to verify that such a reaction occurs in the intact cell in response to insulin. This progression from in vitro to intact cell phosphorylation has been carried out with several putative substrates for the IR kinase (e.g., p120/HA4, PERROTTI et al. 1987; lipocortin 1, KARASIK et al. 1988; calmodulin, COLCA et al. 1987). However, the demonstration of insulin-stimulated Tyr phosphorylation of a protein in an intact cell, while obviously necessary for its continued consideration, is not sufficient to establish the candidate as a physiologically relevant substrate. This critical point is discussed further below.

A second approach to the detection of candidate substrates has been to look directly in intact cells for proteins which undergo an insulin-stimulated increase in phosphotyrosine content. Initially, such experiments involved IEF/SDS–PAGE separation of proteins from ^{32}P-labeled, hormone-treated cells. The much lower abundance of Tyr(P), as compared to Ser(P)/Thr(P) in hormone-stimulated (and even in v-*src* transformed cells), made the detection of ^{32}P-Tyr(P)-containing proteins difficult, even after the bulk of Ser(P) had been hydrolyzed at alkaline pH, under conditions wherein Tyr(P) is quite stable (COOPER and HUNTER 1983). In an effort to enrich for the low abundance Tyr(P)-containing proteins, ROSS et al. (1981) introduced the use of antibodies specific for Tyr(P) residues. The first generation polyclonal and monoclonal (FRACKELTON et al. 1983) Tyr(P) antibodies showed relatively low affinity; however, a series of increasingly more avid reagents have been generated (WANG 1985; KAMPS and SEFTON 1988; GLENNEY et al. 1988) with the concomitant detection of an enlarging list of Tyr(P)-containing proteins. The availability of antibodies capable of detecting Tyr(P) in an immunoblot (WANG 1985) format from extracts of nonradioactive cells, has provided an especially convenient reagent, as the immunoblot does not visualize Ser(P)/Thr(P) which are the dominant signals with ^{32}P-labeling, even after base hydrolysis. Antiphosphotyrosine antibodies, however, do have limitations. It is clear that these reagents vary in their apparent affinity for Tyr(P)-containing proteins. Thus, although calpactin 1 (also called lipocortin 2, the 90 kdalton tetramer of two 36 kdalton and two 10 kdalton polypeptide chains), and calpactin 2 (also called lipocortin 1, the 35 kdalton "EGF receptor substrate" first reported by Fava and Cohen 1984) are both phosphorylated in vitro by the IR, calpactin 1 was not precipitated by anti-Tyr(P) antibodies, whereas calpactin 2 was precipitable under the same conditions (Karasik et al. 1988). In general, the causes for these variations remain unexplored, although some obvious explanations can be envisioned. One example is the likelihood that multiple Tyr(P) on a single polypeptide chain (such as the IR β subunit), by allowing interaction of a single antigen polypeptide chain with multiple antibody combining sites, will cause such polyvalent antigens to be bound by the anti-phosphotyrosine antibodies with very high avidity. There is a considerable likelihood that the specificity of these anti-phosphotyrosine antibodies may depend on the specific phosphotyrosine congener that is used for immunization and/or affinity purification, as well as on the context surrounding the phosphotyrosine-like immunogen; as yet however, no clear-cut example in support of this speculation has appeared. Apparent affinity may be conformation-dependent; selectivity on this basis is of less (theoretical) concern in the immunoblot, inasmuch as the proteins are (at least partially) unfolded. There is however at least one example (p42) of a low abundance (0.002%) cytoplasmic protein which is phosphorylated on Tyr to a considerable stoichiometry in a wide number of cells in response to a diverse set of mitogens (including EGF/PDGF/FGF/TPA) which has been widely reported (Nakamura et al. 1983; Gilmore and Martin 1983; Cooper and Hunter 1985; Kohno 1985) by workers using the ^{32}P-labeling/two-dimensional gel approach, but has not been consistently visualized when anti-Tyr(P) immunoblots or affinity columns are employed (e.g., Kamps and Sefton 1988). Immunoblots also have shortcomings, e.g., the possibility of non-Tyr(P)-cross-reacting species (e.g., His(P) or Tyr SO_4); thus, the identity of all immunoreactive signals requires independent verification by ^{32}P-labeling and analysis of phosphoamino acids after partial protein hydrolysis. Tyr(P)-containing protein which are present constitutively and whose Tyr(P) turns over slowly will be detected by immunoblotting, but will be invisible to the ^{32}P-labeling.

Finally, and perhaps most importantly, a major caveat in the use of anti-Tyr(P) antibodies, as well as the earlier two-dimensional gel methods, is that both approaches are biased toward the detection of the most abundant Tyr(P)-containing proteins, irrespective of their regulatory relevance; thus, a protein of high abundance (say >0.1% of cell protein), may undergo a quantitatively insignificant degree of Tyr phosphorylation (say 0.01–0.1 mol P per polypeptide) in the course of insulin activation of the receptor kinase, and nevertheless, be visualized as a prominent candidate for a physiologic hormone-stimulated Tyr(P) substrate. This situation was encountered in the early studies on RSV transformed cells (Cooper et al. 1983), wherein the major cellular Tyr(P)-containing polypeptides were shown to be the three highly abundant glycolytic enzymes (enolase, lactic dehydrogenase, and phosphoglyceromutase) which were trivially phosphorylated, to an extent which could not influence their overall catalytic activity. This "background noise" in tyrosine phosphorylation is particularly extensive in certain retroviral transformed cells, wherein the retroviral tyrosine kinases are constitutively active and essentially unregulated. This feature is well illustrated in cells infected by the nonmyristolated mutants of v-*src*, which cannot insert in the membrane, and are nontransforming. Such cells exhibit comparable or even higher overall Tyr(P) content as compared to cells infected with wild-type v-*src*; moreover >90% of the Tyr(P) "substrates" seen in infection with wild-type v-*src* are also present in the cells infected with the nonmyristolated, nontransforming variants, distributed in both the cytoplasm and particulate fractions (KAMPS et al. 1986; LINDNER and BURR 1988). Although the general expectation is that the ligand-regulated receptor kinases such as the IR will exhibit a tighter stringency, this inference is not yet supported by direct evidence. In fact, the presence of "background noise" in the insulin stimulation of Ser/Thr phosphorylation is strongly implied by earlier work from this laboratory on the insulin-stimulated phosphorylation ATP citrate lyase, which reflects the (serine-specific) phosphorylation of a very abundant cytosolic protein to a relatively low stoichiometry (PIERCE et al. 1982). *Consequently, even if a protein can be shown to be phosphorylated on tyrosine in response to insulin in the intact cell, this does not establish that a physiologically relevant substrate has been identified. It remains necessary to demonstrate that the modification in situ occurs to a stoichiometry sufficient to alter the function of the protein in a meaningful way, and that the altered function mediates some recognizable biologic response to the hormone.*

A useful variation on the anti-phosphotyrosine antibody/SDS gel approach is to examine immunoprecipitates (or eluates from immunoaffinity columns) for the presence of specific enzymatic or biological activities, rather than ^{32}P-labeled polypeptides. The selective recovery of such activities in increased amounts in anti-phosphotyrosine immunoprecipitates/eluates from extracts of hormone/growth factor-treated cells, suggests that the enzyme may be phosphorylated on tyrosine in response to the hormone/growth factor. Such observations have been reported for a phosphoinositol kinase activity from PDGF-treated cells (WHITMAN et al. 1988), a phospholipase C activity from EGF-treated cells (WAHL et al. 1988), and a particulate (Ser/Thr) protein kinase activity solubilized from insulin-treated adipocytes (YU et al. 1987a). In each case, partial adsorption of the activity was observed, and correlation of the activity with a specific [^{32}P]Tyr-containing polypeptide has not been accomplished, although the PDGF-stimulated phosphoinositol kinase appears to copurify with an 85 kdalton phosphoprotein through several further steps.

IV. Candidate Substrates for the Insulin Receptor Kinase

1. p180

The first polypeptide other than the IR β subunit shown to undergo insulin-stimulated Tyr phosphorylation in an intact cell was p180 (WHITE et al. 1985). Detected initially by immunoprecipitation with anti-Tyr(P) antibodies from extracts of ^{32}P-labeled Fao rat hepatoma cells, p180 was subsequently detected by similar methods in 3T3-L1 cells (GIBBS et al. 1986), normal rat hepatocytes and adipocytes, L6 myocytes, and BC₃H1 cells (WHITE et al. 1987; KADOWAKI et al. 1987; KYRIAKIS and AVRUCH, submitted). In MDCK cells (which lack IR), a 180 kdalton protein is Tyr-phosphorylated in response to IGF-I, whereas EGF does not stimulate the phosphorylation of this 180 kdalton polypeptide (Izumi et al. 1987). In BC$_3$H1 cells (a myocyte line), insulin and IGF-I each stimulate the tyrosine phosphorylation of p180 (as detected by anti-Tyr(P) immunoblot), whereas EGF and FGF do not modify p180, but rather exhibit a set of Tyr(P) protein substrates largely distinct from each other, and from insulin/IGF-I (Kyriakis and Avruch submitted). The similarity (although perhaps not identity: Sahal et al. 1988) in apparent substrate specificity in situ exhibited by the insulin and IGF-I receptor kinases is not surprising in view of the close similarity in receptor structure (Ullrich et al. 1986; Yarden and Ullrich 1988) and the recent demonstration that the short- and long-term biologic responses in CHO cells elicited by stimulation of a transfected type I IGF receptor are indistinguishable from those induced by a stimulation of a transfected IR (Steele-Perkins et al. 1988).

The p180 appears to be a cytosolic protein, at least in its tyrosine-phosphorylated state; it is extractable from cells without the use of detergent, is not labeled by iodination of intact cells under conditions that label the IR α subunit, and does not bind to wheat germ agglutinin Sepharose (White et al. 1987; Kadowaki et al. 1987; Tornqvist et al. 1988). The abundance of p180 and the number of Tyr(P) residues on p180 are unknown. Anti-Tyr(P) immunoblots of Triton extracts from a variety of insulin-treated cells invariably show 3- to 10-fold more Tyr(P) immunoreactivity associated with p180, as compared with the IR. Inasmuch as the minimal Tyr(P) content of the autophosphorylated IR is 5–6 mol $P/(\alpha\beta)_2$, a considerable amount of p180 Tyr(P) is present after insulin stimulation. As suggested above, the high affinity that most anti-Tyr(P) antibodies show for Tyr(P) IR may reflect the multivalent Tyr phosphorylation of the IR β subunit; by extension, the lower avidity for Tyr(P) p180 exhibited by these anti-phosphotyrosine antibodies (indicated by the much lower relative recovery of p180 in anti Tyr(P) immunoaffinity chromatography as well as immunoprecipitations (Tornqvist et al. 1988) may indicate that the Tyr(P) p180 contains fewer Tyr(P) residues than the IR β subunit. If Tyr(P) p180 contains 1–3 Tyr(P) p180 sites per polypeptide chain, then the number of Tyr(P) p180 polypeptide chains would be present in 3- to 30-fold molar excess of the Tyr(P) IR, consistent with a substrate–enzyme relationship. The relative amount of unphosphorylated p180 (i.e., the overall stoichiometry of p180 tyrosine phosphorylation) is unknown.

Kadowaki et al. (1987) and Izumi et al. (1987) reported that insulin-stimulated p180 phosphorylation exhibits a dose-response curve which is superimposable on that for IR β subunit autophosphorylation; this conclusion, however, is probably biased by the contributions of p180 serine phosphorylation (the dominant phosphoamino acid on p180) and by poor recovery of p180 Tyr(P), inasmuch as p180 Tyr(P) content was determined after anti-Tyr(P) immunoprecipitation. By contrast, examination of the dose-response for insulin-stimulated p180 Tyr phosphorylation in H4 hepatoma cells by anti-Tyr(P) immunoblot indicates that half-maximal insulin-stimulated tyrosine phosphorylation of p180 is observed at a 10-fold lower concentration of insulin than is required for half-maximal stimulation of IR autophosphorylation (Fig. 1 a); in addition, half-maximal activation of S6 (serine) kinase occurs at 10-fold lower in-

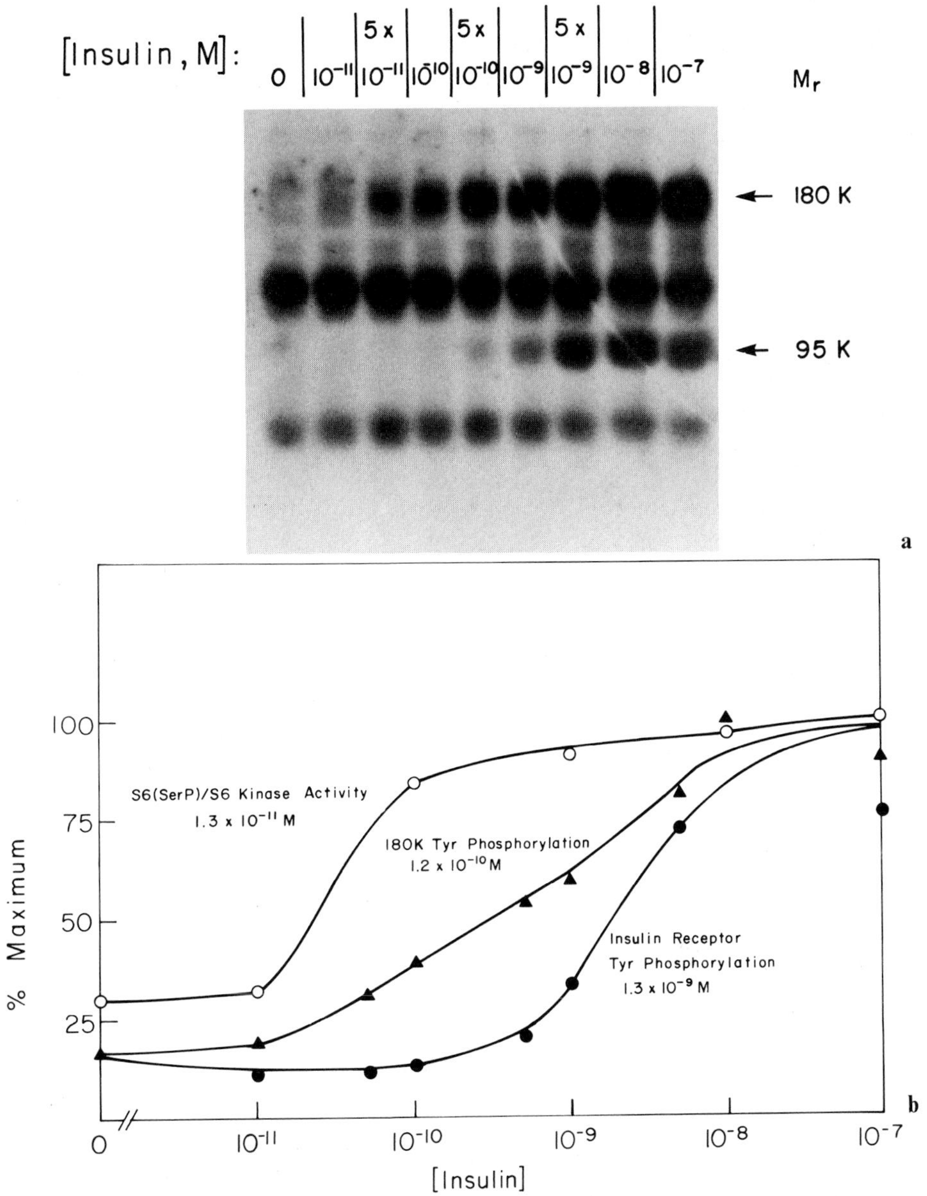

Fig. 1 a, b. Insulin-stimulated protein phosphorylation in hepatoma cells: dose–response. Rat hepatoma cells (H4EIIC) were deprived of serum for 48 h. Insulin was added to each plate at the concentration indicated; 10 min later, the cells were harvested either into the S6 kinase buffer described by NEMENOFF et al. (1986) or into an SDS-containing solution; for SDS-denatured samples, aliquots containing equal protein content were subjected to SDS–PAGE, the proteins were electrophoretically transferred to nitrocellulose, and probed sequentially with an affinity-purified polyclonal autophosphotyrosine antibody, and ^{125}I-labeled protein A as in TORNQVIST et al. (1988). S6 kinase activity was measured as in NEMENOFF et al. (1986). **a** Anti-Tyr(P) immunoblot, **b** activation of the S6 kinase and insulin-stimulated Tyr phosphorylation as a percentage of the maximum

sulin concentrations than p180 Tyr phosphorylation (and 100-fold lower than β subunit autophosphorylation; Fig. 1 b). This progressive leftward shift in K_{act} has several implications: it is possible, for example, that an insulinomimetic agonist (like an anti-IR monoclonal antibody), which can activate distal biologic responses may accomplish this by raising the Tyr(P) phosphorylation of the receptor and p180 to an *undetectably* small, but nevertheless real and biologically significant extent. In addition, the apparent amplification in Tyr(P) content proceeding from the IR β subunit to p180 points to the operation of the receptor kinase in a truly catalytic fashion: if an undetectable number of activated IR, each of which contains multiple Tyr(P), can generate a detectable number of 180 Tyr(P), it is likely that each IR is phosphorylating several p180 molecules, inasmuch as p180 is unlikely to contain a 5- to 10-fold greater Tyr(P) content than that of the IR. Moreover, this leftward shift indicates that H4 cells contain an excess of IR kinase catalytic activity in relation to the cellular content of p180 substrate; recruiting more kinase molecules does not give further Tyr phosphorylation of p180. By contrast, cells such as the BC$_3$H1 myocyte (KYRIAKIS and AVRUCH, manuscript submitted), which express many fewer IR than H4 cells ($\sim 2 \times 10^5$ IR/cell), show superimposed dose–response curves for p180 tyrosine phosphorylation and IR autophosphorylation, as assessed by anti-Tyr(P) immunoblot. This would be anticipated if receptor kinase catalytic activity were limiting in p180 tyrosine phosphorylation. Direct support for the concept that receptor number is limiting to p180 tyrosine phosphorylation is provided by the observation that expression of an additional 10^4–10^6 human IR in CHO (CHOU et al. 1987), or rat 1 cells (IZUMI et al. 1987), which ordinarily exhibit 2000–5000 IR, by transfection of human IR cDNA, leads to a substantial augmentation in the absolute amount of Tyr(P) p180 observed on insulin stimulation.

Insulin-stimulated tyrosine phosphorylation of p180 is very rapid; maximal phosphorylation is achieved by 10–15 s after insulin addition to a variety of cells. This rapidity suggests that prior to its phosphorylation, p180 may be closely apposed to IR; in the limiting case, the IR and p180 may be associated in the basal state, and the receptor-catalyzed tyrosine phosphorylation of p180 may release it into the cytosol. In this regard, the results of MAEGAWA et al. (1988b) are of interest: earlier studies (CHOU et al. 1987; EBINA et al. 1987; McCLAIN et al. 1987) had shown that the expression of mutant IR, whose ATP site has been inactivated by conversion of Lys 1018→Ala, is associated with an inhibition of insulin-stimulated biologic responses, as compared to the parent, untransfected cells. MAEGAWA et al. (1988b) showed that the insulin-stimulated tyrosine phosphorylation of p180 in the Lys 1018→Ala mutant cells is also inhibited, as compared to the parent, untransfected cells. Assuming that the p180 content is not diminished by expression of the inactive receptors, then the inactive receptors, although unable to phosphorylate p180 molecules, may still be capable of sequestering them, perhaps through a specific association. Conceivably, this sequestration might underlie the apparent inhibition of insulin-directed signaling through the endogenous (i.e., active) wild-type receptors, whose apparent kinase activity in these Lys 1018→Ala mutant lines (measured in vitro) appears otherwise largely intact. This speculation is very tentative: IGF-I-stimulated glucose uptake is unaltered in these IR Lys 1018→Ala mutant cells; inasmuch as both IR and IGF-I receptors appear to phosphorylate the same p180, the integrity of the IGF-I response suggests that the inhibitory effect of the mutant IR is exerted through their interaction with the endogenous IR rather than the p180.

All workers report that p180 migrates as a rather broad zone; we find anywhere from one to three distinct insulin-stimulated Tyr(P) polypeptide bands in the $M_r = 180\,000$ region. BC$_3$H1 cells often show all three (KYRIAKIS and AVRUCH, submitted); H4 hepatoma cells and liver cells generally show a doublet, whereas 3T3-L1 cells, a single band. The basis for the size heterogeneity is unknown, but probably not entirely artifactual; it may be due to a variable post-translational modification of a single polypeptide, or to multiple insulin/IGF-I-responsive polypeptides of similar subunit M_r.

Prolonged incubation of cells with insulin is reported to lead, after the initial stimulation of ^{32}P incorporation, to a decrease in the overall ^{32}P incorporation in the p180 region, and the appearance of a doublet of a slightly higher (p210) and slightly lower (p170) apparent M_r (KADOWAKI et al. 1987); the attenuation of the overall ^{32}P-Tyr content may contribute to the visualization of this doublet. WHITE et al. (1987) also observed that treatment of Fao cells with PMA alone led to the appearance of a doublet of similar mobility, as visualized in anti-Tyr(P) immunoprecipitates.

In summary, p180 is the most easily detected insulin-stimulated Tyr(P) protein in all cell types examined thus far, although its actual abundance is unknown. It is phosphorylated to an unknown extent in response to insulin and IGF-I, but not EGF, FGF, or PDGF. Although this is presumed to reflect direct phosphorylation of p180 by the IR kinase, no data on this point are available. The p180 is cytosolic after phosphorylation, and exhibits a modest degree of size heterogeneity, the basis for which is unknown. Cells transfected with mutant IR which exhibit impaired (MAEGAWA et al. 1988b) or absent (WHITE et al. 1988b) p180 phosphorylation, show comparable inhibition of distal biologic responses; p180 tyrosine phosphorylation in intact cells can be stimulated by agents other than insulin (e.g., VO$_4$, protamine, sphingosine; KYRIAKIS and AVRUCH, submitted) which elicit a variety of other cellular responses mimetic of insulin's actions. The p180 is thus an attractive candidate for a physiologic substrate of the IR.

2. p15

FROST and LANE (1985), examining the mechanism of insulin-activated hexose transport in 3T3-L1 cells, observed that the organic arsenical phenylarsine oxide (PAO), a compound with high selectivity for vicinal dithiols, completely inhibited insulin-stimulated glucose uptake with little effect on basal uptake. PAO also exerted a unique action on protein phosphorylation in intact 3T3-L1 cells: PAO (35 μM) permitted the insulin-dependent accumulation of Tyr(P) on a 15 kdalton cytosolic protein with little or no effect on the overall phosphorylation of the IR β subunit (BERNIER et al. 1987; FROST et al. 1987). Insulin did not stimulate detectable Tyr(P) incorporation into p15 in the absence of PAO; IGF-I, although capable of activating glucose transport to the same extent as insulin, did not induce accumulation of Tyr(P) p15 even in the presence of PAO. Prolonged pretreatment of 3T3-L1 cells with VO_4 (1 m$M \times$ 16 h) induces an activation of hexose transport comparable in magnitude to that attained with maximal insulin stimulation, via a dual action: VO_4 inhibition of tyrosine phosphatase also permits the accumulation of autophosphorylated, activated IR kinase molecules, reflected by a twofold increase in overall ^{32}P incorporation into the IR β subunit (exclusively into Tyr, BERNIER et al. 1988). Nevertheless, prolonged incubation with VO_4 does not promote detectable accumulation of Tyr(P) p15, nor does addition of insulin to the VO_4-treated cells (in the absence of PAO) promote the accumulation of Tyr(P) p15. By contrast, addition of PAO to the VO_4-treated cells promotes a substantial accumulation of Tyr(P) p15, as well as a further augmentation (by twofold) of ^{32}P-Tyr(P) incorporation into the IR β subunit. Tyr(P) p15 is susceptible to VO_4-sensitive phosphatases: the dephosphorylation of Tyr(P) p15 which occurs when PAO is inactivated with 2,3-dimercapto-propanol (DMP, a vicinal dithiol), is greatly slowed in the VO_4-treated cells. Thus, inhibition of p15 tyrosine phosphatase by VO_4 does not permit the visualization of IR-mediated tyrosine phosphorylation of p15, indicating that PAO's effect on p15 is not accomplished by inhibition of tyrosine phosphatase (BERNIER et al. 1988). Rather, it appears that PAO must be present in order for p15 to undergo Tyr phosphorylation by an activated IR kinase. The site of action of PAO on tyrosine phosphorylation is not certain, but is probably at the substrate level. PAO does augment the Tyr(P) content of a subset of cellular proteins as assessed by anti-phosphotyrosine immunoblots (KYRIAKIS and AVRUCH, unpublished observations), albeit in a manner quite distinct from that observed with VO_4. Thus, brief exposure (5–20 min) of cells to VO_4 (0.1–2.0 mM) gives a slight increase in "basal" Tyr(P) immunoreactivity; addition of insulin in the presence of VO_4 is associated with a marked augmentation of Tyr(P) content of the p180 and IR β subunit (p15 is not retained on the blots and therefore not evaluated). By contrast, preincubation with PAO (in the absence of hormone) greatly increases the Tyr(P) immunoreactivity of a completely different set of polypeptides, (primarily 100–120 kdalton), but has little effect on the insulin-stimulated tyrosine phosphorylation of IR β subunit or p180 Tyr(P) as occurs after preincubation with VO_4. These features suggest that PAO does not interact primarily with the IR or tyrosine phosphatases, but perhaps with p15 itself, in a manner which alters p15 conformation so as to increase its ability to serve as a substrate for the IR kinase.

FROST et al. (1987) have examined the kinetics of p15 phosphorylation/dephosphorylation in relation to the activation of glucose transport. In the presence of PAO, the insulin-stimulated phosphorylation of p15 proceeds to a maximum more rapidly than the insulin activation of glucose transport (i.e., in the absence of PAO). If Tyr(P) p15 is first allowed to accumulate in the presence of insulin and PAO, inactivation of PAO by addition of DMP induces the rapid dephosphorylation of Tyr(P) p15 and the concomitant activation of glucose transport; the glucose transport activation in this format proceeds more rapidly and without the lag that is observed on direct addition of insulin alone, implying that PAO inhibits transport activation at a relatively late step. Based on these data, FROST et al. (1987) propose that the turnover of Tyr(P) p15 is a critical intermediate step in the activation of glucose transport by insulin, and that perhaps the turnover of p15 is rate-limiting in transport activation. The potential relationship between p15 Tyr(P) turnover and activation of glucose transport is very intriguing, but highly inferential. An equally plausible explanation for the data is that the ability of PAO to inhibit insulin-stimulated hexose transport, and promote the insulin-dependent accumulation of Tyr(P) p15 reflects the dithiol-specific reactivity of PAO acting concomitantly on two independent and functionally unrelated proteins.

HRESKO et al. (1988) have recently found that the amino acid sequence of two proteolytic peptides derived from Tyr(P) p15 corresponds exactly with the sequence deduced from a 3T3-L1-derived cDNA coding for a 14.6 kdalton protein known as p422. This adipocyte protein is highly homologous (62%–64% amino acid identity) with bovine myelin basic protein and rat heart fatty acid-binding protein and is more distantly related to other fatty acid and retinol/retinoic acid-binding proteins (DEMMER et al. 1987). The adipocyte lipid-binding protein (ALBP) was recently purified to homogeneity (MATARESE and BERNLOHR 1988) and shown to bind 1 mol of either oleic acid or retinoic acid at saturation with $K_d = 3 \, \mu M$ and 50 μM, respectively. ALBP/p15 is a highly abundant adipocyte protein, corresponding to 0.5% of total protein, or a cellular content of 270 pmol/10^6 cells ($= 2 \, \mu l$ intracellular water). Based on p15 ^{32}P content, cellular $[\gamma$-^{32}P]ATP-specific activity and assuming 1 mol P/mol p15 (consistent with the presence of a single ^{32}P-labeled tryptic peptide), BERNIER et al. (1988) calculated the content of ^{32}P-Tyr(P) p15, after insulin/VO$_4$ plus PAO as 0.9 pmol/10^6 cells. If these estimates are accurate, then at steady state in the presence of insulin and PAO, about 1 in 300 molecules of p15 bears a Tyr(P). This very low stoichiometry suggests that p15/ALBP tyrosine phosphorylation reflects the incidental modification of a highly abundant cellular protein, perhaps due to binding of PAO by ALBP/p15. The possibility remains that p15 Tyr(P) turnover, rather than phosphorylation, is the critical regulatory event; this outcome would require a reconceptualization of the cellular role of Tyr(P) to that more akin to a phosphoenzyme intermediate. The extensive sequence homology between p15 and the 13–14 kdalton putative growth regulators from hepatocytes (BASSUK et al. 1987) and mammary gland (BOHMER et al. 1987) may simply reflect the ability of these various lipid-binding proteins to bind incidentally hydrophobic growth regulators and/or carcinogens. A more optimistic formulation is that the tyrosine phosphorylation of these polypeptides is not irrelevant, but modulates their (postulated) growth regulatory function.

3. Lipocortin 1

Lipocortin 1, also known as calpactin 2, is a 35 kdalton protein which binds phospholipid and actin in a Ca^{2+}-dependent manner and is avidly phosphorylated by the EGF receptor kinase. It is part of a large family of Ca^{2+}-dependent, phospholipid and cytoskeletal binding proteins (KLEE 1988; CROMPTON et al. 1988); its closest relative is calpactin 1 (lipocortin 2), a 36 kdalton polypeptide which is found monomeric in the cytosol or (predominantly) in a membrane-bound form noncovalently associated with an 11 kdalton light chain as a tetrameric H_2L_2 assembly. The aligned amino acid sequences of calpactin 1 heavy chain and calpactin 2 exhibit $\sim 50\%$ identity overall; the amino terminal 40 residues, which encompass the major sites of Tyr and Ser/Thr phosphorylation, are most dissimilar, while the remaining sequence is composed of homologous repeating units, 70–80 amino acids in length, which fold to form a protease-resistant core which contains the Ca^{2+}/phospholipid and actin binding sites. The RSV tyrosine kinase phosphorylates both calpactin 1 (at Tyr 23) and calpactin 2 (at Tyr 20) avidly, whereas the EGFR kinase phosphorylates only calpactin 2 (at Tyr 20); calpactin 1 lacks a sequence homologous to the major site of EGFR phosphorylation of calpactin 2, i.e., a run of multiple acidic residues amino terminal to the Tyr at residue 20 (ENEEQEY).

The cellular function of the calpactins is unknown; they are relatively abundant cellular proteins, and, based on their biochemical properties, have been suggested to participate in membrane trafficking and cytoskeletal organization. Considerable interest has also focused on a possible role for these proteins as regulators of phospholipase activity. Lipocortins were defined initially as glucocorticoid-induced inhibitors of eicosanoid production (FLOWER et al. 1984), and were thought to act by limiting arachidonic acid precursor availability through the inhibition of phospholipase A_2 (PLA_2); this inhibitory function was reportedly regulated by serine- and tyrosine-specific phosphorylation (HIRATA 1981; HIRATA et al. 1984). HUANG et al. (1986) purified the major PLA_2 inhibitors from human placenta; subsequent analyses employing specific antisera and cDNA clones demonstrated that these two proteins, initially called lipocortin 1 and 2, were identical to calpactin 2 and 1, respectively (SARIS et al. 1986). Recent studies have shown that these proteins serve as PLA_2 inhibitors in vitro by virtue of their interaction with the phospholipid substrate, rather than with the PLA_2 enzyme (HAIGLER et al. 1987; DAVIDSON et al. 1987). This property suggests that the actions of other phospholipases will also prove to be inhibited, at least in vitro.

The physiologic role of the lipocortins as regulators of phospholipase activity in situ remains uncertain; these abundant proteins exhibit a relatively low overall stoichiometry of tyrosine phosphorylation, and a relatively low affinity for Ca^{2+}, if Ca^{2+} is to act as a regulatory ligand. Nevertheless, the cellular content of lipocortin 1 (calpactin 2) is increased by corticosteroid pretreatment, and tyrosine phosphorylation of lipocortin 1 by the EGF receptor in vitro diminishes by fivefold the Ca^{2+} concentration required for binding of lipocortin 1 (calpactin 2) to phosphatidylserine vesicles (SCHLAEPFER and HAIGLER 1987). Thus, the lipocortins remain viable as candidates for physiological regulation through tyrosine phosphorylation.

Lipocortin 1 and 2 are both phosphorylated by the IR kinase in vitro; preliminary peptide maps suggest that IR and EGFR modify the same Tyr residue (Tyr 20) on lipocortin 1 (KARASIK et al. 1988). Employing specific anti-lipocortin 1 and 2 antisera, these workers attempted to immunoprecipitate [^{32}P]lipocortin 1 or 2 from ^{32}P-labelled intact rat hepatocytes; KARASIK et al. (1988) observed that glucocorticoid treatment of the rats induced a large increase in hepatic lipocortin content. Lipocortin 2, although much more abundant than lipocortin 1 (before and after glucocorticoid treatment), did not undergo insulin-stimulated phosphorylation in intact hepatocytes, as would be anticipated by its inability to serve as a receptor kinase substrate in vitro. Lipocortin 1, however, did exhibit insulin-stimulated ^{32}P Tyr(P) incorporation, but only in the hepatocytes from the glucocorticoid-treated rats. The stoichiometry of the insulin-stimulated phosphorylation of lipocortin 1 in hepatocytes is unknown (but probably <0.1 mol P/mol), and the functional consequence obscure, but presumably identical for insulin and EGF, inasmuch as both ligands promote phosphorylation at Tyr 20. KARASIK et al. (1988) consider a role for lipocortin 1 phosphorylation in the usual regulatory program underlying insulin action to be unlikely, in view of the failure to visualize insulin-stimulated lipocortin phosphorylation in the absence of glucocorticoid pretreatment (a maneuver known to induce overall insulin resistance). In fact, they have offered the provocative suggestion that lipocortin 1, as a relatively high affinity IR kinase substrate ($K_m \sim 3 \mu M$) whose abundance increases after glucocorticoid therapy, may serve as a competitive inhibitor of the phosphorylation of the yet-to-be detected native substrates required for insulin action, and thereby underlie, in part, the glucocorticoid-induced insulin resistance.

4. p120

The p120 was detected by REES-JONES and TAYLOR (1985) as an endogenous substrate for insulin-stimulated Tyr phosphorylation in Triton extracts of hepatic membranes purified by wheat germ agglutinin (WGA) Sepharose chromatography; SADOUL et al. (1985) reported a similar observation. Phosphorylation of p120 in these extracts is also stimulated by EGF (PHILLIPS et al. 1987). A p120 IR substrate was detected in similar extracts prepared from liver of a number of species, but not from brain, muscle, heart, kidney, or adipocytes (PHILLIPS et al. 1987). PERROTTI et al. (1987) subsequently showed that insulin caused a 1.6- to 2.5-fold increase in overall ^{32}P content of p120 in intact H35 hepatoma cells, with some increase in [^{32}P]Tyr. The p120 is to be distinguished from a family of 100–120 kdalton Tyr(P)-containing proteins detected by anti-Tyr(P) immunoblotting; the liver-specific p120 glycoprotein described by TAYLOR and colleagues (REES-JONES and TAYLOR 1985; PERROTTI et al. 1987; ACCILI et al. 1986) is not immunoprecipitated by anti-Tyr(P) antibodies which precipitate the IR β subunit (WHITE et al. 1985) and p180 (ACCILI et al. 1986). The immunoactive Tyr(P) at $M_r = 120\,000$ visualized in immunoprecipitates and immunoblots (KYRIAKIS and AVRUCH, submitted) of Triton extracts from several cell types (H4, BC$_3$H1, 3T3-L1, Rat 1 transfected with human IR), are usually readily detectable in the hormone-free state and are distributed in both the cytosol and membrane; the bulk of these proteins are not adsorbed to WGA Sepharose. One or more of these species exhibit marked augmentation in Tyr(P) by EGF, but not in response to insulin. The p120 of MARGOLIS et al. (1988) is precipitated by a monoclonal antibody (HA4) which is reactive with a 120 kdalton rat plasma membrane glycoprotein localized to the bile canalicular face of the hepatocyte. This protein is photolabeled by incubation of liver membranes with a taurocholate derivative, and is thought to participate in bile acid transport (RUETZ et al. 1987). Thus, the liver-specific, 120 kdalton, intrinsic plasma membrane protein is unlikely to be a target of physiological importance to insulin action; it is one minor component of hepatic Tyr(P) 120 kdalton proteins, which are predominantly unidentified membrane and cytosolic protein substrates of the EGF receptor kinase.

5. Calmodulin

Calmodulin is known to bind to the IR (Wong et al. 1988), and recent studies by Sacks and McDonald (1988) and Wong et al. (1988) indicate that the IR can phosphorylate calmodulin in vitro on each of its two Tyr residues (99 and 138). This in vitro phosphorylation was completely dependent on the presence of a basic protein, such as protamine or polylysine, which was not itself phosphory-lated.[1] Another interesting feature of the IR-catalyzed phosphorylation of calmodulin in vitro is the inhibition of the reaction by Ca^{2+} levels $\geq 0.1 \mu M$, presumably through an action on the substrate.

In intact ^{32}P-labeled adipose cells, insulin stimulates ^{32}P incorporation into calmodulin sixfold by 10 min after hormone addition (Colca et al. 1987); consistent stimulation at earlier times was not observed. Estimates of overall insulin-stimulated ^{32}P incorporation into calmodulin were 0.29–0.72 mol P/mol. The incorporated ^{32}P was largely stable to alkaline hydrolysis, but the phosphoamino acid residue was not demonstrated directly (alkali-resistant serine/threonine residues are well recognized). Conversely, inasmuch as calmodulin has been shown to undergo Tyr phosphorylation in RSV transformed cells (Fukami et al. 1986, albeit to an unknown stoichiometry), it is likely that some of the insulin-stimulated phosphorylation will prove to be onto tyrosine. The fundamental role of calmodulin in cell regulation is widely appreciated. Further elucidation of the chemistry and functional significance of insulin-stimulated calmodulin phosphorylation will be of considerable interest.

The four candidate substrates for the IR kinase, reviewed above, are the best characterized thus far, and serve to illustrate the problems encountered in determining the physiologic relevance of hormone-stimulated tyrosine phosphorylation. A number of other in situ substrates (Haring et al. 1987; Kwok and Yip 1987) have been enumerated with little characterization thus far, and more are certain to follow. Among these, one or more substrates of regulatory importance to the cellular program of insulin action seems fairly certain to emerge; however, in view of the limited progress thus far, and perhaps because a positive outcome is not assured, the quest has taken on somewhat the aspect of trench warfare.

---

[1] It is unclear whether these basic polypeptides interact primarily with the IR, the substrate, or both. Stimulation of the partially purified IR kinase activity in vitro by polylysine and protamine has been noted, and we have observed that protamine added to intact cells, in the absence of insulin, promotes the Tyr-specific phosphorylation of p180, suggesting a direct activation of the IR/IGF-I receptor kinase.

C. Insulin-Stimulated Ser/Thr Protein Phosphorylation

The observation that insulin stimulated Ser/Thr phosphorylation in intact adipose cells (BENJAMIN and SINGER 1974, 1975; AVRUCH et al. 1974, 1976), was an unexpected and counterintuitive finding, given the prevailing perception that many metabolic responses elicited by insulin were the consequence of (Ser/Thr) dephosphorylation of key regulatory enzymes (e.g., glycogen synthase and pyruvate dehydrogenase). Nevertheless, the phenomenon of insulin-stimulated (Ser/Thr) protein phosphorylation was rapidly confirmed (HUGHES et al. 1977; FORN and GREENGARD 1976). By analogy to the pivotal regulatory role of Ser/Thr protein phosphorylation in the actions of hormones which elevate cAMP and intracellular calcium, a regulatory role for insulin-stimulated Ser/Thr phosphorylation was immediately envisioned. Several hypothetical mechanisms were proposed (AVRUCH et al. 1979 a, b, 1982, 1985) by which insulin-stimulated Ser/Thr phosphorylation could initiate the regulatory dephosphorylation which occurs in response insulin. Many features of insulin-stimulated (Ser/Thr) protein phosphorylation were compatible with a broad regulatory role; in particular, the ubiquity of the response, its rapid onset (in seconds to minutes of hormone exposure), and reversal when hormone is removed from the intact cell. Direct evidence to support the regulatory role for insulin-stimulated phosphorylation, however, has been slow to appear, and the effort in this area has had much the same character as the search for "native" or "physiologically relevant" insulin-stimulated Tyr(P)-containing proteins. A substantial number of insulin-stimulated Ser/Thr phosphoproteins have now been detected; among the most extensively characterized substrates are ATP citrate lyase (ALEXANDER et al. 1979; RAMAKRISHNA and BENJAMIN 1979), ribosomal 40S subunit protein S6 (HASELBACHER et al. 1979; SMITH et al. 1979), acetyl-CoA carboxylase (BROWNSEY and DENTON 1982; WITTERS 1981), the insulin receptor β subunit and several as yet unidentified proteins (especially an $M_r = 46\,000$ cytosolic protein in liver (AVRUCH et al. 1978), and adipocyte heat/acid-stable $M_r = 22\,000$ cytosolic protein (BELSHAM and DENTON 1980; BLACKSHEAR et al. 1982, 1983), etc. No compelling evidence, however, has been presented to establish that the insulin-stimulated Ser/Thr phosphorylation of any of these proteins is productive of an important change in protein function as measured in vitro, which is relevant to the physiologic functions of that protein in situ, and to the program of insulin action.

The evidence bearing on the physiologic significance of ribosomal S6 phosphorylation in the initiation of protein synthesis, while largely correlative, is, in general, supportive of a positive regulatory role (reviewed in TRAUGH and PENDERGAST 1986; KRIEG et al. 1988). By contrast, sharp disagreement as to the regulatory role of insulin-stimulated phosphorylation on the catalytic function of acetyl-CoA carboxylase is expressed in the literature (BORTHWICK et al. 1987; WITTERS et al. 1988; HAYSTEAD and HARDIE 1986). Treatment of isolated rat adipocytes and hepatocytes with insulin causes activation of acetyl-CoA carboxylase and phosphorylation of the enzyme on a unique peptide segment. The insulin-activated enzyme can be, however, substantially deactivated by gel filtration under conditions that preserve the phosphorylation. The site phosphorylated in insulin-treated cells can be phosphorylated in vitro by casein kinase 2 (HAYSTEAD et al. 1988). A casein kinase 2-like enzyme (i.e., one with similar site specificity) is activated in insulin-treated cells (see below); however, phosphorylation of acetyl-CoA carboxylase by casein kinase 2 in vitro does not alter carboxylase activity. Thus, insulin-stimulated phosphorylation and the activation of carboxylase may well be concomitant but independent events; alternatively, the insulin/casein kinase 2 phosphorylation of carboxylase could alter the interaction of the carboxylase with a yet-to-be identified activating ligand, or somehow promote the dephosphorylation of an inactivating site. WITTERS et al. (1988) have recently observed that the insulin-induced activation of acetyl-CoA carboxylase in Reuber hepatoma cells is actually accompanied by a net dephosphorylation of the carboxylase. At present, the molecular basis for the insulin regulation of this critical lipogenic enzyme and the role of stimulated Ser/Thr phosphorylation remain in doubt.

The lipogenic enzyme, ATP citrate lyase, provided an early and very informative example of the caveats pertaining to the study of protein phosphorylation in intact cells. Insulin and hormones which elevate cAMP regulate de novo fatty acid synthesis in opposite directions (WITTERS et al. 1979); nevertheless, insulin and glucagon (cAMP) each stimulate the phosphorylation of exactly the same serine residue in ATP citrate lyase in intact cells (PIERCE et al. 1981, 1982) which is the dominant and probably sole site of hormone-stimulated phosphorylation. This serine can be phosphorylated by the cAMP kinase in vitro (ALEXANDER et al. 1981), although ATP citrate lyase is rather poor substrate for the cAMP kinase, compared to pyruvate kinase or glycogen synthase (GUY et al. 1980). Stoichiometric phosphorylation of this serine in vitro gives little or no change in the catalytic function of ATP citrate lyase (AVRUCH et al. 1981). These features, together with the relatively low stoichiometry of hormone-stimulated phosphorylation in situ (ALEXANDER et al. 1982) indicate that the phosphorylation of ATP citrate lyase probably is not a "regulatory" modification, but occurs simply because of the very high abundance of the enzyme in lipogenic tissues (3%–5% of cytoplasmic protein in adipocytes and hepatocytes) and the presence of an exposed peptide segment which can be phosphorylated, albeit at a relatively slow rate, by both the cAMP-dependent protein kinase and one or more insulin-regulated, cAMP-independent protein kinases (among which is probably the "S6 kinase," see below). In this view, both the insulin- and glucagon-stimulated phosphorylation of ATP citrate lyase, although occurring in the intact cell, represent the background noise of protein phosphorylation, i.e, events irrelevant to the regulatory program dictated by these hormones and executed by protein kinases. It is important to emphasize that the serious doubts raised about the regulatory significance of the hormone-stimulated phosphorylation of ATP citrate lyase (although see STRALFORS 1987), do not invalidate the more general hypothesis as to the regulatory role of insulin-stimulated (Ser/Thr) phosphorylation. It would be incorrect to conclude, for example, that cAMP-stimulated phosphorylation was not a regulatory response of wide importance, because the glucagon/cAMP-stimulated phosphorylation of ATP citrate lyase proved to be physiologically irrelevant; a similar caveat extends to the general phenomenon of insulin-stimulated Ser/Thr phosphorylation. The more important lesson is that the hormone-stimulated phosphorylation of an abundant cellular protein to a low stoichiometry (whether on Ser/Thr or Tyr) in the intact cell is not likely to be a significant regulatory event, although it may reflect the occurrence of such regulation.

Recently, several low abundance proteins have been detected and purified, which appear quite likely to be true "physiologic substrates" for insulin-stimulated Ser/Thr phosphorylation: a low K_m, cAMP phosphodiesterase and an insulin/growth factor-regulated S6 (protein) kinase; both enzymes will be discussed below. Moreover, it is now clear that insulin-stimulated Ser/Thr phosphorylation reflects the activation of several different Ser/Thr protein kinases. The detection, purification, and characterization of the activation mechanism of such Ser/Thr protein kinases offers a very attractive route back toward an elucidation of the early steps which occur consequent to IR activation. Although some indication of the cellular role of each of these insulin-activated Ser/Thr protein kinases will be provided by an examination of their substrate specificity in vivo, a full understanding is not currently available and will probably require the molecular cloning and expression of cDNA, combined with appropriate mutagenesis, as well as studies aimed at selective inhibition of the expression of endogenous kinase activity.

I. Insulin-Stimulated Ser/Thr Kinases

A number of clearly defined insulin-stimulated Ser/Thr-specific protein kinases have been detected and partially characterized in the past few years; some of these Ser/Thr kinases are also stimulated by growth factor/mitogen treatment of cells. These enzymes include the S6 kinase (NOVAK-HOFER and THOMAS 1984); MAP-2 kinase (STURGILL and RAY 1986), an enzyme detected in extracts of insulin-treated cells by its ability to phosphorylate microtubule-associated protein 2 on Ser/Thr residues; an insulin-stimulated, casein kinase 2-like enzyme (SOMMERCORN et al. 1987); an insulin-stimulated manganese-dependent histone (Ser) kinase (YU et al. 1987 b); and an insulin-stimulated membrane-associated (Ser) histone kinase (YU et al. 1987). Kinase C activity is stimulated by a variety of growth factors, but not by insulin, in most studies. Insulin activation of a novel form of kinase C has been observed (FARESE et al. 1985, 1987; COOPER et al. 1987), although the totality of evidence in support of this idea is not compelling (BLACKSHEAR et al. 1985; SPACH et al. 1986). The accumulating data indicate that a family of insulin/growth factor-regulated (Ser/Thr) protein kinases is emerging, and several considerations indicate that additional insulin/growth factor-regulated (Ser/Thr) protein kinases will be identified: the existence of oncogenic (Ser/Thr) protein kinases (raf, mos) (MOELLING et al. 1984; KLOETZER et al. 1983), whose cellular homologs appear to be participants in growth factor-regulated pathways (SAGATA et al. 1988; MORRISON et al. 1988); the identification of several yeast and Xenopus oocyte proteins involved in cell cycle regulation as Ser/Thr protein kinases (DRAETTA and BEACH 1988; DUNPHY et al. 1988; LOHKA et al. 1988; GAUTIER et al. 1988; ARION et al. 1988; JAZWINSKI 1988; McLEOD and BEACH 1988); and the exponentially growing list of Ser/Thr protein kinases (generated by molecular cloning technology), whose cellular properties and regulatory functions are as yet unknown (see HUNTER 1987; HANKS et al. 1988 for review).

1. S6 Kinase

S6 is a $M_r = 31\,000$ protein which is the sole phosphoprotein of the eukaryotic 40S ribosomal subunit (WOOL 1979); the sequence of the cDNA of rat liver S6 has recently been reported by (CHAN and WOOL 1988). GRESSNER and WOOL (1974a, b, 1976a, b) were the first to demonstrate S6 phosphorylation: they showed that a major increase in hepatic S6 phosphorylation occurred in vivo during liver regeneration or the induction of diabetes, and in response to injection of cycloheximide, puromycin, glucagon, or cAMP; up to five charge forms of S6 of increasing anodal mobility can be discerned on two-dimensional gels. These observations on S6 phosphorylation in intact rats were later extended to cell culture systems, wherein S6 phosphorylation was shown to be diminished by serum starvation and greatly increased by serum, active phorbol esters, insulin, the growth factors, IGF-I, EGF, and PDGF; interestingly, cAMP appeared much less potent in vitro than in vivo (cataloged in GORDON et al. 1982). S6 phosphorylation is also increased by transformation of chick fibroblasts with Rous sarcoma virus, in a manner so as to preempt a further stimulation by serum and/or growth factors (DECKER 1981); mutants of v-src, which are temperature-sensitive for tyrosine kinase activity and transforming ability, stimulate S6 phosphorylation only at the permissive temperature. S6 phosphorylation is also increased during the maturation of *Xenopus* oocytes (GORDON et al. 1982) induced by progesterone or insulin. Interestingly, direct microinjection into oocyte cytoplasms of IR kinase (+ insulin; MALLER et al. 1986), the v-src (SPIVACK et al. 1984), or v-Abelson tyrosine kinase (MALLER et al. 1986), each activates S6 (serine) phosphorylation. These observations gave strong indication that the tyrosine kinase function of these proteins is required for the activation of the S6 (serine) phosphorylation, a conclusion confirmed by the more recent demonstration that transfected IR mutants whose ATP site has been inactivated cannot signal S6 phosphorylation (CHOU et al. 1987). Active phorbol esters (e.g., TPA) also stimulate S6 phosphorylation in many cell lines, through a pathway that requires kinase C, inasmuch as down regulation of kinase C by prolonged exposure to TPA abolishes the ability of TPA to stimulate S6 kinase phosphorylation. Significantly, kinase C downregulation does not prevent S6 phosphorylation by insulin, and is only mildly inhibitory to the S6 phosphorylation stimulated by EGF and PDGF (BLACKSHEAR et al. 1985; SPACH et al. 1986; BLENIS and ERIKSON 1986; PELECH and KREBS 1987).

The presence of an augmented S6 kinase activity in extracts of insulin-treated cells was first reported by Rosen and colleagues (SMITH et al. 1980). A major advance was the demonstration by NOVAK-HOFER and THOMAS (1984) that the inclusion of β-glycerophosphate and high concentrations (>1 mM) of EGTA in the homogenization buffer permitted the detection of 10- to 25-fold stimulation in S6 kinase activity in extracts from EGF and serum-treated 3T3 cells. The utility of these conditions for the detection of insulin-stimulated S6 kinase activity was rapidly confirmed (TABARINI et al. 1985; NEMENOFF et al. 1986; COBB

1986), and they were applied subsequently by many other labs looking at growth factor-activated protein phosphorylation in a wide variety of cell culture systems as well as in the maturing *Xenopus* oocyte. A large number of studies published between 1985 and 1988 provided a fairly cohesive description of some properties of the mitogen-activated S6 kinase activity: treatment of cells with serum/growth factor/insulin/active phorbol esters each stimulated the activity of an S6 kinase in cell extracts which exhibited, for each agent, quite similar properties. The kinase activity was entirely cytosolic, bound to both anion and cation exchange matrices, and (in general) exhibited on gel filtration or density gradient centrifugation $M_r \sim 60$–$70\,000$; an exception is the *Xenopus* S6 kinase II whose molecular mass by SDS-PAGE is about $92\,000$ (ERIKSON and MALLER 1985). The stimulated kinase activity was independent of cAMP, calcium, and phospholipid/diacylglycerol. These extracts from insulin/growth factor-treated cells showed no augmentation in kinase activity toward histone subfractions, casein, and a number of other proteins, although the basal kinase activity for these alternate substrates, as compared to S6, could have easily obscured an increment in absolute activity comparable to that observed using 40S subunits as substrate. The stimulated S6 kinase appears capable of phosphorylating up to five sites on S6. Although the stimulated kinase activity was rather labile, loss of activation was not hastened by dialysis and was stable through several purification steps. Based on size, independence of the activity by known regulators, and an apparently narrow substrate specificity, the S6 kinase detected in these extracts appeared to be different from the many other kinases known to phosphorylate S6. Several studies suggest the existence of two (or more) chromatographically separable, hormone-stimulated S6 kinases, although proteolytically induced heterogeneity is difficult to eliminate.

Efforts at purification have been described by several groups. In general, the lability of the enzyme (both from stimulated and basal cells) and its very low abundance greatly impeded efforts of purification for mammalian and avian sources. BLENIS et al. (1987) reported a $15\,000$-fold purification with 18% recovery of an S6 kinase from 9- to 11-day chicken embryos. The purity of the final product on SDS gels is not shown, although (putative) autophosphorylation of an $M_r = 65\,000$ polypeptide was demonstrated. TABARINI et al. (1987) reported the $54\,000$-fold purification of an S6 kinase from bovine liver; they described an $M_r = 67\,000$ silver-stained polypeptide band and recovered $\sim 120\,\mu g$ protein. The relationship of this enzyme to the mitogen-activated enzymes is uncertain. JENO et al. (1988) purified S6 kinase from vanadate-treated 3T3 cells; a single silver-stained polypeptide band of $M_r = 70\,000$ was detected, and $0.5\,\mu g$ was obtained from 1 g 3T3 cell extract protein.

An S6 kinase from frog eggs has been characterized by ERIKSON and MALLER (1985, 1986); the enzyme, purified 500-fold after an initial DEAE-Sephacel column, exhibits a single polypeptide band of $M_r = 92\,000$ on SDS-PAGE; after excision and "renaturation," about 1% of the original S6 kinase was recovered. An antiserum raised to the excised $M_r = 92\,000$ polypeptide band partially immunoprecipitated $\sim 1\%$–2% of the insulin/progesterone-stimulated S6 kinase activity in oocyte extracts and a comparable fraction of the S6 kinase activity, which is activated in chicken embryo fibroblasts transfected with a temperature-sensitive Rous sarcoma virus, when the cells are shifted to the permissive temperature (ERIKSON et al. 1987). Recently, two cDNA were isolated from a *Xenopus* ovarian library, utilizing synthetic oligonucleotides based on the protein sequence derived from random tryptic peptides of the excised $M_r = 92000$ polypeptide band (JONES et al. 1988). The two cDNAs code for proteins of 733 (α) and 629 (β) residues, respectively. Antisera raised to an SDS-denatured prokaryotic fusion protein containing residues 44–733 of S6 kinase (α) partially immunoprecipitate an autophosphorylated, SDS-denatured S6 kinase II purified from *Xenopus* oocytes. The predicted protein sequence of the putative S6 kinase II cDNAs demonstrate the presence of two internally similar kinase domains; each contains a potential ATP binding site. The amino terminal domain is most closely related to kinase C (31% identity, 59% similarity) whereas the carboxyl terminal segment is comparably similar to phosphorylase b kinase γ (catalytic) subunit. The cDNAs isolated clearly code for two members of a unique, previously undescribed family of protein kinases. However, some questions remain as to the identity of these two protein kinases vis-à-vis the insulin/growth factor-activated S6 kinase(s). The cDNAs code for proteins significantly smaller than the parent polypeptide from which the original tryptic peptides were derived. In addition, of the eight unique tryptic peptide sequences obtained, only four were identified in the amino acid sequence deduced from either the α or β cDNA. This discrepancy may merely reflect the detection of cDNAs coding for two isozymic forms of an S6 kinase gene family. However, the two peptide sequences employed to construct the oligonucleotides used for screening (as well as the other two tryptic peptides identified within the cDNA) come from a region of the protein sequence that is highly conserved among all the Ser/Thr protein kinases (domains VI and VII of HANKS et al. 1988). Thus, the possibility must be considered that the cDNAs do not code for the mitogen-stimulated S6 kinase, and a rigorous proof will require the demonstration that the kinases encoded by these cDNAs exhibit the catalytic and regulatory properties predicted for the insulin/growth factor S6 kinase. In as yet unpublished studies, R. Erikson and colleagues have demonstrated that the α cDNA codes for a regulated S6 protein kinase.

In attempting to identify an abundant mammalian source for the activated S6 kinase, NEMENOFF et al. (1988) harked back to the early observation of GRESSNER and WOOL (1974a), that hepatic S6 phosphorylation increases tenfold during liver regeneration; NEMENOFF et al. (1988) observed that 2 h after 70% hepatectomy in the rat, the specific activity of S6 kinase in the liver remnant had increased sixfold over sham-operated controls, remained at this high level for 36 h, and returned to preoperative levels by 48 h. No increase in hepatic histone H1, H4, and H2B kinase occurred over this interval. More importantly, the S6 kinase activated in regenerating liver eluted as a single (albeit often broad) peak on ion exchange chromatography, whereas *no* dominant (or even distinct) peak of S6 kinase activity was observed on fractionation of comparable extracts of control liver, assayed in the presence of 1 μM PKI and 10 mM EGTA. The S6 kinase which appeared in the regenerating liver comigrated on anion exchange (Mono-Q), cation exchange (Mono-S), and gel filtration chromatography with the activated S6 kinase from insulin-treated H4 hepatoma cells. GRESSNER and WOOL (1974b) had also observed that injection of rats with cycloheximide gave substantially increased phosphorylation of S6. This stimulatory effect of cycloheximide on S6 phosphorylation in vivo was confirmed later by NIELSEN et al. (1982). PRICE and AVRUCH (submitted) showed that intraperitoneal injection of cycloheximide produces a stimulation of hepatic S6 kinase activity, comparable in magnitude to the increase that occurs during liver regeneration; as with the activity engendered during liver regeneration, the cycloheximide-induced S6 kinase also copurifies with the insulin-stimulated enzyme from H4 hepatoma cells.

The enzyme from the liver of cycloheximide-treated rats has been purified to near homogeneity, approximately over 10^5-fold (PRICE et al. 1989). The purified hepatic S6 kinase exhibits two polypeptide bands, a dominant species of $M_r = 70\,000$ and a minor band of $M_r = 92\,000$; both undergo autophosphorylation, and coelute on gel filtration at an apparent molecular weight of $70\,000$. The larger polypeptide is presumed to be an isozyme because of its similar size to the *Xenopus* S6 kinase II; alternative possibilities include a proteolytic precursor, regulatory subunit, substrate, or another protein kinase (perhaps S6 kinase-kinase). The rat liver S6 kinase is unreactive with antisera raised against the purified Xenopus S6 kinase II, or the recombinant Xenopus S6 kinase α. The highly purified hepatic S6 exhibits a broader substrate specifity than anticipated by the earlier published studies, which primarily employed much less highly purified enzymes: when compared at roughly equal "S6" phosphorylating activity to the catalytic subunit of the cAMP-dependent protein kinase, the hepatic S6 kinase exhibits 5- to 10-fold *higher* activity toward ATP citrate lyase, a comparable rate of phosphorylation of skeletal muscle glycogen synthase, and perhaps one-third the activity toward histone 2B. Whereas the cAMP kinase also phosphorylates histone H1, H4, and acetyl-CoA carboxylase, the S6 kinase is devoid of such activity, and is also unable to phosphorylate purified skeletal muscle phosphatase inhibitor 2, partially purified preparations of the adipocyte heat-stable $M_r = 22\,000$ polypeptide, and the human placental IR. The hepatic S6 kinase thus phosphorylates two of the in situ substrates for insulin-stimulated Ser/Thr phosphorylation, S6 and ATP citrate lyase, but not acetyl-CoA carboxylase and others. The phosphorylation of ATP citrate lyase by the S6 kinase may well underlie the insulin-stimulated phosphorylation of ATP citrate lyase in intact cells; analysis of the site phosphorylated by S6 kinase on ATP citrate lyase will be necessary to confirm this conclusion. Inasmuch as the insulin-stimulated phosphorylation of ATP citrate lyase in intact cells is viewed as "background noise" in the program of insulin action, these observations cannot be taken to support the conclusion that S6 kinase operates as a protein kinase of "broad" substrate specificity in situ. Nevertheless, the ability of the S6 kinase to phosphorylate proteins other than S6 (such as glycogen synthase and histone 2B) in vitro at rates comparable to those catalyzed by kinases such as the cAMP-dependent kinase, which is known to operate on these proteins as a physiologic regulator in situ, raises for consideration the possibility that other physiologic substrates for the S6 kinase may exist. In view of the mitogen activation of S6 kinase, an examination of the ability of this enzyme to phosphorylate transcriptional regulatory proteins, as well as the initiation and elongation factors involved in translational regulation, appears warranted. In this regard, we have observed that S6 kinase avidly phosphorylates a number of polypeptides in NaCl extracts of liver and adipocyte nuclei, whereas the catalytic subunit of the cAMP-dependent protein kinase, added at similar S6 phosphorylating activity, yields much less extensive phosphorylation of these extracts.

The widespread interest in S6 kinase, an enzyme whose cellular role is largely unknown, arises primarily from the desire to understand the mechanism of its activation by insulin and growth factors. The S6 kinases purified thus far appear to be single polypeptide chains; an important prediction from the deduced protein sequence of the putative *Xenopus* S6 kinase cDNA (keeping in mind the reservations concerning the identity of this kinase as the insulin/mitogen-stimulated S6 kinase) is the lack of a potential "regulatory" domain as seen in protein kinase C or myosin light chain kinase, other regulated kinases encoded on a single polypeptide chain (JONES et al. 1988). Considerable attention has focused on the possibility that the stable activation of S6 kinase induced by growth factors may be due to phosphorylation of the S6 kinase itself. Certainly, activation/deactivation in the intact cell is rapidly responsive to the addition/removal of hormones (NEMENOFF et al. 1988). BALLOU and THOMAS (1988) observed that their purified 3T3 S6 kinase can be deactivated by protein phosphatase I and IIA, especially the latter. BALLOU et al. (1988) have partially ($\sim$5000-fold) purified the S6 kinase activity from ^{32}P-labeled 3T3 cells: a 70 kdalton polypeptide is recovered as the dominant ^{32}P-labeled polypeptide from the serum-treated cells, whereas this ^{32}P-peptide is not present in extracts from serum-starved cells. The partially purified 70 kdalton ^{32}P-labeled polypeptide undergoes (presumed) autophosphorylation in vitro, and is dephosphorylated by phosphatase IIA, concomitant with the loss of S6 kinase activity. After brief (1 h) acid hydrolysis, the excised ^{32}P-labeled 70 kdalton polypeptide exhibits only ^{32}P-Ser and ^{32}P-Tyr no ^{32}P-Tyr is detected. These data strongly indicate that activation of S6 kinase is accompanied by and due to its phosphorylation on Ser and/or Thr residues. PRICE and AVRUCH (submitted) have observed that the hepatic S6 kinase purified from cycloheximide-treated rats can be fully deactivated by incubation with protein phosphatase IIA; deactivation is completely prevented by phosphatase inhibitors such as *p*-nitrophenyl phosphate. Moreover, antisera raised against the rat liver S6 kinase purified from cycloheximide treated rats immunoprecipitate an insulin stimulated S6 kinase activity from H_4 hepatoma cells. Immunoprecipitates prepared from ^{32}P labelled H_4 cells contain a 70 000 dalton ^{32}P-peptide whose ^{32}P content is increased after insulin treatment of the intact cells. Thus, the S6 kinase itself has become the best documented candidate for a physiologically important substrate for insulin-stimulated Ser/Thr phosphorylation. Most recently, STURGILL et al. (1988) have reported that the *Xenopus* S6 kinase II, which had first been deactivated with either protein phosphatase I or IIA, could be partially reactivated by phosphorylation with a preparation of MAP-2 kinase partially purified from insulin-treated 3T3-L1 cells, whereas kinase C and kinase A were ineffective. This observation provides for a potential cascade of insulin-stimulated serine kinases. Although its premature based on available data to conclude that MAP-2 kinase is the enzyme mediating the insulin/growth factor-induced activation of S6 kinase in situ, this idea clearly merits close consideration.

2. MAP-2 Kinase

Microtubule-associated proteins (MAPs) copurify with cytoplasmic micro-tubules; when added to tubulin in vitro, MAPs stimulate microtubule assembly and enhance microtubule stability (OLMSTED 1986). Microtubule disassembly is necessary for DNA synthesis and cell division (FRIEDKIN and ROZENGURT 1980; THYBERG 1984). Agents which destabilize microtubules potentiate the mitogenic effects of growth factors, and can, in some instances, apparently serve to initiate DNA synthesis in the absence of growth factors. Conversely, agents which stabilize microtubules can effectively block the response to a variety of mitogens. MAPs are highly phosphorylated in situ, and serve as substrate for a variety of Ser/Thr and Tyr kinases in vitro. Ser/Thr-specific phosphorylation of MAPs in vitro reduces their affinity for microtubules; by inference, MAP (Ser/Thr) phosphorylation would be expected to destabilize microtubules in situ and thereby contribute to the mitogenic program initiated by growth factors. In fact, exposure of quiescent cells to growth factors leads to an enhanced ^{32}P incorporation into a variety of MAPs, which thereupon appear to redistribute into the soluble (unassembled) fraction (SHAW et al. 1988). Thus, the detection of a stimulated kinase activity directed at MAPs in extracts of insulin/growth factor-treated cells would be consistent with existing concepts on the regulation of microtubule assembly.

In the course of examining the phosphorylation of phosphatase inhibitor 2 by extracts of 3T3-L1 cells, STURGILL and RAY (1986) noted that extracts from insulin-treated cells showed a considerably enhanced kinase activity directed at a high molecular weight contaminant of their I-2 preparation, but not I-2 itself. This high molecular weight contaminant cross-reacted with antibodies raised against MAP-2. Utilizing as substrate partially purified brain MAP-2, RAY and STURGILL (1987, 1988a, b, c) subsequently showed that an insulin-stimulated MAP-2 kinase is reproducibly observed as an entirely cytosolic activity, which phosphorylates Ser/Thr, but not Tyr residues on MAP-2; the enzyme does not detectably phosphorylate ATP citrate lyase, acetyl-CoA carboxylase, S6, and has very low activity versus casein and histone. This insulin-stimulated MAP-2 kinase is further distinguishable from the S6 kinase by a more rapid activation following insulin treatment, a more rapid inactivation during incubation in vitro after cell disruption, and by its distinct chromatographic properties. Insulin-stimulated MAP-2 kinase activity in extracts from 3T3-L1 cells, BC$_3$-H1 cells, freshly isolated rat adipocytes, and hepatocytes, exhibited, in each case, an apparent molecular weight of 33000 by gel filtration (RAY and STURGILL 1988c). In addition, a portion of the insulin-stimulated MAP-2 kinase absorbs strongly to phenyl-Sepharose and coelutes with an $M_r = 40000$ ^{32}P-labeled phosphoprotein, which contains both ^{32}P-Tyr and ^{32}P-Thr (RAY and STURGILL 1988a). Approximately 14% of the MAP-2 kinase activity, which has been partially purified from insulin-treated cells by hydrophobic chromatography, can be absorbed to polyclonal anti-phosphotyrosine antibodies. These findings suggest that the MAP-2 kinase may serve as direct substrate for the IR tyrosine kinase, and, in turn, may phosphorylate and activate the S6 kinase and perhaps other (as yet unidentified) substrates of regulatory significance. HOSHI et al. (1988) described a very similar MAP-2 kinase activity in extracts of human embryonic lung fibroblasts treated with EGF, PDGF, FGF, IGF-I, insulin, and TPA. This MAP 2 kinase activity was almost completely inhibited by low concentrations of free Ca (2 μM).

Among the many unanswered questions concerning the insulin/mitogen-activated Ser/Thr protein kinases is the mechanism by which signals initiated by diverse receptors converge so as to lead to the activation of a single Ser/Thr protein kinase (e.g., S6 kinase or MAP-2 kinase). A simple explanation is that several Tyr kinases and (Ser/Thr) kinases (such as kinase C), despite their fundamentally different substrate specificities, share one or more common targets. The detection of ^{32}P-Tyr(P) and ^{32}P(Thr) on MAP-2 kinase suggests this enzyme as one potential locus for integration of these signals.

3. Casein Kinase 2

Casein kinase 2 is a ubiquitous enzyme first detected as one of two dominant casein/phosvitin phosphorylating activities in several tissues (both of which are distinct from the protein kinase responsible for casein phosphorylation in mammary tissues, in situ (HATHAWAY and TRAUGH 1982). The enzyme, purified from a variety of sources, exhibits an $\alpha_2\beta_2$ structure, and both subunits have been sequenced (TAKIO et al. 1987). The β subunit, 24–26 kdalton undergoes an autophosphorylation reaction, but has no known function, and is not homologous in protein sequence to other proteins; the α subunit, 40–44 kdalton, contains the catalytic activity and exhibits strong homology in its carboxyl terminal two-thirds, to CDC 28, a yeast protein kinase critical to cell division. Although predominantly cytosolic, casein kinase 2 is found in the nucleus and can phosphorylate in vitro a wide array of proteins found in both sites including acetyl-CoA carboxylase, glycogen synthase, protein phosphatase inhibitor 2, the type II regulatory subunit of the cAMP kinase, a variety of protein synthesis initiation factors, RNA polymerase I and II, DNA topoisomerase 1 and 2, and a number of nucleolar and nonhistone nuclear proteins thought to participate in the regulation of gene expression (HATHAWAY and TRAUGH 1982; SOMMERCORN and KREBS 1987a).

As with all Ser/Thr protein kinases studied thus far, the substrate specificity of casein kinase 2 is determined in large part by the amino acid sequence surrounding the site of phosphorylation. Casein kinase 2 is one of a small family of enzymes which phosphorylates Ser (preferentially) or Thr residues situated in a cluster of acidic residues. Studies with synthetic peptides indicate that acidic residues (Asp preferred to Glu) carboxyl terminal to the phosphorylation site are the primary recognition determinants; e.g., in the peptide, RRREEESEEE, the carboxyl terminal Glu is by far the most potent determinant of recognition (KUENZEL et al. 1987). Peptides with carboxyl terminal extensions of 5–6 acidic residues show even faster phosphorylation (MEGGIO et al. 1984), and most protein substrates exhibit clusters of 5–6 acidic residues carboxyl terminal to the phosphorylation site. Amino terminal acidic residues also serve as positive determinants in synthetic peptide substrates, but are not necessary, and are lacking entirely in some of the protein substrates.

Casein kinase 2 is strongly inhibited competitively by a variety of polyanions, e.g., polyaspartic acid/polyglutamic acid, heparin, and 2,3-DPG and is stimulated by polyamines (at least toward several substrates) and by polybasic proteins such as polylysine and histone. In contradistinction from most protein kinases, casein kinase 2 can utilize GTP $(K_m \sim 30 \, \mu M)$ nearly as well as ATP $(K_m \sim 14 \, \mu M)$. These novel features of casein kinase 2 have been utilized to design assays for the enzyme in relatively crude extracts: the phosphorylation of the peptide, RRREEETEEE, in the presence of $[\gamma\text{-}^{32}P]$GTP or $[\gamma\text{-}^{32}P]$ATP, which can be inhibited by low concentrations of heparin, is taken to reflect the activity of casein kinase 2 (KUENZEL and KREBS 1985; SOMMERCORN and KREBS 1987b; SOMMERCORN et al. 1987; KLARLUND and CZECH 1988). The specificity of this assay for classical casein kinase 2 is not entirely certain; although this peptide cannot be phosphorylated by casein kinase 1, a number of other protein kinases have been described which are "casein kinase 2-like" in their general substrate specificity, ability to utilize GTP, and inhibition by heparin, but which clearly differ in molecular structure from classical casein kinase 2 (KISHIMOTO et al. 1987). Whether such casein kinase 2-like activities are subjected to rapid regulation and whether they can phosphorylate the synthetic peptide substrates is not known.

Nevertheless, several labs employing such peptide phosphorylation assays have observed the stimulation of a "casein kinase 2" activity by insulin and/or growth factors. SOMMERCORN et al. (1987) observed that addition of insulin to differentiated 3T3-L1 cells gave a 1.3- to 2.5-fold stimulation of casein kinase 2 activity; the stimulation was maximal by 10 min after insulin addition and half-maximally stimulated by 0.15 nM insulin; EGF gave a stimulation of similar magnitude. The insulin-stimulated activation was unaffected by gel filtration, and was characterized by an increase in V_{max} for the peptide substrate, with no change in K_m. Overall content of casein kinase 2 judged by immunoblot was not altered by insulin, and brief preincubation with cycloheximide did not diminish (and perhaps enhanced) the basal and insulin-stimulated activities. KLARLUND and CZECH (1988) examined casein kinase 2 activity in extracts of growth factor-treated 3T3 fibroblasts after an initial batch adsorption/elution from phosphocellulose. The desalted eluate (which contained $\sim 47\%$ of the activity in the initial cell extract) exhibited a 1.6- to 2.5-fold higher activity when the extracts were prepared from cells treated with insulin or IGF-I, but no increase in activity when cells were treated with EGF or PDGF, although the latter agents effectively stimulated $[^3H]$thymidine incorporation. The insulin/IGF-I stimulation was half-maximal at 100 nM/1 nM, respectively, and fully developed by 15 min after hormone addition.

The mechanism of hormone activation is obscure at present. The persistence of the hormonal activation through desalting and phosphocellulose chromatography favors certain activation mechanisms, such as hormone-induced covalent modification or the inactivation/dissociation of an inhibitor. Nevertheless, the possibility that hormone treatment generates an activating ligand/protein of sufficiently high affinity to remain associated through partial enzyme purification cannot be eliminated. Purified casein kinase 2 contains Ser(P)/Thr(P), but not Tyr(P). Dephosphorylation of purified casein kinase 2 in vitro with phosphatase IIA and I can activate the enzyme up to 2- to 3-fold, to an extent that is inversely related to the casein kinase 2 concentration (AGOSTINIS et al. 1987). The significance and mechanism of this activation, its relation to casein kinase 2 polymerization (GLOVER 1986) and autophosphorylation are not clear. We have been unable to phosphorylate and/or activate purified casein kinase 2 in vitro with purified placental IR (WITTERS and AVRUCH, unpublished observations). Phosphorylation of casein kinase 2 by other protein kinases, or the state of casein kinase 2 in ^{32}P-labeled, insulin-treated cells, has not been reported. Thus, despite the extensive knowledge of the molecular structure of casein kinase 2, virtually nothing is known of its regulation in situ; moreover, the coidentity of the hormone-stimulated casein kinase 2 activity vis-à-vis classical casein kinase 2 must be viewed as not yet securely established.

In addition to the uncertainty surrounding the regulation of casein kinase 2, the precise cellular roles of the enzyme are unknown, in large part, because many of the phosphorylations catalyzed by the kinase in vitro do not directly alter the catalytic function of the protein substrate (e.g., glycogen synthase, acetyl-CoA carboxylase). Casein kinase 2, however, is capable of modifying enzyme activity indirectly in certain instances; phosphorylation of glycogen synthase (at site 5; PICTON et al. 1982a, b; DEPAOLI-ROACH et al. 1984), and phosphatase inhibitor 2 (DEPAOLI-ROACH et al. 1981, 1984) (at Ser 86, 120, 121; HOLMES et al. 1986, 1987) by casein kinase 2 creates a recognition site which permits another protein kinase, variously known as glycogen synthase kinase 3 (GSK-3) or F_A (the activating factor of the Mg-ATP-dependent protein phosphatase) to interact with the casein kinase 2-modified substrate, and phosphorylate it at a Ser/Thr located several residues to the amino terminal side of the casein kinase 2 site. In the case of glycogen synthase, a sequential set of four additional GSK-3 phosphorylation sites is formed by the subsequent action of GSK-3 itself, (Ser-XXX-Ser(P)) with each phosphorylation generating the recognition motif for the next GSK-3 phosphorylation which occurs further amino terminal (FIOL et al. 1987). Phosphorylation of glycogen synthase by GSK-3/F_A at the three most amino terminal sites (3a, b, c) is functionally significant, producing a profound inhibition of synthase activity (COHEN 1986). Similarly, GSK-3/F_A phosphorylation of phosphatase inhibitor 2 (I-2) at Thr 72 (a GSK-3/F_A recognition site created by the casein kinase 2 phosphorylation of I-2), results in the activation of the Mg-ATP-dependent phosphatase (INGEBRITSEN and COHEN 1983; BALLOU and FISCHER 1986). Inasmuch as the sites phosphorylated on glycogen synthase by GSK-3 are avidly dephosphorylated by the Mg-ATP-dependent protein phosphatase, a surprising potential outcome of the linked effects of casein kinase 2 and GSK-3/F_A is that the insulin activation of casein kinase 2 is theoretically capable of leading either to the activation or inactivation of glycogen synthase. Inactivation would proceed by the insulin-augmented phosphorylation of synthase (at sites 5 and 3a, b, c); activation of glycogen synthase would occur through the casein kinase 2/GSK-3 phosphorylation of phosphatase inhibitor 2, activation of the Mg-ATP-dependent phosphatase and dephosphorylation of glycogen synthase (at sites 3a, b, c, 2, 1a, b, and possibly others).

All workers agree that insulin action is associated with a net dephosphorylation of sites 3a, b, c on glycogen synthase, and probably sites 2 and 1b as well (PARKER et al. 1983; COHEN 1986; POULTER et al. 1988; LAWRENCE et al. 1986; SMITH et al. 1988). This outcome could be rationalized with an insulin activation of casein kinase 2 if site 5 on glycogen synthase was largely phosphorylated in the "basal" state prior to insulin addition (as appears to be the case in skeletal muscle in vivo, COHEN 1986; POULTER et al. 1988), so that further phosphorylation of site 5 (and in turn, the inactivating sites 3 a, b, c) were not engendered by an increase of casein kinase 2 activity; rather, the casein kinase 2/GSK-3 phosphorylation of phosphatase inhibitor 2 served to activate the cytosolic Mg-ATP-dependent (type I) phosphatase and tipped the balance of kinase/phosphatase action on glycogen synthase toward net dephosphorylation. Consistent with this rather byzantine formulation are several reports (see below) of an insulin-induced increase in protein phosphatase activity, and the preliminary observation of an insulin-induced increase in the ^{32}P content of I-2 in intact adipose cells (LAWRENCE et al. 1988). However, the role of I-2 phosphorylation in vitro in the activation and maintenance of phosphatase activity is complex; although F_A/GSK-3-mediated I-2 phosphorylation is necessary to initiate phosphatase activation, the activated state of the phosphatase is maintained despite the autodephosphorylation of Thr 72 on I-2 by the phosphatase catalytic subunit (BALLOU and FISCHER 1986). The regulation of this phosphatase in situ, and especially the role of I-2 phosphorylation, is largely unknown. Clearly, the role of insulin-stimulated casein kinase 2 activity in glycogen synthase/protein phosphatase regulation will require much further study.

In summary, it is now well established that one or more casein kinase 2-like enzymes are activated by insulin. Much is known of the molecular structure and potential substrates of classical casein kinase 2. Nevertheless, the molecular structure of the insulin-regulated kinase, the proximate mechanism of insulin activation, and the role of the kinase in cell regulation are all uncertain.

4. Other Possible Insulin-Activated Ser/Thr Protein Kinases

YU et al. (1987 a, b) have recently described two insulin-activated Ser/Thr protein kinase activities detected in extracts of isolated rat adipocytes: one is a histone kinase associated with a high density microsomal fraction, the second is a cytosolic enzyme which is Mn^{2+}-dependent, and phosphorylates histine, Kemptide (LRRASLG), as well as ATP citrate lyase. Few data on the molecular properties of these enzymes are available as yet.

A number of groups have described the insulin-stimulated phosphorylation of serine residues on the β subunit of partially purified IR (GAZZANO et al. 1983; BALLOTI et al. 1986). It is not yet clear whether this phenomenon is due to insulin stimulation of the activity of a (Ser/Thr) protein kinase which copurifies with the IR, or to an insulin-induced alteration in receptor conformation with a concomitant alteration in the susceptibility of the IR β subunit to a contaminating protein kinase (perhaps analogous to the β-adrenergic receptor kinase, BENOVIC et al. 1986–1988).

A brief report (YANG et al. 1988) described the insulin-induced translocation of an F_A/GSK-3 activity in intact human platelets from a membrane localization, where the enzyme was latent and presumably inactive in situ, to the cytosol, where the enzyme was fully active. This very provocative observation could provide an attractive mechanism for concomitantly enhancing phosphorylation of some proteins while dephosphorylating others.

A series of reports from Farese et al. (1985, 1987) and Cooper et al. (1987) have proposed that insulin treatment of BC_3H1 cells can activate a form of kinase C with unusual features. Blackshear et al. (1985) and Spach et al. (1986), however, have shown in this and other cells that insulin does not stimulate phosphorylation of ubiquitous 80 kdalton polypeptide substrate of kinase C; moreover, extensive downregulation of kinase C by prolonged exposure to active phorbol esters does not modify the insulin-stimulated phosphorylation of S6 or the activation of S6 kinase. Nevertheless, a number of protein kinases related to kinase C, but atypical in their response to Ca^{2+}, diglyceride, or phospholipids have been described (Gonzatti-Haces and Traugh 1986; Klemm and Elias 1987; Ohno et al. 1988; Nishizuka 1988). The idea that insulin could activate one or more such enzymes is quite plausible. The relation of kinase C and its many cousins to the so-called protease-activated kinases described by Gonzatti-Haces and Traugh (1986), Lubben and Traugh (1983), De la Houssaye et al. (1983), and Masaracchia et al. (1988) is not clear. Protease-activated kinase (PAK)II is a 50 kdalton species generated by light tryptic/chymotryptic treatment of an inactive, 80 kdalton precursor; PAKII is moderated/stimulated by diglyceride and phospholipid, but inhibited by calcium. Insulin treatment of 3T3-L1 cells has been reported to augment the PAKII activity which is generated by proteolytic treatment of partially purified extracts (Perisic and Traugh 1983).

D. Insulin-Induced Dephosphorylation

Insulin alters the activity of a number of critical metabolic pathways by promoting the dephosphorylation of rate-limiting enzymes, e.g., inhibition of lipolysis (dephosphorylation of hormone-sensitive triglyceride lipase; Belfrage et al. 1986a; Honnor and Londos 1986) and gluconeogenesis (dephosphorylation of pyruvate kinase and fructose-2,6-diphosphatase; Claus et al. 1986), activation of glycogenesis (dephosphorylation of glycogen synthase; Cohen 1986) and de novo fatty acid synthesis (Denton et al. 1986; several steps are facilitated, e.g., glucose transport, pyruvate oxidation, mitochondrial transport of pyruvate and citrate, malonyl-CoA synthesis, fatty acid activation and esterification; only pyruvate dehydrogenase is known with certainty to be dephosphorylated). The molecular mechanisms underlying the dephosphorylations appear to be multiple and diverse and will be considered only in brief.

I. Inhibition of Protein Kinase Activity

Perhaps the most clearly delineated mechanism underlying certain dephosphorylations is the insulin inhibition of the cAMP-dependent protein kinase. This is the predominant site of insulin's action to dephosphorylate hormone-sensitive triglyceride lipase (Belfrage et al. 1986), pyruvate kinase, and fructose-2,6-diphosphatase (Claus et al. 1986). The hormonal regulation of the activity of these enzymes is defined in vivo by their phosphorylation at a single dominant site which is catalyzed primarily by the cAMP-dependent protein kinase. In the absence of a hormone which stimulates cAMP accumulation, the extent of phosphorylation of these enzymes at this regulatory site is low, and usually not detectably altered by insulin addition (AVRUCH et al. 1976, 1978). A large number of "catabolic" hormones induce an increase in cAMP synthesis and activation of the cAMP-dependent protein kinase (HONNOR et al. 1985a, b). Insulin antagonizes this accumulation of cAMP primarily by activating a specific low K_m cAMP phosphodiesterase (PDE), and accelerating cAMP breakdown (see Chap. 17; DEGERMAN and BELFRAGE 1987). The elevated levels of cAMP fall, and the kinase is concomitantly deactivated (LONDOS et al. 1985). Cyclase activation in response to hormones can yield cAMP levels far above those required for maximal activation of the kinase; if the initial cAMP levels are so high that the lower level attained after insulin activation of PDE remains supramaximal for kinase activation, no insulin-induced inhibition of the kinase is detected. Under these conditions, at least in isolated adipocytes, a net dephosphorylation/deactivation of hormone-sensitive triglyceride lipase can still be engendered by insulin, through a mechanism which (by inference) is thought to involve an increase in protein phosphatase activity (LONDOS et al. 1985; see below).

The mechanism by which insulin stimulates PDE activity has been difficult to ascertain, because of the slow progress in identification and purification of the specific PDE responsive to insulin. The PDE activity responsive to insulin also appears to be the species which is activated by hormones which increase cAMP; as expected, the cAMP-induced increase in PDE activity is mediated via protein phosphorylation (MACPHEE et al. 1988; GETTYS et al. 1988; GRANT et al. 1988). Recently, DEGERMAN and BELFRAGE (1987) have purified this enzyme to apparent homogeneity over 50000-fold from adipose tissue. It is a 130 kdalton polypeptide, a peripheral membrane protein. Immunoprecipitation of the enzyme from ^{32}P-labeled adipocytes reveals that insulin and cAMP each are associated with an increase in ^{32}P incorporation into the enzyme polypeptide, concomitant with activation (DEGERMAN et al. 1989a, b). The insulin activation of cAMP PDE via augmented Ser phosphorylation of the enzyme, if verified, provides a molecular basis for several of insulin's most important actions on energy metabolism in vivo, and will serve to establish insulin-stimulated Ser phosphorylation as a central regulatory reaction in insulin action. In order to establish that the insulin-stimulated phosphorylation underlies the activation, it will be necessary to: (a) deactivate the PDE with protein phosphatase in vitro; (b) reactivate the PDE by phosphorylation with an insulin-responsive protein kinase; and (c) demonstrate that the site phosphorylated in vivo in response to insulin corresponds to that phosphorylated in vitro by the insulin-responsive kinase.

A number of reports conclude that insulin can modify the activity of the cAMP-dependent protein kinase by altering the ability of the kinase to be activated by cAMP. Much of the evidence presented is indirect (GABBAY and LARDY 1984, 1987), although some direct evidence is available (WALKENBACK et al. 1978; MOR et al. 1981); most published work on this point does not detect such an insulin-induced alteration in the cAMP sensitivity of the cAMP kinase (e.g., HONNOR et al. 1985c and references therein). The molecular basis for such an alteration in kinase activity is unknown.

Insulin can antagonize the metabolic responses to non-cAMP-linked hormones, e.g., α_1-catecholamines acting in liver (BLACKMORE et al. 1979;

THOMAS et al. 1985): these agents act through accelerating phosphatidyl inositide turnover and are thought to utilize Ca^{2+} (and DAG) as their intracellular mediator(s). A number of actions of insulin on intracellular Ca^{2+} homeostasis have been described which could be marshaled to explain either an increase or decrease in cytosolic Ca^{2+}. It seems likely, however, that the insulin antagonism of the gluconeogenic and glycogenolytic responses to α_1-catecholamines in liver will be due in part to an insulin-induced inhibition in the elevation of cytoplasmic Ca^{2+} engendered by these hormones, with a consequent inhibition of Ca^{2+}-regulated protein kinases (phosphorylase b kinase and/or Ca^{2+} calmodulin kinase 2).

Insulin may inhibit certain protein kinases by mechanisms other than lowering the levels of an activating ligand. RAMAKRISHNA and BENJAMIN (1985) have purified a $cAMP/Ca^{2+}$-independent protein kinase which they have called "multifunctional protein kinase." This enzyme is recovered as a single polypeptide chain of 36 kdaltons; it phosphorylates ATP citrate lyase, at a site distinct from the site of insulin/cAMP-stimulated phosphorylation and only after the insulin/cAMP site has first been phosphorylated (RAMAKRISHNA et al. 1983). The designation of "multifunctional" is based on the ability of this kinase to phosphorylate, in vitro, glycogen synthase (at sites 2 and 3a, b, c; SHEORAIN et al. 1985), acetyl-CoA carboxylase (at a site distinct from that phosphorylated by the cAMP kinase), phosphatase inhibitor 2 (with a resultant increase in the activity of the Mg-ATP-dependent protein phosphatase), ribosomal S6, and eIF-2 β chain. Somewhat similar enzymes have been described by HEGAZY et al. (1987), YANG et al. (1987a, b). RAMAKRISHNA and BENJAMIN (1988) have recently observed that insulin treatment of adipose tissue leads to 50%–60% decrease in this activity, as assayed in extracts subjected to sequential DEAE/phosphocellulose chromatography. Inasmuch as this inhibition was no longer observed when phosphatase inhibitors were omitted from the extraction medium, the authors speculate that regulation of this kinase may be mediated by an insulin-induced alteration in its phosphorylation state.

II. Activation of Protein Phosphatase

The ability of insulin to stimulate protein phosphatase activity has been inferred from a variety of indirect evidence, some quite compelling. In addition to work already cited, the effects of insulin on the phosphorylation state and activity of protein phosphatase inhibitor 1 indicate strongly that an increase in type I phosphatase activity should accompany insulin action (FOULKES et al. 1980; KHATRA et al. 1980; NEMENOFF et al. 1983). The difficulties in reliable assay of protein phosphatase activity in crude extracts have, however, impeded efforts at the direct demonstration of insulin activation of protein phosphatase activity (INGEBRITSEN and COHEN 1983; BALLOU and FISCHER 1986). Protein phosphatase apparent specific activity, for example, increases with dilution of extracts; this is presumably the conseqence of the dilution of endogenous inhibitors which include proteins (e.g., phosphatase inhibitors 1 and 2, phosphatase substrates) and ligands (which act both on the added ^{32}P-labeled substrate, and on the phosphatase). The action of endogenous proteases is a further confounding element, capable of both increasing and decreasing apparent phosphatase activity. Redistribution of phosphatase may occur, from a latent membrane-bound state to an active state as a cytosolic component. Endogenous glycogen can interact with the relevant enzymes and substrates. The results obtained may vary with the substrate employed: [^{32}P] phosphorylase a from rabbit skeletal muscle is a preferred substrate because it is simple to prepare, has a single site of phosphorylation, and is susceptible to essentially all known phosphatases; yet the latter two features may obscure contributions of specific phosphatases, or site-site interactions at the substrate level that facilitate dephosphorylation. In spite of these difficulties, a number of reports demonstrating an insulin-induced increase in cellular/cytosolic protein phosphatase activity have appeared, summarized by BRAUTIGAN and KUPLIC (1988). Recently, CHAN et al. (1988) reported that insulin, EGF, and PDGF, each of which can act on serum-starved 3T3 cells to activate glycogen synthase, also stimulate [^{32}P] phosphorylase phosphatase activity, assayed in cell extracts. The magnitude of insulin-stimulated phosphatase activity was a modest 1.3- to 1.5-fold, and required the inclusion of EDTA and glycogen in the homogenization (sonication) medium. The hormone stimulation was half-maximal at ~ 5 nM insulin, rapid in onset, and transient, and completely abolished by addition of protein phosphatase inhibitor 2 to the extract, indicating that a type I phosphatase was responsible. Potential mechanisms for the stimulation of type I phosphatase include a translocation of phosphatase catalytic subunit to the cytosol from membranes of glycogen particles, by reactions analogous to those uncovered by HIRAGA and COHEN (1986). Alternatively, the observed stimulation could reflect an activation of the cytosolic type I phosphatase, mediated by casein kinase 2 and GSK-3/F$_A$, as discussed earlier, or by entirely novel, undiscovered mechanisms. These observations on protein phosphatase regulation in serum-starved 3T3 cells are buttressed by several reports of insulin activation of type I phosphatase activity in rat liver, skeletal muscle, and adipose tissue (BRAUTIGAN and KUPLIC 1988; TOTH et al. 1988). In addition, insulin action appears to be associated with activation of pyruvate kinase phosphatase, a Mg-dependent (type IIC) phosphatase (FELIU 1986), as well as the intramitochondrial PDH phosphatase (THOMAS et al. 1986), an enzyme entirely distinct in its regulation from the cytoplasmic type I and type II enzymes. Given the clear-cut evidence now available that insulin regulates the activity of *at least* three and perhaps six or more different Ser/Thr protein kinases, it is not inappropriate to entertain the possibility that insulin regulates the activity of multiple, independent protein phosphatases. A detailed examination of the cellular protein phosphatases is beyond the scope of this discussion.

E. Conclusion

Regulatory protein phosphorylation is the critical mediator of the cellular response to insulin, and operates at virtually every step in the intracellular transmission and dissemination of the signal generated by the insulin–receptor interaction. The apex of this system of intracellular regulation is the IR kinase; the ability of the IR to catalyze tyrosine phosphorylation is necessary for all responses to the hormone (although see DEBANT et al. 1988). The only physiologic target of this reaction identified thus far, however, is the IR β subunit itself: the tyrosine-specific, intramolecular autophosphorylation promoted by insulin binding is absolutely required in order for the receptor to phosphorylate other protein substrates in vitro, and presumably in situ. The identity (and therefore the existence) of these native substrates remains in doubt. Although a variety of candidate substrates have been detected, the biochemical function of these candidates and the effects of tyrosine phosphorylation is unknown (e.g., p180) or of unresolved (lipocortin 1, calmodulin) or highly doubtful (p15, p120/HA4) significance to the program of insulin action. The absence of certifiable "physiologic" substrates for the IR kinase is not compelling evidence against the existence of these targets, but indicates the inefficiency and/or insensitivity of current approaches to their detection and isolation, and perhaps the scarcity of such molecules.

Insulin/growth factor-stimulated Ser/Thr-specific protein phosphorylation, mediated by a family of activated Ser/Thr protein kinases, is now emerging as an important and ubiquitous mechanism for the dissemination and amplification of the early intracellular signals generated by activation of the tyrosine kinase function. Little is known about how these multiple, different Ser/Thr kinases are regulated, although preliminary evidence suggests the importance of direct phosphorylation, an a cascade of protein kinases. The existence of novel nonenzymatic activators and inhibitors acting on these kinases, either polypeptides (perhaps kinase targets themselves) or other low molecular weight ligands (ALEMANY et al. 1987; LARNER 1988), remains an ever present, though as yet unvalidated consideration. Very little is known about the identity of the physiologically important substrates for these activated Ser/Thr kinases, although recently, the S6 kinase itself, and a low K_m cAMP phosphodiesterase have emerged as strong candidates for regulatory targets of insulin-activated Ser/Thr kinases. S6 kinase (through the phosphorylation of the 40S ribosomal subunit) may regulate the overall efficiency of protein synthesis at the initiation step of translation, and integrate the signals generated by diverse hormones, analogous to the role of phosphorylase b kinase in the regulation of glycogenolysis. The cAMP/insulin-regulated cAMP PDE acts to autoregulate the response to elevations of cAMP, and serves as a locus for cross-regulation of the cAMP pathway by other signaling pathways.

Although extrapolation from two examples is premature, it is striking that these enzyme substrates are themselves intermediates in the regulatory network. The regulatory superstructure provided by the protein kinase network is very complex, and its complexity is reflected by the large numbers of Tyr and Ser/Thr protein kinases expressed by cells; the only known function of these enzymes is to regulate protein function via posttranslational (and perhaps cotranslational) phosphorylation. Current knowledge as to the number, regulation, and organization of the subset of Ser/Thr protein kinases responsive to insulin and other growth factors is embryonic, in comparison to the cyclic nucleotide-, Ca^{2+}-, and lipid-regulated protein kinases (COHEN 1988). Virtually nothing is known about how insulin-regulated Ser/Thr kinases regulate fundamental processes of the cell, i.e., nutrient and ion transport, energy metabolism, protein synthesis, gene transcription, cell division, etc. Even less is known concerning the ways in which the insulin/growth factor-regulated (Ser/Thr) kinases interact with analogous effectors in other signaling pathways. Appreciation of the complexity of the protein kinase network has emerged rather slowly. The organizational model provided by the cAMP system suggested that a single kinase catalytic entity served to couple the signal generated at the cell surface (i.e., cAMP) rather directly to the metabolic enzymes of the cells, e.g., hormone-sensitive lipase, glycogen synthase, with the exceptional occurrence of an intervening kinase (i.e., phosphorylase b kinase). It now seems likely that will prove to be the simplest case. Already, a minimum of five Ca^{2+}/calmodulin-regulated protein kinases have been detected, some exhibiting a broad substrate specificity (e.g., Ca/calmodulin kinase II), others a more narrow range (e.g., phosphorylase b kinase, myosin light chain kinase). The kinase family of Ca/diglyceride/phospholipid enzymes exhibits an enlarging complexity, and examples of enzymes with related substrate specificity, but slightly altered regulation (in response to Ca or lipid) are now emerging. In addition, it appears that the regulation of kinase activity through phosphorylation by another kinase, first exemplified by phosphorylase b kinase, may prove to be, together with ligand regulation, a ubiquitous control mechanism for protein kinases, just as these twin control mechanisms are known to interact in the regulation of metabolic enzymes themselves such as glycogen synthase, PDH, glycogen phosphorylase, and acetyl-CoA carboxylase. The very large and exponentially growing number of protein kinases of itself indicates that an accurate blueprint of this network, for example as it operates in control of critical, multistep cell functions such as mitosis, is likely to be very complicated. The regulatory kinase/phosphatase network which encompasses glycogen metabolism exemplifies the constantly enlarging complexity revealed by intense and incisive scrutiny (COHEN 1983, 1987). Finally, the narrow view presented in this chapter concerning the regulation of protein phosphatases, which suggests that the activity of these enzymes is directed by protein kinases, is very incomplete. The recent demonstration that the catalytic subunit of the protein tyrosine phosphatase is highly homologous to a transmembrane receptor (CHARBONNEAU et al. 1988) suggests that, as with the protein kinases, direct control of protein phosphatases by extracellular signals acting via transmembrane mechanisms, as well as intracellular messengers, acting directly (e.g., Ca^{2+}/phosphatase IIB) and indirectly (e.g., cAMP/phosphatase inhibitor 1), will all be important regulatory mechanisms.

The elucidation of the insulin-regulated network of Tyrosine and Ser/Thr protein kinases presents an exceptional challenge. Among the immediate biochemical problems to be solved are: (a) identification of the reactions through which the tyrosine kinases regulate serine kinase activity; (b) enumeration of the full array of insulin-regulated Ser/Thr kinases, and their organization relative to the receptor tyrosine kinase and each other; (c) elucidation of the substrates for each of the insulin-regulated Ser/Thr kinases; and (d) elucidation of the insulin-activated protein kinases which participate in the regulation of protein phosphatase activity.

Acknowledgments. The authors wish to thank numerous colleagues who kindly provided manuscripts relevant to this review. Support for the work cited from the author's lab was provided by the Howard Hughes Medical Institute and NIH grant DK17776. Martha Chambers is thanked for excellent editorial and secretarial assistance.

H.T. wishes to acknowledge support from the Swedish Medical Research Council (grant 8689) and the Nordic Insulin Foundation.

References

Accili D, Perrotti N, Rees-Jones R, Taylor SI (1986) Tissue distribution and subcellular localization of an endogenous substrate (pp120) for the insulin receptor-associated tyrosine kinase. Endocrinology 119:1274–1280

Agostinis P, Goris J, Pinna LA, Merlevede W (1987) Regulation of casein kinase 2 by phosphorylation/dephosphorylation. Biochem J 248:785–789

Alexander MC, Kowaloff EM, Witters LA, Dennihy DT, Avruch J (1979) Purification of a hepatic 123 000-dalton hormone-stimulated ^{32}P-peptide and its identification as ATP-citrate lyase. J Biol Chem 254:8052–8056

Alexander MC, Palmer JL, Pointer RH, Koumjian L, Avruch J (1981) The role of cAMP-dependent protein kinase in the glucagon-stimulated phosphorylation of ATP-citrate lyase. Biochim Biophys Acta 674:37–47

Alexander MC, Palmer JL, Pointer RH, Kowaloff EM, Koumjian L, Avruch J (1982) Insulin-stimulated phosphorylation of ATP-citrate lyase in isolated hepatocytes. Stoichiometry and relation to the phosphoenzyme intermediate. J Biol Chem 257:2049–2055

Allemany S, Mato JM, Stralfors P (1987) Phospho-dephospho-control by insulin is mimicked by a phospho-oligosaccharide in adipocytes. Nature 330:77–79

Arion D, Meijer L, Brizuela L, Beach D (1988) cdc2 is a component of the M phase-specific histone H1 kinase: evidence for identity with MPF. Cell 55:371–378

Avruch J, Leone G, Martin DB (1974) Effects of insulin and epinephrine on protein phosphorylation in adipocytes. Diabetes 23:348

Avruch J, Leone G, Martin DB (1976) Effect of epinephrine and insulin on phosphopeptide metabolism in adipocytes. J Biol Chem 251:1511–1515

Avruch J, Witters LA, Alexander MC, Bush MA (1978) Effects of glucagon and insulin on cytoplasmic protein phosphorylation in hepatocytes. J Biol Chem 253:4754–4761

Avruch J, Witters LA, Alexander MC, Bush MA, Crapo LM (1979a) Insulin and phosphorylation of intracellular proteins. In: Bitensky M, Collier RJ, Steiner OF, Fox CF (eds) Transmembrane signalling. Prog Clin Biol Res 31:621–628

Avruch J, Witters LA, Alexander MC, Kowaloff EM, Palmer JL (1979b) Insulin-stimulated protein phosphorylation. In: Waldhausl W (ed) Proceedings of the 10th congress of the International Diabetes Federation. Excerpta Medica, Amsterdam, pp 185–190

Avruch J, Alexander MC, Palmer JL, Pointer RH, Nemenoff R, Pierce MW (1981) Hormone-stimulated phosphorylation of ATP-citrate lyase: mechanisms and significance. In: Rosen O and Krebs E (eds) Cold Spring Harbor conference on cell proliferation. Cold Spring Harbor, New York, pp 759–770

Avruch J, Alexander MC, Palmer JL, Pierce MW, Nemenoff RA, Blackshear PJ, Tipper JP, Witters LA (1982) The role of insulin-stimulated protein phosphorylation in insulin action. Fed Proc 41:2629–2633

Avruch J, Nemenoff RA, Pierce MP, Kwok YC, Blackshear PJ (1985) Protein phosphorylation as a mode of insulin action. In: Czech M (ed) Molecular basis of insulin action. Plenum, New York, pp 263–296

Balloti R, Kowalski A, Le Marchand-Brustel Y, van Obberghen E (1986) Presence of an insulin-stimulated serine kinase in cell extracts from IM-9 cells. Biochem Biophys Res Commun 139:179–185

Ballou L, Fischer E (1986) Protein phosphatases. In: Boyer P, Krebs E (eds) The enzymes. Academic, New York, pp 312–355

Ballou LM, Thomas G (1988) Protein phosphatase 2A inactivates the mitogen-stimulated S6 kinase from Swiss mouse 3T3 cells. J Biol Chem 263:1188–1194

Ballou LM, Siegmann M, Thomas G (1988) S6 kinase in quiescent Swiss mouse 3T3 cells is activated by phosphorylation in response to serum treatment. Proc Natl Acad Sci USA 85:7154–7158

Bassuk JA, Tsichlis PN, Solof S (1987) Liver fatty acid binding protein is the mitosis-associated polypeptide target of a carcinogen in rat hepatocytes. Proc Natl Acad Sci USA 84:7547–7551

Belfrage P, Donner J, Eriksson H, Stralfors P (1986a) Mechanisms for the control of lipolysis by insulin and pituitary growth hormone. In: Belfrage P, Donner J, Stralfors P (eds) Mechanisms of insulin action. Elsevier, New York, pp 323–340

Belfrage P, Donner J, Stralfors P (1986b) Mechanisms of insulin action. Fernstrom Found Ser 7:1–409

Belsham GJ, Denton RM (1980) The effect of insulin and adrenalin on the phosphorylation of a 22000 molecular weight protein within isolated fat cells. Biochem Soc Trans 8:383

Benjamin WB, Singer I (1974) Effect of insulin on the phosphorylation of adipose tissue protein. Biochim Biophys Acta 351:28–41

Benjamin WB, Singer I (1975) Actions of insulin, epinephrine, and dibutyryl cyclic adenosine 5′-monophosphate on fat cell protein phosphorylations. Cyclic adenosine 5′-monophosphate dependent and independent mechanisms. Biochemistry 14:3301–3309

Benkovic SJ, Fierke CA, Naylor AN (1988) Insights into enzyme function from studies on mutants of dihydrofolate reductase. Science 239:1105–1110

Benovic JL, Strasser RH, Caron MG, Lefkowitz RJ (1986) Beta-adrenergic receptor kinase: identification of a novel protein kinase that phosphorylates the agonist-occupied form of the receptor. Proc Natl Acad Sci USA 83:2797–2801

Benovic JL, Mayor F Jr, Staniszewski C, Lefkowitz R, Caron M (1987) Purification and characterization of the beta-adrenergic receptor kinase. J Biol Chem 262:9026–9032

Benovic J, Staniszewski C, Mayor F Jr, Caron MG, Lefkowitz RJ (1988) Beta-adrenergic receptor kinase. Activity of partial agonists for stimulation of adenylate cyclase correlates with ability to promote receptor phosphorylation. J Biol Chem 263:3893–3897

Bernier M, Laird D, Lane M (1987) Insulin-activated tyrosine phosphorylation of a 15-kilodalton protein in intact 3T3-L1 adipocytes. Proc Natl Acad Sci USA 84:1844–1848

Bernier M, Laird DM, Lane MD (1988) Effect of vanadate on the cellular accumulation of pp15, an apparent product of insulin receptor tyrosine kinase action. J Biol Chem 263:13626–13634

Blackmore PF, Assimacopoulos-Jeannet F, Chan TM, Exton JH (1979) Studies on alpha-adrenergic activation of hepatic glucose output. Insulin inhibition of alpha-adrenergic and glucagon actions in normal and calcium depleted hepatocytes. J Biol Chem 254:2828–2834

Blackshear PJ, Nemenoff RA, Avruch J (1982) Preliminary characterization of a heat-stable protein from rat adipose tissue whose phosphorylation is stimulated by insulin. Biochem J 204:817–824

Blackshear PJ, Nemenoff RA, Avruch J (1983) Insulin and growth factors stimulate the phosphorylation of a M_r-22000 protein in 3T3-L1 adipocytes. Biochem J 214:11–19

Blackshear PJ, Witters LA, Girard PR, Kuo JF, Quamo SN (1985) Growth factor-stimulated protein phosphorylation in 3T3-L1 cells. Evidence for protein kinase C-dependent pathways. J Biol Chem 260:13304–13315

Blenis J, Erikson RL (1986) Stimulation of ribosomal protein S6 kinase activity by pp60v-src or by serum: dissociation from phorbol ester-stimulated activity. Proc Natl Acad Sci USA 83:1733–1737

Blenis J, Kuo CJ, Erikson RL (1987) Identification of a ribosomal protein S6 kinase regulated by transformation and growth-promoting stimuli. J Biol Chem 262:14373–14376

Bohmer FD, Kraft R, Otto A, Wernstedt C, Hellman U, Kurtz A, Muller T et al. (1987) Identification of a polypeptide growth inhibitor from bovine mammary gland. Sequence homology to fatty acid- and retinoid-binding proteins. J Biol Chem 262:15137–15143

Borthwick AC, Edgell NJ, Denton RM (1987) Use of rapid gel-permeation chromatography to explore the inter-relationships between polymerization, phosphorylation and activity of acetyl-CoA carboxylase. Biochem J 241:773–782

Boyer PD, Krebs EG (1986) The enzymes, vol XVII, part A. Academic, Orlando

Brautigan DL, Kuplic JD (1988) Proposal for a pathway to mediate the metabolic effects of insulin. Int J Biochem 20:349–356

Brownsey RW, Denton RM (1982) Evidence that insulin activates fat-cell acetyl-CoA carboxylase by increased phosphorylation at a specific site. Biochem J 202:77–86

Cahill GF Jr (1971) Physiology of insulin in man. Diabetes 20:785–799

Cahill GF Jr, Steiner D (1972) The endocrine pancreas. William and Wilkins, Philadelphia (Handbook of physiology: Endocrinology I)

Carpenter G, King L Jr, Cohen S (1979) Rapid enhancement of protein phosphorylation in A-431 cell membrane preparations by epidermal growth factor. J Biol Chem 254:4884–4891

Chan CP, McNall SJ, Krebs EG, Fischer EH (1988) Stimulation of protein phosphatase activity by insulin and growth factor in 3T3 cells. Proc Natl Acad Sci USA 85:6257–6261

Chan Y-L, Wool IG (1988) The primary structure of rat ribosomal protein S6. J Biol Chem 263:2891–2896

Charbonneau H, Tonks NK, Walsh KA, Fischer EH (1988) The leukocyte common antigen (CD45): a putative receptor-linked protein tyrosine phosphatase. Proc Natl Acad Sci USA 85:7182–7186

Chou CK, Dull TJ, Russell DS, Gherzi R, Lebwohl D, Ullrich A, Rosen OM (1987) Human insulin receptors mutated at the ATP-binding site lack protein tyrosine kinase activity and fall to mediate postreceptor effects of insulin. J Biol Chem 262:1842–1847

Claus TH, Raafat El-Maghrabi M, Regen DM, Pilkis SJ (1986) Mechanism of the acute action of insulin on hepatic gluconeogenesis. In: Belfrage P, Donner J, Stralfors P (eds) Mechanisms of insulin action. Elsevier, New York, pp 305–321

Cobb MH (1986) An insulin-stimulated ribosomal protein S6 kinase in 3T3-L1 cells. J Biol Chem 261:12 994–12 999

Cohen P (1983) Protein phosphorylation and the control of glycogen metabolism in skeletal muscle. Philos Trans R Soc Lond [Biol] 302:13–25

Cohen P (1986) Muscle glycogen synthase. In: Boyer P, Krebs E (eds) The enzymes. Academic, New York, pp 461–497

Cohen P (1987) Molecular mechanisms involved in the control of glycogenolysis in skeletal muscle by calcium and cyclic AMP. Biochem Soc Trans 15:999–1001

Cohen P (1988) Protein phosphorylation and hormone action. Proc R Soc Lond [Biol] 234:115–144

Cohen S, Carpenter G, King L Jr (1980) Epidermal growth factor-receptor-protein kinase interactions. Co-purification of receptor and epidermal growth factor-enhanced phosphorylation activity. J Biol Chem 255:4832–4842

Colca JR, DeWald DB, Pearson JD, Palazuk BJ, Laurino JP, McDonald JM (1987) Insulin stimulates the phosphorylation of calmodulin in intact adipocytes. J Biol Chem 262:11 399–11 402

Cooper DA, Konda TS, Standaert ML, Davis JL, Pollet RJ, Farese RV (1987) Insulin increases membrane and cytosolic protein kinase C activity in BC3H1 myocytes. J Biol Chem 262:3633–3639

Cooper JA, Hunter T (1983) Regulation of cell growth and transformation by tyrosine-specific protein kinases: The search for important cellular substrate proteins. Curr Top Microbiol Immunol 107:125–182

Cooper JA, Hunter T (1985) Major substrate for growth factor-activated protein-tyrosine kinases is a low-abundance protein. Mol Cell Biol 5:3304–3309

Cooper JA, Reiss NA, Schwartz RJ, Hunter T (1983) Three glycolytic enzymes are phosphorylated at tyrosine in cells transformed by Rous sarcoma virus. Nature 302:218–223

Cooper JA, Esch F, Taylor SS, Hunter T (1984) Phosphorylation sites in enolase and lactate dehydrogenase utilized by tyrosine protein kinases in vivo and in vitro. J Biol Chem 259:7835–7841

Crompton MR, Moss SE, Crumpton MJ (1988) Diversity in the lipocortin/calpactin family. Cell 55:1–3

Czech MP (ed) (1985) Molecular basis of insulin action. Plenum, New York

Czech MP, Klarlund JV, Yagaloff KA, Bradford AP, Lewis RE (1988) Insulin receptor signalling: activation of multiple serine kinases. J Biol Chem 263:11 017–11 020

Davidson FF, Dennis EA, Powell M, Glenney J Jr (1987) Inhibition of phospholipase A2 by "lipocortins" and calpactins. An effect of binding to substrate phospholipids. J Biol Chem 262:1698–1705

Debant A, Caluser E, Ponzio G, Filloux, Auzan C, Contreres J-O, Rossi B (1988) Replacement of insulin receptor tyrosine residues 1162 and 1163 does not alter the mitogenic effect of the hormone. Proc Natl Acad Sci USA 85:8032–8036

Decker S (1981) Phosphorylation of ribosomal protein S6 in avian sarcoma virus-transformed chicken embryo fibroblasts. Proc Natl Acad Sci USA 78:4112–4115

Degerman E, Belfrage P (1987) Purification of the putative hormone-sensitive cAMP phosphodiesterase from rat adipose tissue using a derivative of cilostamide as a novel affinity ligand. J Biol Chem 262:5797–5807

Degerman E, Olsson H, Stralfors P, Smith CJ, Manganiello VC, Belfrage P (1989 a) Phosphorylation/dephosphorylation of the hormone-sensitive, low K_m cAMP phosphodiesterase from adipose tissue by cAMP-dependent protein kinase and protein phosphatases. (submitted)

Degerman E, Smith CJ, Tornqvist H, Vasta V, Belfrage P, Manganiello VC (1989 b) Evidence that insulin and isoprenaline activates the cGMP-inhibited low K_m cAMP phosphodiesterase in rat fat cells by phosphorylation. (submitted)

De la Houssaye BA, Eckols TK, Masaracchia RA (1983) Activation of a cyclic AMP-dependent protein kinase by an endogenous ATP-requiring protease from lymphosarcoma cells. J Biol Chem 258:4272–4278

Demmer LA, Birkenmeier EH, Sweetser DA, Levin MS, Zollerman S, Sparkes RS, Mohandas T et al. (1987) The cellular retinol binding protein II gene. Sequence analysis of the rat gene, chromosomal localization in mice and humans, and documentation of its close linkage to the cellular retinol binding protein gene. J Biol Chem 262:2458–2467

Denton D (1986) Early events in insulin actions. Adv Cyclic Nucleotide Protein Phosphorylation Res 20:293–341

Denton R, Thomas AP, Tavare JM, Borthwick AC, Brownsey RW, Hopkirk TJ, McCormack JG (1986) Mechanisms involved in the stimulation of fatty acid synthesis by insulin. In: Belfrage P, Donner J, Stralfors P (eds) Mechanisms of insulin action. Elsevier, New York, pp 283–304

DePaoli-Roach AA (1984) Synergistic phosphorylation and activation of ATP-Mg-dependent phosphoprotein phosphatase by FA/GSK-3 and casein kinase II (PCO.7). J Biol Chem 259:12 144–12 152

DePaoli-Roach AA, Ahmad Z, Roach P (1981) Characterization of a rabbit skeletal muscle protein kinase (PCO.7) able to phosphorylate glycogen synthase and phosvitin. J Biol Chem 256:8955–8962

DePaoli-Roach AA, Ahmad Z, Camici M, Lawrence JJ, Roach PJ (1983) Multiple phosphorylation of rabbit skeletal muscle glycogen synthase. Evidence for interactions among phosphorylation sites and the resolution of electrophoretically distinct forms of the subunit. J Biol Chem 258:10 702–10 709

Draetta G, Beach D (1988) Activation of cdc2 protein kinase during mitosis in human cells: cell cycle-dependent phosphorylation and substrate rearrangements. Cell 54:17–26

Dunphy WG, Brizuela L, Beach D, Newport J (1988) The Xenopus cdc2 protein is a component of MPF, a cytoplasmic regulator of mitosis. Cell 54:423–431

Ebina Y, Ellis L, Jarnagin K, Edery M, Graf L, Clauser E, Ou J-H, Masiarz F et al. (1985 a) The human insulin receptor cDNA: the structural basis for hormone-activated transmembrane signalling. Cell 40:747–758

Ebina Y, Edery M, Ellis L, Standring D, Beaudoin J, Roth RA, Rutter WJ (1985 b) Expression of a functional human insulin receptor from a cloned cDNA in Chinese hamster ovary cells. Proc Natl Acad Sci USA 82:8014–8018

Ebina Y, Araki E, Taira M, Shimada F, Mori M, Craik C, Siddle K et al. (1987) Replacement of lysine residue 1030 in the putative ATP-binding region of the insulin receptor abolishes insulin- and antibody-stimulated glucose uptake and receptor kinase activity. Proc Natl Acad Sci USA 84:704–708

356   J. AVRUCH et al.

Edelman AM, Blumenthal DK, Krebs EG (1987) Protein serine/threonine kinases. Annu Rev Biochem 56:567–613

Ellis L, Clauser E, Morgan DO, Edery M, Roth RA, Rutter WJ (1986) Replacement of insulin receptor tyrosine residues 1162 and 1163 compromises insulin-stimulated kinase activity and uptake of 2-deoxyglucose. Cell 45:721–732

Ellis L, Morgan DO, Jong S-M, Wang L-H, Roth RA, Rutter WJ (1987a) Heterologous transmembrane signaling by a human insulin receptor-v-*ros* hybrid in Chinese hamster ovary cells. Proc Natl Acad Sci USA 84:5101–5105

Ellis L, Morgan DO, Clauser E, Roth RA, Rutter WJ (1987b) A membrane-anchored cytoplasmic domain of the human insulin receptor mediates a constitutively elevated insulin-independent uptake of 2-deoxyglucose. Mol Endocrinol 1:15–24

Erikson E, Maller JL (1985) A protein kinase from *Xenopus* eggs specific for ribosomal protein S6. Proc Natl Acad Sci USA 82:742–746

Erikson E, Maller JL (1986) Purification and characterization of a protein kinase from *Xenopus* eggs highly specific for ribosomal protein S6. J Biol Chem 261:350–355

Erikson E, Stefanovic D, Blenis J, Erikson RL, Maller JL (1987) Antibodies to *Xenopus* eggs S6 kinase II recognize S6 kinase from progesterone- and insulin-stimulated *Xenopus* oocytes and from proliferating chicken embryo fibroblasts. Mol Cell Biol 7:3147–3155

Farese RV, Suman Konda T, Davis JS, Standaert ML, Pollet RJ, Cooper DR (1985) Insulin rapidly increases diacylglycerol by activating de novo phosphatidic acid synthesis. Science 236:586–589

Farese RV, Davis JS, Barnes DE, Standaert ML, Babischkin JS, Hock R, Rosic NK, Pollet RJ (1987) The de novo phospholipid effect of insulin is associated with increases in diacylglycerol, but not inositol phosphates or cytosolic Ca^{2+}. Biochem J 231:269–287

Fava RS, Cohen S (1984) Isolation of a calcium-dependent 35-kilodalton substrate for the epidermal growth factor receptor/kinase from A-431 cells. J Biol Chem 259:2636–2645

Feliu JE (1986) Adv Protein Phosphorylation 3:163–185

Fiol CJ, Mahrenholz AM, Wang Y, Roeske RW, Roach PJ (1987) Formation of protein kinase recognition sites by covalent modification of the substrate. J Biol Chem 262:14042–14048

Flower RK, Wood JN, Parante L (1984) Macrocortin and the mechanism of action of the glucocorticoids. Adv Infl Res 7:61–70

Forn J, Greenbard P (1976) Regulation of lipolytic and antilipolytic compounds of the phosphorylation of specific proteins in isolated intact fat cells. Arch Biochem Biophys 176:721–733

Forsayeth JR, Montemurro A, Maddux BA, DiPirro R, Goldfine ID (1987a) Effect of monoclonal antibodies on human insulin receptor autophosphorylation, negative cooperativity, and down-regulation. J Biol Chem 262:4134–4140

Forsayeth JR, Caro JF, Sinha MK, Maddux BA, Goldfine ID (1987b) Monoclonal antibodies to the human insulin receptor that activate glucose transport but not insulin receptor kinase activity. Proc Natl Acad Sci USA 84:3448–3451

Foulkes JG, Jefferson LS, Cohen P (1980) The hormonal control of glycogen metabolism: dephosphorylation of protein phosphatase inhibitor-1 in vivo in response to insulin. FEBS Lett 112:21–24

Foulkes JG, Cohen P, Strada SJ, Everson WV, Jefferson LS (1982) Antagonistic effects of insulin and beta-adrenergic agonists on the activity of protein phosphatase inhibitor-1 in skeletal muscle of the perfused rat hemicorpus. J Biol Chem 257:12493–12496

Frackelton AR Jr, Ross AH, Eisen HN (1983) Characterization and use of monoclonal antibodies for isolation of phosphotyrosyl proteins from retrovirus-transformed cells and growth factor-stimulated cells. Mol Cell Biol 3:1343–1352

Friedkin M, Rozengurt E (1980) The role of cytoplasmic microtubules in the regulation of the activity of peptide growth factors. Adv Enzyme Regul 19:39–59

Frost SC, Lane MD (1985) Evidence for the involvement of vicinal sulfhydryl groups in insulin-activated hexose transport by 3T3-L1 adipocytes. J Biol Chem 260:2646–2652

Frost SC, Kohanski RA, Lane MD (1987) Effect of phenylarsine oxide on insulin-dependent protein phosphorylation and glucose transport in 3T3-L1 adipocytes. J Biol Chem 262:9872–9876

Fukami Y, Nakamura T, Nakayama A, Kanehisa T (1986) Phosphorylation of tyrosine residues of calmodulin in Rous sarcoma virus-transformed cells. Proc Natl Acad Sci USA 83:4190–4193

Gabbay RS, Lardy HA (1984) Site of insulin inhibition of cAMP-stimulated glycogenolysis. J Biol Chem 259:6052–6055

Gabbay RA, Lardy HA (1987) Insulin inhibition of hepatic cAMP-dependent protein kinase: decreased affinity of protein kinase for cAMP and possible differential regulation of intrachain sites 1 and 2. Proc Natl Acad Sci USA 84:2218–2222

Gallis B, Edelman AM, Casnellie JE, Krebs EG (1983) Epidermal growth factor stimulates tyrosine phosphorylation of the myosin regulatory light chain from smooth muscle. J Biol Chem 258:13089–13093

Gautier J, Norbury C, Lohka M, Nurse P, Maller J (1988) Purified maturation-promoting factor contains the product of a *Xenopus* homolog of the fission yeast cell cycle control gene cdc2. Cell 54:433–439

Gazzano H, Kowalski A, Fehlmann M, van Obberghen E (1983) Two different protein kinase activities are associated with the insulin receptor. Biochem J 216:575–582

Gettys TW, Vine AJ, Simonds MF, Corbin JD (1988) Activation of the particulate low K_m phosphodiesterase of adipocytes by addition of cAMP-dependent protein kinase. J Biol Chem 263:10359–10363

Gibbs EM, Allard WJ, Lienhard GE (1986) The glucose transporter in 3T3-L1 adipocytes is phosphorylated in response to phorbol ester but not in response to insulin. J Biol Chem 261:16597–16603

Gilmore T, Martin GS (1983) Phorbol ester and diacylglycerol induce protein phosphorylation at tyrosine. Nature 306:487–490

Glenney JR, Lokas L, Kamps MP (1988) Monoclonal antibodies to phosphotyrosine. J Immunol Methods 109:277–285

Glover CVC (1986) A filamentous form of *Drosophila* casein kinase II. J Biol Chem 261:14349–14354

Gonzatti-Haces M, Traugh JA (1986) Ca^{2+}-independent activation of protease-activated kinase II by phospholipids/diolein and comparison with the Ca^{2+}/phospholipid-dependent protein kinase. J Biol Chem 261:15266–15272

Gordon J, Nielsen PJ, Manchester KL, Towbin H, Jimenez de Asua L, Thomas G (1982) Criteria for establishment of the biological significance of ribosomal protein phosphorylation. Curr Top Cell Regul 21:89–99

Goren HJ, White MF, Kahn CR (1987) Separate domains of the insulin receptor contain sites of autophosphorylation and tyrosine kinase activity. Biochemistry 26:2374–2382

Goren HJ, Hollenberg MD, Roncari DAK (1988) Insulin action and diabetes. Raven, New York

Grant PG, Mannarino AF, Colman RW (1988) cAMP-mediated phosphorylation of the low-K_m cAMP phosphodiesterase markedly stimulates its catalytic activity. Proc Natl Acad Sci USA 85:9071–9075

Graves CB, Goewert RR, McDonald JM (1985) The insulin receptor contains a calmodulin-binding domain. Science 230:827–829

Gressner AM, Wool IG (1974a) The phosphorylation of liver ribosomal proteins in vivo. Evidence that only a single small subunit protein (S6) is phosphorylated. J Biol Chem 249:6917–6925

Gressner AM, Wool IG (1974b) The stimulation of the phosphorylation of ribosomal protein S6 by cycloheximide and puromycin. Biochem Biophys Res Commun 60:1482–1490

Gressner AM, Wool IG (1976a) Effect of experimental diabetes and insulin on phosphorylation of rat liver ribosomal protein S6. Nature 259:148–150

Gressner AM, Wool IG (1976b) Influence of glucagon and cyclic adenosine 3′:5′-monophosphate on the phosphorylation of rat liver ribosomal protein S6. J Biol Chem 251:1500–1504

Guy PS, Cohen P, Hardie DG (1980) Rat mammary gland ATP citrate lyase is phosphory-lated by cyclic AMP-dependent protein kinase. FEBS Lett 109:205–208

Haigler HT, Schaepfer DD, Burgess WH (1987) Characterization of lipocortin I and an immunologically unrelated 33-kDa as epidermal growth factor receptor/kinase sub-strates and phospholipase A2 inhibitors. J Biol Chem 262:6921–6930

Hanks SK, Quinn MA, Hunter T (1988) The protein kinase family: conserved features and deduced phylogeny of the catalytic domains. Science 241:42–52

Haring HU, White MF, Machicao F, Ermel B, Obermaier B (1987) Insulin rapidly stimu-lates phosphorylation of a 46-kDa membrane protein in tyrosine residues as well as phosphorylation of several soluble proteins in intact fat cells. Proc Natl Acad Sci USA 84:113–117

Haselbacher GK, Humbel RE, Thomas G (1979) Insulin-like growth factor: insulin or serum increase phosphorylation of ribosomal protein S6 during transition of stationary chick embryo fibroblasts into early G_i phase of the cell cycle. FEBS Lett 100:185–190

Hathaway GM, Traugh JA (1982) Casein kinases: multipotential protein kinases. Curr Top Cell Regul 21:101–127

Haystead TAJ, Hardie DG (1986) Evidence that activation of acetyl-CoA carboxylase by insulin in adipocytes is mediated by a low-M_r effector and not by increased phosphorylation. Biochem J 240:99–106

Haystead TAJ, Campbell DG, Hardie DG (1988) Analysis of sites phosphorylated on acetyl-CoA carboxylase in response to insulin in isolated adipocytes: comparison with sites phosphorylated by casein kinase-2 and the calmodulin-dependent multiprotein kinase. Eur J Biochem 175:347–354

Hegazy MG, Thysseril TJ, Schlender KK, Reimann EM (1987) Characterization of GSK-M, a glycogen synthase kinase from rat skeletal muscle. Arch Biochem Biophys 258:470–481

Herrera R, Rosen OM (1986) Autophosphorylation of the insulin receptor in vitro. Desig-nation of phosphorylation sites and correlation with receptor kinase activation. J Biol Chem 261:11980–11985

Herrera R, Petruzzelli L, Thomas N, Bramson H, Kaiser E, Rosen OM (1985) An antipep-tide antibody that specifically inhibits insulin receptor autophosphorylation and protein kinase activity. Proc Natl Acad Sci USA 82:7899–7903

Herrera R, Garcia de Herreros A, Kallen RG, Rosen OM (1988) Synthesis, purification and characterization of the cytoplasmic domain of the human insulin receptor using a baculovirus expression system. J Biol Chem 263:5560–5568

Hiraga, Cohen P (1986) Phosphorylation of the glycogen binding subunit of protein phosphatase: 1G by cyclic AMP-dependent protein kinase promotes translocation of the phosphatase from glycogen to cytosol in rabbit skeletal muscle. Eur J Biochem 161:763–769

Hirata F (1981) The regulation of lipomodulin, a phospholipase inhibitory protein, in rab-bit neutrophils by phosphorylation. J Biol Chem 256:7730–7733

Hirata F, Matsuda K, Notsu Y, Hattori T, del Carmine R (1984) Phosphorylation at a tyrosine residue of lipomodulin in mitogen-stimulated murine thymocytes. Proc Natl Acad Sci USA 81:4717–4721

Holmes CFB, Kuret J, Chisholm AAK, Cohen P (1986) Identification of the sites on rabbit skeletal muscle protein phosphatase inhibitor-2 phosphorylated by casein kinase-II. Biochim Biophys Acta 870:408–416

Holmes CFB, Tonks NK, Major H, Cohen O (1987) Analysis of the in vivo phosphoryla-tion state of protein phosphatase inhibitor-2 from rabbit skeletal muscle by fast-atom bombardment mass spectrometry. Biochim Biophys Acta 929:208–219

Honnor RS, Londos C (1986) New insights into the antilipolytic action of insulin. In: Bel-frage P, Donner J, Stralfors OP (eds) Mechanisms of insulin action. Elsevier, New York, pp 341–363

Honnor RP, Dhillon GS, Londos C (1985a) cAMP-dependent protein kinase and lipolysis in rat adipocytes. I. Cell preparation, manipulation and predictability in behavior. J Biol Chem 260:15122–15129

Honnor RS, Dhillon GS, Londos C (1985b) cAMP-dependent protein kinase and lipolysis in rat adipocytes. II. Definition of steady state relationship with lipolytic and antilipolytic modulators. J Biol Chem 260:15130–15138

Hoshi M, Nishida E, Sakai H (1988) Activation of a Ca^{2+}-inhibitable protein kinase that phosphorylates microtubule-associated protein-2 in vitro by growth factors, phorbol esters, and serum in quiescent cultured human fibroblasts. J Biol Chem 263:5396–5401

Hresko RC, Bernier M, Hoffman R, Flores-Riveros JR, Liao K, Laird DM, Lane MD (1988) Identification of phosphorylated 422(aP2) protein as pp15, the 15-kilodalton target of the insulin receptor tyrosine kinase in 3T3-L1 adipocytes. Proc Natl Acad Sci USA 85:8835–8839

Huang KS, Wallner BP, Mattaliano RJ, Tizard R, Burne L, Frey A, Hession C et al. (1986) Two human 35 Kd inhibitors of phospholipase A2 are related to substrate pp60^{v-src} and of the epidermal growth factor receptor/kinase. Cell 46:191–199

Hughes WA, Brownsey R, Denton RM (1977) The effects of insulin on the phosphorylation of pyruvate dehydrogenase and other proteins in adipocyte tissue. In: Pinna L (ed) Phosphorylated proteins and their related enzymes. Information Retrieval, London, pp 7–31

Hunter T (1982) Synthetic peptide substrates for a tyrosine protein kinase. J Biol Chem 257:4843–4848

Hunter T (1987) A thousand and one protein kinases. Cell 50:823–829

Hunter T, Cooper JA (1985) Protein-tyrosine kinases. Annu Rev Biochem 54:897–930

Ingebritsen TS, Cohen P (1983) Protein phosphatases: properties and protein cellular regulation. Science 221:331–338

Izumi T, White MF, Kadowaki T, Takaku F, Akanuma Y, Kasuga M (1987) Insulin-like growth factor I rapidly stimulates tyrosine phosphorylation of a M_r 185000 protein in intact cells. J Biol Chem 262:1282–1287

Jazwinski SM (1988) cdc7-dependent protein kinase activity in yeast replicative-complex preparations. Proc Natl Acad Sci USA 85:2101–2105

Jeno P, Ballou LM, Novak-Hofer I, Thomas G (1988) Identification and characterization of a mitogen-activated S6 kinase. Proc Natl Acad Sci USA 85:406–410

Jones SW, Erikson E, Blenis J, Maller JL, Erikson RL (1988) A *Xenopus* ribosomal protein S6 kinase has two apparent kinase domains that are each similar to distinct protein kinases. Proc Natl Acad Sci USA 85:3377–3381

Kadowaki T, Koyasu S, Nishida E, Tobe K, Izumi T, Takaku F, Sakai H et al. (1987) Tyrosine phosphorylation of common and specific sets of cellular proteins rapidly induced by insulin, insulin-like growth factor I, and epidermal growth factor in an intact cell. J Biol Chem 262:7342–7350

Kahn CR, Baird K, Jarrett DB, Flier JS (1978) Direct demonstration that cross linking or aggregation is important in insulin action. Proc Natl Acad Sci USA 74:4209–4213

Kamps MP, Sefton BM (1988) Identification of multiple novel polypeptide substrates of the v-*src*, v-*yes*, v-*fps*, v-*ros*, and v-*erb*-B oncogenic tyrosine protein kinases utilizing antisera against phosphotyrosine. Oncogene 2:305–315

Kamps MP, Buss JE, Sefton BM (1986) Rous sarcoma virus transforming protein lacking myristic acid phosphorylates known polypeptide substrates without inducing transformation. Cell 45:105–112

Karasik A, Pepinsky RB, Shoelson SE, Kahn CR (1988) Lipocortins 1 and 2 as substrates for the insulin receptor kinase in rat liver. J Biol Chem 263:11862–11867

Kasuga M, Karlsson FA, Kahn CR (1982) Insulin stimulates the phosphorylation of the 95000-dalton subunit of its own receptor. Science 215:185–187

Khatra BS, Chiasson J-L, Shikama H, Exton JH, Soderling TR (1980) Effect of epinephrine and insulin on the phosphorylation of phosphorylase phosphatase inhibitor 1 in perfused rat skeletal muscle. FEBS Lett 114:253–256

Kishimoto A, Goldstein JL, Brown MS (1987) Purification of catalytic subunit of low density lipoprotein receptor kinase and identification of heat-stable activator protein. J Biol Chem 262:9367–9373

Klarlund JK, Czech MP (1988) Insulin-like growth factor I and insulin rapidly increases casein kinase II activity in BALB/c 3T3 fibroblasts. J Biol Chem 263:15872–15875

Klee CA (1988) Ca^{2+}-dependent phospholipid-(and membrane-)binding proteins. Biochemistry 27:6645–6653

Klein HH, Friedenberg GR, Cordera R, Olefsky JM (1985) Substrate specificities of insulin and epidermal receptor kinases. Biochem Biophys Res Commun 127:254–263

Klein HH, Freidenberg GR, Kladde M, Olefsky JM (1986) Insulin activation of insulin receptor tyrosine kinase in intact rat adipocytes. An in vitro system to measure histone kinase activity of insulin receptors activated in vivo. J Biol Chem 261:4691–4697

Klemm DJ, Elias L (1987) A distinctive phospholipid-stimulated protein kinase of normal and malignant murine hemopoietic cells. J Biol Chem 262:7580–7585

Kloetzer WS, Maxwell SA, Arlinghaus RB (1983) p85$^{gag-mos}$ encoded by ts110 Moloney murine sarcoma virus has an associated protein kinase activity. Proc Natl Acad Sci USA 80:412–416

Kohanski RA, Lane MD (1986) Kinetic evidence for activating and non-activating components of autophosphorylation of the insulin receptor protein kinase. Biochem Biophys Res Commun 134:1312–1318

Kohno M (1985) Diverse mitogenic agents induce rapid phosphorylation of a common set of cellular proteins at tyrosine in quiescent mammalian cells. J Biol Chem 260:1771–1779

Korn LJ, Siebel CW, McCormick F, Roth RA (1987) Rasp21 as a potential mediator of insulin action in *Xenopus* oocytes. Science 236:840–843

Krieg J, Hofstennge J, Thomas G (1988) Identification of the 40S ribosomal protein S6 phosphorylation sites induced by cycloheximide. J Biol Chem 263:11 473–11 477

Krupinski J, Rajaram R, Lakonishok M, Benovic JL, Cerione RA (1988) Insulin-dependent phosphorylation of GTP-binding proteins in phospholipid vesicles. J Biol Chem 263:12 333–12 341

Kuenzel EA, Krebs EG (1985) Synthetic peptide substrate specific for casein kinase II. Proc Natl Acad Sci USA 82:737–741

Kuenzel EA, Mulligan JA, Sommercorn J, Krebs EG (1987) Substrate specificity determinants for casein kinase II as deduced from studies with synthetic peptides. J Biol Chem 262:9136–9140

Kwok YC, Yip CP (1987) Tyrosine phosphorylation of two cytosolic proteins of 50 kDa and 35 kDa in rat liver by insulin-receptor kinase in vitro. Biochem J 248:27–33

Kwok YC, Nemenoff RA, Powers AC, Avruch J (1986) Kinetic properties of the insulin receptor tyrosine protein kinase: activation through an insulin-stimulated tyrosine-specific intramolecular autophosphorylation. Arch Biochem Biophys 244:102–113

Larner J (1988) Insulin signalling mechanisms – lessons from the Old Testament of glycogen metabolism and from the New Testament of molecular biology. Diabetes 37:262–275

Lawrence JC, James C, Hiken JF (1986) Control of glycogen synthase by insulin and isoproterenol in rat adipocytes. Changes in the distribution of phosphate in the synthase subunit in response to insulin and beta-adrenergic activation. J Biol Chem 261:669–677

Lawrence JC Jr, Hiden JF, Burnette B, DePaoli-Roach AA (1988) Phosphorylation of phosphoprotein phosphatase inhibitor-2 (I-2) in rat fat cells. Biochem Biophys Res Commun 150:197–203

Linder ME, Burr JG (1988) Nonmyristoylated p60^{v-src} fails to phosphorylate proteins of 115-120 kDa in chicken embryo fibroblasts. Proc Natl Acad Sci USA 85:2608–2612

Lohka MJ, Hayes MK, Maller JL (1988) Purification of maturation-promoting factor, an intracellular regulator of early mitotic events. Proc Natl Acad Sci USA 85:3009–3013

Londos C, Honnor RC, Dhillon GS (1985) cAMP-dependent protein kinase and lipolysis in rat adipocytes. III. Multiple modes of insulin regulation of lipolysis and regulation of insulin responses by adenylate cyclase regulators. J Biol Chem 260:15 139–15 145

Lubben TH, Traugh JA (1983) Cyclic nucleotide independent protein kinases from rabbit reticulocytes. Purification and characterization of protease-activated kinase II. J Biol Chem 258:13 992–13 997

Macphee CH, Reifsnyder DH, Moore TA, Lerea KM, Beavo JA (1988) Phosphorylation results in activation of a cAMP phosphodiesterase in human platelets. J Biol Chem 263:10 353–10 358

Maegawa H, McClain DA, Freidenberg G, Olefsky JM, Napier M, Lipari T, Dull TJ et al. (1988a) Properties of a human insulin receptor with a COOH-terminal truncation. II. Truncated receptors have normal kinase activity but are defective in signaling metabolic events. J Biol Chem 263:8912–8917

Maegawa H, Olefsky J, Thies S, Boyd D, Ullrich A, McClain DA (1988b) Insulin receptors with defective tyrosine kinase inhibit normal receptor function at the level of substrate phosphorylation. J Biol Chem 263:12629–12637

Maller JL, Foulkes JG, Erikson E, Baltimore D (1985) Phosphorylation of ribosomal protein S6 on serine after microinjection of the Abelson murine leukemia virus tyrosine-specific protein kinase in Xenopus oocytes. Proc Natl Acad Sci USA 82:272–276

Maller JL, Pike LJ, Freidenberg GR, Cordera R, Stith BJ, Olefsky JM, Krebs EG (1986) Increased phosphorylation of ribosomal protein S6 following microinjection of insulin receptor-kinase into Xenopus oocytes. Nature 320:459–461

Margolis RV, Taylor SI, Seminara D, Hubbard AL (1988) Identification of pp120, an endogenous substrate for the hepatocyte insulin receptor tyrosine kinase, as an integral membrane glycoprotein of the bile canalicular domain. Proc Natl Acad Sci USA 85:7256–7260

Masaracchia RA, Mallick S, Murdoch FE (1988) Regulation of S6 protein kinase from human placenta: evidence for activation by an activating enzyme and the insulin receptor. In: Goren H, Hollenberg M, Roncari D (eds) Insulin action and diabetes. Raven, New York, pp 169–172

Matarese V, Bernlohr DA (1988) Purification of murine adipocyte lipid-binding protein. Characterization as a fatty acid- and retinoic acid-binding protein. J Biol Chem 263:14544–14551

McClain DA, Maegawa H, Lee J, Dull TJ, Ullrich A, Olefsky JM (1987) A mutant insulin receptor with defective tyrosine kinase displays no biologic activity and does not undergo endocytosis. J Biol Chem 262:14663–14571

McClain DA, Maegawa H, Levy J, Huecksteadt T, Dull TJ, Lee J, Ullrich A, Olefsky JM (1988) Properties of a human insulin receptor with a C terminal truncation. I. Insulin binding, autophosphorylation and endocytosis. J Biol Chem 263:8904–8911

McLeod M, Beach D (1988) A specific inhibitor of the ran1+ protein kinase regulates entry into meiosis in Schizosaccharomyces pombe. Nature 332:509–514

Meggio F, Marchiori F, Borin G, Ghessa G, Pinna LA (1984) Synthetic peptides including acidic clusters as substrates and inhibitors of rat liver casein kinase TS (type-2). J Biol Chem 259:14576–14579

Moelling K, Heimann B, Beimling P, Rapp UR, Sander T (1984) Serine- amd threonine-specific protein kinase activities of purified gag-mil and gag-raf proteins. Nature 312:558–561

Mor MA, Vila J, Ciudad CJ, Guinouart J (1981) Insulin inactivation of rat hepatocyte cyclic AMP-dependent protein kinase. FEBS Lett 136:131–134

Morgan DO, Roth RA (1981) Acute insulin action requires insulin receptor kinase activity: introduction of an inhibitory monoclonal antibody into mammalian cells blocks the rapid effects of insulin. Proc Natl Acad Sci USA 84:41–45

Morgan DO, Ho L, Korn LJ, Roth RA (1986a) Insulin action is blocked by a monoclonal antibody that inhibits the insulin receptor kinase. Proc Natl Acad Sci USA 83:328–332

Morgan DO, Roth RA (1986b) Mapping surface structures of the human insulin receptor with monoclonal antibodies: localization of main immunogenic regions to the receptor kinase domain. Biochemistry 25:1364–1371

Morrison BD, Pessin JE (1987) Insulin stimulation of the insulin receptor kinase can occur in the complete absence of beta subunit autophosphorylation. J Biol Chem 262:2861–2868

Morrison DK, Kaplan DR, Rapp U, Roberts TM (1988) Signal transduction from membrane to cytoplasm: growth factors and membrane-bound oncogene products increase Raf-1 phosphorylation and associated protein kinase activity. Proc Natl Acad Sci USA 85:8855–8859

Nakamura KP, Martinez R, Weber MJ (1983) Tyrosine phosphorylation of specific proteins after mitogen stimulation of chicken embryo fibroblasts. Mol Cell Biol 3:380–390

Nemenoff RA, Blackshear PJ, Avruch J (1983) Hormonal regulation of protein dephosphorylation. Identification and hormonal regulation of inhibitor-1 in rat adipose tissue. J Biol Chem 258:9537–9543

Nemenoff RA, Kwok YC, Shulman GI, Blackshear PJ, Osathanondh R, Avruch J (1984) Insulin-stimulated tyrosine protein kinase: characterization and relation to the insulin receptor. J Biol Chem 259:5058–5065

Nemenoff RA, Gunsalus JR, Avruch J (1986) An insulin-stimulated (ribosomal S6) protein kinase from soluble extracts of H4 hepatoma cells. Arch Biochem Biophys 245:196–203

Nemenoff RA, Mendelsohn JM, Gunsalus JR, Carter E, Avruch J (1988) An S6 kinase in H4 hepatoma cells. J Biol Chem 263:19 455–19 560

Nielsen PJ, Manchester KL, Towbin H, Gordon J, Thomas G (1982) The phosphorylation of ribosomal protein S6 in rat tissues following cycloheximide injection, in diabetes, and after denervation of diaphragm. J Biol Chem 257:12 316–12 321

Nishizuka Y (1988) The molecular heterogeneity of protein kinase and its implications for cellular regulation. Nature 334:661–665

Novak-Hofer I, Thomas G (1984) An activated S6 kinase in extracts from serum- and epidermal growth factor-stimulated Swiss 3T3 cells. J Biol Chem 259:5995–6000

Ohno S, Akita Y, Konno Y, Imajoh S, Suzuki K (1988) A novel phorbol ester receptor/protein kinase, nPKC, distantly related to the protein kinase C family. Cell 53:731–741

Olmsted JB (1986) Microtubule-associated proteins. Annu Rev Cell Biol 2:421–457

Pang DT, Sharma BB, Shafer JA (1985) Purification of the catalytic active phosphorylated form of insulin receptor kinase by affinity chromatography with O-phosphotyrosyl-binding antibodies. Arch Biochem Biophys 242:176–186

Parker PJ, Caudwell B, Cohen P (1983) Glycogen synthase from rabbit skeletal muscle: effect of insulin on the state of phosphorylation of the seven phosphoserine residues in vivo. Eur J Biochem 130:227–234

Pelech SL, Krebs EG (1987) Mitogen-activated S6 kinase is stimulated via protein kinase C-dependent and independent pathways in Swiss 3T3 cells. J Biol Chem 262:11 598–11 606

Perisic O, Traugh JA (1983) Protease-activated kinase II as the potential mediator of insulin-stimulated phosphorylation of ribosomal protein S6. J Biol Chem 258:9589–9592

Perrotti N, Accili D, Marcus-Samuels B, Rees-Jones RW, Taylor SI (1987) Insulin stimulates phosphorylation of a 120-kDa glycoprotein substrate (pp120) for the receptor-associated protein kinase in intact H-35 hepatoma cells. J Biol Chem 84:3137–3140

Petruzzelli L, Herrera R, Rosen OM (1984) Insulin receptor is an insulin-dependent tyrosine protein kinase: copurification of insulin-binding activity and protein kinase activity to homogeneity from human placenta. Proc Natl Acad Sci USA 81:3327–3331

Phillips SA, Perrotti N, Taylor SI (1987) Rat liver membranes contain a 120 kDa glycoprotein which serves as a substrate for the tyrosine kinases of the receptors for insulin and epidermal growth factor. FEBS Lett 212:141–144

Picton C, Woodgett J, Hemmings B, Cohen P (1982a) Multisite phosphorylation of glycogen synthase from rabbit skeletal muscle. Phosphorylation of site 5 by glycogen synthase kinase-5 (casein kinase-II) is a prerequisite for phosphorylation of sites 3 by glycogen synthase kinase-3. FEBS Lett 150:191–196

Picton C, Aitken A, Bilham T, Cohen P (1982b) Multiple phosphorylation of glycogen synthase from rabbit skeletal muscle. Organisation of the seven sites in the polypeptide chain. Eur J Biochem 124:37–45

Pierce MW, Palmer JL, Keutmann HT, Avruch J (1981) ATP-citrate lyase. Structure of a tryptic peptide containing the phosphorylation site directed by glucagon and the cAMP-dependent protein kinase. J Biol Chem 256:8867–8870

Pierce MW, Palmer JL, Keutmann HT, Hall TA, Avruch J (1982) The insulin-directed phosphorylation site of ATP-citrate lyase is identical to the site phosphorylated by the cAMP-dependent protein kinase in vitro. J Biol Chem 257:10681–10685

Pike LJ, Kuenzel AE, Casnellie JE, Krebs EG (1984) A comparison of the insulin and epidermal growth factor-stimulated protein kinases from human placenta. J Biol Chem 259:9913–9921

Poulter L, Ang S-G, Gibson BW, Williams DH, Holmes CFB, Caudwell FB, Pitcher J, Cohen P (1988) Analysis of the in vivo phosphorylation state of rabbit skeletal muscle glycogen synthase by fast-atom-bombardment mass spectrometry. Eur J Biochem 175:497–510

Price DJ, Nemenoff RA, Avruch J (1989) Purification of a hepatic S6 kinase from cycloheximide-treated rats. Biol Chem 264:13825–13833

Ramakrishna S, Benjamin WB (1979) Fat cell protein phosphorylation: identification of phosphoprotein-2 as ATP-citrate lyase. J Biol Chem 254:9232–9237

Ramakrishna S, Benjamin WB (1985) Cyclic nucleotide-independent protein kinase from rat liver. Purification and characterization of a multifunctional protein kinase. J Biol Chem 260:12280–12286

Ramakrishna S, Benjamin WB (1988) Insulin action rapidly decreases multifunctional protein kinase activity in rat adipose tissue. J Biol Chem 263:12677–12681

Ramakrishna S, Pucci DL, Benjamin WB (1983) Dependence of ATP-citrate lyase kinase activity on the phosphorylation of ATP-citrate lyase by cyclic AMP-dependent protein kinase. J Biol Chem 258:4950–4956

Ray LB, Sturgill TW (1987) Rapid stimulation by insulin of a serine/threonine kinase in 3T3-L1 adipocytes that phosphorylates microtubule-associated protein 2 in vitro. Proc Natl Acad Sci USA 84:1502–1506

Ray LB, Sturgill TW (1988a) Insulin-stimulated microtubule-associated protein kinase is phosphorylated on tyrosine and threonine in vivo. Proc Natl Acad Sci USA 85:3753–3757

Ray LB, Sturgill TW (1988b) Characterization of insulin-stimulated microtubule-associated protein kinase. Rapid isolation and stabilization of a novel serine/threonine kinase from 3T3-L1 cells. J Biol Chem 263:12721–12727

Ray LB, Sturgill TW (1988c) Insulin-stimulated microtubule-associated protein kinase is detectable by analytical gel chromatography as a 35-kDa protein in myocytes, adipocytes, and hepatocytes. Arch Biochem Biophys 262:307–313

Rees-Jones RW, Taylor SI (1985) An endogenous substrate for the insulin receptor-associated tyrosine kinase. J Biol Chem 260:4461–4467

Rosen OM (1987) After insulin binds. Science 237:1452–1458

Rosen OM, Herrera R, Olowe Y, Petruzzelli LM, Cobb MH (1983) Phosphorylation activates the insulin receptor tyrosine protein kinase. Proc Natl Acad Sci USA 80:3237–3240

Ross AH, Baltimore D, Eisen HN (1981) Phosphotyrosine-containing proteins isolated by affinity chromatography with antibodies to a synthetic hapten. Nature 294:654–656

Rothenberg PL, Kahn CR (1988) Insulin inhibits pertussis toxin-catalyzed ATP-ribosylation of G-proteins. Evidence for a novel interaction between insulin receptors and G-proteins. J Biol Chem 263:15546–15552

Ruetz S, Fricker G, Hugentobler G, Winterhalter K, Kurz G, Meier PJ (1987) Isolation and characterization of the putative canalicular bile salt transport system of rat liver. J Biol Chem 262:11324–11330

Sacks DB, McDonald JM (1988) Insulin-stimulated phosphorylation of calmodulin by rat liver insulin receptor preparations. J Biol Chem 263:2377–2383

Sadoul S-L, Peyron J-F, Ballotti R, Debant A, Fehlmann M, van Obberghen E (1985) Identification of a cellular 110000-Da protein substrate for the insulin receptor kinase. Biochem J 227:887–892

Sagata N, Oskarsson M, Copeland T, Brumbaugh J, Vande Woude GF (1988) Function of c-mos proto-oncogene product in meiotic maturation in Xenopus oocytes. Nature 335:519–525

Sahal D, Ramachandran J, Fugita-Yamaguchi Y (1988) Specificity of tyrosine protein kinases of the structurally related receptors for insulin and insulin-like growth factor I: Tyr containing synthetic polymers as specific inhibitor substrates. Arch Biochem Biophys 260:416–426

Saris CJM, Tack B, Kristensen T, Glenney JR Jr, Hunter T (1986) The cDNA sequence for the protein-tyrosine kinase substrate (calpactin heavy chain) reveals a multidomain protein with internal repeats. Cell 46:201–212

Schlaepfer DD, Haigler HT (1987) Characterization of Ca^{2+}-dependent phospholipid binding and phosphorylation of lipocortin I. J Biol Chem 262:6931–6937

Shaw JF, Chou I-N, Anand B (1988) Rapid phosphorylation of microtubule-associated proteins through distinct mitogenic pathways. J Biol Chem 263:1459–1466

Sheorain VS, Ramakrishna S, Benjamin WB, Soderling TR (1985) Phosphorylation of sites 3 and 2 in rabbit skeletal muscle glycogen synthase by multifunctional protein kinase (ATP-citrate lyase kinase). J Biol Chem 260:12287–12292

Siddle K, Soos MA, O'Brien RM, Ganderton RH, Taylor R (1986) Monoclonal antibodies as probes of the structure and function of insulin receptors. Biochem Soc Trans 15:47–51

Smart JE, Oppermann H, Czernilofsky AP, Purchio A, Erikson RL, Bishop JM (1981) Characterization of sites for tyrosine phosphorylation in the transforming protein of Rous sarcoma virus ($pp60^{v-src}$) and its normal cellular homologue ($pp60^{c-src}$). Proc Natl Acad Sci USA 78:6013–6017

Smith CJ, Wejksnora PJ, Warner JR, Rubin CS, Rosen OM (1979) Insulin-stimulated protein phosphorylation in 3T3 preadipocytes. Proc Natl Acad Sci USA 76:2725–2729

Smith CJ, Rubin CS, Rosen OM (1980) Insulin-treated 3T3-L1 adipocytes and cell-free extracts derived from them incorporated ^{32}P into ribosomal protein S6. Proc Natl Acad Sci USA 77:2641–2645

Smith RL, Roach PJ, Lawrence JC Jr (1988) Insulin resistance in denervated skeletal muscle. Inability of insulin to stimulate dephosphorylation of glycogen synthase in denervated rat epitrochlearis muscle. J Biol Chem 263:658–665

Sommercorn J, Krebs EG (1987a) Classification of protein kinases into messenger-dependent and independent kinases. The regulation of independent kinases. In: Zappia V (ed) Post-translational modifications of proteins and ageing. Plenum, New York

Sommercorn J, Krebs EG (1987b) Induction of casein kinase II during differentiation of 3T3-L1 cells. J Biol Chem 262:3839–3943

Sommercorn J, Mulligan JA, Lozeman FJ, Krebs EG (1987) Activation of casein kinase-II in response to insulin and to epidermal growth factor. Proc Natl Acad Sci USA 84:8834–8838

Spach DH, Nemenoff RA, Blackshear PJ (1986) Protein phosphorylation and protein kinase activities in BC3H-1 myocytes. Differences between the effects of insulin and phorbol esters. J Biol Chem 261:12750–12753

Spivack JA, Erikson RL, Maller JL (1984) Microinjection of $pp60^{v-src}$ into *Xenopus* oocytes increases phosphorylation of ribosomal protein S6 and accelerates the rate of progesterone-induced meiotic maturation. Mol Cell Biol 4:1631–1634

Stadtmauer L, Rosen OM (1986) Phosphorylation of synthetic insulin receptor peptides by the insulin receptor kinase and evidence that the preferred sequence containing Tyr-1150 is phosphorylated in vivo. J Biol Chem 261:10000–10005

Steele-Perkins G, Turner J, Edman JC, Hari J, Pierce SBV, Stover C, Rutter WJ, Roth RA (1988) Expression and characterization of a functional human insulin-like growth factor I receptor. J Biol Chem 263:11486–11492

Stralfors P (1987) Isoproterenol and insulin control the cellular localization of ATP citrate-lyase through its phosphorylation in adipocytes. J Biol Chem 262:11486–11489

Sturgill TW, Ray LB (1986) Muscle proteins related to microtubule associated protein-2 are substrates for an insulin-stimulatable kinase. Biochem Biophys Res Commun 134:565–571

Sturgill TW, Ray LB, Erikson E, Maller JL (1988) Insulin-stimulated MAP-2 kinase phosphorylates and activates ribosomal protein S6 kinase II. Nature 334:715–718

Tabarini D, Heinrich J, Rosen IM (1985) Activation of S6 kinase activity in 3T3-L1 cells by insulin and phorbol esters. Proc Natl Acad Sci USA 82:4369–4373

Tabarini D, Garcia de Herreros A, Heinrich J, Rosen OM (1987) Purification of a bovine liver S6 kinase. Biochem Biophys Res Commun 144:891–899

Takio K, Kuenzel EA, Walsh KA, Krebs EG (1987) Amino acid sequence of the beta subunit of bovine lung casein kinase II. Proc Natl Acad Sci USA 84:4851–4855

Taylor R, Soos MA, Wells A, Argyraki M, Siddle K (1987) Insulin-like and insulin-inhibitory effects of monoclonal antibodies for different epitopes on the human insulin receptor. Biochem J 242:123–129

Thomas AP, Martin-Requero A, Williamson JR (1985) Interactions between insulin and alpha-adrenergic agents in the regulation of glycogen metabolism in isolated adipocytes. J Biol Chem 260:5963–5973

Thomas AP, Diggle TA, Denton RM (1986) Sensitivity of pyruvate dehydrogenase phosphate phosphatase to magnesium ions. Similar effects of spermine and insulin. Biochem J 238:83–91

Thyberg J (1984) The microtubule cytoskeleton and the initiation of DNA synthesis. Exp Cell Res 155:1–8

Tornqvist HE, Avruch J (1988) Relationship of site-specific beta subunit tyrosine autophosphorylation to insulin activation of the insulin receptor (tyrosine) protein kinase activity. J Biol Chem 263:4593–4601

Tornqvist HE, Pierce MW, Frackelton AR, Nemenoff RA, Avruch J (1987) Identification of insulin receptor tyrosine residues autophosphorylated in vitro. J Biol Chem 262:10212–10219

Tornqvist HE, Gunsalus JR, Nemenoff RA, Frackelton AR, Pierce MW, Avruch J (1988) Identification of the insulin receptor tyrosine residues undergoing insulin-stimulated phosphorylation in intact rat hepatoma cells. J Biol Chem 263:350–359

Toth B, Bollen M, Stalmans W (1988) Acute regulation of hepatic protein phosphatases by glucagon, insulin, and glucose. J Biol Chem 263:14061–14066

Traugh JA, Pendergast AM (1986) Regulation of protein synthesis by phosphorylation of ribosomal protein S6 and aminoacyl-tRNA synthetases. Prog Nucleic Acid Res 33:195–230

Ullrich A, Bell JR, Chen EY, Herrera R, Petruzzelli LM, Dull TG, Gray A et al. (1985) Human insulin receptor and its relationship to the tyrosine kinase family of oncogenes. Nature 313:756–761

Ullrich A, Gray A, Tam AW, Yang-Feng T, Tsubokawa M, Collins C, Henzel W et al. (1986) Insulin-like growth factor I receptor primary structure: comparison with insulin receptor suggests structural determinants that define functional specificity. EMBO J 5:2503–2512

Ushiro H, Cohen S (1980) Identification of phosphotyrosine as a product of epidermal growth factor-activated protein kinase in A-431 cell membranes. J Biol Chem 255:8363–8365

Wahl MI, Daniel TO, Carpenter D (1988) Antiphosphorylation recovery of phospholipase C activity after EGF treatment of A-431 cells. Science 241:968–970

Walkenbach RJ, Hazen R, Larner J (1978) Reversible inhibition of cyclic AMP-dependent protein kinase by insulin. Mol Cell Biochem 19:31–41

Wang JYC (1985) Isolation of antibodies for phosphotyrosine by immunization with a v-abl oncogene-encoded protein. Mol Cell Biol 5:3640–3643

White MF, Maron R, Kahn CR (1985) Insulin rapidly stimulates tyrosine phosphorylation of a M_r-185000 protein in intact cells. Nature 318:183–186

White MF, Stegmann EW, Dull TJ, Ullrich A, Kahn CR (1987) Characterization of an endogenous substrate of the insulin receptor in cultured cells. J Biol Chem 262:9769–9777

White MF, Shoelson SE, Keutmann H, Kahn CR (1988a) A cascade of tyrosine autophosphorylation in the beta-subunit activates the phosphotransferase of the insulin receptor. J Biol Chem 263:2969–2930

White MF, Livingston JJ, Backer JM, Lauris V, Dull T, Ullrich A, Kahn CR (1988b) Mutation of the insulin receptor at tyrosine 960 inhibits signal transmission but does not affect its tyrosine kinase activity. Cell 54:641–649

Whitman M, Downes CP, Keeler M, Keller T, Cantley L (1988) Type I phosphatidylinositol kinase makes a novel inositol phospholipid, phosphatidylinisitol-3-phosphate. Nature 332:644–646

Witters LA (1981) Insulin stimulates the phosphorylation of acetyl-CoA carboxylase. Biochem Biophys Res Commun 100:872–878

Witters LA, Moriarity D, Martin DB (1979) Regulation of hepatic acetyl-CoA carboxylase by changes in the polymeric state of the enzyme. J Biol Chem 254:6644–6649

Witters LA, Watts TD, Daniels DL, Evans JL (1988) Insulin stimulates the dephosphorylation and activation of acetyl-CoA carboxylase. Proc Natl Acad Sci USA 85:5473–5477

Wong ECC, Sacks DB, Laurino JP, McDonald JM (1988) Characteristics of calmodulin phosphorylation by the insulin receptor kinase. Endocrinology 123:1830–1836

Wool IG (1979) The structure and function of eukaryotic ribosomes. Annu Rev Biochem 48:719–754

Yang S-D, Fong Y-L, Yu J-S, Liu J-S (1987a) Identification and characterization of a phosphorylation-activated cyclic AMP and Ca^{2+}-independent protein kinase in the brain. J Biol Chem 262:7034–7040

Yang S-D, Chang S-Y, Soderling TR (1987b) Characterization of an autophosphorylation-dependent multifunctional protein kinase from liver. J Biol Chem 262:9421–9427

Yang S-D, Ho L-T, Fung TJ (1988) Insulin induces activation and translocation of protein kinase F_A (multifunctional protein phosphatase activator) in human platelet. Biochem Biophys Res Commun 151:61–69

Yarden Y, Ullrich A (1988) Growth factor receptor tyrosine kinases. Annu Rev Biochem 57:443–478

Yu K-T, Czech MP (1984) Tyrosine phosphorylation of the insulin receptor beta subunit activates the receptor-associated tyrosine kinase activity. J Biol Chem 259:5277–5286

Yu K-T, Czech MP (1986) Tyrosine phosphorylation of insulin receptor beta subunit activates the receptor tyrosine kinase in intact H-35 hepatoma cells. J Biol Chem 261:4715–4722

Yu K-T, Khalaf N, Czech MP (1987a) Insulin stimulates a membrane-bound serine kinase that may be phosphorylated on tyrosine. Proc Natl Acad Sci USA 84:3972–3976

Yu K-T, Khalaf N, Czech MP (1987b) Insulin stimulates a novel Mn^{2+}-dependent cytosolic serine kinase in rat adipocytes. J Biol Chem 262:16677–16685

Zick Y, Grunberger G, Podskalny JM, Moncada V, Taylor SI, Gordon P, Roth J (1983) Insulin stimulates phosphorylation of serine residues in soluble insulin receptors. Biochem Biophys Res Commun 116:1129–1135

Zick Y, Grunberger G, Rees-Jones RW, Comi R (1985) Use of tyrosine-containing polymers to characterize the substrates specificity of insulin and other hormone-stimulated tyrosine kinases. Eur J Biochem 148:177–182

CHAPTER 16

Effects of Insulin on Glycogen Metabolism

J. LARNER

A. Introduction

I. Historical

Since its discovery by Claude BERNARD (1857), glycogen and its synthesis have become closely intertwined with insulin action. In 1921, insulin was purified based on its effect to decrease blood glucose. Two years later Banting and Best reported that insulin increased both glycogen and fat in several tissues, together with increasing the respiratory quotient (see BEST 1963). This seminal early work clearly established that insulin promoted intracellular glucose disposal via two anabolic reactions and that the required energy was derived from increased glucose oxidation. Today, increased glycogen synthesis is recognized as a universal intracellular metabolic marker of insulin action while increased glucose transport and fat synthesis occur only in specific insulin-sensitive cells.

The discovery of phosphorolysis, glucose 1-P, and phosphorylase by PARNAS (1938), HANES (1940), and CORI and CORI (1936) demonstrated that a macromolecule resembling glycogen ("blue" glycogen) could be synthesized in vitro in the presence of a primer, glucose 1-P and phosphorylase. However, from metabolite measurements, it became clear that phosphorylase could not function biosynthetically in vivo (LARNER et al. 1960). The high millimolar concentrations of P_i compared with the low micromolar concentrations of glucose 1-P in tissues would drive the phosphorylase equilibrium reaction toward glycogen degradation. A separate pathway bypassing P_i was clearly required for synthesis. LELOIR and CARDINI (1957) were the first to identify glycogen synthase, an enzyme which utilized UDPG as glucosyl donor in liver extracts. VILLAR-PALASI and LARNER (1960a) next established the complete biosynthetic pathway by identifying UDPG pyrophosphorylase as the precursor-forming enzyme together with glycogen synthase. They described a glycogen cycle composed of separate biosynthetic and degradative reactions (VILLAR-PALASI and LARNER 1960a). From metabolite measurements and balance studies, LARNER et al. (1960) proposed an intracellular action of insulin to promote glycogen synthesis directly, i.e, independent of glucose transport, and located the activation step between glucose 1-P and glycogen. Direct activity measurements of the glycogen cycle enzymes in extracts of diaphragm muscle pretreated with insulin demonstrated that insulin had activated only one enzyme of four tested in a stable manner; namely, glycogen synthase (VILLAR-PALASI and LARNER 1960b). This activation occurred with no change in "total" enzyme activity, i.e, the activity measured in the presence of the allosteric activator glucose 6-P. Therefore, an interconversion between inactive and active forms was postulated. The interconversion by phosphorylation and dephosphorylation was next demonstrated by FRIEDMAN and LARNER (1963) who identified a protein kinase and a phosphoprotein phosphatase catalyzing the two interconversion reactions. The intracellular mechanism of insulin's activation of glycogen synthase was fully established by direct chemical analysis of phosphate in the homogeneous muscle enzyme. Glycogen synthase isolated from rabbits treated with insulin in vivo had a decreased covalent phosphorylation state associated with increased activity ROACH et al. (1977; see Sect. B. II).

Interestingly, the control of glycogen synthase by phosphorylation was opposite in directionality to the two previously known examples, phosphorylase and phosphorylase b kinase. Subsequently, a similar dephosphorylation control mechanism was found by LINN et al. (1969) for the rate-limiting enzyme of fat synthesis, mitochondrial pyruvate dehydrogenase. The insulin activation of both glycogen and fat synthesis by a similar covalent dephosphorylation mechanism provided a biochemical explanation of the very first findings of Banting and Best in 1923 (see BANTING et al. 1963).

II. Research Approach

Our overall approach to gain an understanding of the mechanism by which insulin controls glycogen synthesis was to trace backward from our marker enzyme, dephosphorylated glycogen synthase, the biochemical steps to the cell membrane and the set of events initiated by insulin's binding to its receptor. We had already described the kinase and the phosphatase responsible for phosphorylating and dephosphorylating glycogen synthase. The kinase was rapidly shown to be distinct from phosphorylase b kinase (FRIEDMAN and LARNER 1966) and stimulated by cAMP in the micromolar range in a dose-dependent manner (HUIJING and LARNER 1966). It is now known as cAMP-dependent protein kinase (WALSH et al. 1968). We carefully demonstrated in many published experiments that insulin activated glycogen synthase without detectably decreasing basal tissue cAMP concentrations (see LARNER 1988). Next, we demonstrated that the cAMP-dependent protein kinase itself was a target of insulin action. Insulin decreased the activity (SHEN et al. 1970) and the sensitivity (WALKENBACH et al. 1980) of the cAMP-dependent protein kinase to existing cAMP concentrations. Experiments with anti-insulin receptor antibody demonstrated that this immunoglobulin activated intracellular glycogen synthase indistinguishably from insulin, implicating a receptor-initiated signaling mechanism (LAWRENCE et al. 1978). Based upon the cAMP second messenger model, we set out to seek a possible mediator with an assay designed to detect inhibition of the cAMP-dependent protein kinase in an insulin-selective manner. Two insulin mediators have now been found, purified, and their generation and action studied (see Sects. B. IX and B. X). One inhibits the cAMP-dependent protein kinase and adenylate cyclase and thus leads presumably to activation of glycogen synthase. The other activates mitochondrial pyruvate dehydrogenase by stimulating PDH phosphatase. Both act in vitro in a manner similar to insulin. The identification of the two mediators and their sites of action further illuminates the signaling mechanism used by insulin to activate these two distinct anabolic pathways via dephosphorylation.

B. Experimental

I. Phosphorylation Sites of Glycogen Synthase

We first reported that glycogen synthase contained multiple sites of serine phosphorylation (SMITH et al. 1971) an observation based on direct chemical analysis of the enzyme. These results indicated that a stoichiometry of 7 alkali-labile P/100000 or 6/85000, subunit molecular weight existed. These phosphates then were conclusively proven to be present in a single polypeptide chain (TAKEDA et al. 1975). Subsequently, amino acid sequence studies have provided a working map of the phosphorylated sites of rabbit muscle glycogen synthase (SODERLING et al. 1977; PICTON et al. 1982a).

As shown in Fig. 1 all the phosphorylated sites are contained on two cyanogen bromide fragments CB1 and CB2. CB1 contains sites 2A and 2B which derive from the NH_2 terminal end. CB2 which derives from the COOH terminal end contains sites 1A and B, 3A, B, and C, 4, and 5. Thus, a total of nine sites have been identified and these are capable of being phosphorylated by at least ten known kinases. Table 1 lists the kinases and their preferred phosphorylation sites. While there is some overlap in specificity, clearly casein kinase 1 is the least specific, glycogen synthase kinase 2, GSK-3, the most specific, and the other kinases, including cAMP-dependent protein kinase, have a moderate, but not absolute degree of specificity.

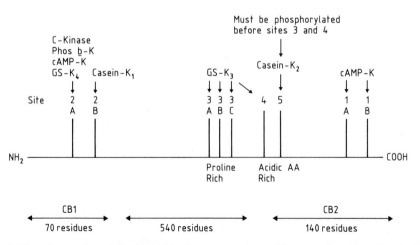

Fig. 1. Phosphorylation sites on muscle glycogen synthase. The nine sites (1A, 1B, 2A, 2B, 3A, 3B, 3C, 4, and 5) are located in the two cyanogen bromide fragments (*CB1* and *CB2*). Kinases responsible for phosphorylation of respective sites: *cAMP-K*, cAMP-dependent protein kinase; *GS-K₃*, glycogen synthase kinase 3; *C-Kinase*, Ca^{2+}, diacylglycerol, and phospholipid-stimulated C kinase; *Phos b-K*, phosphorylase *b* kinase; *GS-K₄*, glycogen synthase kinase 4; *Casein-K₁* casein kinase 1; *Casein-K₂*, casein kinase 2: Casein-K_1 can phosphorylate a number of other sites in the molecule as well

Table 1. Ten protein kinases phosphorylating glycogen synthase in vitro

Kinase	Molecular weight	Activators	Sites in synthase phosphorylated	Degree of inactivation of synthase
cAMP-dependent	170 000	cAMP	1A, 1B, 2, 4	+
cGMP-dependent	160 000	cGMP	1A, 1B, 2	+
Phosphorylase b kinase	1 300 000	Ca^{2+} Glycogen	2	+
Ca^{2+} (calmodulin-dependent)	275 000–850 000	Ca^{2+} Calmodulin	2, 1B	+
Glycogen synthase kinase 3	57 000	?	3A, 3B, 3C	+ + +
Glycogen synthase kinase 4	100 000	?	2	+
Casein kinase 2	150 000	Polyamines	5	− − −
Casein kinase 1	30 000	?	At least 9	+ + + +
Protein kinase C	82 000	Diacylglycerol Phorbol esters	1A	+
Heparin-activated	70 000	Heparin	?	+

II. Decreased Phosphorylation of Glycogen Synthase with Insulin Action

While there is general agreement that glycogen synthase becomes dephosphorylated with insulin action, there is no agreement on which site or sites are dephosphorylated. We demonstrated a decrease of 0.5 mol P/mol subunit with insulin action in vivo in rabbit skeletal muscle glycogen synthase (ROACH et al. 1977), a magnitude of decrease that has been observed by all subsequent workers. Cohen's laboratory has concluded that sites 3A, B, C are the *only* sites dephosphorylated in the enzyme with insulin action (PARKER et al. 1983), while LAWRENCE et al. (1983) and SHEORAIN et al. (1984) have noted a "spread" of dephosphorylation to include sites in both CB1 and CB2. Careful examination of data from Cohen's laboratory reveals smaller decreases in phosphorylation in sites other than 3A, B, C; however, these were not considered to be statistically significant. Thus, while the evidence is still incomplete and the question controversial, the possibility of multiple site dephosphorylation with insulin action must be kept in mind. In vivo experiments with epinephrine have clearly established a pattern of increased multisite phosphorylation (PARKER et al. 1982). Since epinephrine acts via adenylate cyclase, cAMP, and cAMP-dependent protein kinase activation, and insulin is known to cause both inhibition and desensitization of the cAMP-dependent protein kinase to cAMP, how can the spread of increased phosphorylation and the potential spread of decreased phosphorylation on glycogen synthase with epinephrine and insulin action be explained?

While many complex interlocking mechanisms can be envisioned, one simple mechanism involving control of the phosphoprotein phosphatases can be proposed involving the initial activation or inhibition of the cAMP-dependent protein kinase and a subsequent "feed forward" effect on the phosphoprotein phosphatase via the kinase acting on inhibitor 1. This inhibitor which is active only when phosphorylated by the cAMP-dependent protein kinase, would thus become an active inhibitor with epinephrine action. Conversely, with insulin action, inhibitor 1 would remain unphosphorylated and inactive, allowing the phosphatase activity to increase. The activity state of the phosphatase would thus be modified by the activity state of the cAMP-dependent protein kinase and thus allow a spread of the phosphorylation or dephosphorylation sites to occur since the phosphatases are recognized as having less site specificity than kinases (INGEBRITSEN and COHEN 1983). For an additional insulin-cAMP kinase mechanism enhancing phosphatase activity see Sect. B. VII.

III. Integration of Covalent and Allosteric Controls of Phosphorylase and Glycogen Synthase

What have we learned about the coordination of allosteric and covalent controls of phosphorylase and glycogen synthase? It has become clear with both enzymes that the active forms are not as sensitive to the allosteric controls as the inactive forms are. This was first established for phosphorylase by MORGAN and PARMEG-GIANI (1964) and for glycogen synthase by PIRAS and STANELONI (1969). Thus, enzyme phosphorylation state alters enzyme sensitivity to allosteric modifiers. Phosphorylase b is stimulated by AMP and this stimulation is reversed by glucose 6-P. Glycogen synthase D is stimulated by glucose 6-P and this stimulation is reversed by nucleotides including UMP or AMP. Thus, both inactive forms are markedly sensitive to the allosteric controls. The two active forms on

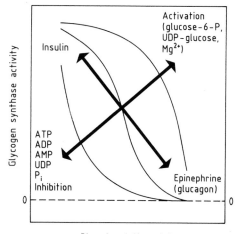

Fig. 2. Model for the interaction of hormonal and small molecule effectors in the regulation of glycogen synthase. The diagram shows schematically a summary of the small molecule effects on glycogen synthase in relation to the phosphorylation state of the enzyme. Effectors are considered to shift the position of the enzyme activity between the phosphorylation curves in the sense indicated. The effects of hormones are considered to act through phosphorylation, causing progress along the activity–phosphorylation curve as indicated

the other hand, phosphorylase a, and glycogen synthase I are at best minimally sensitive to the allosteric controls. With graded phosphorylation of the multiple sites in glycogen synthase, a graded response and sensitivity to allosteric modifiers is achieved (ROACH and LARNER 1976). A model of the inter-relationship of phosphorylation state controlled by hormones and sensitivity to allosteric modifiers is shown in Fig. 2. The hormones act along the full lines, the small molecules shift the enzyme activity between the full lines. With plentiful energy (upper curve), epinephrine or glucagon would be required to phosphory-late the enzyme markedly to decrease its activity. However, activation of the en-zyme could be obtained even at a relatively high phosphorylation state. Con-versely with diminished energy (lowest curve), insulin would be required to decrease phosphorylation state considerably more to activate glycogen synthesis.

IV. Emerging Significance of Multiple Phosphorylation

In addition to producing a more graded response than single phosphorylation, multiple phosphorylation has been shown to be used for a rather unexpected on–off regulatory property. In three separate cases of multiple phosphorylation, it has been shown that one specific site *must* be phosphorylated prior to the phosphorylation of adjoining sites in the molecule. This was first observed with glycogen synthase where site 5 was shown to be obligatorily phosphorylated by casein kinase 2 prior to phosphorylation of sites 4, and 3A, B, C by GSK-3 (DePaoli-Roach et al. 1983; Picton et al. 1982 b). A similar set of circumstances occurs for type 2 regulatory subunit of cAMP-dependent protein kinase and for phosphoprotein phosphatase inhibitor 2. Very recently using a synthetic peptide containing glycogen synthase sites 5, 4, 3C, B, and A, Fiol et al. (1987) demon-strated that the regularly spaced serine phosphorylation sites are phosphorylated in a distinct order beginning with site 5 phosphorylated by casein kinase 2 and proceeding regularly through sites 4, 3C, 3B, and 3A in the NH_2 terminal direc-tion, catalyzed by GSK-3.

V. Insulin and Tissue cAMP Concentrations

It is now well established that insulin can and does act without altering basal tis-sue cAMP concentrations (see LARNER 1988). On the other hand, insulin can decrease elevated concentrations of cAMP (see LARNER 1988). Insulin has also been shown to delay cAMP accumulation in tissue treated with epinephrine (CRAIG et al. 1969). All of these effects can now be explained by the generation of a mediator which inhibits the catalytic subunit of adenylate cyclase (see LARNER 1988).

VI. Insulin and cAMP-Dependent Protein Kinase: A Marker for Mediator

The first demonstration of insulin's effect to inactivate the cAMP-dependent protein kinase was by VILLAR-PALASI and WENGER (1967) in skeletal muscle in vivo. This was confirmed and extended by SHEN et al. (1970) and by WALAAS et

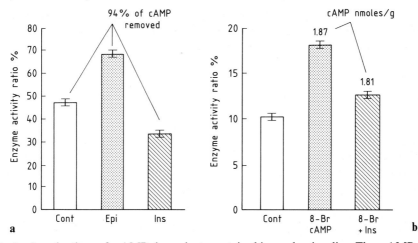

Fig. 3 a, b. Inactivation of cAMP-dependent protein kinase by insulin. The cAMP-dependent protein kinase activity ratio is measured as ±cAMP and expressed as a percentage, 8Br-cAMP, 8-bromo-cAMP; CONT, control; EPI, epinephrine; Ins, insulin. **a** from Shen et al. (1970), **b** from Gabbay and Lardy (1984)

al. (1972) in rat diaphragm. Most recently, experiments in fat and liver have produced similar results. Figure 3 compares the inactivation of the enzyme in diaphragm (Shen et al. 1970) with that in hepatocytes (Gabbay and Lardy 1984). The "total" enzyme activity measaured with excess cAMP was not affected by insulin while the activity ratio ($-$cAMP/$+$cAMP) was decreased. Further analytical data on gels demonstrate directly that the enzyme was in-

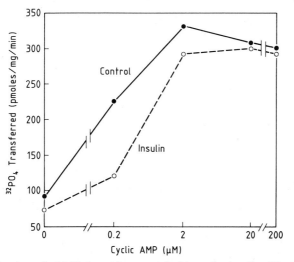

Fig. 4. Desensitization of cAMP-dependent protein kinase by insulin. Diaphragm extracts were prepared from control and insulin-treated tissue and assayed for cAMP-dependent protein kinase with various concentrations of added cAMP as described by Walkenbach et al. (1980)

activated by insulin by conversion to its inactive holoenzyme form (CUIDAD et al. 1987). In addition, it is clear that the enzyme has also been desensitized to the activating effect of cAMP (Fig. 4; WALKENBACH et al. 1980) an effect also recently demonstrated by GABBAY and LARDY (1987). All of the data suggest that a modified holoenzyme has been produced by insulin action. Perhaps this modification resides in an altered phosphorylation state. The insulin-specific inhibition of the cAMP-dependent protein kinase was of great significance since it became the first bioassay for the putative insulin mediator.

VII. Effects of Insulin on Other Kinases: A Potential Phosphorylation Cascade

Effects of insulin on three other kinases will be discussed which may be related to receptor inactivation, increased phosphorylation via a cascade, and control of dephosphorylation. The interaction of insulin and protein kinase C has evoked considerable interest. Through the work of FARESE et al. (1984), SALTIEL et al. (1986) and others, it is now recognized that insulin action leads to rapid increases in tissue diacylglycerol. Within 0.5–1 min there is a "burst" increase, followed by a decrease and a subsequent slower increase. LUTTRELL et al. (1988) have shown that pertussis toxin pretreatment abolishes the effect, implicating a G-like protein in the mechanism. The diacylglycerol could potentially originate from: (a) cleavage of phospholipids by phospholipase C activation; (b) de novo synthesis; (c) cleavage of glycophospholipids by an inositol-specific phospholipase C; and (d) cleavage of protein-linked glycophospholipids by an inositol-specific phospholipase C. Whatever the source(s), the diacylglycerol produced can potentially activate one or more of the several protein kinase C isozymes which then can phosphorylate sensitive substrates, including the insulin and EGF receptors (BOLLAG et al. 1986; COCHET et al. 1984), glycogen synthase (CUIDAD et al. 1984), and other polypeptides. Phosphorylation of these receptors can lead to an attenuation of hormone action by a decrease in receptor tyrosine kinase activity (BOLLAG et al. 1986; COCHET et al. 1984).

A recent development has been the discovery of an insulin-activated kinase which phosphorylates microtubule-associated protein 2 (MAP-2). MAP-2 kinase is activated by insulin prior to the activation of a separate kinase which phosphorylates the ribosomal protein S6, known as S6 kinase (RAY and STURGILL 1987). MAP-2 kinase can itself phosphorylate and activate S-6 kinase. Further work has demonstrated that MAP-2 kinase is active when it is phosphorylated on tyrosine and threonine. This suggests the existence of a cascade of MAP-2 kinase, possibly by the insulin receptor tyrosine kinase. MAP-2 kinase in turn phosphorylates and activates S6 kinase. MAP-2 phosphorylation may be involved in the control of cell shape or mobility, while S6 phosphorylation may be involved in the control of protein synthesis (RAY and STURGILL 1988).

Effects of insulin have also been recently reported on casein kinase 1 and casein kinase 2. SOMMERCORN et al. (1987) have demonstrated a "stable" activation of casein 2 with insulin and EGF treatment. As noted in a previous section casein kinase 2 must obligatorily phosphorylate glycogen synthase (site 5), phosphatase inhibitor 2, and the regulatory subunit of cAMP-dependent protein kinase (type 2) prior to phosphorylation occurring by other independent protein kinases (DePaoli-Roach et al. 1983; Picton et al. 1982b). Presumably as a result of insulin-activated casein kinase 2, the phosphorylation of phosphatase inhibitor 2 is increased (Lawrence et al. 1988). This would in turn lead to more rapid phosphorylation of inhibitor 2 by GSK-3, and activation of phosphoprotein phosphatase 1, resulting in dephosphorylation of glycogen synthase. Since glycogen synthase usually is essentially fully phosphorylated in site 5, its activity would not be directly affected by activation of casein kinase 2 (Parker et al. 1983). Thus, by this mechanism, activation of a kinase would lead to activation of a phosphatase.

Insulin and vanadate, an insulin-like agent which acts through the insulin receptor (Tamura et al. 1984a) also produce activation of casein kinase 1 (Villar-Palasi et al. 1988). The significance of this effect is not clear at present. Diabetes, insulin injection, and glucagon have also been shown to affect the kinetic characteristics of casein kinase 1 and 2 in sensitive tissues (Grande et al. 1987). Further results in this rapidly developing area are awaited with interest.

VIII. Effects of Insulin on Phosphoprotein Phosphatases: A Mechanism for Dephosphorylation

Since the original report of Bishop on a "stable" phosphatase activation by insulin (Bishop 1970) a number of confirmatory reports have appeared (Nuttall et al. 1976; Gold et al. 1979). None has led to a definitive study of the mechanism of the activation. Very recently, further reports have appeared detailing the conditions necessary to stabilize the effect, most importantly homogenate concentration (Toth et al. 1987). Since the molecular characteristics and subunit structures of the various classes of phosphoprotein phosphatases are now being established (Ingebritsen and Cohen 1983), it is anticipated that future work will begin to elucidate the biochemical mechanisms of the effects of insulin on this class of enzymes. Of interest will be the possible relationship of the "stable" phosphatase activations previously observed to the more recent findings of decreased phosphorylation of inhibitor 1 and increased phosphorylation of inhibitor 2 with insulin action. It is already clear that insulin has profound effects to activate phosphatase biosynthesis in long-term experiments (Goheer et al. 1987).

IX. Purification and Action of Two Mediators: A Mechanism for Control of Dephosphorylation

We have employed two general approaches in purifying two insulin mediators to homogeneity – molecular sizing and ion exchange. We originally demonstrated that one mediator which inhibited the cAMP-dependent protein kinase was present in a specific fraction (fraction 2) on a Sephadex G-25 column with an approximate molecular weight of 1500 (Larner et al. 1974, 1979). This fraction when sent to Jarett's laboratory was shown to activate pyruvate dehydrogenase phosphatase (Jarett and Seals 1979). Subsequently the two activities were

Table 2. Carbohydrate constituents of two insulin mediators

Pyruvate dehydrogenase phosphate (stimulator)	Adenylate cyclase/cAMP kinase (inhibitor)
D-*chiro*-inositol	*Myo*-inositol
Galactosamine	Glucosamine
Mannose	Galactose

separated on a Sephadex G-15 column when it was shown in our laboratory that the cAMP kinase inhibitor sized slightly smaller than the PDH phosphatase activator (CHENG et al. 1985).

More recently a separation on the basis of ion exchange has also been achieved. When acid- and heat-inactivated, charcoal-treated extracts are adsorbed onto the anion exchange resin AG1 and eluted with dilute HCl, the PDH phosphatase activator elutes at pH 2, whereas the cAMP kinase inhibitor elutes at pH 1.3–1.5. It has been shown in our laboratory that the cAMP kinase inhibitor also inhibited the catalytic unit of adenylate cyclase (MALCHOFF et al. 1987). Thus, this mediator neutralizes not only the production, but the action of cAMP as well. Both of the mediators have been purified to essential homogeneity and their carbohydrate constituents shown to be chemically distinct. The composition data are shown in Table 2. Both have symmetrical, but nonidentical carbohydrate compositions, indicating that they arise from similar, but nonidentical precursors (LARNER et al. 1988).

The kinetics of the mediator which inhibits the cAMP-dependent protein kinase and adenylate cyclase for ATP is mixed, with effects on both K_m and V_{max}. The kinetics of this mediator action on the cAMP-dependent protein kinase is purely noncompetitive for ATP and cAMP and interestingly purely competitive for histone (MALCHOFF et al. 1987). The PDH phosphatase stimulatory mediator has been shown to act on the partially purified phosphatase by NEWMAN et al. (1985), but no kinetic data have been reported.

X. Formation of Insulin Mediators

As discussed in the previous section the compositions of the two insulin mediators including the possible presence of amino acids and ethanolamine clearly suggest that they arise from the glycophospholipid anchors of a set of exofacial-linked cell membrane proteins (see LOW 1987).

A general model depicting the involvement of PI-glycan lipids in the generation of insulin mediators has been proposed by SALTIEL et al. (1986). According to this model, the interaction of insulin with its receptor stimulates a PI-glycan-specific phospholipase C (PLC) which in turn hydrolyzes a PI-glycan precursor presumably located in the inner face of the cell membrane, thus releasing insulin modulator in the cytoplasm. In support of this model SALTIEL (1987) and MATO et al. (1987) have isolated a PI-glycan from rat liver membranes which is hydrolyzed by bacterial PLC to generate modulator. The stimulation of tissue PLC by insulin has yet to be demonstrated.

The involvement of proteolytic enzymes in the mechanisms of insulin action has been hypothesized for many years. Most of the evidence leading to this hypothesis arises from the effects of proteolytic enzyme inhibitors on several insulin-regulated cell metabolic functions (LARNER et al. 1982; MUCHMORE et al. 1982; CHERQUI et al. 1985). Direct stimulation of proteolysis by insulin has also been shown (CHERQUI et al. 1985). However, there are no unified hypotheses by which the involvement of proteolysis in the regulatory mechanisms of insulin may be explained. On one hand, certain proteases such as trypsin stimulate insulin receptor autophosphorylation (LEEF and LARNER 1987) and certain proteolysis inhibitors inhibit the insulin-stimulated autophosphorylation of the receptor (TAMURA et al. 1984b). On the other hand, recent work by RINAUDO et al. (1987) as well as previous work by BEGUM et al. (1983) suggests that pretreatment with certain protease inhibitors blocks the insulin-stimulated release of PDH activator from isolated rat liver membranes, suggesting that an insulin-sensitive proteolytic step may be involved in the generation of insulin mediators. Thus, there is evidence to suggest the involvement of more than one hydrolytic step in the generation of insulin mediators.

We have now shown that alkaline phosphatase and PDH mediator release to the extracellular medium are stimulated by insulin, following a kinetic profile which clearly suggests a direct relationship between both events (Fig. 5) (ROMERO et al. 1988). BC3H-1 myocytes were cultured in Dulbecco's modified Eagle's medium supplemented with 20% fetal bovine serum in collagen-coated Petri dishes (100 mm) or in 12-well plates as described by ROMERO et al. (1988). Cells were serum starved for 20 h prior to the addition of insulin. Insulin stimulation was initiated by the addition of serum-free medium containing 0.1% bovine serum albumin and 100 nM porcine insulin. All incubations were carried out at 37° C. Figure 5a shows the release of alkaline phosphatase to the incubation medium in the presence (full circles) and absence (open circles) of insulin. Cells were cultured in 100-mm Petri dishes, serum starved, and stimulated by addition of 10 ml insulin-containing culture medium as described above. At the indicated times, 0.4 ml aliquots were withdrawn and rapidly frozen in liquid nitrogen. The activity of alkaline phosphatase in the medium was estimated from the rate of hydrolysis of p-nitrophenylphosphate. The alkaline phosphatase assay was carried out at 37° C in 0.1 M Tris (pH 9.2) containing 2 mg/ml substrate and 2 mM MgCl$_2$ (0.5 ml final volume). The reaction was followed spectrophotometrically at 410 nm and the rate of hydrolysis of the substrate was determined from the slope of the time course of the reaction. The error bars represent the standard error of the estimate ($n=4$).

Figure 5b shows the release of alkaline phosphatase from pertussis toxin-treated BC3H-1 cells in the presence (full circles) and absence (open circles) of insulin. Cells were cultured as described above and exposed to toxin (100 ng/ml) for 20 h before insulin stimulation. Treatment with insulin and alkaline phosphatase assays were carried out as described above. The error bars represent the standard error of the estimate ($n=4$).

Figure 5c shows the effect of p-aminobenzamidine on the insulin-stimulated release of alkaline phosphatase (full circles). Controls in the absence of insulin are shown by the open circles. The inhibitor (0.8 mM) was added to the culture

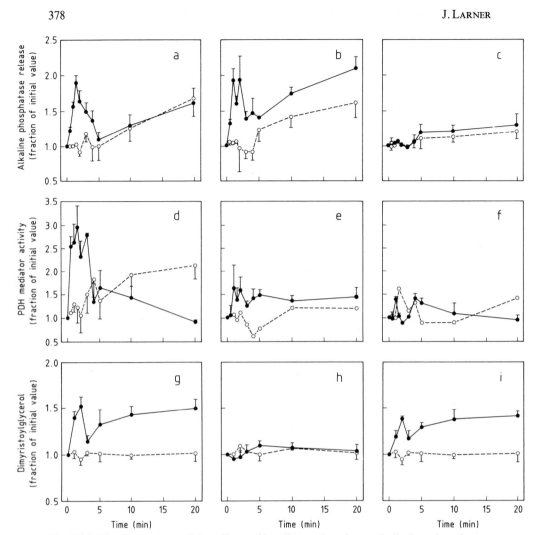

Fig. 5 a–i. The time course of the effects of insulin on the release of alkaline phosphatase (**a–c**), release of insulin mediator to the incubation medium (**d–f**), and generation of dimyristoylglycerol(**g–i**)

medium 2–4 min before exposure to insulin. The error bars represent the standard error of the estimate ($n = 3$). Figure 5d shows the release of insulin mediator to the extracellular medium by BC3H-1 myocytes upon stimulation with insulin (*full circles*). The *open circles* show the controls in the absence of insulin. Cells were grown in 100-mm plates and stimulated with insulin as described above. Aliquots of the incubation medium were collected and frozen in liquid nitrogen as described. Mediator activity was determined in 0.2-ml aliquots of the collected samples using a pyruvate dehydrogenase stimulation assay as described elsewhere (LARNER et al. 1988).

Since the cell culture medium contains a number of substances which interfere with the PDH assay, the samples were chromatographed in 2-ml AG-1 columns eluted successively with 10 ml distilled water and three solutions of aqueous HCl adjusted at pH 3, 2, and 1.3, respectively. The eluates were concentrated by lyophilization, redissolved in water, and relyophilized. Mediator activity was usually found in the pH 2 and pH 3 fractions. The chromatographic behavior of the PDH-activating factor released by the myocytes was identical to that of a mediator purified from liver cells in our laboratory. This purified mediator has been shown to contain inositol and other sugars (LARNER et al. 1988). These data and those presented in Fig. 5e and f correspond to the average of three separate experiments, each carried out in duplicate. The error bars represent the standard error of the estimate. Figure 5e shows the effect of pertussis toxin treatment on the insulin-stimulated release of mediator to the culture medium (*full circles*). Cells were pretreated with toxin as described above and stimulated with insulin. Mediator activity was determined as described. The *full circles* show the results obtained upon addition of insulin; the *open circles* show the controls in the absence of insulin. Figure 5f shows results for cells pretreated with *p*-aminobenzamidine (0.8 mM) for 2–4 min and incubated in the presence (*full circles*) or absence (*open circles*) of insulin (100 nM). Mediator activity was determined in the culture medium as described. Figure 5g shows the production of dimyristoylglycerol by BC3H-1 myocytes in the presence (*full circles*) and absence (*open circles*) of insulin. Cells were cultured in 12-well plates, serum starved as described, and labeled with [^{3}H] myristate for 20 h before treatment with insulin. Dimyristoylglycerol production was determined by extraction of lipid with chloroform–methanol–HCl (200:100:1, v/v) and subsequent chromatography on silica gel plates using petroleum ether–diethyl ether–acetic acid (70:30:2, v/v) for development. Lipids were scraped and radioactivity determined by liquid scintillation counting. The data shown correspond to the average of five independent experiments ($\pm$ standard error).

Figure 5h shows the effect of pertussis toxin on the insulin-stimulated production of dimyristoylglycerol. The insulin stimulation experiments are shown by the *full circles*. The *open circles* are controls in the absence of insulin. The data show the average of three separate experiments. Figure 5i shows the effect of *p*-aminobenzamidine on the insulin-stimulated generation of dimyristoyl-glycerol. The *full circles* show the effects of insulin treatment. The *open circles* represent the controls with no additions. The error bars represent the standard error of the estimate ($n = 3$).

The time course again is consistent with the kinetics of release of both alkaline phosphatase and PDH mediator, suggesting a relationship between all three events. However, while treatment of cells with pertussis toxin clearly abolished the insulin-stimulated increase in diacylglycerol and PDH mediator generation, it had only minimal effects on release of alkaline phosphatase. Thus, while each blocking agent inhibits mediator generation, each blocking agent does not act on the alternate release process – strongly indicating that two separate hydrolytic events are required to generate mediator, but that only one event is necessary to release each separate product, i.e., diacylglycerol or alkaline phosphatase. A model summarizing these data is shown in Fig. 6 with the accompanying requirements for a transporter to carry the mediator formed extracellularly to its required intracellular site of action.

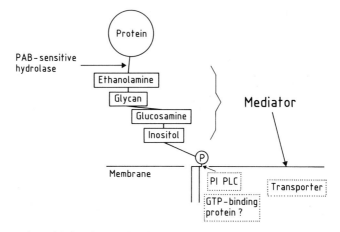

Fig. 6. Proposed model for the mechanism of PDH mediator release from BC3H-1 cell membranes. The model proposes that the glycophospholipid anchor of a membrane protein is a primary source of insulin mediator. Insulin activates, possibly by a protein phosphorylation process, two hydrolytic enzymes. The first of these is a protease-like enzyme which is inhibited by p-aminobenzamidine and hydrolyzes the protein or the anchor in a position distal to inositol. This enzyme releases the protein from its anchor and generates a glycophospholipid similar to those described by SALTIEL et al. (1986) and MATO et al. (1987). The second is a phospholipase C, the activity of which is apparently modulated by a GTP binding protein sensitive to pertussis toxin. This enzyme cleaves the inositol–phosphate bond releasing diacylglycerol and mediator. The mediator is thus released to the extracellular medium and is transported into the cell by yet another component of the system

XI. Insulin Receptor Activation and Mediator Formation

We have obtained clear evidence linking an insulin receptor-initiated phosphorylation event in the generation/release/activation of insulin mediator (SUZUKI et al. 1987). ATP-Mn^{2+}, the specific preferred substrate requirement for insulin receptor tyrosine kinase autophosphorylation and activation stimulates mediator formation/release/activation 10- to 20-fold in liver cell membranes in the presence of insulin. Insulin alone stimulates this event 2- to 4-fold. The stimulation is specific both for ATP and for Mn^{2+}. ATP-Mn^{2+} could be required to phosphorylate either one of the hydrolytic enzymes involved in forming mediator, or of the substrate from which mediator is formed, or of the mediator itself, activating it. While the detailed significance of this ATP-Mn^{2+} requirement is as yet unclear, it serves as an important probe for investigating the detailed steps in generation, transport and/or metabolism of mediator.

C. Summary

The enhanced phosphorylations via cAMP, Ca^{2+} mobilization, and diacylglycerol formation via the activation of the respective kinases is now classical. The decreased phosphorylation via inhibition of adenylate cyclase via the adrenergic receptor is also becoming understood. What the insulin studies have taught us, via the novel P-P-1G mediators, is that these compounds act to decrease phosphorylation state, particularly of glycogen synthase and pyruvate dehydrogenase, activating these two key rate-limiting enzymes leading to the accumulation of glycogen and fat in sensitive tissue. While much more needs to be done to work out the details, the basic mechanisms are now being established and they point out the novel and unexpected nature of the mechanism of insulin action and explain in part why it has taken so long to work out the mechanisms.

Acknowledgments. I am deeply grateful and indebted to co-workers and colleagues including Laura Huang, Carlos Villar-Palasi, Charles F. W. Schwartz, Guillermo Romero, Guozhi Tang, Susumu Suzuki, T. Y. Shen, A. Shawn Oswald, Zeses Roulidis, Erik Hewlett, Louis Luttrell, Alan Rogol, Konrad Zeller, and Douglas Heimark whose work contributed to and is cited in this chapter. Discussions with Rodney Biltonen are also gratefully acknowledged. The work described was supported in part by grants from USPHS NIH DK14334 and the University of Virginia Diabetes Research and Training Center.

References

Banting FG, Best CH, Collip JB, Macleod JJR, Noble EC (1963) The effect of insulin on the percentage amounts of fat and glycogen in the liver and other organs of diabetic animals. In: Selected papers of Charles H. Best. University of Toronto Press, Toronto, pp 84–87

Begum N, Tepperman HM, Tepperman J (1983) Effects of high fat and high carbohydrate diets on liver pyruvate dehydrogenase and its activation by a chemical mediator released from insulin-treated liver particulate fraction. Endocrinology 112:50–59

Bernard C (1877) Leçons sur le diabéte et la glycogénése animale. Bailliére, Paris

Bishop JS (1970) Inability of insulin to activate liver glycogen transferase D phosphatase in the diabetic pancreatectomized dog. Biochim Biophys Acta 208:208–218

Bollag GE, Roth RA, Beaudoin J, Mochly-Rosen D, Koshland DE Jr (1986) Protein kinase C directly phosphorylates the insulin receptor in vitro and reduces its protein-tyrosine kinase activity. Proc Natl Acad Sci USA 83:5822–5824

Cheng K, Thompson M, Schwartz C, Malchoff C, Tamura S, Craig J, Locher E, Larner J (1985) Multiple intracellular peptide mediators of insulin action. In: Czech MP (ed) Molecular basis of insulin action. Plenum, New York, pp 171–182

Cherqui G, Caron M, Capeau J, Picard J (1985) Further evidence for the involvement of a membrane proteolytic step in insulin action. Biochem J 227:137–147

Cochet C, Gill GN, Meisenhelder J, Cooper JA, Hunter T (1984) C-kinase phosphorylates the epidermal growth factor receptor and reduces its epidermal growth factor-stimulated tyrosine protein kinase activity. J Biol Chem 259:2553–2558

Cori CF, Cori GT (1936) Mechanism of formation of hexose monophosphate in muscle and isolation of a new phosphate ester. Proc Soc Exp Biol Med 34:702–705

Craig JW, Rall TW, Larner J (1969) The influence of insulin and epinephrine on adenosine 3′5′phosphate and glycogen transferase in muscle. Biochim Biophys Acta 177:213–219

Cuidad CJ, Camici M, Ahmad Z, Wang Y, DePaoli-Roach A, Roach PJ (1984) Eur J Biochem 142:511–520

Cuidad CJ, Vila J, Mor MA, Guinovart JJ (1987) Effects of glucagon and insulin on the cyclic AMP binding capacity of hepatocyte cyclic AMP-dependent protein kinase. Mol Cell Biochem 73:37–44

DePaoli-Roach AA, Ahmad Z, Camici M, Lawrence JC, Roach PJ (1983) Multiple phosphorylation of rabbit skeletal muscle glycogen synthase. Evidence for interactions among phosphorylation sites and the resolution of electrophoretically distinct forms of the enzyme. J Biol Chem 258:10701–10709

Farese R, Barnes DE, Davis JS, Standaert ML, Pollet RJ (1984) Effects of insulin and protein synthesis inhibitors on phospholipid metabolism, diacylglycerol levels and pyruvate dehydrogenase activity in BC_3H-1 cultured myocytes. J Biol Chem 259:7094–7100

Fiol CJ, Mahrenholz AM, Wang Y, Roeske RW, Roach PJ (1987) Formation of protein kinase recognition sites by covalent modification of the substrate. J Biol Chem 262:14042–14048

Friedman DL, Larner J (1963) Interconversion of two forms of muscle UDPG-glycan transglucosylase by a phosphorylation-dephosphorylation reaction sequence. Biochemistry 2:669–675

Friedman DL, Larner J (1966) Studies on UDPGα-1,4, glycanα-4-glucosyl transferase VIII catalysis of the phosphorylation of muscle phosphorylase and transferase by separate enzymes. Biochemistry 4:2261–2264

Gabbay RA, Lardy HA (1984) Site of insulin inhibition of cAMP-stimulated glycogenolysis, cAMP changes. J Biol Chem 259:6052–6055

Gabbay RA, Lardy HA (1987) Insulin inhibition of hepatic cAMP-dependent protein kinase: decreased affinity of protein kinase for cAMP and possible differential regulation of intrachain sites 1 and 2. Proc Natl Acad Sci USA 84:2218–2222

Goheer MA, Larner J, Curnow RT (1987) Long-term regulation of glycogen metabolizing enzymes by insulin in H_4 hepatoma cells. Mol Cell Biochem 75:137–149

Gold AH, Dickemper D, Haverstick DM (1979) Insulin sensitivity of liver glycogen synthase b into a conversion. Mol Cell Biochem 25:47–59

Grande J, Perez M, Itarte E (1987) Hepatic casein kinases 1 and 2 are under control by insulin and glucagon. European Association for the Study of Diabetes, Brussels

Hanes GS (1940) The breakdown and synthesis of starch by an enzyme system from pea seeds. Proc R Soc Lond [Biol] 128:421–450

Huijing F, Larner J (1966) On the effect of adenosine 3′5′cyclophosphate on the kinase of UDPG: α-1,4-glycan α-4-glucosyl transferase. Biochem Biophys Res Commun 23:259–263

Ingebritsen TS, Cohen P (1983) Protein phosphatases: properties and role in cellular regulation. Science 221:331–338

Jarett L, Seals JR (1979) Pyruvate dehydrogenase activation in adipocyte mitochondria by an insulin-generated mediator from muscle. Science 206:1406–1408

Larner J (1988) Insulin-signaling mechanisms: lessons from the Old Testament of glycogen metabolism and the New Testament of molecular biology. Banting lecture. Diabetes 37:262–275

Larner J, Villar-Palasi C, Richman DJ (1960) Insulin-stimulated glycogen formation in rat diaphragm. Levels of tissue intermediates in short-time experiments. Arch Biochem Biophys 86:56–60

Larner J, Huang LC, Broker G, Murad F, Miller TB (1974) Inhibitor of protein kinase formed in insulin treated muscle. Fed Proc 33:261

Larner J, Galasko G, Cheng K, DePaoli-Roach AA, Huang L, Daggy F, Kellogg J (1979) Generation by insulin of a chemical mediator that controls protein phosphorylation and dephosphorylation. Science 206:1408–1410

Larner J, Cheng K, Schwartz C, Kikuchi K, Tamura S, Creacy S, Dubler R et al. (1982) A proteolytic mechanism for the action of insulin via oligopeptide mediator formation. Fed Proc 41:2724–2729

Larner J, Huang LC, Schwartz CFW, Oswald AS, Shen FY, Kinter M, Tang G, Zeller K (1988) Rat liver insulin mediator which stimulates pyruvate dehydrogenase phosphatase contains galactosamine and D-chiroinositol. Biochem Biophys Res Commun 151:1416–1426

Lawrence JC, Larner J, Kahn CR, Roth J (1978) Autoantibodies to the insulin receptor activate glycogen synthase in rat adipocytes. Mol Cell Biochem 22:153–157

Lawrence JC, Hiken JF, DePaoli-Roach AA, Roach PJ (1983) Hormonal control of glycogen synthase in rat hemidiaphragms. Effects of insulin and epinephrine on the distribution of phosphate between cyanogen bromide fragments. J Biol Chem 258:10710–10719

Lawrence JC Jr, Hiken J, Burnette B, DePaoli-Roach AA (1988) Phosphorylation of phosphoprotein phosphatase inhibitor-2 (I-2) in rat fat cells. Biochem Biophys Res Commun 150:197–203

Leef JW, Larner J (1987) Insulin-mimetic effect of trypsin on the insulin receptor tyrosine kinase in intact adipocytes. J Biol Chem 262:14837–14842

Leloir LF, Cardini CE (1957) Biosynthesis of glycogen from uridine diphosphate glucose. J Am Chem Soc 79:6340–6341

Linn TC, Pettit FH, Reed LJ (1969) α-Keto acid dehydrogenase complexes. X. Regulation of the activity of the pyruvate dehydrogenase complex from beef kidney mitochondria by phosphorylation and dephosphorylation. Proc Natl Acad Sci USA 62:234–241

Low MG (1987) Biochemistry of the glycosyl-phosphatidylinositol membrane protein anchors. Biochem J 244:1–13

Luttrell LM, Hewlett EL, Romero G, Rogol AD (1988) Pertussis toxin treatment attenuates some effects of insulin in BC_3H_1 murine myocytes. J Biol Chem 263:6134–6141

Malchoff CD, Huang L, Gillespie N, Villar-Palasi C, Schwartz CFW, Cheng K, Hewlett EL, Larner J (1987) A putative mediator of insulin action which inhibits adenylate cyclase and adenosine 3'5'-monophosphate-dependent protein kinase; partial purification from rat livers. Site and kinetic mechanism of action. Endocrinology 120:1327–1337

Mato JM, Kelley KL, Abler A, Jarett L (1987) Identification of a novel insulin-sensitive glycophospholipid from H35 hepatoma cells. J Biol Chem 262:2131–2137

Morgan HE, Parmeggiani A (1964) Regulation of glycogenolysis in muscle. J Biol Chem 239:2440–2445

Muchmore DB, Raess BU, Bergstrom RW, deHaen C (1982) On the mechanisms of inhibition of insulin action by small-molecular weight trypsin inhibitors. Diabetes 31:976–984

Newman JD, Armstrong JM, Bornstein J (1985) Assay of insulin mediator with soluble pyruvate dehydrogenase phosphatase. Endocrinology 116:1912–1919

Nuttall FQ, Gannon MC, Corbett VA, Wheeler MP (1976) Insulin stimulation of heart glycogen synthase D phosphatase (protein phosphatase). J Biol Chem 251:6724–6729

Parker PJ, Embi N, Caudwell FB, Cohen P (1982) Glycogen synthase from rabbit skeletal muscle. State of phosphorylation of the seven phosphoserine residues in vivo in the presence and absence of adrenaline. Eur J Biochem 124:47–55

Parker PJ, Caudwell FB, Cohen D (1983) Glycogen synthase from rabbit skeletal muscle: effect of insulin on the state of phosphorylation of the seven phosphoserine residues in vivo. Eur J Biochem 130:227–234

Parnas JK (1938) Über die enzymatischen Phospholierungen in der alkoholischen Gärung und in der Muskelglykogenolyse. Enzymologia 5:166–184

Picton C, Aitken A, Bilham T, Cohen P (1982a) Multisite phosphorylation of glycogen synthase from rabbit skeletal muscle. Organization of the seven sites in the polypeptide chain. Eur J Biochem 124:37–45

Picton C, Woodgett J, Hemmings B, Cohen P (1982b) Multisite phosphorylation of glycogen synthase from rabbit skeletal muscle. FEBS Lett 150:191–195

Piras R, Staneloni R (1969) In vivo regulation of rat muscle glycogen synthetase activity. Biochemistry 8:2153–2160

Ray LB, Sturgill TW (1987) Rapid stimulation by insulin of a serine/threonine kinase in 3T3-L1 adipocytes that phosphorylates microtubule-associated protein 2 in vitro. Proc Natl Acad Sci USA 84:1502–1506

Ray LB, Sturgill TW (1988) Insulin-stimulated MAP (microtubule-associated protein) kinase is phosphorylated on tyrosine and threonine in vivo. Proc Natl Acad Sci USA 85:3753–3757

Rinaudo MT, Curto M, Bruno R, Marino C, Rossetti V, Mostert M (1987) Evidence of an insulin generated pyruvate dehydrogenase stimulating factor in rat brain plasma membranes. Int J Biochem 19:909–913

Roach PJ, Larner J (1976) Rabbit skeletal muscle glycogen synthetase. Enzyme phosphorylation state and effector concentrations as interacting control parameters. J Biol Chem 251:1920–1925

Roach PJ, Rosell-Perez M, Larner J (1977) Muscle glycogen synthase in vivo state: effects of insulin administration on the chemical and kinetic properties of the purified enzyme. FEBS Lett 80:95–98

Romero G, Luttrell L, Rogol A, Zeller K, Hewlett E, Larner J (1988) The phosphatidyl-glycan anchors of membrane proteins: potential precursors of insulin mediators. Science 240:509–511

Saltiel AR (1987) Insulin generates an enzyme modulator from hepatic plasma membranes: regulation of adenosine 3'5'-monophosphate phosphodiesterase, pyruvate dehydrogenase, and adenylate cyclase. Endocrinology 120:967–972

Saltiel AR, Fox JA, Sherline P, Cuatrecasas P (1986) Insulin-stimulated hydrolysis of a novel glycolipid generates modulators of cAMP phosphodiesterase. Science 233:967–971

Shen LC, Villar-Palasi C, Larner J (1970) Hormonal alteration of protein kinase sensitivity to 3'5'-cyclic AMP. Physiol Chem Phys 2:536–44

Sheorain VS, Juhl H, Bass M, Soderling TR (1984) Effect of epinephrine diabetes and insulin on rabbit skeletal muscle glycogen synthase. J Biol Chem 259:7024–7030

Smith CH, Brown NE, Larner J (1971) Molecular characteristics of the totally dependent and independent forms of glycogen synthase of rabbit skeletal muscle. Biochim Biophys Acta 242:81–88

Soderling TR, Jett MF, Hutson NJ, Khatra BS (1977) Regulation of glycogen synthase. J Biol Chem 252:7517–7524

Sommercorn J, Mulligan JA, Lozeman FJ, Krebs EG (1987) Activation of casein kinase II in response to insulin and to epidermal growth factor. Proc Natl Acad Sci USA 84:8834–8838

Suzuki S, Toyota T, Tamura S, Kikuchi K, Tsuiki S, Huang L, Villar-Palasi C et al. (1987) ATP-Mn^{2+} stimulates the generation of a putative mediator of insulin action. J Biol Chem 262:3199–3204

Takeda Y, Brewer HB Jr, Larner J (1975) Structural studies on rabbit muscle glycogen synthase. J Biol Chem 250:8943–8950

Tamura S, Brown TA, Whipple JH, Fujita-Yamaguchi Y, Dubler RE, Chang K, J (1984a) A novel mechanism for the insulin-like effect of vanadate on glycogen synthase in rat adipocytes. J Biol Chem 259:6650–6658

Tamura S, Schwartz CFW, Whipple JH, Dubler RE, Fujita-Yamaguchi Y, Larner J (1984b) Selective inhibition of the insulin-stimulated phosphorylation of the 95000 dalton subunit of the insulin receptor by TAME or BAEE. Biochem Biophys Res Commun 119:465–472

Toth B, Bollen M, Stalmans W (1987) Short term effects of glucagon and insulin on cytosolic fraction of protein phsophatase in rat liver. Arch Int Physiol Biochem 95:B242

Villar-Palasi C, Larner J (1960a) A uridine coenzyme-linked pathway of glycogen synthesis in muscle. Biochim Biophys Acta 30:449

Villar-Palasi C, Larner J (1960b) Levels of activity of the enzymes of the glycogen cycle in rat tissues. Arch Biochem Biophys 86:270–273

Villar-Palasi C, Wenger JI (1967) In vivo effect of insulin on muscle glycogen synthetase, identification of the action pathway. Fed Proc 26:563

Villar-Palasi C, Gomez-Foix AM, Rodriguez-Gil JE, Guinovart JJ, Bosch F (1989) Activation of casein kinase 1 by vanadate in rat hepatocytes. J Biochem (in press)

Walaas O, Walaas E, Gronnerod O (1972) Effect of insulin and epinephrine on cyclic AMP-dependent protein kinase in rat diaphragm. Isr J Med Sci 8:353–57

Walkenbach RJ, Hazen R, Larner J (1980) Hormonal regulation of glycogen synthase: insulin decreases protein kinase sensitivity to cyclic AMP. Biochem Biophys Acta 629:421–430

Walsh DA, Perkins JP, Krebs EG (1968) An adenosine 3'5' monophosphate dependent protein kinase from rabbit skeletal muscle. J Biol Chem 243:3763–3765

CHAPTER 17

Insulin-Sensitive cAMP Phosphodiesterase

T. KONO

A. Introduction

Cyclic nucleotide phosphodiesterase (simply referred to as phosphodiesterase in this chapter) is an enzyme that hydrolyzes 3′,5′-cyclic nucleotide to 5′-AMP. This enzyme was first described by SUTHERLAND and RALL (1958) and subsequently purified by BUTCHER and SUTHERLAND (1962) from beef hearts. However, it was soon discovered that so-called phosphodiesterase is not a single enzyme, but a group of enzymes having significantly different properties (see, e.g., WELLS and HARDMAN 1977 for review). Although classification of phosphodiesterases is yet to be completed, they have often been divided into four groups (STRADA and THOMPSON 1984): type I, Ca^{2+}- and calmodulin-sensitive phosphodiesterases; type II, cGMP-sensitive enzymes; type III, rhodopsin-sensitive cGMP phosphodiesterases; and type IV, cAMP phosphodiesterases that include the hormone-sensitive enzymes. Although this classification is now somewhat obsolete, it clearly indicates that there are different types of phosphodiesterases. In general, more than one species of phosphodiesterases coexist within a cell; for example, WEBER and APPLEMAN (1982) reported that phosphodiesterases from rat adipocytes are separable into four discrete peaks by DEAE chromatography, and that only one of them is insulin-sensitive.

Although the number of insulin-sensitive phosphodiesterases is not large, it is still impossible to cite all the relevant literature in the limited number of pages that are allocated to this chapter. Additional information on phosphodiesterase in general may be obtained in articles written by WELLS and HARDMAN (1977) and BEAVO et al. (1982). Insulin-sensitive phosphodiesterases in particular have been reviewed previously by FRANCIS and KONO (1982), APPLEMAN et al. (1984), and MAKINO (1987).

B. Localization of Insulin-Sensitive Phosphodiesterase

Insulin-sensitive phosphodiesterases are present in the liver (ALLAN and SNEYD 1975; TRIA et al. 1977), muscle (GAIN and APPLEMAN 1978), and fat cells (LOTEN and SNEYD 1970; MANGANIELLO and VAUGHAN 1973). They commonly exhibit low/K_m activities ($K_m \sim 0.1$–0.7 μM) towards cAMP and are associated with cellular membranes. When the particulate fraction is obtained from rat adipocytes by mild homogenization of cells in isotonic sucrose solution, and is subjected to a linear sucrose density gradient centrifugation, most of the phosphodiesterase activity in the fraction is concentrated in the endoplasmic reticulum fraction, and the rest is recovered in the plasma membrane fraction (KONO et al. 1975). The enzyme activities in the two fractions are both increased 2- to 3-fold when cells are exposed to insulin prior to homogenization (KONO et al. 1975; MAKINO and KONO 1980). Nevertheless, the data obtained by MACAULAY et al. (1983) suggest that the plasma membrane and the endoplasmic reticulum are associated with different types of phosphodiesterases. Thus, these investigators found that the enzyme in the plasma membrane fraction was stimulated up to 36% with lysophosphatidylglycerol > lysophosphatidylcholine > lysophosphadtidylserine > phosphatidylserine > phosphatidylglycerol in the order listed, whereas the enzyme in the endoplasmic reticulum was activated up to 78% with the same agents in the reverse order.

As first reported by MANGANIELLO and VAUGHAN (1973), insulin-sensitive phosphodiesterase in fat cells can easily be enriched into a crude microsomal fraction by a two-step differential centrifugation of the cell homogenate. The fraction thus obtained is often referred to as fraction P-2 and widely used in studies on the potential mechanism of insulin action (see Sect. D). HEYWORTH et al. (1983 b) found that the insulin-sensitive liver enzymes are associated with two types of membranes: the plasma membrane and the "dense vesicles." The morphological identity of the latter is not clear. Earlier, LOTEN et al. (1978) reported that the liver enzyme was associated with certain vesicles that were apparently distinct from either the plasma membrane or the endoplasmic reticulum. According to the data obtained by MARCHMONT et al. (1981) and PYNE et al. (1987), the liver enzymes associated with the plasma membrane and the dense vesicles are apparently different (see Sect. C). Earlier, WOO and MANERY (1973) noted that a type of insulin-sensitive phosphodiesterase exists on the external surface of skeletal muscle cells. Likewise, SMOAKE et al. (1981) found that phosphodiesterase in the liver plasma membrane is an ectoenzyme and hydrolyzes extracellular cAMP. However, HEYWORTH et al. (1984) contended that the insulin-sensitive, liver "periphery" enzyme is involved in the regulation of the cAMP concentration in the cell. MARCHMONT and HOUSLAY (1980 b) and MARCHMONT et al. (1981) noted that the liver peripheral enzyme can be solubilized with a concentrated salt solution (e.g., 0.2–0.4 M NaCl). On the other hand, LOTEN (1983) reported that the liver "particulate" enzyme was not solubilized with salt solutions. The fat cell enzyme in fraction P-2 (see above) is not solubilized with 0.2–1.0 M NaCl or KCl (MAKINO et al. 1980).

C. Solubilization, Purification, and Characterization

LOVELL-SMITH et al. (1977) were the first to attempt to purify insulin-sensitive phosphodiesterase. They solubilized the adipocyte enzyme in fraction P-2 with triton X-100 and obtained two peaks of the enzyme or enzymes by chromatography on DEAE. Unfortunately, a significant fraction of the enzyme activity was lost during the process, and the effect of insulin was considerably diminished. MAKINO and KONO (1980) reported that the hormonal effect was largely preserved when the enzyme was solubilized with a mixture of diluted Zwittergent 3–14 and Lubrol WX. The Stokes radii of the enzyme solubilized by this method were 8.7 nm (in the basal state) and 9.4 nm (in the plus insulin state), suggesting that the enzyme had a large mass: either $M_r = 720\,000$ (basal) or $M_r = 820\,000$ (plus insulin) (MAKINO and KONO 1980). SALTIEL and STEIGERWALT (1986) solubilized the adipocyte enzyme with triton X-100 and obtained an apparently homogeneous preparation after a mere 54-fold purification. The molecular weight of their enzyme was approximately 60\,000. DEGERMAN et al. (1987) solubilized the fat cell enzyme by sonication in the presence of a non-ionic detergent, NaBr, and glycerol. They obtained an apparently homogeneous preparation after more than 65\,000-fold enrichment; their purification steps included chromatography on DEAE, gel filtration, and affinity chromatography on immobilized cilostamide. Cilostamide is an agent thought to be a specific inhibitor of membrane-bound phosphodiesterase (ELKS and MANGANIELLO 1984). The molecular weight of their purified enzyme was determined to be approximately 64\,000 by SDS–PAGE, and 100\,000–110\,000 by gel filtration (DEGERMAN et al. 1987). Therefore, the enzyme may exist in the dimeric form in nature.

DEGERMAN et al. (1987) further reported that their purified enzyme exhibited normal Michaelis–Menten kinetics, and hydrolyzed both cAMP ($K_m = 0.38\ \mu M$, $V_{max} = 8.5\ \mu mol\ min^{-1}mg^{-1}$) and cGMP ($K_m = 0.28\ \mu M$, $V_{max} = 2.0\ \mu mol\ min^{-1}mg^{-1}$); the hydrolysis of cAMP was competitively inhibited by cGMP ($K_i = 0.15\ \mu M$). The liver peripheral enzyme was purified by MARCHMONT et al. (1981) after solubilization with 0.4 M NaCl. They obtained an apparently homogeneous preparation after a 9500-fold enrichment; the molecular weight of the enzyme was estimated to be 52\,000 by both SDS–PAGE and gel filtration. It would appear, therefore, that this enzyme is in the monomeric form; nevertheless, MARCHMONT et al. (1981) found that the enzyme exhibits a non-Michaelis-type kinetics towards hydrolysis of cAMP (apparent $K_m = 0.7\ \mu M$, $V_{max} = 9.1$ units/mg, and Hill coefficient $= 0.62$). Their enzyme was poorly active to cGMP ($K_m = 120\ \mu M$, V_{max} 0.4 units/mg). The catalytic domain (see below) of the liver enzyme associated with the dense vesicles was purified by PYNE et al. (1987). They clipped off the catalytic domain of the enzyme by a mild proteolysis of the holoenzyme (see below) and purified the former to apparent homogeneity after 2000- to 3000-fold enrichment. Their "enzyme" (or the catalytic domain of phosphodiesterase) appeared to be a dimer ($M_r = 112\,000$) of identical monomers ($M_r = 57\,000$). The purified catalytic domain hydrolyzed both cAMP and cGMP, and its action on cAMP was negatively cooperative (Hill coefficient $= 0.43$) while its action on CGMP was the normal Michaelis–Menten type (PYNE et al. 1987).

Using a crude fat cell enzyme, LOTEN et al. (1980) concluded that phosphodiesterase may have two distinct domains; one for the catalytic activity and the other for its regulation (see Fig. 1 later). This conclusion is based on their observations: (a) that upon mild trypsin treatment of phosphodiesterase in a particulate fraction (fraction P-2), a portion of the enzyme is solubilized; and (b) the solubilized portion of the enzyme possesses the full catalytic activity of the holoenzyme regardless of whether it is derived from the basal or insulin-stimulated states of fraction P-2. Independently, MAKINO et al. (1980) found that the catalytic activity of phosphodiesterase is solubilized when fraction P-2 from adipocytes is incubated overnight with dithiothreitol at 4° C. Again, the maximum enzyme activity is solubilized from both basal and insulin-stimulated states of fraction P-2. Later, MAKINO et al. (1982) concluded that this dithiothreitol-dependent solubilization/stimulation was secondary to the activation of sulfhydryl proteinases in fraction P-2 since the reaction was blocked by various inhibitors of proteolysis. Earlier, LOTEN et al. (1978) found that (the catalytic domain of) liver phosphodiesterase was rapidly solubilized when the enzyme in the particulate fraction was washed with hypotonic aqueous solution. They suggested that this reaction was secondary to "activation" of the proteolytic enzyme in the fraction, possibly as a result of hypotonic rupture of lysosomes.

It should be noted at this point that no information is currently available about the "activation state" of the aforementioned purified enzyme preparations. In other words, it is not clear whether the purified enzymes were in the basal state or in the stimulated state. In fact, it is even obscure whether some of the purified "enzymes" were the holoenzymes or the catalytic domains of the enzymes. Since the basal and insulin-stimulated states of phosphodiesterase are clearly distinguishable only prior to solubilization (or before complete solubilization?) of the enzyme, the hormonal regulation of phosphodiesterase has been studied, in most cases, by measuring the enzyme activity either in the cell homogenate or in fraction P-2, as described in the next section.

D. Hormonal Stimulation and its Physiological Significance

Initially, LOTEN and SNEYD (1970) reported that insulin stimulated a high K_m phosphodiesterase in adipocytes by changing its K_m value, and a low K_m enzyme by increasing its V_{max} value. In subsequent studies, however, only the latter part of their observation was confirmed (e.g., MANGANIELLO and VAUGHAN 1973). ZINMAN and HOLLENBERG (1974) found that when fat cells are exposed to insulin, the cellular phosphodiesterase activity is stimulated to the maximum level in approximately 5 min at 37° C and remains at the same level as long as the incubation is continued. On the other hand, the dose–response curve of the insulin effect is biphasic; the enzyme is stimulated maximally when the hormone concentration is 1–3 nM (KONO et al. 1975). The action of insulin on phosphodiesterase is ATP-dependent; the hormone-dependent stimulation is blocked almost completely in the presence of 1 mM 2,4-dinitrophenol, 1 mM dicumarol, 1–2 mM KCN, or 10 mM sodium azide (KONO et al. 1977). When insulin-treated cells are deprived of ATP, the hormone-stimulated enzyme activity is reduced to the basal level in 5 min (KONO et al. 1977). This last observation may not be surprising; nevertheless, it is of considerable interest as ATP is involved in both insulin-dependent activation and subsequent deactivation of glucose transport activity in the same cell (VEGA et al. 1980). Apparently, the effect of insulin on phosphodiesterase is secondary to the binding of the hormone to its receptor since the hormonal effect is abolished by proteolytic modification of the cellular insulin receptor (SUZUKI et al. 1987).

As mentioned earlier, insulin-dependent stimulation of phosphodiesterase in adipocytes is detectable in the cell homogenate, provided that a low concentration of cAMP (e.g., 0.1 µM) is used as the substrate (MANGANIELLO and VAUGHAN 1973). However, a considerably larger hormonal effect is observable if the enzyme activity is determined in fraction P-2 (the crude microsomal fraction; MANGANIELLO and VAUGHAN 1973). The insulin-dependent stimulation of phosphodiesterase detected in fraction P-2 is approximately 2.5–3.0 (KONO et al. 1975). In some cell preparations, a larger insulin effect is recorded when cells are treated with the hormone in the presence of 5 mM glucose or sodium pyruvate (KONO, unpublished data). On the other hand, the apparent insulin effect is rendered considerably smaller if cells are homogenized in the presence of EDTA or dithiothreitol (KONO and BARHAM 1973). The latter observations have been interpreted as indicating that phosphodiesterase may exist in its reduced form in the cell, and that its insulin-stimulated form is unstable unless the enzyme is oxidized with air at the time of cell homogenization (KONO et al. 1975). Also, the apparent effect of insulin observed in fraction P-2 is significantly smaller if cells are homogenized at pH 7.4–7.5 rather than pH 7.0, because the basal activity is significantly elevated in the cell homogenate at pH 7.4–7.5 (VEGA et al. 1980). Nevertheless, the enzyme that has been partially purified through the step of fraction P-2 is most stable and most active at pH 7.5, regardless of whether it is in the basal or insulin-stimulated state (MAKINO and KONO 1980).

As was first reported by ZINMAN and HOLLENBERG (1974), phosphodiesterase in adipocyte is stimulated not only by insulin, but also by catecholamines and other so-called lipolytic agents that would increase the cellular level of cAMP. Although PAWLSON et al. (1974) reported that the actions of insulin and lipolytic agents were additive, MAKINO and KONO (1980) found it otherwise. The latter investigators concluded the same species of phosphodiesterase may be stimulated by both insulin and isoproterenol (a catecholamine) since the enzymes stimulated by the two effectors showed identical responses to heat, pH, salts, and several inhibitors. It must be noted, however, that different mechanisms may be involved in the actions of the two effectors since the action of catecholamines is biphasic with respect to time (ZINMAN and HOLLENBERG 1974) while that of insulin is not (see above). The catecholamine-dependent stimulation is maximum at 5–10 min of incubation (ZINMAN and HOLLENBERG 1974) when the cellular level of cAMP is also maximum (MANGANIELLO et al. 1971). It is suggested, therefore, that cAMP is involved in the stimulation of phosphodiesterase in response to catecholamines or other lipolytic agents. For some unknown reason, the maximum level of stimulation induced by lipolytic agents is significantly and consistently lower than that brought about by insulin (MAKINO and KONO 1980).

The action of insulin or catecholamines to stimulate intracellular phosphodiesterase is mimicked by 1 mM $MnSO_4$ + A-23187 (UEDA et al. 1984), 10 mM $MnSO_4$ (UEDA et al. 1984), 1 mM hydrogen peroxide (MAKINO et al. 1980), sulfonylureas (e.g., 1 mM tolubutamide, acetohexamide, and carboxytolbutamide (OSEGAWA et al. 1982), concanavalin A (SUZUKI et al. 1984), anti-insulin receptor antibody (MAKINO 1987), and insulin-like growth factor I (IGF-I) (MAKINO 1987). The enzyme is transiently stimulated by extracellular ATP (HASHIMOTO et al. 1987) or adenosine (WONG and OOI 1985).

LOTEN et al. (1978) found that liver phosphodiesterase is stimulated not only with insulin, but also with glucagon, which increases the cAMP concentration in hepatocytes. In contrast, HEYWORTH et al. (1983b) reported that while liver phosphodiesterase in the dense vesicles is stimulated with both insulin and glucagon, the insulin-sensitive enzyme associated with the plasma membrane is not activated with glucagon.

The physiological significance of insulin action on phosphodiesterase is still controversial. The problem is not in the validity of the observations that insulin stimulates phosphodiesterase and concomitantly lowers the cellular level of cAMP (KONO and BARHAM 1973). Instead, it is yet to be determined whether the action of insulin on phosphodiesterase is (a) the only, (b) the primary, or (c) a nonessential action of the hormone for the manifestation of its effects on physiological activities, such as lipolysis. In theory, insulin could modulate the physiological activity of a cAMP system by (i) inhibiting adenylate cyclase; (ii) stimulating phosphodiesterase; or (iii) inhibiting cAMP-dependent protein kinase. The possibility that adenylate cyclase is the primary site of insulin action (i above) was first proposed by HEPP et al. (1969) and ILLIANO and CUATRECASAS (1972). Although their results are yet to be confirmed by others, it was recently reported by HEYWORTH and HOUSLAY (1983) that insulin inhibited adenylate cyclase when the system is fortified with GTP. Their data suggest that a certain GTP-binding protein might be involved in the action of insulin. The possibility that insulin inhibits cAMP-dependent protein kinase (iii above) is advocated by GABBAY and LARDY (1984, 1987), who found that insulin inhibited lipolysis that was stimulated by 8-bromo-cAMP without significantly reducing the concentration of the nucleotide. In contrast, ELKS and MANGANIELLO (1985) and BEEBE et al. (1985) contend that the primary site of insulin action is on phosphodiesterase (ii above). Thus, ELKS and MANGANIELLO (1984, 1985) found that insulin can block lipolysis induced by Ro-20-1720 (which inhibits soluble phosphodiesterase that is insulin-insensitive), but not lipolysis stimulated by cilostamide (which inhibits membrane-bound, insulin-sensitive phosphodiesterase). BEEBE et al. (1985) found that there are rough correlations among the relative effects of 20 species of cAMP-analogs on phosphodiesterase and those on lipolysis in adipocytes as well as on phosphorylase in hepatocytes. Since insulin exerts a clear-cut effect on phosphodiesterase, I feel that the effect must have some physiological meaning. At the same time, I would not be surprised if phosphodiesterase is not the only site of insulin action in the cAMP system since the optimum insulin concentration for inhibition of lipolysis is 0.1 nM (KONO and BARHAM 1973) while that for the stimulation of phosphodiesterase is 1–3 nM (KONO et al. 1975).

As for the cAMP-dependent stimulation of phosphodiesterase, most investigators seem to agree that its physiological role is in the reduction of the overly produced cAMP (ZINMAN and HOLLENBERG 1974). As is well known, when cells are exposed to a lipolytic agent, e.g., isoproterenol, the cellular level of cAMP is elevated very rapidly during the first 5–10 min, and then reduced precipitously during the next 20–30 min (MANGANIELLO et al. 1971).

E. Cell-Free Activation and Deactivation

As mentioned earlier (see Sect. D), both basal and insulin-stimulated phosphodiesterases are highly stable if they are partially purified through the step of fraction P-2 and suspended in buffer A (0.25 M sucrose, 10 mM 2-[{2-hydroxy-1,1-bis(hydroxymethyl)-ethyl}amino]ethanesulfonic acid (TES) buffer, pH 7.5) at 0° C or below. Nevertheless, the enzyme activity is rapidly and profoundly altered if fraction P-2 is incubated at 37° C in the presence of certain agents.

I. Effects of Salts

When fraction P-2 in buffer A (see above) is incubated at 37° C, not only the insulin-stimulated, but also the basal phosphodiesterase activities are rapidly reduced. In contrast, when fraction P-2 is heated to 37° C in buffer A supplemented with 0.15 M KCl (or NaCl) plus 5 mM MgSO$_4$, the basal enzyme activity is rapidly increased more than 100%, while the plus-insulin activity is slightly elevated (MAKINO et al. 1980). These salt effects are minimum when salt concentration is modest; e.g., when buffer A is mixed with 0.1 M NaCl or KCl. Under the latter conditions, the plus-insulin activity is slightly reduced while the basal activity is slightly increased (MAKINO et al. 1980).

II. Effects of Dithiothreitol at 37° C

When phosphodiesterase is incubated with 1 mM dithiothreitol (a reductant) at 37° C, the plus-insulin activity is reduced to the basal level in 5 min, while the basal activity is not significantly affected (MAKINO et al. 1980). In addition, the salt-dependent stimulation (see above) is either reversed or prevented in the presence of 1 mM dithiothreitol (MAKINO et al. 1980). The underlying mechanism of this dithiothreitol effect is still obscure; however, it is currently assumed: (a) that, as mentioned earlier in Sect. D, phosphodiesterase can take either a reduced (SH) form or an oxidized (SS) form; and (b) that the oxidized form is salt-sensitive while the reduced form is not (ROBINSON et al. 1988). Since phosphodiesterase maintains a low activity in the basal state of adipocytes, it is suggested that the enzyme assumes a reduced form in the cell (ROBINSON et al. 1989). These dithiothreitol effects, which are observed at 37° C, should not be confused with the aforementioned stimulatory effect of the same agent observed at 4° C (see Sect. C).

III. Effects of SH Blocking Agents

In the presence of 0.1 M NaCl (which stabilizes the enzyme activity as described above), both basal and insulin-stimulated phosphodiesterases (in fraction P-2) from adipocytes are strongly inhibited with p-chloromercuribenzenesulfonic acid (PCMBS), but not with iodoacetamide (IAA) or N-ethylmaleimide (NEM) (MAKINO et al. 1980). However, LOTEN and REDSHAW-LOTEN (1986) reported that the liver enzyme is stimulated by low concentrations of SH blocking agents.

IV. Effects of a Mild Proteolysis

As described earlier (see Sect. C), the catalytic domain of phosphodiesterase is separated from its regulatory domain by a mild proteolysis of the holoenzyme. The enzymatically active catalytic domain can be obtained even from the enzymes that have been inactivated by treatment with either low salt buffer or dithiothreitol (MAKINO et al. 1980). In contrast, no active fragment is generated from the enzyme that has been inactivated with PCMBS (MAKINO et al. 1980). It would appear, therefore, that both low salt buffer (Sect. E. I) and dithiothreitol (Sect. E. II) deactivate the enzyme by modulating its regulatory domain, whereas PCMBS (Sect. E. III) may do so by attacking the catalytic domain of the enzyme.

V. Effects of cGMP and cAMP

Cyclic GMP is a potent competitive inhibitor of hormone-sensitive phosphodiesterase ($K_i = 0.15 \ \mu M$) (DEGERMAN et al. 1987). Nevertheless, the same agent at 10 μM stimulates the enzyme more than 100% when fraction P-2 is: (a) incubated with the agent for 5–20 min at 37° C in the presence of 154 mM KCl + 5 mM MgSO$_4$ + 1 mM dithiothreitol; (b) diluted with excess low salt buffer (containing 1 mM sodium tetrathionate for oxidation of thiol residues); (c) pelleted by centrifugation; and (d) resuspended in the standard buffer for determination of phosphodiesterase activity (ROBINSON et al. 1989). It is suggested, therefore: (a) that phosphodiesterase has two sites for interaction with cGMP, one for stimulation and the other for competitive inhibition; and (b) that the inhibitory effect of cGMP may disappear while the stimulatory effect may persist when the cGMP-treated enzyme is diluted with a buffer free from the agent (ROBINSON et al. 1989). Under these conditions, the enzyme is also stimulated with 10 μM cAMP, although the stimulatory effect of cAMP is significantly less than that of cGMP (ROBINSON et al. 1989). These observations suggest that insulin-sensitive phosphodiesterase is closely related to so-called cGMP-sensitive phosphodiesterase often classified as type II (see Sect. E. I). The possible mechanism of cGMP-dependent stimulation and inhibition of phosphodiesterase is schematically illustrated in Fig. 1.

VI. Effects of GTP and ATP

HEYWORTH et al. (1983 a) reported that the liver phosphodiesterase is stimulated with several guanosine derivatives (e.g., GTP, GTP-γ-S, etc.) in the presence of ATP. Their data imply that G protein might be involved in the stimulation of phosphodiesterase. However, no such effect of GTP plus ATP has been observed in the fat cell system (ROBINSON et al. 1988).

VII. Effects of Phosphodiesterase-Specific Inhibitors

As is well known, phosphodiesterase is inhibited with methylxanthines, especially with 3-isobutyl-1-methylxanthine (IBMX) (BEAVO et al. 1970). ELKS and MANGANIELLO (1984) found that: (a) whereas IBMX inhibits both soluble and

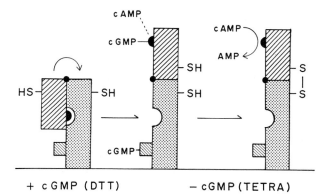

$$+ \ c\,G\,M\,P \ (D\,T\,T) \qquad\qquad - \ c\,G\,M\,P\,(T\,E\,T\,R\,A)$$

Fig. 1. Schematic illustration of the possible mechanism of cGMP-dependent stimulation and inhibition of phosphodiesterase. The figure suggests that phosphodiesterase has a catalytic domain (*hatched area*) and a regulatory domain (*stippled area*), and is associated with cellular membranes at the regulatory domain. The reduced, basal form of the enzyme (far left) changes its conformation when it is interacted with cGMP at the regulatory domain (center). However, at this stage, the enzyme is still inactive, since its catalytic activity is competitively inhibited by cGMP. When the cGMP-treated enzyme is washed with a cGMP-free buffer that contains sodium tetrathionate (an oxidant), the enzyme is freed from the inhibitory effect of cGMP and, at the same time, is locked into the activated conformation by oxidation (far right). It is further suggested that the "hinge" between the catalytic and regulatory domains is trypsin-sensitive, and that the catalytic domain is easily separated from the holoenzyme by mild proteolysis

particulate phosphodiesterases; (b) Ro-20-1724 preferentially inhibits soluble phosphodiesterase; and (c) cilostamide specifically blocks the activity of the particulate enzyme. Cilostamide is a synthetic compound, originally reported to inhibit phosphodiesterase from platelets with a K_i value as low as 5 nM (HIDAKA et al. 1979). Another inhibitor, griseolic acid, is an antibiotic; it has a chemical structure similar to that of cAMP and shows different potencies towards phosphodiesterases from different tissues (NAKAGAWA et al. 1985; TAKAHASHI et al. 1985). Its K_i value to insulin-stimulated phosphodiesterase from rat adipocytes is 0.19 μM (ROBINSON et al. 1988).

VIII. Effects of TPA (a Phorbol Ester)

IRVINE et al. (1986) reported that TPA inhibited phosphodiesterase in hepatocytes. However, the agent appears to have no significant effect on the fat cell enzyme or its insulin-dependent stimulation (KONO, unpublished data).

IX. Effects of Phosphorylation

Since the action of insulin on phosphodiesterase is blocked by depletion of intracellular ATP (KONO et al. 1977), we previously suggested that phosphorylation might be involved in the hormonal stimulation of the enzyme (MAKINO and KONO 1980; KONO 1983). In agreement with this view, MARCHMONT and HOUSLAY (1980a, 1981) have reported that the liver enzyme was stimulated by

phosphorylation with ATP in the presence of cAMP and insulin. More recently
Corbin and his associates observed that the fat cell enzyme in fraction P-2 is
stimulated by incubation with ATP plus the catalytic subunit of cAMP-
dependent protein kinase (Gettys et al. 1988). Likewise, Macphee et al. (1987)
reported that cGMP-inhibitable, low K_m phosphodiesterase from human
platelets is stimulated by phosphorylation with cAMP-dependent protein kinase.

F. Possible Mechanisms of Insulin Action

The molecular mechanism by which insulin stimulates phosphodiesterase is still
largely unknown. Nevertheless, it has been well established that the enzyme is
stimulated in situ (Makino and Kono 1980) without being translocated as the
glucose transport system. Presumably, phosphodiesterase is stimulated by a conformational change, as suggested above in Sect. E.V (Fig. 1). The possible
candidates for the hormonal signal that would induce such a conformational
change have been discussed before (Kono 1983; Makino 1987), and some new
theories are considered below.

I. Phosphorylation Theory

As described at the end of the previous section, at least three laboratories now
claim that phosphodiesterase is stimulated by cAMP-dependent phosphorylation. These observations may explain how phosphodiesterase is stimulated by
various lipolytic agents that would increase the cellular concentration of cAMP
(see Sect. D). However, they fail to delineate the mechanism by which insulin
stimulates the enzyme since insulin lowers, rather than elevates, the intracellular
level of cAMP (Kono and Barham 1973). Since the insulin receptor is a
hormone-sensitive tyrosine kinase (Kasuga et al. 1982), it is of interest to know
whether phosphodiesterase is stimulated by the receptor-mediated phosphorylation.

II. Mediator Theory

This subject is discussed in detail elsewhere in this volume (see Chap. 14); therefore, it is not considered here.

III. G Protein Theory

A number of hormonal signals are now known to be transmitted from the surface
receptor to intracellular enzymes by way of GTP binding (or G) proteins (see,
e.g., Dohlman et al. 1987; Levitzki 1987). Ui and his associates (e.g., Katada et
al. 1982) have shown that biological activities of certain G proteins are abolished
by the action of pertussis toxin. The effects of pertussis toxin on the actions of insulin are controversial; Moreno et al. (1983), Kather et al. (1983), and Weber et
al. (1987) concluded that the toxin has no significant effects on the system. In

contrast, GOREN et al. (1985) found that pertussis toxin apparently blocked the hormonal actions on both glucose transport and lipolysis. HEYWORTH et al. (1986) reported that pertussis toxin inhibited the action of insulin on liver phosphodiesterase associated with the dense vesicles, but not that of the enzyme in the plasma membrane. Our own data indicate that pertussis toxin may have no direct effect on the actions of insulin to stimulate phosphodiesterase and glucose transport. Our own data are consistent with the interpretation that the apparent effects of pertussis toxin on the system are secondary to the toxin-dependent accumulation of fatty acids in the incubation mixture (SHIBATA and KONO, unpublished). It should be noted, however, that pertussis toxin does not inhibit all kinds of G proteins. HOUSLAY et al. (1984) suggest that at least some of the actions of insulin might be mediated by a specific G protein; their idea is intriguing, but it is yet to be verified with solid experimental evidence.

Acknowledgments. Our original studies cited in this chapter were supported by NIH Grant DK 07625. I am grateful to my colleagues, research associates, research assistants, and students who have worked with me in our studies on insulin-sensitive phosphodiesterase. Their names appear in the individual articles cited in this chapter. I am also thankful to Mrs. Frances W. Robinson, Miss Jo Ellen Flanagan, and Mrs. Patsy Raymer for their help in the preparation of this chapter. I am also grateful to Dr. Daryl Granner, the Chairman of our department, for his interest in our studies and for his continued encouragement.

References

Allan EH, Sneyd JGT (1975) An effect of glucagon on 3′,5′-cyclic AMP phosphodiesterase activity in isolated rat hepatocytes. Biochem Biophys Res Commun 62:594–601

Appleman MM, Allan EH, Ariano MA, Ong KK, Tusang CA, Weber HW, Whitson RH (1984) Insulin control of cyclic AMP phosphodiesterase. Adv Cyclic Nucleotide Protein Phosphorylation Res 16:149–158

Beavo JA, Rogers NL, Crofford OB, Hardman JG, Sutherland EW, Newman EN (1970) Effects of xanthine derivatives on lipolysis and on adenosine 3′,5′-monophosphate phosphodiesterase activity. Mol Pharmacol 6:597–603

Beavo JA, Hansen RS, Harrison SA, Hurwitz RL, Martins TJ, Mumby MC (1982) Identification and properties of cyclic nucleotide phosphodiesterases. Mol Cell Endocrinol 28:387–410

Beebe SJ, Redmon JB, Blackmore PF, Corbin JD (1985) Discriminative insulin antagonism of stimulatory effects of various cAMP analogs on adipocyte lipolysis and hepatocyte glycogenolysis. J Biol Chem 260:15781–15788

Butcher RW, Sutherland EW (1962) Adenosine 3′,5′-phosphate in biological materials. I. Purification and properties of cyclic 3′,5′-nucleotide phosphodiesterase and use of this enzyme to characterize adenosine 3′,5′-phosphate in human urine. J Biol Chem 237:1244–1250

Degerman E, Belfrage P, Newman AH, Rice KC, Manganiello VC (1987) Purification of the putative hormone-sensitive cyclic AMP phosphodiesterase from rat adipose tissue using a derivative of cilostamide as a novel affinity ligand. J Biol Chem 262:5797–5807

Dohlman HG, Caron MG, Lefkowitz RJ (1987) A family of receptors coupled to guanine nucleotide regulatory proteins. Biochemistry 26:2657–2664

Elks ML, Manganiello VC (1984) Selective effects of phosphodiesterase inhibitors on different phosphodiesterases, adenosine 3′,5′-monophosphate metabolism, and lipolysis in 3T3-L1 adipocytes. Endocrinology 115:1262–1268

Elks ML, Manganiello VC (1985) Antilipolytic action of insulin: role of cAMP phosphodiesterase activation. Endocrinology 116:2199–2121

Francis SH, Kono T (1982) Hormone-sensitive cAMP phosphodiesterase in liver and fat cells. Mol Cell Biochem 42:109–116

Gabbay RA, Lardy HA (1984) Site of insulin inhibition of cAMP-stimulated glycogenolysis. J Biol Chem 259:6052–6055

Gabbay RA, Lardy HA (1987) Insulin inhibition of hepatic cAMP-dependent protein kinase: decreased affinity of protein kinase for cAMP and possible differential regulation of intrachain sites 1 and 2. Proc Natl Acad Sci USA 84:2218–2222

Gain KR, Appleman MM (1978) Distribution and regulation of the phosphodiesterases of muscle tissues. Adv Cyclic Nucleotide Res 9:221–231

Gettys TW, Vine AJ, Simonds MF, Corbin JD (1988) Activation of the particulate low K_m phosphodiesterase of adipocytes by addition of cAMP-dependent protein kinase. J Biol Chem 263:10359–10363

Goren HJ, Northrup JK, Hollenberg MD (1985) Action of insulin modulated by pertussis toxin in rat adipocytes. Can J Physiol Pharmacol 63:1017–1022

Hashimoto N, Robinson FW, Shibata Y, Flanagan JE, Kono T (1987) Diversity in the effects of extracellular ATP and adenosine on the cellular processing and physiologic actions of insulin in rat adipocytes. J Biol Chem 262:15026–15032

Hepp KD, Menahan LA, Wieland O, Williams RH (1969) Studies on the action of insulin in isolated adipose tissue cells. II. 3′,5′-Nucleotide phosphodiesterase and antilipolysis. Biochim Biophys Acta 184:554–565

Heyworth CM, Houslay MD (1983) Insulin exerts actions through a distinct species of guanine nucleotide regulatory protein: inhibition of adenylate cyclase. Biochem J 214:547–552

Heyworth CM, Rawal S, Houslay MD (1983a) Guanine nucleotides can activate the insulin-stimulated phosphodiesterase in liver plasma membranes. FEBS Lett 154:87–91

Heyworth CM, Wallace AV, Houslay MD (1983b) Insulin and glucagon regulate the activation of two distinct membrane-bound cyclic AMP phosphodiesterases in hepatocytes. Biochem J 214:99–110

Heyworth CM, Wallace AV, Wilson SR, Houslay MD (1984) An assessment of the ability of insulin-stimulated cyclic AMP phosphodiesterase to decrease hepatocyte intracellular cyclic AMP concentrations. Biochem J 222:183–187

Heyworth CM, Grey A-M, Wilson SR, Hanski E, Houslay MD (1986) The action of islet activating protein (pertussis toxin) on insulin's ability to inhibit adenylate cyclase and activate cyclic AMP phosphodiesterase in hepatocytes. Biochem J 235:145–149

Hidaka H, Hayashi H, Kohri H, Kimura Y, Hosokawa T, Igawa T, Saitoh Y (1979) Selective inhibitor of platelet cyclic adenosine monophosphate phosphodiesterase, cilostamide, inhibits platelet aggregation. J Pharmacol Exp Ther 211:26–30

Houslay MD, Wallace AV, Cooper ME, Pyne NJ, Wilson SR, Heyworth CM (1984) Does insulin exert certain of its actions through a distinct species of guanine nucleotide regulatory protein? Biochem Soc Trans 12:766–768

Illiano G, Cuatrecasas P (1972) Modulation of adenylate cyclase activity in liver and fat cell membranes by insulin. Science 175:906–908

Irvine F, Pyne NJ, Houslay MD (1986) The phorbol ester TPA inhibits cyclic AMP phosphodiesterase activity in intact hepatocytes. FEBS Lett 208:455–459

Kasuga M, Zick Y, Blithe DL, Crettaz M, Kahn CR (1982) Insulin stimulates tyrosine phosphorylation of the insulin receptor in a cell-free system. Nature 298:667–669

Katada T, Amano T, Ui M (1982) Modulation by islet-activating protein of adenylate cyclase activity in C6 glioma cells. J Biol Chem 257:3739–3746

Kather H, Aktories K, Schulz G, Jakobs KH (1983) Islet-activating protein discriminates the antilipolytic mechanism of insulin from that of other antilipolytic compounds. FEBS Lett 161:149–152

Kono T (1983) Actions of insulin on glucose transport and cAMP phosphodiesterase in fat cells: involvement of two distinct molecular mechanisms. Recent Prog Horm Res 39:519–557

Kono T, Barham FW (1973) Effects of insulin on the levels of adenosine 3′:5′-monophosphate and lipolysis in isolated rat epidermal fat cells. J Biol Chem 248:7417–7426

Kono T, Robinson FW, Sarver JA (1975) Insulin-sensitive phosphodiesterase. J Biol Chem 250:7826–7835

Kono T, Robinson FW, Sarver JA, Vega FV, Pointer RH (1977) Actions of insulin in fat cells. J Biol Chem 252:2226–2233

Levitzki A (1987) Regulation of adenylate cyclase by hormones and G-proteins. FEBS Lett 211:113–118

Loten EG (1983) Detergent solubilization of rat liver particulate cyclic AMP phosphodiesterase. Int J Biochem 15:923–928

Loten EG, Redshaw-Loten JC (1986) Stimulation of low K_m cyclic AMP phosphodiesterase by sulphydryl modification. Int J Biochem 18:847–851

Loten EG, Sneyd JGT (1970) An effect of insulin on adipose tissue adenosine 3′:5′-cyclic monophosphate phosphodiesterase. Biochem J 120:187–193

Loten EG, Assimacopoulos-Jeannet FD, Exton JH, Park CR (1978) Stimulation of a low K_m phosphodiesterase from liver by insulin and glucagon. J Biol Chem 253:746–757

Loten EG, Francis SH, Corbin JD (1980) Proteolytic solubilization and modification of hormone-sensitive cyclic nucleotide phosphodiesterase. J Biol Chem 255:7838–7844

Lovell-Smith CJ, Manganiello VC, Vaughan M (1977) Solubilization and characterization of hormone-responsive phosphodiesterase activity of rat fat cells. Biochim Biophys Acta 497:447–458

Macauly SL, Kiechle FL, Jarett L (1983) Comparison of phospholipid effects on insulin-sensitive low K_m cyclic AMP phosphodiesterase in adipocyte plasma membranes and microsomes. Biochim Biophys Acta 760:293–299

Macphee CH, Reifsnyder DH, Moore TA, Beavo JA (1987) Intact cell and cell-free phosphorylation and concomitant activation of a low K_m, cAMP phosphodiesterase found in human platelets. J Cyclic Nucleotide Protein Phosphor Res 11:487–496

Makino H (1987) Action of insulin and insulin-sensitive phosphodiesterase (in Japanese). In: Kosaka K, Akanuma Y (eds) Tonyobyo gaku. Shindan To Chiryo Sha, Tokyo, pp 115–150

Makino H, Kono T (1980) Characterization of insulin-sensitive phosphodiesterase in fat cells. II. Comparison of enzyme activities stimulated by insulin and by isoproterenol. J Biol Chem 255:7850–7854

Makino H, de Buschiazzo PM, Pointer RH, Jordan JE, Kono T (1980) Characterization of insulin-sensitive phosphodiesterase in fat cells. I. Effects of salts and oxidation-reduction agents. J Biol Chem 255:7845–7849

Makino H, Kanatsuka A, Osegawa M, Kumagai A (1982) Effects of dithiothreitol on insulin-sensitive phosphodiesterase in rat fat cells. Biochim Biophys Acta 704:31–36

Manganiello VC, Vaughan M (1973) An effect of insulin on cyclic adenosine 3′:5′-monophosphate phosphodiesterase activity in fat cells. J Biol Chem 248:7164–7170

Manganiello VC, Murad F, Vaughan M (1971) Effects of lipolytic and antilipolytic agents on cyclic 3′,5′-adenosine monophosphate in fat cells. J Biol Chem 246:2195–2202

Marchmont RJ, Houslay MD (1980a) Insulin triggers cyclic AMP-dependent activation and phosphorylation of a plasma membrane cyclic AMP phosphodiesterase. Nature 286:904–906

Marchmont RJ, Houslay MD (1980b) A peripheral and an intrinsic enzyme constitute the cyclic AMP phosphodiesterase activity of rat liver plasma membranes. Biochem J 187:381–392

Marchmont RJ, Houslay MD (1981) Characterization of the phosphorylated form of the insulin-stimulated cyclic AMP phosphodiesterase from rat liver plasma membranes. Biochem J 195:653–660

Marchmont RJ, Ayad SR, Houslay MD (1981) Purification and properties of the insulin-stimulated cyclic AMP phosphodiesterase from rat liver plasma membranes. Biochem J 195:645–652

Moreno FJ, Mills I, Garcia-Sáinz JA, Fain JN (1983) Effects of pertussis toxin treatment on the metabolism of rat adipocytes. J Biol Chem 258:10938–10943

Nakagawa F, Okazaki T, Naito A (1985) Griseolic acid, an inhibitor of cyclic adenosine 3′,5′-monophosphate phosphodiesterase. I. Taxonomy, isolation and characterization. J Antibiot (Tokyo) 38:823–829

Osegawa M, Makino H, Kanatsuka A, Kumagai A (1982) Effects of sulfonylureas on membrane-bound low K_m cyclic AMP phosphodiesterase in rat fat cells. Biochim Biophys Acta 721:289–296

Pawlson LG, Lovell-Smith CJ, Manganiello VC, Vaughan M (1974) Effects of epinephrine, adrenocorticotrophic hormone, and theophylline on adenosine 3′,5′-monophosphate phosphodiesterase activity in fat cells. Proc Natl Acad Sci USA 71:1639–1642

Pyne NJ, Cooper ME, Houslay MD (1987) The insulin- and glucagon-stimulated "dense-vesicle" high-affinity cyclic AMP phosphodiesterase from rat liver. Biochem J 242:33–42

Robinson FW, Smith CJ, Flanagan JE, Shibata H, Kono T (1989) Cyclic GMP-dependent stimulation of the membrane-bound insulin-sensitive, low K_m cAMP phosphodiesterase from rat adipocytes. J Biol Chem 264:16458–16464

Saltiel AR, Steigerwalt RW (1986) Purification of putative insulin-sensitive cAMP phosphodiesterase or its catalytic domain from rat adipocytes. Diabetes 35:698–704

Smoake JA, McMahon KL, Wright RK, Solomon SS (1981) Hormonally sensitive cyclic AMP phosphodiesterase in liver cells. J Biol Chem 256:8531–8535

Strada SJ, Thompson WJ (1984) Nomenclature recommendations (cyclic nucleotide phosphodiesterases). Adv Cyclic Nucleotide Protein Phosphorylation Res 16:vi

Sutherland EW, Rall TW (1958) Fractionation and characterization of a cyclic adenine ribonucleotide formed by tissue particles. J Biol Chem 232:1077–1091

Suzuki T, Makino H, Kanatsuka A, Osegawa M, Yoshida S, Sakamoto Y (1984) Activation of insulin-sensitive phosphodiesterase by lectins and insulin-dextran complex in rat fat cells. Metabolism 33:572–576

Suzuki T, Makino H, Kanatsuka A, Yoshida S (1987) Changes in insulin binding and action in fat cells treated with trypsin (in Japanese, Abstract in English). J Jpn Diabetes Soc 30:133–139

Takahashi S, Nakagawa F, Kawazoe K, Furukawa Y, Sato S, Tamura C, Naito A (1985) Griseolic acid, an inhibitor of cyclic adenosine 3′,5′-monophosphate phosphodiesterase. II. The structure of griseolic acid. J Antibiot (Tokyo) 38:830–834

Tria E, Scapin S, Cocco C, Luly P (1977) Insulin-sensitive adenosine 3′,5′-cyclic monophosphate phosphodiesterase of hepatocyte plasma membrane. Biochim Biophys Acta 496:77–87

Ueda M, Robinson FW, Smith MM, Kono T (1984) Effects of divalent cations on the regulation of insulin-sensitive glucose transport and cAMP phosphodiesterase in adipocytes. J Biol Chem 259:9520–9525

Vega FV, Key RJ, Jordan JE, Kono T (1980) Reversal of insulin effects in fat cells may require energy for deactivation of glucose transport, but not for deactivation of phosphodiesterase. Arch Biochem Biophys 203:167–173

Weber HW, Appleman MM (1982) Insulin-dependent and insulin-independent low K_m cyclic AMP phosphodiesterase from rat adipose tissue. J Biol Chem 257:5339–5341

Weber HW, Chung F-Z, Day K, Appleman MM (1987) Insulin stimulation of cyclic AMP phosphodiesterase is independent from the G-protein pathways involved in adenylate cyclase regulation. J Cyclic Nucleotide Protein Phosphor Res 11:345–354

Wells JN, Hardman JG (1977) Cyclic nucleotide phosphodiesterases. Adv Cyclic Nucleotide Res 8:119–143

Wong EH-A, Ooi S-O (1985) The action of adenosine in relation to that of insulin on the low-K_m cyclic AMP phosphodiesterase in rat adipocytes. Biochem J 227:815–821

Woo Y-T, Manery JF (1973) Cyclic AMP phosphodiesterase activity at the external surface of intact skeletal muscles and stimulation of the enzyme by insulin. Arch Biochem Biophys 154:510–519

Zinman B, Hollenberg CH (1974) Effect of insulin and lipolytic agents on rat adipocyte low K_m cyclic adenosine 3′:5′-monophosphate phosphodiesterase. J Biol Chem 249:2182–2187

CHAPTER 18

Regulation of Gene Expression by Insulin

J. L. MESSINA

A. Introduction

Insulin exerts a variety of biological effects – some of which occur within seconds
to minutes, such as glucose transport and the regulation of enzyme activity.
Other effects of insulin occur over hours to days, including the "growth effects"
of insulin which have traditionally included RNA and DNA synthesis. Until the
last few years, insulin's effects on RNA synthesis were studied by measuring
uridine incorporation into total cellular RNA or by measuring the production of
specific immunoprecipable proteins. The study of overall RNA synthesis does
not determine specific effects of insulin on the expression of a single mRNA. The
study of total protein synthesis, precipitated from either in vitro translation assays or tissue or serum samples also does not necessarily describe the effects of insulin on expression of a specific mRNA. Besides affecting mRNA concentrations, insulin may alter the ability to translate an mRNA into protein and can
alter the stability of a protein. Since the last volume on insulin in this series
(HASSELBLATT and BRUCHHAUSEN 1975) molecular biological techniques have become widely used to study insulin's regulation of specific genes. This has led to
the design of experiments analyzing the effects of insulin on cellular concentrations of specific mRNAs using specific cDNA probes that hybridize with only
one mRNA species. This chapter will review the ever-expanding group of genes
regulated by insulin at the level of mRNA transcription, or at a step following
transcription, but prior to translation of the protein.

When reviewing the literature in this area there are three primary issues to
consider before defining a gene as one under the control of insulin. It is first
necessary to determine whether the effects of insulin on mRNA production are
due specifically to insulin's action through its own receptor or whether insulin is
acting by its ability to cross-react at high concentrations with receptors for
insulin-like growth factor I or II (IGF-I, IGF-II; see Chaps. 9 and 13). Since insulin can interact with these other receptors, it is important that low hormonal
concentrations prove effective. Alternatively, using specific antibodies to block
the IGF-I and IGF-II receptors, it can be ascertained that insulin is working
through its own receptor.

A second question to address is whether a gene is induced directly by insulin
or whether the gene is induced only in response to insulin's action in promoting
DNA synthesis and cell division, or as a consequence of insulin's action in
regulating sugar and amino acid transport. Certain genes are cyclically regulated
in the cell cycle and are not under specific control of insulin – even when insulin

stimulates a cell to progress through the cycle. Other genes are regulated by changes in intracellular availability of specific metabolic substrates. Insulin may be regulating availability of the metabolic substrates, but have no direct effect on expression of these genes. When insulin affects transcription of an mRNA within minutes, these possibilities are lessened, although not completely discounted. However, when the effects of insulin are studied over many hours, it is often difficult to be sure that insulin is directly regulating expression of the mRNA in question.

The third criterion to address when studying insulin's regulation of gene expression is whether effects of insulin are due directly to insulin, or whether insulin is altering the sensitivity of cells to other hormones and factors. In vivo, it is often impossible to determine whether insulin is acting directly, or is acting indirectly through it effects on other hormones or alterations in systemic availability of metabolic substrates. Determining whether insulin is acting directly or by altering the availability of other factors is important for our understanding of the intracellular mechanism by which insulin acts to regulate a gene. To determine whether insulin specifically regulates a gene, it is preferable to study effects of insulin: (a) in a cell culture system or in isolated explants or slices of tissue; (b) in the absence of other hormones or under other strictly defined conditions; and (c) in situations where effects of insulin can be rapidly measured. The eventual study of specific sequences within a gene that alter transcription of a gene in response to insulin, and isolation of proteins involved in regulation of these genes, is the next exciting step in understanding of insulin's action on gene expression. However,

Table 1. Genes regulated by insulin

Induced by insulin	Inhibited by insulin	Inhibited and induced by insulin
p33	Phosphoenolpyruvate	Albumin
c-fos	carboxykinase	Tyrosine
c-myc	Carbamoyl phosphate	Aminotrans-
Glyceraldehyde-3-phosphate	synthetase I[a]	ferase
dehydrogenase	Serine dehydratase[a]	
Fatty acid synthetase[b]	Tryptophan-2,3-dioxygenase	
Malic enzyme[b]		
Pyruvate kinase[a]		
Ornithine decarboxylase[a]		
Metallothionein[a]		
Apolipoprotein A–IV[a]		
Apolipoprotein A–I[a]		
Glucokinase		
Glucose-6-phosphate dehydrogenase[a]		
α_{2u}-Globulin[a]		
Casein[a]		
Amylase[a]		
Ovalbumin[a]		
δ-Crystallin[a]		

[a] A direct effect of insulin on the transcription of this gene has not been demonstrated.
[b] Insulin probably works solely at a post transcriptional site.

there are other possible effects of insulin that can alter cytoplasmic concentrations of an mRNA. These include regulation of nuclear or cytoplasmic stability of mRNA, nuclear processing and polyadenylation of primary transcripts, and transport of mRNA out of the nucleus. This chapter will discuss the diverse effects of insulin on expression of a variety of specific mRNAs. Genes inhibited by insulin will be discussed first, followed by genes induced by insulin. The last category, genes both inhibited and induced by insulin, depending on the cell lines used or other variables, will be discussed in the third section (see Table 1).

B. Genes Inhibited by Insulin

I. Phosphoenolpyruvate Carboxykinase

Hepatic phosphoenolpyruvate carboxykinase (PEPCK) is the best documented gene regulated by insulin and has been studied extensively over the last decade, mainly by the research groups of R. W. Hanson and D. K. Granner. Many excellent reviews are available on the regulation of this gene by insulin as well as by glucocorticoids and hormones that stimulate production of cyclic AMP. Due to space limitations, I will not attempt to encompass the complete literature on this well-researched system, but refer you to several of these reviews (see LOOSE et al. 1985; GRANNER and ANDREONE 1985). The following discussion is a short summary of this reviewed literature and of work that has emerged over the last several years.

PEPCK is mainly a cytosolic enzyme, distributed in the liver, kidney cortex, adipose tissue, and small intestine (LOOSE et al. 1985; GRANNER and ANDREONE 1985). It catalyzes the conversion of oxaloacetate to phosphoenolpyruvate, the rate-limiting step in hepatic and renal gluconeogenesis. Glucagon, epinephrine, thyroxine, glucocorticoids, and insulin all regulate expression of this gene. Feeding of normal, fasted animals and insulin treatment of diabetic chow-fed animals leads to a decrease in PEPCK synthesis and PEPCK mRNA (LOOSE et al. 1985; GRANNER and ANDREONE 1985). It is now well established that in both intact rats and in cultured hepatoma cells insulin acts to inhibit the transcription of PEPCK mRNA. Insulin does not affect PEPCK mRNA stability nor its transport out of the nucleus. Insulin's inhibition of PEPCK transcription counteracts the actions of both glucocorticoids and hormones that work through increasing cAMP levels.

In recent work, MAGNUSON et al. (1987) constructed fusion genes containing the 5'promoter region of the PEPCK gene and the coding sequence of the chloramphenicol acetyltransferase (CAT) gene, a bacterial gene not expressed in mammalian cells. These fusion genes were regulated in a fashion analogous to the endogenous PEPCK gene when transfected into H4IIE (H4) hepatoma cells. When a fusion gene containing approximately 2 kb of the 5' nontranscribed region of the PEPCK gene was used, CAT activity was reduced by 25%–50% following addition of 10 nM insulin for 18 h (Fig. 1). Treatment of H4 cells with dexamethasone (Dex), a synthetic glucocorticoid, for 18 h led to a 2.5-fold increase in CAT. Simultaneous addition of insulin blocked Dex's stimulation of

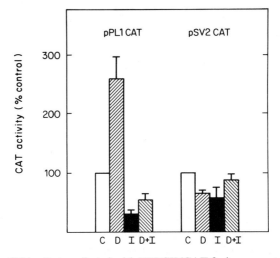

Fig. 1. Response of H4 cells transfected with PEPCK/CAT fusion genes to dexamethasone and insulin. Cultures of H4IIE cells were transfected in suspension with either pPL1CAT, which contains the 5' fragment of the PEPCK gene spliced to the CAT coding sequence or pSV2CAT which contains the CAT gene, but no PEPCK sequences. Transfected cells were treated for 18 h with: *C* untreated control; *D* 100 n*M* Dex; *I* 10 n*M* insulin. The mean $\pm$ standard deviations are indicated for four separate experiments. (From Magnuson et al. 1987)

CAT activity. This effect was significant with concentrations of insulin as low as 10 p*M*. Only 600 base pairs immediately 5' to the transcriptional start site were necessary for regulation of the fusion genes by insulin. Thus, within this 600 base pairs was a site (or sites) specifically sensitive to long-term insulin addition to H4 cells.

II. Other Genes Inhibited by Insulin

1. Carbamoyl Phosphate Synthetase I

In addition to PEPCK, there are other genes whose expression are inhibited by insulin. Carbamoyl phosphate synthetase I is a mitochondrial matrix enzyme catalyzing the first step of the urea cycle. Dex treatment of H35 hepatoma cell cultures for 13 h induced the carbamoyl phosphate synthetase I gene (Kitagawa et al. 1985). Insulin added along with Dex suppressed induction of carbamoyl phosphate synthetase I mRNA by 50%.

2. Serine Dehydratase

Immunoprecipitable serine dehydratase translated in vitro from primary cultures of rat hepatocytes was increased 5- to 10-fold following 12 h of incubation with Dex and glucagon (Noda et al. 1984). Addition of insulin along with Dex and glucagon suppressed this induction of serine dehydratase mRNA activity by 60%–70%.

3. Tryptophan 2,3-Dioxygenase

Tryptophan 2,3-dioxygenase is a key enzyme in tryptophan metabolism. Culture of rat hepatocytes with dex and glucagon for 12 h induced translatable mRNA activity, hybridizable mRNA levels, and transcription of tryptophan 2,3-dioxygenase mRNA 10- to 40-fold (Niimi et al. 1983; Nakamura et al. 1987). When added along with these other hormones, 100 nM insulin suppressed this induction by 70%–85%.

C. Genes Induced by Insulin

I. *p33*

Insulin induces the expression of a gene termed *p33*. When in vitro translations of purified *p33* mRNA were performed, and isolated by SDS-PAGE, a 53 kdalton peptide was obtained. The cDNA was originally cloned from rat livers by Kenney and his coworkers [Lee et al. (1985)] from a group of mRNAs that were induced in livers of hydrocortisone-treated adrenalectomized rats. Both glucocorticoids and insulin were found to increase cytoplasmic concentrations and transcription of *p33* mRNA following administration to adrenalectomized rats, reaching a maximum of a fivefold increase after 1 h and returning to baseline by 2 h. However, it was unclear whether insulin treatment in vivo was acting directly or through altering plasma levels of other hormones and metabolic substrates.

In cultured, serum-starved rat H4IIEC3 hepatoma cells, insulin caused a 10- to 15-fold increase in *p33* mRNA (Messina et al. 1985a), reaching a peak within 1–2 h which was maintained for 4–6 hours (Table 2). Insulin at concentrations as low as 1 pM induced this mRNA in the absence of any other hormones or serum.

Table 2. Regulation of *p*33 mRNA[a]

Additions	Fold induction of		Stability of mRNA $t_{1/2}$ (min)
	mRNA	Transcription	
Control	1	1	34
Insulin	12	3	38
WGA	28	ND	40
Con A	21	ND	47
Dex	16	18	32

[a] H4 hepatoma cells were incubated with insulin (5 nM), WGA (100 μg/ml), Con A (10 μg/ml), or Dex (50 nM) prior to preparation of mRNA for determination of cellular *p*33 mRNA concentrations or transcription rate. The different agents had slightly different time courses and transcription increased more rapidly than did mRNA accumulation. Thus, the values expressed are the peak values attained with each agent. The half-life ($t_{1/2}$) was determined following addition of actinomycin D (5 μg/ml) to the cells in the presence of the other agents. The values are the means of 4–8 separate experiments for each agent in each category. ND, not determined.

Lectins (concanavalin A, Con A, and wheat germ agglutinin, WGA) which often mimic the effect of insulin on isolated or cultured cells (Cuatrecasas 1973; Cuatrecasas and Tell 1973; Smith and Liu 1981), also increased the cellular accumulation of *p33* mRNA (Messina et al. 1985b, 1987a). Neither insulin nor either of the lectins altered the cytoplasmic stability of *p33* mRNA (Table 2; Messina et al. 1985a, b, 1987a).

In recent work Dex also induced the cytoplasmic accumulation of *p33* mRNA in H4IIEC3 cells (Messina 1989a). The 10- to 20-fold effect of Dex on *p33* mRNA accumulation was entirely accounted for by a 10- to 20-fold increase in *p33* mRNA transcription. However, insulin administration resulted in only a threefold increase in the transcription of *p33* mRNA (Table 2). The insulin induction of transcription was time- and dose-dependent and was blocked by addition of α-amanitin (Messina 1989a). Since the cellular concentration of *p33* mRNA was induced to a greater extent than was transcription, and since insulin did not alter the cytoplasmic stability of *p33* mRNA, insulin must have been regulating at least one other step in the sequence between RNA synthesis and RNA stability. Possible posttranscriptional effects of insulin include decreased nuclear degradation of nascent transcripts, increased processing of the nascent transcripts, or increased transport of processed, mature *p33* mRNA out of the nucleus. This is the only example to date of both transcriptional and posttranscriptional regulation of an mRNA by insulin. As more genes are studied, other instances of complex regulation by insulin may become apparent.

II. c-*fos* and c-*myc*

The proto-oncogenes c-*fos* and c-*myc* appear to be part of a growth or differentiation response of cells to various growth factors. Insulin was also found to regulate these genes in several cell lines. In 3T3-L1 adipocytes, and to a lesser extent in 3T3-L1 fibroblasts, insulin induced accumulation of c-*fos* mRNA (Stumpo and Blackshear 1986). Effects of insulin were rapid (reached a maximum by 30 min), but not exclusive since other growth factors, including platelet-derived growth factor (PDGF), fibroblast growth factor (FGF), and bombesin all increased accumulation of c-*fos* mRNA. The minimum concentration of insulin that induced c-*fos* mRNA was 7 nM, with 70 nM giving a maximum effect. Insulin-like growth factor II was ineffective in inducing this mRNA while IGF-I induced the accumulation of c-*fos* mRNA, but at tenfold higher concentrations. Insulin also increased the accumulation of both c-*fos* and c-*myc* in rat H35 and H4IIEC3 hepatoma cells. In parallel experiments, insulin (1–5 nM) induced the accumulation of c-*myc* mRNA 8- to 15-fold following 2 h of addition to serum-starved cells (Taub et al. 1987; Messina 1989c, d). Accumulation of c-*fos* was also induced in hepatoma cells, reaching a tenfold induction within 15–30 min. Serum was as effective as insulin in inducing the accumulation of both proto-oncogenes whereas PDGF, FGF, IGF-I, IGF-II, and epidermal growth factor (EGF) were ineffective (Taub et al. 1987; Messina 1989c, d).

In recent experiments insulin addition (5 nM) to serum-starved H4 hepatoma cells for 15 min was found to stimulate a sevenfold induction of c-*fos* transcrip-

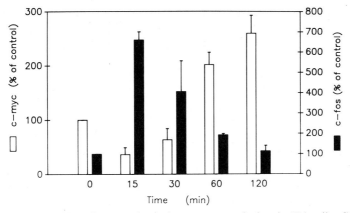

Fig. 2. Time course of insulin on c-*fos* and c-*myc* transcription in H4 cells. Cultures of H4IIE cells were incubated with insulin (5 nM) for the indicated times before measuring the transcription of c-*fos* (*hatched bars*) and c-*myc* (*open bars*) mRNAs. The effect of insulin is expressed as a percentage of control, untreated cells. The scales are different for c-*fos* and c-*myc*. The mean ± standard errors are indicated for four separate experiments

tion (Fig. 2; MESSINA 1989c). This effect of insulin was sufficient to explain the increase in cellular accumulation of c-*fos* mRNA. Effects of insulin were evident at 5 min and at concentrations as low as 0.05 nM. Treatment of H4 cells with the protein synthesis inhibitor cycloheximide, alone or in combination with insulin, led to a much larger increase in c-*fos* transcription (50- to 70-fold) than did treatment with insulin alone (MESSINA 1989c).

Insulin also regulates the transcription of c-*myc* mRNA, but in a more complex manner. Following 15 and 30 min of insulin addition to H4 hepatoma cells, there was a 40%–60% suppression of c-*myc* transcription (Fig. 2; MESSINA 1989d). This suppression reversed after 30 min, and by 60 min there was a 2- to 3-fold induction of transcription. This increase in c-*myc* transcription was not sufficient to account for the much larger increase in c-*myc* mRNA levels. Cycloheximide addition mimicked the initial insulin effect in suppressing c-*myc* transcription, but instead of reversing after 30 min, transcription kept declining to less than 20% of original levels by 60 min.

In a recent study, a sequence in the c-*fos* 5′ flanking sequence was found to be necessary for induction of c-*fos* gene expression. A fusion gene was produced containing the 5′ flanking sequence of the c-*fos* gene preceding coding sequences of the CAT gene. Insulin (70 nM) induced accumulation of this fusion gene about threefold following 45 min of treatment when this gene was transfected into an insulin-responsive chinese hamster ovary (CHO) fibroblast cell line (STUMPO et al. 1988). There was a region of the c-*fos* gene, referred to as the serum-responsive element, that was necessary for the induction of c-*fos* gene expression in response to serum and growth factors such as PDGF. If four base pairs within the serum-responsive element were altered, the transfected gene was no longer responsive to insulin. Thus, sequences in the serum-responsive element were necessary for the insulin-induced accumulation of c-*fos* mRNA in CHO cells.

III. Glyceraldehyde-3-Phosphate Dehydrogenase

Insulin, at concentrations as low as 0.2 nM, was able to induce mRNA levels of the glycolytic enzyme glyceraldehyde-3-phosphate dehydrogenase (GAPDH) in differentiated 3T3F442A adipocytes (ALEXANDER et al. 1985). Effects of insulin (2 nM) on GAPDH mRNA levels were twofold after 2 h of treatment and reached a maximum eightfold induction by 24 h. In preliminary studies, insulin induced the transcription of the GAPDH gene. In addition a 600 base pair, 5′ flanking sequence GADPH fragment – CAT fusion gene was induced 2- to 3-fold following 4 h of insulin treatment when transfected into 3T3 cells (ALEXANDER-BRIDGES 1987). This suggests the presence of an insulin-sensitive regulatory sequence in the 5′ flanking region of the GADPH gene.

IV. Fatty Acid Synthase and Malic Enzyme

The rate of hepatic fatty acid synthesis was decreased in starved or diabetic animals and was increased by refeeding or insulin treatment (see GOODRIDGE 1985 for a review). Coincident with increased fatty acid synthesis was an increase in malic enzyme and fatty acid synthase activity. In vivo and in vitro insulin and thyroid hormone increased and glucagon decreased the activity of these two enzymes. Alloxan-diabetic rats had substantially less hybridizable malic enzyme mRNA, but following 12 h of insulin treatment there was a 50% increase in this mRNA, which continued to increase (up to fivefold) following 60 h of continued insulin treatment (DRAKE and MUCENSKI 1985). In other studies, the in vitro translation of malic enzyme mRNA increased 15-fold after 24 h of insulin treatment (KATSURADA et al. 1983). However, in cultured chicken hepatocytes, insulin had no effect on malic enzyme mRNA levels, although hepatocyte response to triiodothyronine was increased (GOODRIDGE 1985). Thus, insulin's effects on malic enzyme gene expression, both in vivo and in vitro are probably dependent upon the presence of other factors.

Fatty acid synthase mRNA activity decreased over 95% in livers of streptozotocin-treated rats and increased to normal following 12 h of insulin treatment (PRY and PORTER 1981). However, in cultured chicken hepatocytes insulin had no effect on fatty acid synthase mRNA levels, whereas insulin enhanced the stimulatory effect of triiodothyronine (GOODRIDGE 1985). Thus, as with malic enzyme mRNA, insulin may not be a primary regulator of fatty acid synthase gene expression.

V. Pyruvate Kinase

Activity of the hepatic, glycolytic enzyme pyruvate kinase (L type) is regulated by dietary conditions and hormonal status (for review see CIMBALA et al. 1985). Streptozotocin- or alloxan-induced diabetes depressed hepatic L-pyruvate kinase enzyme activity and mRNA activity by 75%–90% (NOGUCHI et al. 1982; CIMBALA et al. 1985). Insulin treatment of diabetic animals restored both these activities within 24 h with the first increase in mRNA activity evident at 10 h. Hybridizable L-pyruvate kinase mRNA was also reduced in diabetic animals and increased to normal levels 48 h after insulin administration (MUNNICH et al. 1984). There was a considerable lag since no effect was measurable before 10 h of insulin treatment (NOGUCHI et al. 1985). A 12-fold increase was evident by 16 h and was paralleled by a 6-fold increase in transcription. Blocking protein synthesis by injection of cycloheximide 4 h after insulin treatment inhibited the insulin-induced accumulation of L-pyruvate kinase mRNA and transcription of this gene. This suggests that insulin may not have acted directly to regulate the pyruvate kinase gene, but altered synthesis of some other protein.

VI. Ornithine Decarboxylase

Insulin regulates ornithine decarboxylase (ODC) activity in some, but not all insulin-sensitive cells (HOVIS et al. 1986). In rat H4 hepatoma cells, insulin increased ODC activity and ODC mRNA (BLACKSHEAR et al. 1987). The effects of insulin on ODC activity were maximal by 4 h while its effects on ODC mRNA levels were maximal between 8 and 12 h. If insulin was acting to increase ODC activity by increasing ODC mRNA, then the increase in ODC mRNA should have preceded the increase in activity. Thus, the initial activation of ODC activity may not be due to an insulin-induced increase in ODC mRNA, but due to another effect of insulin. Insulin induced DNA synthesis after 12 h and cell division after 24 h of treatment in rat H35 hepatoma cells, the parent cell line of the H4 cells (KOONTZ and IWAHASHI 1981; KOONTZ 1984). Since activation of ODC, the rate-limiting enzyme in polyamine synthesis, has been correlated with cell cycling (RUSSELL and SNYDER 1968; CRESS and GERNER 1980), one must consider whether effects of insulin on ODC mRNA were due to cell cycle-dependent alterations in the patterns of ODC mRNA expression, or indeed represented a direct effect of insulin.

VII. Other Hepatic Genes Induced by Insulin

1. Metallothionein

In addition to the genes discussed above, many other genes are induced by insulin, but with little detailed information concerning insulin's exact role in this regulation. One such example is the metallothionein gene(s) which code for heavy metal binding proteins. Insulin induced metallothionein mRNA levels following 8 h of treatment of human HepG2 hepatoma cells, but not following treatment of HeLa cells (IMBRA and KARIN 1987).

2. Apolipoproteins A-I and A-IV

Apolipoproteins are components of chylomicrons, very low density lipoproteins, and high density lipoproteins. A 4-day incubation of cultured rat hepatocytes with insulin (10 nM) induced a sevenfold increase in apolipoprotein A-IV mRNA and a twofold increase in apolipoprotein A-I mRNA, with no significant change in apolipoprotein-E mRNA (ELSHOURBAGY et al. 1985).

3. Glucokinase

Starvation or streptozotocin-induced diabetes decreased hepatic glucokinase mRNA activity compared to normal rats (Spence 1983; Sibrowski and Seitz 1984). In both studies, insulin treatment of diabetic rats increased glucokinase mRNA activity from 3- to 13-fold within 1.5 h with effects of insulin evident by as little as 20 min.

4. Glucose-6-Phosphate Dehydrogenase

Glucose-6-phosphate dehydrogenase (G6PDH) is a rate-limiting enzyme in the pentose phosphate pathway which provides reducing equivalents for a variety of biosynthetic reactions. Yoshimoto et al. (1983) found that 12 h of incubation with insulin increased the G6PDH translational activity of hepatocyte poly-adenylated mRNA. G6PDH mRNA activity increased twofold at 1 nM and a maximum of fivefold at 10 nM insulin. In similar experiments, 2 days of culture of rat hepatocytes with 300 nM insulin increased the G6PDH mRNA activity and hybridizable mRNA 1.5- to 2-fold (Stumpo and Kletzien 1984; Fritz et al. 1986).

5. α_{2u}-Globulin

The major urinary protein of mature male rats, α_{2u}-globulin is regulated by adrenal and testicular steroids, growth hormone, and thyroid hormone. Insulin deficiency in streptozotocin-induced diabetes led to 80% reduction in hepatic concentration of this protein (Roy et al. 1980). This depressed rate of synthesis was increased threefold within 4 days of continuous insulin treatment. The diabetes- and insulin-induced changes in α_{2u}-globulin were caused by equivalent changes in translatable mRNA activity and hybridizable mRNA specific for α_{2u}-globulin.

VIII. Other Nonhepatic Genes Induced by Insulin

1. Amylase

In rats made diabetic with streptozotocin, mRNA for the pancreatic digestive enzyme, amylase, progressively decreased to less than 10% of original values within 6 days, and to undetectable levels by 13 days of diabetes (Korc et al. 1981). Pancreatic amylase mRNA increased 2- to 3-fold after 2 days of treatment of diabetic rats, had returned to normal levels by 7 days, and exceeded prediabetic levels by 2- to 6-fold (depending on the method of measurement) following 10 days of insulin treatment. Essentially similar results were found in streptozotocin-diabetic mice (Draginis et al. 1984).

2. Casein

Dispersed murine mammary epithelial cells responded to insulin with a threefold increase in rapidly labeled nuclear RNA synthesis (Turkington 1970). In murine mammary gland explants, in the presence of cortisol and prolactin, insulin (8 nM) increased the accumulation of casein mRNA 14-fold within 48 h. Serum, EGF, IGF-I, and IGF-II were ineffective in duplicating insulin's induction of casein mRNA (BOLANDER et al. 1981; KULSKI et al. 1983). When rat mammary explants were cultured for 2–5 days with serum, cortisol, and prolactin, insulin addition was also necessary to support casein mRNA accumulation and casein gene transcription (CHOMCZYNSKI et al. 1984).

3. Ovalbumin

When chicken oviduct explant cultures were initially cultured for 2 h in hormone-free medium, followed by the addition of estradiol and insulin for 8 h, there was a fourfold increase in the mRNA for the egg white protein, ovalbumin (EVANS and MCKNIGHT 1984). Half-maximal effects of insulin were evident at 1 nM, and insulin was 5–10 times more potent than IGF-I and IGF-II. In similar studies, primary cultures of chicken oviduct cells required the presence of insulin in addition to estradiol and cortisol for the induction of ovalbumin mRNA (SANDERS and MCKNIGHT 1985). By 2 days of incubation with insulin and the two steroid hormones, there was a 300-fold increase in the accumulation of ovalbumin mRNA and a 35-fold increase in conalbumin mRNA.

4. δ-Crystallin

Insulin (170 nM) added to embryonic chick lens epithelia cells for 24 h caused a 3.6-fold increase in mRNA for the lens-specific protein, δ-crystallin (MILSTONE and PIATIGORSKY 1977). A minimum of 5 h with these high concentrations of insulin was necessary to induce δ-crystallin mRNA, followed by a linear increase between 5 and 25 h.

D. Genes Inhibited or Induced by Insulin

I. Albumin

Albumin is synthesized and secreted by the liver. Hepatic secretion of albumin decreased by over 90% in rats made diabetic by alloxan and returned towards normal following insulin treatment. Likewise, hepatic albumin mRNA decreased over 70% and was restored to normal levels by insulin treatment (PEAVY et al. 1978, 1985; DRAKE and MUCENSKI 1985). Similar results were obtained in spontaneously diabetic BB/W rats maintained on insulin therapy (JEFFERSON et al. 1983). Albumin mRNA transcription was reduced approximately 50% in nuclei isolated from alloxan-diabetic rats compared to normal controls (LLOYD et al. 1987). Diabetes in both models also resulted in a 50%–75% decrease in the synthesis of total secreted proteins and a 40% decrease in the synthesis of hepatic nonsecreted proteins.

In primary cultures of rat hepatocytes, maintained in chemically defined, serum-free media containing Dex and glucagon, removal of insulin from the media for 6 days resulted in a 60% decrease in albumin secretion, and a 40% decrease in cellular albumin mRNA concentrations (FLAIM et al. 1985). There

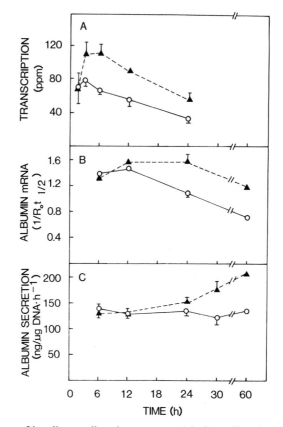

Fig. 3. Time course of insulin on albumin gene transcription, albumin mRNA concentration, and albumin secretion rate in cultured rat hepatocytes. After initial plating in presence of insulin, Dex, and glucagon, hepatocytes were maintained in primary culture for 40 h in hormone-deficient medium. At this time (0 h), half of the cells received insulin (100 nM; *full triangles*); controls received no insulin (*open circles*). Medium was changed and insulin was added every 24 h thereafter. Cells were harvested at indicated times and albumin transcription, mRNA, and secretion was measured. Each point represents the mean of 2–8 independent experiments. (From LLOYD et al. 1987)

were also reductions in total protein synthesis, but the decline in albumin was greater than that of total or other secreted proteins. Insulin treatment (10 nM) returned albumin secretion and albumin mRNA levels to that of hepatocytes continually maintained with insulin (Fig. 3). In similar experiments, chick embryo hepatocytes maintained in the absence of any hormones lost over 99% of their albumin mRNA within 3 days of culture. Insulin addition (35 nM) slowed the loss of albumin mRNA by 2- to 3-fold during the third day of culture (PLANT et al. 1983).

In recent experiments albumin transcription was measured in nuclei isolated from hepatocytes cultured for 40 h in the absence of any hormones. Insulin induced transcription twofold by 3–6 h (no induction at 1.5 h) compared to untreated hepatocytes, and this effect was maintained for 24 h (Fig. 3; LLOYD et al.

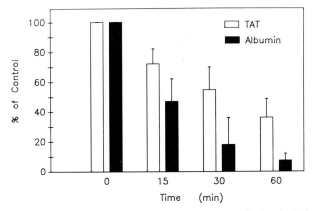

Fig. 4. Time course of insulin on TAT and albumin transcription in H4 cells. Cultures of H4IIE cells were incubated with insulin (5 n*M*) for the indicated times before measuring the transcription of TAT (*open bars*) and albumin (*hatched bars*) mRNAs. The effect of insulin is expressed as a percentage of control, untreated cells. The mean ± standard errors are indicated for four separate experiments

1987). However, these effects were detected upon a background of continually diminishing albumin transcription in untreated hepatocytes.

In contrast to these studies, albumin mRNA was found to increase when insulin was removed from rat hepatocytes cultured in hormonally defined, serum-free medium (JEFFERSON et al. 1984). Insulin was also found to decrease the cellular concentration of albumin mRNA by 95% or more following 12–28 h of treatment in serum-starved rat H4 hepatoma cells (STRAUS and TAKEMOTO 1987). Following 24 h of insulin treatment there was an 80% reduction in albumin mRNA transcription. However, this decrease in transcription was measured many hours after decreases in cellular albumin mRNA concentrations were first evident. Recent work has indicated that there is a rapid effect of insulin on albumin mRNA transcription (MESSINA 1989b). Transcription of albumin mRNA was reduced 90% within 1 h of insulin addition with a 40%–50% reduction discernible after only 15 min of insulin treatment (Fig. 4). The reduction was dose dependent and was maximal at physiologic concentrations of insulin (1–5 n*M*). Since following transcription the transcript is processed before being transported out of the nucleus, any reductions of transcription should precede changes in cellular mRNA concentrations. The inhibition of albumin gene transcription found in this study (MESSINA 1989b) occurred long before the decrease in mRNA found in the study by STRAUS and TAKEMOTO (1987), suggesting that the changes in transcription were probably a major cause of the decrease in cellular albumin mRNA levels.

The differences between the work with hepatocytes and that with hepatoma cells described above is indicative of a problem evident in much of the work on the regulation of hepatic gene expression. Many liver-specific functions are altered when hepatocytes are prepared, with radical alterations in the expression of liver-specific genes (BISSELL and GUZELIAN 1981; CLAYTON and DARNELL 1983; ENAT et al. 1984; JEFFERSON et al. 1984; NAKAMURA et al. 1984). It is therefore difficult to determine whether an effect of insulin or any other hormone is primarily due to a direct effect on a gene, or is due to a modification of the rate of change in gene expression due to the culturing of the cells. The hormonal control of specific genes may differ in hepatoma cells from that in livers of intact animals. However, Reuber H35 hepatoma cells and their clonal derivatives, the H4IIEC3 and Fao cell lines, have been successfully used to study many of insulin's actions. These include insulin binding, regulation of the insulin receptor tyrosine kinase, and insulin-induced changes in enzyme activity and macromolecular synthesis (HOFMANN et al. 1980; CRETTAZ and KAHN 1983). Thus, while not ideal, cultured hepatoma cells do provide a stable, homogeneous source of cells in which to study regulation of genes in the absence of other serum hormones and growth factors.

II. Tyrosine Aminotransferase

Another hepatic protein regulated by insulin is the enzyme tyrosine aminotransferase (TAT), the rate-limiting enzyme for tyrosine catabolism. Rat liver TAT is under the control of glucocorticoids, glucagon, insulin, and growth hormone (see GRANNER and HARGROVE 1983 for a review). However, studies examining the regulation of TAT by insulin have often been contradictory. Insulin acutely induced TAT activity in rat liver and rat H35 hepatoma cells, probably due to an increase in TAT synthesis (HOLTEN and KENNEY 1967; REEL et al. 1970), and induced the enzyme by retarding its degradation in HTC hepatoma cells (GELEHRTER and TOMKINS 1970; SPENCER et al. 1978). In contrast, cultured hepatocytes treated with insulin for 24 h resulted in a decrease in TAT activity and a decrease in Dex-stimulated TAT activity (NAKAMURA et al. 1981).

Actinomycin D, an inhibitor of RNA synthesis, did not completely block insulin's induction of TAT at concentrations that inhibited glucocorticoid effects on TAT synthesis in rat H35 hepatoma cells (GELEHRTER and TOMKINS 1970; LEE et al. 1970). Data obtained using an in vitro translation assay indicated that activity of TAT mRNA increased 2- to 3-fold following 1–1.5 h of insulin treatment in adrenalectomized rats. Coincident with this increase was a 35%–45% augmentation of the translational activity of total hepatic polyadenylated RNA (HILL et al. 1981). The authors interpreted this to indicate a generalized ability of insulin to increase the translational efficiency of hepatic mRNA, and a particular sensitivity of TAT mRNA to this effect of insulin.

Since the isolation of a cDNA for TAT, both glucocorticoids and cAMP have been found to induce TAT transcription and to stimulate the accumulation of TAT mRNA (SCHERER et al. 1982; HASHIMOTO et al. 1984). The role of insulin in the regulation of TAT mRNA levels has not been as well established. SCHUBART (1986) showed that insulin treatment did not alter the levels of hybridizable TAT mRNA in cultured rat hepatocytes following 24 h of Dex or 3 h of cAMP treatment. Insulin also had no effect on the cellular concentration of TAT mRNA in untreated or 24-h Dex-stimulated rat FTO-2B hepatoma cells. However, insulin did reduce TAT mRNA in hepatoma cells induced by cAMP.

In recent work in my laboratory, insulin was shown to rapidly inhibit transcription of TAT mRNA (MESSINA 1989b). Insulin reduced TAT gene transcription by over 60% within 30 min of addition to serum-starved H4 hepatoma cells with effects being evident by 15 min (Fig. 4). Concentrations of insulin as low as 10 pM were sufficient to inhibit TAT mRNA transcription by 20% with maximal inhibition at 5 nM.

While it is difficult to compare many of the experiments described above due to differences in experimental conditions and designs, it appears that insulin has several effects on the expression of the TAT gene. One effect of insulin is to cause a reduction in TAT synthesis by inhibiting its transcription. Since TAT is a gluconeogenic enzyme, inhibition by insulin is not unexpected. However, TAT activity may also be necessary following a meal since it may prevent the cellular accumulation of toxic concentrations of tyrosine (GRANNER and HARGROVE 1983). Thus, insulin may be able to acutely increase the synthesis of TAT, by either transiently extending the half-life of TAT mRNA or by increasing the translational efficiency of existing TAT mRNA. Additional studies will be required to further elucidate these complex effects of insulin.

E. Protein Kinase C and Insulin's Regulation of Gene Expression

Phorbol esters such as phorbol 12-myristate 13-acetate (PMA) are tumor-promoting agents which activate protein kinase C (CASTAGNA et al. 1982). Phorbol esters mimic several of insulin's actions, such as increasing pyruvate dehydrogenase activity, amino acid uptake, glucose transport, and ribosomal S6 protein phosphorylation (FARESE et al. 1985; WU and BYUS 1981; TREVILLYAN et al. 1985). The incubation of rat H4 cells with high concentrations of PMA for 16 h reduced protein kinase C activity 93% in supernatant fractions and 97% in detergent-extracted particulate fractions (BLACKSHEAR et al. 1987). Hepatoma cells pretreated with phorbol esters as described above, to render them deficient in protein kinase C activity, were still capable of responding to insulin with an increase in ODC mRNA to near normal levels. CHU et al. (1987) have shown that both insulin and PMA suppress PEPCK gene expression in H4 hepatoma cells. The insulin suppression was unaffected by pretreating the cells with high concentrations of PMA whereas the PMA effect was blocked by prior pretreatment with PMA. Thus, protein kinase C did not appear to play a role in modulating the effect of insulin on PEPCK and probably ODC in H4 cells.

The results of these two studies differ from our work, since we found that protein kinase C activity was necessary for insulin to exert its full effect on *p33* mRNA transcription and cellular mRNA accumulation (WEINSTOCK and MESSINA 1988). When H4 cells were pretreated with PMA for 24 h, and then exposed to insulin, *p33* transcriptional activity decreased by 37% and the cytoplasmic accumulation of *p33* mRNA decreased by 38% (Fig. 5). In addition, pretreated cells were unable to respond with an induction of *p33* transcription in response to additional PMA. This suggests that, at least for the *p33* gene, the effects of insulin were in part modulated by protein kinase C. Alternatively, part of the pathway by which phorbol esters induce *p33* mRNA (distal to protein kinase C activation) and by which insulin induces *p33* mRNA accumulation are shared. For example, there may be a common substrate acted upon by both protein kinase C and an

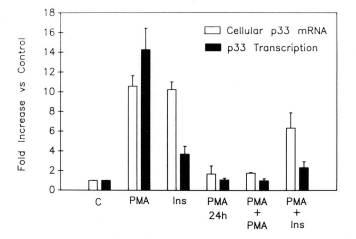

Fig. 5. Effect of 24-h pretreatment with PMA on the induction of *p33* mRNA accumulation and transcription by insulin and PMA in H4 cells. Cultures of H4IIE cells were treated with PMA (1 µg/ml) or untreated for 24 h. Then either insulin (5 n*M*) or PMA (1 µg/ml) was added and cellular mRNA concentrations (*open bars*) and transcription rates (*full bars*) of *p33* mRNA were measured. *C* control; *PMA* no pretreatment, PMA added for 1 h; Ins no pretreatment, insulin added for 1 h (cellular mRNA) or 30 min (transcription); *PMA 24 h* PMA pretreatment for 24 h; *PMA + PMA* PMA pretreatment, then PMA added for 1 h; *PMA + Ins* PMA pretreatment, then insulin added for 1 h (cellular mRNA) or 30 min (transcription). The mean ± standard errors are indicated for four separate experiments. (Adapted from Weinstock and Messina 1988)

insulin-induced protein kinase. If this substrate is reduced due to pretreatment with phorbol esters, then the actions of insulin may also be reduced. Since insulin treatment of liver cells stimulates a selective phospholipase C activity that generates inositol–glucosamine-containing compounds and diacylglycerol (DAG), an endogenous activator of protein kinase C, the role of DAG and protein kinase C must be considered in the evaluation of insulin's actions (see Chap. 14). Our data concerning the regulation of the *p33* gene suggests that each gene regulated by insulin must be studied individually to determine whether protein kinase C is involved in its regulation by insulin. Consistent with this hypothesis is the work of Imbra and Karin (1987) which showed that both PMA and insulin induced the levels of metallothionein mRNA in HepG2 hepatoma cells. Treatment of HepG2 cells with a protein kinase inhibitor having a preference for protein kinase C completely blocked the induction of metallothionein mRNA by PMA, while the response to insulin was partially inhibited. These data suggest a role for protein kinase C, or a related kinase, in at least some of insulin's actions.

F. Summary and Conclusions

Recent advances in the molecular cloning of probes for distinct mRNAs has led to an ability to measure the effects of insulin on specific mRNAs. In several instances (PEPCK, *p33*, c-*fos*, c-*myc*, GAPDH, albumin, TAT) insulin has been shown to have rapid effects on gene transcription (see Table 1). Evidently, insulin can regulate the expression of certain genes within a matter of minutes and this effect may not be dependent on de novo protein synthesis. This suggests that insulin acts directly to regulate the transcription of these genes. The search for specific genomic sequences sensitive to insulin addition to cells is under way and the purification of specific transcriptional regulatory factors, both stimulatory and inhibitory, is now within our grasp.

It is also becoming apparent that insulin may not always regulate the synthesis of an mRNA, but can have several other actions on an mRNA. Insulin can act to stabilize presynthesized mRNA or increase the translational efficiency of the available mRNA. At times insulin may function by increasing the availability of other factors or regulate the cellular sensitivity to other hormones. Systems now exist which will allow determination of insulin's transcriptional and posttranscriptional effects. As with many of insulin's actions, insulin's regulation of gene expression will probably not fit a simple model. Instead, insulin may be able to exert multiple effects on the transcription and metabolism of mRNA.

Acknowledgments. I am especially indebted to my colleagues Ruth S. Weinstock and Phillip H. Smith for their many helpful discussions and their reading of this chapter. This work was supported by grants from the Juvenile Diabetes Foundation and the American Diabetes Association.

References

Alexander M, Curtis G, Avruch J, Goodman HM (1985) Insulin regulation of protein biosynthesis in differentiated 3T3 adipocytes. J Biol Chem 260:11978–11985

Alexander-Bridges M (1987) A human GAPDH promoter chloramphenicol acetyl transferase (CAT) gene fusion is regulated by insulin in 3T3 adipocytes. Diabetes 36:8A

Bissell DM, Guzelian PS (1981) Phenotypic stability of adult rat hepatocytes in primary monolayer culture. Ann NY Acad Sci 349:85–98

Blackshear PJ, Nemenoff RA, Hovis JG, Halsey DL, Stumpo DJ, Huang JK (1987) Insulin action in normal and protein kinase C-deficient rat hepatoma cells. Effects of protein phosphorylation, protein kinase activities, and ornithine decarboxylase activities and messenger ribonucleic acid levels. Mol Endocrinol 1:44–52

Bolander FF Jr, Nicholas KR, van Wyk JJ, Topper YJ (1981) Insulin is essential for accumulation of casein mRNA in mouse mammary epithelial cells. Proc Natl Acad Sci USA 78:5682–5684

Castagna M, Takai Y, Kaibuchi K, Sano K, Kikkawa U, Nishizuka Y (1982) Direct activation of calcium activated, phospholipid-dependent protein kinase by tumor-promoting phorbol esters. J Biol Chem 257:7847

Chomczynski P, Qasba P, Topper YJ (1984) Essential role of insulin in transcription of the rat 25,000 molecular weight casein gene. Science 226:1326–1328

Chu DTW, Stumpo DJ, Blackshear PJ, Granner DK (1987) The inhibition of PEPCK gene expression by insulin is not mediated by protein kinase C. Mol Endocrinol 1:53

Cimbala MA, Lau D, Daigneault JF (1985) Regulation of pyruvate kinase gene expression by hormones and developmental factors. In: Czech MP (ed) Molecular basis of insulin action. Plenum, New York, p 385

Clayton DF, Darnell JE Jr (1983) Changes in liver-specific compared to common gene transcription during primary culture of mouse hepatocytes. Mol Cell Biol 3:1552–1561

Cress AE, Gerner EW (1980) Ornithine decarboxylase induction in cells stimulated to proliferate differs from that in continuously dividing cells. Biochem J 188:375–380

Crettaz M, Kahn RC (1983) Analysis of insulin action using differentiated and dedifferentiated hepatoma cells. Endocrinology 113:1201–1209

Cuatrecasas P (1973) Interaction of concanavalin A and wheat germ agglutinin with the insulin receptor of fat cells and liver. J Biol Chem 248:3528–3534

Cuatrecasas P, Tell GPE (1973) Insulin-like activity of concanavalin A and wheat germ agglutinin – direct interaction with insulin receptors. Proc Natl Acad Sci USA 70:485–489

Drake RL, Mucenski CG (1985) Insulin mediates the asynchronous accumulation of hepatic albumin and malic enzyme messenger RNA's. Biochem Biophys Res Commun 130:317–324

Dranginis A, Morley M, Nesbitt M, Rosenblum BB, Meisler M (1984) Independent regulation of nonallelic pancreatic amylase genes in diabetic mice. J Biol Chem 259:12216–12219

Elshourbagy NA, Boguski MS, Liao WSL, Jefferson LS, Gordon JI, Taylor JM (1985) Expression of rat apolipoprotein A-IV and A-I genes: mRNA induction during development and in response to glucocorticoids and insulin. Proc Natl Acad Sci USA 82:8242–8246

Enat R, Jefferson DM, Ruiz-Opazo N, Gatmaitan Z, Leinwand LA, Reid LM (1984) Hepatocyte proliferation in vitro: its dependence on the use of serum-free, hormonally defined medium and substrata of extracellular matrix. Proc Natl Acad Sci USA 81:1411–1415

Evans MI, McKnight GS (1984) Regulation of the ovalbumin gene: effects of insulin, adenosine 3′,5′-monophosphate, and estrogen. Endocrinology 115:368–377

Farese RV, Standaert ML, Barnes DE, David JS, Pollet RJ (1985) Phorbol ester provokes insulin-like effects on glucose transport, amino acid uptake and pyruvate dehydrogenase activity in BC3H-1 cultured myocytes. Endocrinology 116:2650–2655

Flaim KE, Hutson SM, Lloyd CE, Taylor JM, Shiman R, Jefferson LS (1985) Direct effect of insulin on albumin gene expression in primary cultures of rat hepatocytes. Am J Physiol 249:E447–E453

Fritz RS, Stumpo DJ, Kletzien RF (1986) Glucose-6-phosphate dehydrogenase mRNA sequence abundance in primary cultures of rat hepatocytes. Biochem J 237:617–619

Gelehrter TD, Tomkins GM (1970) Posttranscriptional control of tyrosine aminotransferase synthesis by insulin. Proc Natl Acad Sci USA 66:390–397

Goodridge AG (1985) Hormonal regulation of the expression of the genes for malic enzyme and fatty acid synthase. In: Czech MP (ed) Molecular basis of insulin action. Plenum, New York, p 369

Granner DK, Andreone TL (1985) Insulin modulation of gene expression. Diabetes Metab Rev 12:139–170

Granner DK, Hargrove JL (1983) Regulation of the synthesis of tyrosine aminotransferase: the relationship to mRNA^{TAT}. Mol Cell Biochem 53/54:113–128

Hashimoto S, Schmid W, Schutz G (1984) Transcriptional activation of the rat liver tyrosine aminotransferase gene by cAMP. Proc Natl Acad Sci USA 81:6637–6641

Hasselblatt A, Bruchhausen FV (eds) (1975) Insulin II. Springer, Berlin Heidelberg New York

Hill RE, Lee K-L, Kenney FT (1981) Effects of insulin on messenger RNA activities in rat liver. J Biol Chem 256:1510–1513

Hofmann C, Marsh JW, Miller B, Steiner DF (1980) Cultured hepatoma cells as a model system for studying insulin processing and biologic responsiveness. Diabetes 29:865–874

Holten D, Kenney FT (1967) Regulation of tyrosine α-ketoglutarate transaminase in rat liver. J Biol Chem 242:4372–4377

Hovis JG, Stumpo DJ, Halsey DL, Blackshear PJ (1986) Effects of mitogens on ornithine decarboxylase activity and messenger RNA levels in normal and protein kinase C-deficient NIH-3T3 fibroblasts. J Biol Chem 261:10380–10386

Imbra RJ, Karin M (1987) Metallothionein gene expression is regulated by serum factors and activators of protein kinase C. Mol Cell Biol 7:1358–1363

Jefferson DM, Clayton DF, Darnell JE Jr, Reid LM (1984) Posttranscriptional modulation of gene expression in cultured rat hepatocytes. Mol Cell Biol 4:1929–1934

Jefferson LD, Liao WSL, Peauy DE, Miller TB, Appel MC, Taylor JM (1983) Diabetes-induced alterations in liver protein synthesis: changes in the relative abundance of mRNA for albumin and other plasma proteins. J Biol Chem 258:1369–1375

Katsurada A, Iritani N, Fukuda H, Noguchi T, Tanaka T (1983) Induction of rat liver malic enzyme messenger RNA activity by insulin and by fructose. Biochem Biophys Res Commun 112:176–182

Kenney FT, Lee K-L, Pomato N, Nickol JM (1979) Multiple hormonal control of enzyme synthesis in liver and hepatoma cells. In: Sato GH, Ross R (eds) Cold Spring Harbor conferences on cell proliferation. Cold Spring Harbor, New York, p 905

Kitagawa Y, Ryall J, Nguyen M, Shore GC (1985) Expression of carbamoyl-phosphate synthetase I mRNA in Reuber hepatoma H-35 cells. Regulation by glucocorticoids and insulin. Biochim Biophys Acta 825:148–153

Koontz JW (1984) The role of the insulin receptor in mediating the insulin-stimulated growth response in Reuber H-35 cells. Mol Cell Biochem 58:139–146

Koontz J, Iwahashi M (1981) Insulin as a potent, specific growth factor in a rat hepatoma cell line. Science 211:947–949

Korc M, Owerbach D, Quinto C, Rutter WJ (1981) Pancreatic islet-acinar cell interaction: amylase messenger RNA levels are determined by insulin. Science 213:351–353

Kulski JK, Nicholas KR, Topper YJ, Qasba P (1983) Essentiality of insulin and prolactin for accumulation of rat casein mRNAs. Biochem Biophys Res Commun 116:994–999

Lee K-L, Reel JR, Kenney FT (1970) Regulation of tyrosine α-ketoglutarate transaminase in rat liver. J Biol Chem 245:5806–5812

Lee K-L, Isham KR, Stringfellow L, Rothrock R, Kenney FT (1985) Molecular cloning of cDNAs cognate to genes sensitive to hormonal control in rat liver. J Biol Chem 260:16433–16438

Lloyd CE, Kalinyak JE, Hutson SM, Jefferson LS (1987) Stimulation of albumin gene transcription by insulin in primary cultures of rat hepatocytes. Am J Physiol 252:C205–C214

Loose DS, Wynshaw-Boris A, Meisner HM, Hod Y, Hanson RW (1985) Hormonal regulation of phosphoenolpyruvate carboxykinase gene expression. In: Czech MP (ed) Molecular basis of insulin action. Plenum, New York, p 347

Magnuson MA, Quinn PG, Granner DK (1987) Multihormonal regulation of phosphoenolpyruvate carboxykinase-chloramphenicol acetyltransferase fusion genes. J Biol Chem 262:14917–14920

Messina JL (1989a) Insulin and dexamethone regulation of a rat hepatoma ribonucleic acid: insulin has a transcriptional and a post-transcriptional effect. Endocrinology 124:754–761

Messina JL (1989b) Rapid regulation of albumin and tyrosine aminotransferase transcription by insulin and dexamethasone in rat hepatoma cells (submitted)

Messina JL (1989c) Insulin's regulation of c-fos gene transcription in hepatoma cells (submitted)

Messina JL (1989d) Inhibition and stimulation of c-myc gene transcription by insulin (submitted)

Messina JL, Hamlin J, Larner J (1985a) Effects of insulin alone on the accumulation of a specific mRNA in rat hepatoma cells. J Biol Chem 260:16418–16423

Messina JL, Hamlin J, Azizkahn J, Larner J (1985b) The effects of insulin and concanavalin A on the accumulation of a specific mRNA in rat hepatoma cells. Biochem Biophys Res Commun 133:1168–1174

Messina JL, Hamlin J, Larner J (1987a) Insulin-mimetic actions of wheat germ agglutinin and concanavalin A on specific mRNA levels. Arch Biochem Biophys 254:110–115

Messina JL, Hamlin J, Larner J (1987b) Positive interaction between insulin and phorbol esters on the regulation of a specific messenger ribonucleic acid in rat hepatoma cells. Endocrinology 121:1227–1232

Milstone LM, Piatigorsky J (1977) δ-Crystallin gene expression in embryonic chick lens epithelia cultured in the presence of insulin. Exp Cell Res 105:9–14

Munnich A, Marie J, Reach G, Vaulont S, Simon M-P, Kahn A (1984) In vivo hormonal control of L-type pyruvate kinase gene expression. J Biol Chem 259:10228–10231

Nakamura T, Noda C, Ichihara A (1981) Two phase regulation of tyrosine aminotransferase activity by insulin in primary cultured hepatocytes of adult rats. Biochem Biophys Res Commun 99:775–780

Nakamura T, Nakayama Y, Teramoto H, Nawa K, Ichihara A (1984) Loss of reciprocal modulations of growth and liver function of hepatoma cells in culture by contact with cells or cell membranes. Proc Natl Acad Sci USA 81:6398–6402

Nakamura T, Niimi S, Nawa K, Noda C, Ichihara A, Takagi Y, Anai M, Sakaki Y (1987) Multihormonal regulation of transcription of the tryptophan 2,3-dioxygenase gene in primary cultures of adult rat hepatocytes with special reference to the presence of a transcriptional protein mediating the action of glucocorticoids. J Biol Chem 262:727–733

Niimi S, Nakamura T, Nawa K, Ichihara A (1983) Hormonal regulation of translatable mRNA of tryptophan 2,3-dioxygenase in primary cultures of adult rat hepatocytes. J Biochem 94:1697–1706

Noda C, Tomomura M, Nakamura T, Ichihara A (1984) Hormonal control of serine dehydratase mRNA in primary cultures of adult rat hepatocytes. J Biochem 95:37–45

Noguchi T, Inoue H, Tanaka T (1982) Regulation of rat liver L-type pyruvate kinase mRNA by insulin and by fructose. Eur J Biochem 128:583–588

Noguchi T, Inoue H, Tanaka T (1985) Transcriptional and post-transcriptional regulation of L-type pyruvate kinase in diabetic rat liver by insulin and dietary fructose. J Biol Chem 260:14393–14397

Peavy DE, Taylor JM, Jefferson LS (1978) Correlation of albumin production rates and albumin mRNA levels in livers of normal, diabetic and insulin-treated diabetic rats. Proc Natl Acad Sci USA 75:5879–5883

Peavy DE, Taylor JM, Jefferson LS (1985) Time course of changes in albumin synthesis and mRNA in diabetic and insulin-treated diabetic rats. Am J Physiol 248:E656–E663

Plant PW, Deeley RG, Grieninger G (1983) Selective block of albumin gene expression in chick embryo hepatocytes cultured without hormones and its partial reversal by insulin. J Biol Chem 258:15355–15360

Pry TA, Porter JW (1981) Control of fatty acid synthetase mRNA levels in rat liver by insulin, glucagon, and dibutyryl cyclic AMP. Biochem Biophys Res Commun 100:1002–1009

Reel JR, Lee K-L, Kenney FT (1970) Regulation of tyrosine α-ketoglutarate transaminase in rat liver. J Biol Chem 245:5800–5805

Roy AK, Chatterjee B, Prasad MSK, Unakar N (1980) Role of insulin in the regulation of the hepatic messenger RNA for α_{2u}-globulin in diabetic rats. J Biol Chem 255:11614–11618

Russell D, Snyder SH (1968) Amine synthesis in rapidly growing tissues: ornithine decarboxylase activity in regenerating rat liver, chick embryo, and various tumors. Biochemistry 60:1420–1427

Sanders MM, McKnight GS (1985) Chicken egg white genes: multihormonal regulation in a primary cell culture system. Endocrinology 116:398–405

Scherer G, Schmid W, Strange CM, Rowekamp W, Schutz G (1982) Isolation of cDNA clones coding for rat tyrosine aminotransferase. Proc Natl Acad Sci USA 79:7205–7208

Schubart UK (1986) Regulation of gene expression in rat hepatocytes and hepatoma cells by insulin: quantitation of messenger ribonucleic acid's coding for tyrosine aminotransferase, tryptophan oxygenase, and phosphoenolpyruvate carboxykinase. Endocrinology 119:1741–1749

Sibrowski W, Seitz H (1984) Rapid action of insulin and cyclic AMP in the regulation of functional messenger RNA coding for glucokinase in rat liver. J Biol Chem 259:343–346

Smith JD, Liu AYC (1981) Lectins mimic insulin in the induction of tyrosine aminotransferase. Science 214:799–800

Spence JT (1983) Levels of translatable mRNA coding for rat liver glucokinase. J Biol Chem 258:9143–9146

Spencer CJ, Heaton JH, Gelehrter TD, Richardson KI, Garwin JL (1978) Insulin selectively slows the degradation rate of tyrosine aminotransferase. J Biol Chem 253:7677–7682

Straus DS, Takemoto CD (1987) Insulin negatively regulates albumin mRNA at the transcriptional and post-transcriptional level in rat hepatoma cells. J Biol Chem 262:1955–1960

Stumpo DJ, Blackshear PJ (1986) Insulin and growth factor effects on c-*fos* expression in normal and protein kinase C-deficient 3T3-L1 fibroblasts and adipocytes. Proc Natl Acad Sci USA 83:9453–9457

Stumpo DJ, Kletzien RF (1984) Regulation of glucose-6-phosphate dehydrogenase mRNA by insulin and the glucocorticoids in primary cultures of rat hepatocytes. Eur J Biochem 144:497–502

Stumpo DJ, Stewart TN, Gilman MZ, Blackshear PJ (1988) Identification of c-*fos* sequences involved in induction by insulin and phorbol esters. J Biol Chem 263:1611–1614

Taub R, Roy A, Dieter R, Koontz J (1987) Insulin as a growth factor in rat hepatoma cells. J Biol Chem 262:10893–10897

Trevillyan JM, Perisic O, Traugh JA, Byus CV (1985) Insulin and phorbol ester-stimulated phosphorylation of ribosomal protein S6. J Biol Chem 260:3041–3044

Turkington RW (1970) Stimulation of RNA synthesis in isolated mammary cells by insulin and prolactin bound to sepharose. Biochem Biophys Res Commun 41:1362–1366

Weinstock RS, Messina JL (1988) Transcriptional regulation of a rat hepatoma gene by insulin and protein kinase C. Endocrinology 123:366–372

Wu VS, Byus CV (1981) The induction of ornithine decarboxylase by tumor promoting phorbol ester analogues in Reuber H35 hepatoma cells. Life Sci 29:1855–1863

Yoshimoto K, Nakamura T, Niimi S, Ichihara A (1983) Hormonal regulation of translatable mRNA of glucose-6-phosphate dehydrogenase in primary cultures of adult rat hepatocytes. Biochim Biophys Acta 741:143–149

CHAPTER 19

Insulin, Membrane Polarization, and Ionic Currents

K. ZIERLER

A. Insulin Effect on Membrane Potential

I. Some Pertinent Electrophysiology

Membrane potential is measured, conventionally, by placing one electrode in the extracellular, or outside, solution, a second electrode inside the cell, and connecting the two electrodes to an electrometer. If the aqueous solutions separated by the cell membrane had the same composition there would be no transmembrane electrical potential difference. In fact, however, there are great differences in ion concentration on the two sides. Osmolality of the two solutions must be identical, otherwise cell volume could not be constant. It is crucial to note carefully the composition of solutions in which investigators bathe cells and tissues. They need to be isosmotic with the cell's internal solution, and ionic composition and contents are critical. The major intracellular cation is K^+, normally in mammals at a concentrations of about 155 mM in many cells, compared to about 4 mM outside. It is not unusual to read that investigators use as their "normal" outside K^+ concentration as high as 6 mM, a concentration which a clinician finds worrisome in life. Outside Na^+ concentration is about 140 mM; inside Na^+ is difficult to measure, but has been estimated from 1–2 mM in some mammalian skeletal muscle to 10–20 mM. Outside Cl^- concentration is about 100 mM in situ, but most "extracellular" solutions in the laboratory have a higher Cl^- concentration, depending on the buffer. It is important to note what that concentration is because in some cells most of the transmembrane current is carried by Cl^-, and the abnormally high Cl^- concentration gradient may obscure results. In many cell types there is a Cl^--pump (to be defined later), and the equilibrium potential (to be defined later) for Cl^- is often more positive than the observed potential. In other cells, notably in amphibian and mammalian skeletal muscle, although there are discrepancies in the literature and alternative interpretations, the distribution of Cl^- between the two aqueous phases is as though it is passive, as though there is no Cl^--pump. The distribution, then, is determined by the membrane potential. If the membrane potential is, for example, -90 mV, the ratio of outside/inside Cl^- concentration is about 30. Outside Ca^{2+} concentration is a few millimolar; inside is a few nanomolar. It is critical for membrane integrity that there be an adequate Ca^{2+} concentration in the bathing solution; for skeletal muscle this is at least 100 µM. Inside H^+ ion concentration is greater than outside. For most ions, then, there is disequilibrium. This disequilibrium is the source of the transmembrane electrical potential difference, the membrane potential.

At some point in the life of a cell, ion pumps, for example a Na^+, K^+, a Na^+, H^+, or a Ca^{2+}, Na^+ exchange pump, had to appear in order to initiate and maintain transmembrane gradients in electrolyte concentration. An electrogenic pump is one which pumps more ionic charge in one direction across the membrane than it does in the opposite direction. Disequilibrium can be established and maintained by an ion pump even if the pump is not electrogenic.

No matter how the steady state gradient of electrolyte concentration was created, the fact that there is a gradient between two aqueous phases separated by a poorly permeable membrane means that there is potential energy in the system, just as there is when a dam raises the upstream water level. Work can be done when sluice gates are opened and water flows. GOLDMAN (1943) presented the constant field equation, which provided the necessary background for HODGKIN and KATZ (1949) to express the steady state membrane potential in terms of steady state electrolyte distribution and permeability of a given cell membrane to each specific electrolyte

$$V_D = -\frac{RT}{F} \ln \frac{P_K(K^+)_i + P_{Na}(Na^+)_i + P_{Cl}(Cl^-)_o}{P_K(K^+)_o + P_{Na}(Na^+)_o + P_{Cl}(Cl^-)_i}, \tag{1}$$

where V_D is the diffusive component of the steady state membrane potential, R is the gas constant, T the temperature, ln the natural logarithm, the P are permeability coefficients describing the ability of the ion specified by the subscript to pass through the given membrane, and the parentheses indicate activities of the specified ion inside, i, or outside, o, the cell. Because we do not usually measure activities and because it has been demonstrated in some types of cells by use of ion-specific electrodes, that the activity coefficients for Na^+ and K^+ are approximately the same inside and outside the cell, Eq. (1) is usually expressed in terms of electrolyte concentration, indicated by brackets; e.g. [K^+].

P can be viewed as ionic diffusivity D per unit distance. D has dimensions cm^2 s^{-1}; therefore, P has dimensions of velocity, $cm\ s^{-1}$. P is proportional to the mobility of the ion in the membrane and the "solubility" of the ion in the membrane, and inversely proportional to the thickness of the membrane. Only the monovalent ions are considered. [H^+] is excluded from Eq. (1) because its concentration is negligible compared to the millimolar and 100 millimolar concentrations of the three specified ions. Divalent ions are excluded from the equation because, for many cell types, P for divalent ions are negligible compared to the three specified. There are cells for which these conditions are not true; contributions of divalent ions, particularly Ca^{2+}, need to be introduced into the equation. These conditions will not concern us here.

The normal cell membrane is very poorly permeable to all ions except those that have substantial lipid solubility. Electrical currents across cell membranes are carried by flow of ions through the membrane. Electrical resistance of animal cell membranes in the resting steady state is about the same as is expected of a piece of glass of the same thickness. Ions are translocated across cell membranes by one of two processes. They are pumped by ion exchange metabolic processes, or they pass through the membrane as though the driving force is the thermodynamic chemical potential gradient. The latter translocation is passive or diffusive, although the word "diffusive" is unlikely to describe the details of the actual translocation. The observed membrane potential V_m then is the weighted sum of the contributions of the pump component V_P and the diffusive component V_D

$$V_m = T_P V_P + T_D V_D, \qquad (2)$$

where the T are transference or transport numbers, the fractional contribution to total current through the membrane made by ions driven by the process indicated by the subscript. Equation (2) is the statement that two potential sources are arranged in parallel.

In normal rat skeletal muscle, ouabain, at a concentration and over a time period demonstrated to completely block β-adrenergic stimulation of the Na^+,K^+-pump in that tissue, had no effect on V_m; that is, in normal rat skeletal muscle at rest the pump contribution in Eq. (2) is indistinguishable from zero; it is less than 0.5%.

Given the observed gradient in, say, (K^+) across the membrane, the work necessary to maintain that gradient is

$$E_K = - \frac{RT}{F} \ln \frac{[K^+]_i}{[K^+]_o},$$

where E_K is the equilibrium potential, or the Nernst potential, for K^+. It is expressed properly in terms of activity, but is given here in terms of measured concentrations on the assumption that the activity coefficients are identical in the two solutions. There is a similar definition for E_{Na} and E_{Cl}.

Current carried by a given ion through the membrane is proportional to the electrical potential driving that current. The driving potential is the difference between the observed potential V_m and the equilibrium potential for that ion. The contribution to current density through the membrane made by K^+, for example, is

$$I_K = g_K(V_m - E_K)$$

where g is electrical conductance. Total membrane current density, then, is the sum of all the individual ionic currents. The Nernst or equilibrium potential is often called the reversal potential, the null potential, or the zero-current potential in the electrophysiologic literature. The relationship between ion currents and voltage is obtained from voltage-clamp experiments. At a certain applied voltage, current just equals the ionic equilibrium potential, $V_m = E_K$, for example, and current is zero, or null. In practice applied voltages are more likely to bracket the null voltage, being a little smaller and a little larger than the null voltage, so that current reverses direction, from outward to inward; hence, reversal potential.

It is obvious at once that if there is no pump for the ion specified, as may be true for Cl^- in some cells, $V_m = E_{Cl}$ and $I_{Cl} = 0$. Total membrane current density is zero at the observed V_m.

$$I_m = g_m V_m - g_K E_K - g_{Na} E_{Na} + g_{Cl} E_{Cl} = 0$$

On rearrangement and division by total conductance,

$$V_m = T_K E_K + T_{Na} E_{Na} - T_{Cl} E_{Cl}$$

where the T are transference numbers. Comparison with Eq. (1) shows that the relationship between g_K and P_K is nonlinear. Nevertheless, if a condition decreases P_K it also decreases g_K. The value of g is measured electrically; P is measured by flux studies or calculated from Eq. (1) with measurements of V_m and the indicated ion concentrations.

Ionic currents of the diffusive component pass through "channels" in the membrane, which are proteins, probably arranged to have a modified cavity-like structure penetrating the membrane. Channels are more or less specific for one ion; that is, there is a recognition site. There are one or more energy barriers, which may be ion binding sites, through the channel, and there is a "gate" which is either open or closed, which admits or prevents ion passage. The open or closed state may be controlled by voltage, by a receptor, by concentration of some other ion (e.g., Ca^{2+}-activated K^+ channels), by a GTP binding protein system, or by other means. Once a channel has undergone transition to the open state, current through it is proportional to the activity gradient of that ion; that is, conductance through all ion channels of that type is virtually identical. Total current attributed to a single kind of ion is the sum of all currents through single channels. A condition which alters current across the whole cell membrane may do so by altering single-channel conductance, total channel number, or driving force.

II. Insulin Hyperpolarizes Skeletal Muscle

In 1957 I thought that it ought to be possible to get insights into mechanisms of insulin action by studying insulin effects on potassium metabolism, because we knew that insulin caused hypokalemia and that the shift of K from extracellular space to cells was more sensitive to insulin than was stimulated glucose uptake. I predicted, on the basis of Eq. (1) that because, as I then thought, insulin decreased serum $[K^+]$ there would be a larger negative membrane potential, i.e., hyperpolarization.

The predicted result occurred, but not for the predicted reason (ZIERLER 1957). Experiments were carried out on excised skeletal muscle, extensor digitorum longus, of rats. Muscles were bathed in a volume of solution orders of magnitude larger than muscle volume. This relationship ensured that $[K^+]_o$ remained effectively constant so that transfer of K from outside to inside was indicated, not by decreased $[K]_o$, as in the whole-body observations, but as increased intracellular K. However, measurement of the time course of changes in intracellular $[K]$ showed that there was no significant increase in $[K]_i$ during the first hour, and that even after 3 h the increase in $[K]_i$ was inadequate to account for the observed hyperpolarization. Hyperpolarization preceded increased $[K]_i$.

We have repeatedly demonstrated that insulin hyperpolarizes rat striated muscle (extensor digitorum longus, soleus, caudofemoralis, diaphragm, and primary culture of rat embryonic hindlimb muscle; ZIERLER 1987, 1972 for citation of earlier reports), and we have also found it in mouse diaphragm.

Caudofemoralis responds maximally to an insulin concentration of 1 mU/ml (about 6 nM) with an 8- to 10-mV hyperpolarization from a control value of -78 mV in less than 1 min, and half-maximally to 100 μU/ml (600 pM) (ZIERLER and ROGUS 1981 a). By pressure-puffing insulin near the surface of a muscle fiber already impaled with the microelectrode it was possible to demonstrate insulin-induced hyperpolarization (IIH) in less than 1 s, sometimes in less than 500 ms (ZIERLER and ROGUS 1981a). IIH does not require the presence of glucose in the bathing solution (ZIERLER 1959).

Since 1957 IIH has been reported by a number of investigators for a variety of striated muscles from several species: KERNAN (1961), DEMELLO (1967), MOORE and RABOVSKY (1979), MARUNAKA (1986, 1987) in frog skeletal muscle; FLATMAN and CLAUSEN (1979), FROL'KIS et al. (1977), and FROL'KIS (1980) in rat skeletal muscle; OTSUKA and OHTSUKI (1965, 1970), BOLTE and LÜDERITZ (1968), DENGLER et al. (1979) and TAKAMORI and TSUJIHATA (1981) in rat diaphragm, ZEMKOVA et al. (1982) in mouse diaphragm, and HOFMANN et al. (1983) in human intercostal muscle. There have also been reports of hyperpolarization of skeletal muscle in situ in response to insulin injection, but, because insulin reduces serum $[K^+]$ so profoundly these studies are not as unambiguous as are studies of excised muscle in which $[K^+]_o$ is held constant. However, there is an interesting and pertinent study of muscle in vivo by FLATMAN and CLAUSEN (1979). Under pentobarbital anesthesia, rats received glucose intravenously. Serum insulin concentration rose as expected. Serum [K] fell only from 4.2 to 3.9 mequiv./l, and the soleus muscle, in situ, hyperpolarized from -70.3 to -75.8 mV. The reader can calculate, from a modified version of Eq. (1) in which Cl^- terms are dropped, using 155 mM for $[K^+]_i$, 140 mM for $[Na^+]_o$, and the reported value of control V_m, that the ratio P_{Na}/P_K was 0.045. If that ratio was not changed by insulin (but evidence will be presented later that it is reduced by insulin) and if $[K]_i$ remained the same, then the predicted V_m for the observed decrease in $[K]_o$ is -71.1 mV; that is, the observed hyperpolarization cannot be accounted for only on the basis of the observed reduction in serum [K].

There are two reports of failure to find IIH in striated muscle (STARK and O'DOHERTY 1982; KLIP et al. 1986). STARK et al. (1980) used conventional intracellular electrodes in excised rat skeletal muscle placed in a chamber superfused continuously. Detailed criticism of their experimental design and enumeration of what appeared to be possible technical flaws have been published, together with a response from the first author (ZIERLER and MOORE 1984). KLIP et al. (1986) used a voltage-dependent fluorescent dye to assess V_m in L6 cells, a cell line from transformed rat embryo thigh muscles. It is not very sensitive to insulin with respect to glucose uptake (0.2 μM, ~ 30 mU/ml, increased uptake by only about 45%), and one might not expect as large a response with respect to IIH as in normal muscle. Although the authors state that there was no fluorescence intensity response to insulin (0.1 μM), the illustration shows a slow change in fluorescence intensity compatible with hyperpolarization by a few millivolts. The technique is not the one of choice for measuring V_m; it requires a number of assumptions and allied measurements in order to calibrate fluorescence intensity, so that the accumulated uncertainty may be too large for the task.

III. Insulin Hyperpolarizes Adipocytes

BEIGELMAN and HOLLANDER (1962) reported that insulin hyperpolarizes cells of rat epididymal fat pads. It is technically difficult to impale white fat cells to measure V_m because their water content is only about 4%, of their volume, and it was quite a feat to have accomplished it. The first records were quite noisy, reflecting technical difficulty, with a wide spread among responses to insulin, and a question was raised as to whether the probe was recording from adipocytes or from fibroblasts in the fat pad. Accordingly, investigators turned to indirect methods to estimate insulin's effect on V_m in suspensions of adipocytes from white fat.

The first report was based on use of a voltage-sensitive fluorescent dye (PETROZZO and ZIERLER 1976) and suffered from the fact that background fluorescence was so large. Nevertheless, it was clear that insulin hyperpolarized, although there was no calibration and results were not quantitative. A better method was used by CHENG et al. (1981) and by DAVIS et al. (1981). Both groups used a largely nonpolar molecule containing a radiolabeled ionizable group. Hydrophobicity confers membrane permeability. Its ionized form is distributed at equilibrium so that its Nernst potential is whatever V_m happens to be, and so V_m can be calculated. Both groups reported IIH comparable quantitatively to that produced in striated muscle. Davis et al. carried the matter further. They pointed out that the amount of the probe inside the cell is not distributed homogeneously because the same kinds of forces that distribute it across the cell membrane operate to distribute it across membranes of intracellular organelles. Of these, on account of their volume and their transmembrane electrical potential gradient, mitochondria are likely to be important. With appropriate ancillary measurements, Davis et al. calculated that V_m of white adipocytes was -75 mV, substantially greater than had been estimated previously, and that a supramaximal concentration of insulin hyperpolarized by 9 mV. At the same time they calculated that insulin depolarized mitochondria by 19 mV. The same laboratory from which there was reported failure to find IIH in skeletal muscle also failed to find it in rat epididymal fat pad (STARK et al. 1980).

KRISHNA et al. (1970) were able to impale rat excised interscapular fat pads (brown fat) with microelectrodes. Brown fat participates in thermoregulation. Fat globules are not coalesced as in white fat, and cell water content is greater. They reported V_m was only -33 mV. Norepinephrine depolarized. Insulin, up to 1 mU/ml (about 600 pM), had no effect alone, but inhibited depolarization produced otherwise by norepinephrine.

IV. Insulin Hyperpolarizes Myocardium

Insulin has been reported to hyperpolarize dog heart muscle (IMANAGA 1978), cultured chick myocardial cells (LANTZ et al. 1980), and canine false tendons and feline and rat papillary muscles (LAMANNA and FERRIER 1981). There is a negative report (MALINOW 1958), but it neglects to show that insulin was biologically active under the conditions of the experiment.

There are a number of reports of insulin effects on other electrical phenomena in the heart, many of them stimulated by a series of papers by SODI-PALLARES et al. (1962). They infused intravenously a solution containing glucose–insulin–potassium in patients and in experimental animals with myocardial infarction. SODI-PALLARES argued that because it had been shown that insulin hyperpolarizes muscle, the solution he used, which he called a polarizing solution, might restore K to damaged heart muscle, and, with that, restore myocardial function. His first reports, and those of others, noted favorable electrocardiographic and dynamic responses. Later reports were less consistent and, if there are benefits, they do not appear to be dramatic (see KINOSHITA and YATANI 1982; ANDERSON et al. 1983 for experimental studies).

V. Insulin and Liver Membrane Potential

FRIEDMANN and DAMBACH (1982) found no effect of insulin alone on V_m of perfused rat liver. Glucagon hyperpolarized, and insulin prevented hyperpolarization by glucagon. WONDERGEM (1983) observed that insulin produced a slow depolarization of cultured hepatocytes. FROL'KIS et al. (1977) and FROL'KIS (1980) reported that insulin hyperpolarized rat liver, but this was in response to large doses of insulin injected into the animal, with its associated hypokalemia, which might have been responsible for the result. Based on experiments in which $[K^+]_o$ was controlled, by Friedman et al. and by Wondergem, there is no evidence that insulin hyperpolarizes hepatic cells, and, under conditions used by WONDERGEM (1983) insulin depolarized. We are reminded that in response to insulin, skeletal muscle, heart, and adipose tissue all increase glucose uptake, but liver does not.

VI. Insulin and V_m of Other Cells

Insulin hyperpolarizes toad bladder and toad colon mucosa (CRABBÉ 1969), rat gastric mucosa (REHM et al. 1961), rabbit ciliary epithelium (MILLER and CONSTANT 1960), and cultured toad kidney cells (FIDELMAN et al. 1982). Some of these tissues are not ordinarily thought of as insulin responsive and most required large concentrations of (mammalian) insulin to demonstrate the effect. There is, however, increasing evidence that the serosal, or antiluminal, or basolateral surface has insulin receptors or is the site of initial responsiveness. Cultured toad kidney cells respond to small concentrations of insulin and exhibit typical metabolic responses when insulin is added to the basal, not the apical, surface (FIDELMAN et al. 1982).

WILLIAMS et al. (1982) reported that insulin had no effect on V_m of pancreatic acinar cells in pancreatic fragments from mice diabetic by streptozotocin. KANNO and SAITO (1976) studied effects of insulin on pancreozymin-induced hyperpolarization and amylase release from perfused pancreatic acinar cells. Insulin alone had no effect. Insulin increased pancreozymin-induced hyperpolarization by about 10 mV and increased pancreozymin-induced amylase release by about 80%, maximally. SINGH (1985) reported that a very large concentration of insulin (1 μM, about 1.6 U/ml) hyperpolarized mouse pancreatic acinar cells by about 2 mV, a highly significant effect. The report adds that insulin alone had no effect on amylase secretion, but that it potentiated secretion stimulated by acetylcholine. Acetylcholine depolarized by about 5 mV whether or not insulin was present, so that it is difficult to link amylase secretion quantitatively to V_m under these circumstances.

VII. A Tentative Conclusion About Insulin-Induced Hyperpolarization

The evidence is secure that insulin hyperpolarizes mammalian and amphibian striated muscle, white adipocytes, and heart muscle under conditions in which the bathing solution is reasonably physiologic and tissues have been excised from normal animals. We should include in this group the basal surface of toad kidney cells. The best evidence is that insulin does not hyperpolarize liver cells. With these observations the following tentative generalization is made: insulin hyperpolarizes cell membranes of those target cells that respond to insulin by increasing D-glucose uptake; insulin does not hyperpolarize cell membranes of target cells in which it does not stimulate D-glucose uptake. This leads one to suspect that hyperpolarization may be a step in the transduction chain leading to glucose uptake.

B. Some Other Electrical Effects, Not on Striated Muscle

Insulin increases short-circuit current (active Na^+ transport) in frog and toad skin and toad bladder and kidney cells (CRABBÉ 1969; COX and SINGER 1977; FIDELMAN et al. 1982) in an Ussing chamber. Insulin is effective only if added to the serosal side. In toad kidney cell culture, the effect does not require glucose, is visible at an insulin concentration of 10 µU/ml, and is half maximum at 100 µU/ml, which are small concentrations, particularly for an amphibian response to mammalian insulin. The increase in short-circuit current and the increase in absolute value of the elctrical potential difference across the membrane forming the window of the Ussing chamber are reflections of the same process, an electrogenic increase in Na^+ transport. There are some electrical effects of insulin which we tentatively assign to the miscellaneous category until a common thread is found, if ever.

MORRILL et al. (1985) found that insulin (0.1–1 µM) hyperpolarized frog oocytes by 14–20 mV, from −43 mV, and initiated meiotic division. It is interesting to note for later reference when we consider the mechanisms by which insulin hyperpolarizes striated muscle, that these effects occurred whether or not strophanthin was present, and that Na^+ conductance decreased by 40%–50%. When $[Na^+]_0$ was reduced to 2.5 mM V_m became −75 mV and there was no insulin effect. These observations, along with insulin-induced increased ^{22}Na uptake and increased intracellular pH, led to the reasonable suggestion under the circumstances that insulin hyperpolarized by stimulating an Na^+,H^+ exchange pump. Major difficulties with accepting the interpretation as a physiological effect of insulin are the enormous concentrations of insulin that had to be used and the lack of test of the hypothesis by such experiments as use of inhibitors of Na^+,H^+ exchange.

VAN DER KLOOT and VAN DER KLOOT (1986) studied effects of several agents, including insulin, on quantal size and frequency of miniature end-plate potentials at the neuromuscular junction of frog sartorius. Porcine insulin (10–100 mU/ml, about 6–60 nM) increased quantal size, but not quantal number per unit time. It did not alter the density of α-bungarotoxin binding sites (density of acetylcholine receptors), reversal potentials, or time constants of the membrane (product of membrane capacitance and membrane resistance).

There is a report, difficult to interpret because it includes intravenous injection of insulin into the animal, with its attendant alterations in metabolism, of the migrating myoelectric complex of the jejunum in dogs and sheep in situ (BUÉNO and RUCKEBUSCH 1976). Under thiopental anesthesia electrodes were inserted at three sites along the jejunum through serosal and muscular layers. A migrating electrical complex was replaced by a continuous pattern of electrical activity after injection of insulin, as it was after feeding. The response to feeding was diminished in the alloxan diabetic dog.

ENOMOTO et al. (1986) studied effects of various hormones on some electrical properties of cultured colonies of mouse mammary epithelial cells, which were electrically coupled. In the absence of added hormone, V_m was constant at about -33 mV. The hormones had no effect on the resting level of membrane potential, but, after a delay of 30 min to hours, insulin produced slow depolarizing potentials to a peak of about 5 mV lasting for about 1 s. The more hyperpolarized at rest, the larger the insulin-induced depolarization. These were not action potentials. One suspects that the depolarization may have been due to inward Ca^{2+} current and that the repolarization may have been due to K^+ current, perhaps Ca^{2+}-activated. Indeed, the reversal potential of the hyperpolarizing response was about -90 mV, suggesting that it was due to K^+ current.

Finally, COKER et al. (1981) measured zeta potentials of chick erythrocytes in the presence and absence of insulin (1 µU/ml, about 6 pM, an extraordinarily small concentration, if it is not a typographical error). A zeta potential is an index of surface charge density and can be measured, as it was in this case, in a cell microelectrophoresis apparatus. Insulin produced a significantly more negative zeta potential by about 30%. Addition of Mg^{2+} or Ca^{2+} incrased both the control and the insulin-induced negative zeta potential. Conclusion: insulin increased cell surface charge density.

C. Insulin's Effect on V_m of Striated Muscle when K^+ Distribution Has Been Altered

OTSUKA and OHTSUKI (1965), BOLTE and LÜDERITZ (1968), and DENGLER et al. (1979) measured V_m of rat diaphragms from rats maintained on a K-deficient diet. Details of the diet protocols varied, which may account for some of the observed differences. When the diaphragms from K-deficient rats were incubated in what each investigator called "normal" $[K]_0$, OTSUKA and OHTSUKI (1965) observed about 10 mV hyperpolarization from a V_m of about -67 mV (insulin 1 U/ml, $[K]_0$ 5.9 mM), BOLTE and LÜDERITZ (1968) observed a depolarization by 11 mV from V_m of -95 mV (insulin 100 mU/ml, $[K]_0$ 4.7 mM), and DENGLER et al. (1979) observed no change in V_m from -77 mV (insulin 400 mU/ml, $[K]_0$ 3.5 mM). The latter two groups compared their data to those from diaphragms from normally fed rats. For BOLTE and LÜDERITZ (1968) insulin hyperpolarized by 8 mV from -79 mV. Notice that V_m was much more polarized in K-deficient diaphragms, and almost certainly exceeded E_K in that case. For DENGLER et al.

(1979) insulin hyperpolarized normal diaphragm by 5 mV from V_m of -83 mV. Notice that in this case V_m was less polarized in K-deficient diaphragms.

OTSUKA and OHTSUKI (1965) observed that IIH of K-deficient diapragms decreased as $[K]_o$ increased, and became insignificant at 30 mM $[K]_o$. They and DENGLER et al. (1979) found that insulin depolarized when K-deficient diaphragms were bathed in solutions in which $[K]_o$ was 0.5 mM or less.

It is difficult to rationalize these discrepant observations on K-deficient diaphragms. We (ZIERLER et al. 1985) found that in caudofemoralis muscle from normal rats IIH decreased with increasing $[K]_o$, confirming similar observations by OTSUKA and OHTSUKI (1965). The decreasing insulin effect depended quantitatively on how $[K]_o$ was increased. When it was increased by equimolar substitution of KCl for NaCl, IIH decreased more rapidly and vanished at about 20 mM $[K]_o$. When it was increased at the expense of $[Na]_o$, but the product $[K]_o \cdot [Cl]_o$ was held constant, the decrease in IIH was more gradual and did not disappear even at 80 mM $[K]_o$. The mechanisms by which insulin alters V_m are considered in the next section. In anticipation of this, note that diminished IIH with increasing $[K]_o$ is consistent with the hypothesis that insulin hyperpolarizes by decreasing Na permeability, P_{Na}.

D. The Immediate Mechanism by which Insulin Hyperpolarizes Skeletal Muscle

I. List of Candidates

The list follows from Eq. (1) and (2). From Eq. (2), hyperpolarization can be the result of increasing either the pump potential term or the diffusion potential term. The pump potential term might be augmented in several ways: by increasing the density of pumps in the cell membrane, by increasing the coupling ratio (pumping out even more Na^+ than it pumps in K^+), by accelerating turnover so that the T_p factor increases.

The diffusion potential term might be augmented in several ways revealed by examination of Eq. (1). First, the distribution of ions might be altered so that their steady state concentrations define a more polarized potential. Because the experiments we are considering were all carried out in vitro under conditions in which external concentrations were held constant, only changes in intracellular concentration are candidates. Of these, $[Na^+]_i$ and P_{Na} are already so small that decreasing $[Na]_i$ cannot account for hyperpolarization. If it is true that $E_{Cl} = V_m$ in skeletal muscle, then it does not matter what happens to $[Cl^-]_i$. This leaves only $[K^+]_i$, with the caveat that one must be sure that the role of Cl^- can be dismissed.

The remaining candidates, from Eq. (1), are the ion permeabilities or conductances. If the Cl^- terms can be omitted, then hyperpolarization could occur if P_K increased, if P_{Na} decreased, or with any combination resulting in decreased ratio P_{Na}/P_K. If Cl^- terms cannot be neglected, then, because E_{Cl}, if it is not V_m, is likely to be less polarized than V_m, P_{Cl} would have to decrease. The directional changes considered above for the P apply also to conductances.

II. The Pump as Candidate

1. Insulin Effects on Na$^+$ Transport Across Amphibian Epithelial Cells

a) Experiments

There are a number of reports interpreted to indicate that insulin stimulates an electrogenic Na$^+$,K$^+$-pump, and in many of these it is reported that ouabain or strophantin prevented or abolished either IIH or other insulin-induced effects attributed to stimulation of the pump. Much of this evidence has been reviewed by MOORE (1983) and by CLAUSEN and FLATMAN (1987), to whom the reader is referred for more details.

CREESE and NORTHOVER (1961) observed that the apparent Na content of excised rat diaphragm increased as it was kept in a buffered physiological salt solution and that this increase was prevented by insulin. They suggested that insulin might affect an Na$^+$,K$^+$ exchange process. KERNAN (1961) found that insulin increased ^{22}Na efflux from frog muscle. HERRERA et al. (1963); HERRERA (1965) reported that insulin stimulated the short-circuit current across frog skin and toad bladder in Ussing chambers. As described above, this current is the electrical manifestation of active Na$^+$ transport.

Although very large insulin concentrations were used generally in the early experiments, a fact that might have been attributed to poor responsiveness of amphibia to mammalian insulin, and very large mammalian insulin concentrations are still often used in studies of amphibian tissues, COX and SINGER (1977) reported that 1 mU/ml increased ^{24}Na transport by about 15% in normal [K]$_o$ and by about 40% in high [K]$_o$ across toad bladder in an Ussing chamber. Even this relatively large concentration, which would be maximal in mammalian muscle or adipocytes, is small compared to those others have had to use. CRABBÉ (1981) reviewed the literature and his own experience (toad bladder, colon, and abdominal skin) with insulin effects on ion transport across epithelia. There were no effects at 1 mU/ml, some effects at 10 mU/ml, and more (doubling the short-circuit current) at 100 mU/ml. Insulin stimulation of short-circuit current develops slowly, over several hours, and is prevented by ouabain, 50 μM, which concentration also nearly abolishes the short-circuit current in the absence of insulin.

SCHOEN and ERLIJ (1987) extended electrophysiological studies of ion transport across amphibian epithelia. With conventional microelectrodes impaling frog skin, they obtained data for current–voltage, I–V, curves. Porcine insulin, 100 mU/ml, increased short-circuit current by 40%. It depolarized. Depolarization is explained on the basis of increased Na$^+$ conductance (from analysis of I-V plots). P_{Na} increased.

b) Summary of Insulin Effect on Short-Circuit
Current Across Amphibian Epithelia

Insulin increases short-circuit current by about 15%–100% in different preparations. It requires a large concentration, usually at least 100 mU/ml, of mammalian insulin, to obtain this effect. Insulin is effective only when added to the serosal or basolateral surface, where, presumably, insulin receptors are located. The weight of evidence is that the short-circuit current, which, within error, correlates one-to-one with active Na transport, is the result of an electrogenic pump. In most experiments this pump is ouabain inhibitable whether or not insulin is present. Stimulation of the pump by insulin develops very slowly, over hours, a time course very different from that observed for IIH of mammalian striated muscle. Insulin does not stimulate the Na^+ pump in amphibian epithelial cells by the same mechanism as does aldosterone because, unlike the case for aldosterone, cyclohexamide does not abolish the insulin effect (COX and SINGER 1977).

Although it may be simply that amphibian epithelial cells respond poorly to mammalian insulin, and might respond to far smaller concentrations of amphibian insulin if that were available, the fact remains that, because insulin concentrations were of the order of 1000 times those clearly effective in mammalian striated muscle and adipocytes, one ought to keep in mind the possibility that some or all of the observed effects were not initiated by insulin binding to an insulin receptor, but by insulin binding to some other receptor, perhaps a growth receptor, or that the effects were due to a contaminating protein.

2. Evidence that Insulin Stimulates a Ouabain-Inhibitable Electrogenic Na^+,K^+-Pump in Striated Muscle

In 1965 MOORE suspected that the hyperpolarization insulin produced in rat skeletal muscle, which our laboratory had reported, and the reported stimulation of active Na^+ transport across amphibian epithelia might be related, and that the hyperpolarization of skeletal muscle was the result of insulin stimulation of an electrogenic Na^+,K^+-pump. MOORE and RABOVSKY (1979) studied frog sartorius muscle. Porcine insulin, 250 mU/ml, hyperpolarized frog sartorius in about 20 min from a resting level of -91 mV to about -95 mV. That value was maintained for another 30 min. Ouabain, 10^{-4} M, was added and V_m was depolarized 20–25 min later by 6 mV. It is important to note that E_K in these experiments was about -103 mV; IIH was far short of E_K. It is also important to note for comparison with estimates to be discussed later that Moore and Rabovsky calculated that insulin did not change the P_{Na}/P_K ratio in these experiments.

FLATMAN and CLAUSEN (1979) reported that insulin, 100 mU/ml (about 640 nM), hyperpolarized soleus muscles from rats weighing 200–400 g by about 4 mV after 15 min. Either salbutamol or epinephrine, each at 10^{-5} M, hyperpolarized to a greater extent, up to 9 mV. Ouabain, 10^{-3} M, was added to bathing solutions 30 min before either insulin or epinephrine was added. Another 15 min elapsed before measurements of V_m began, in the case of insulin. Under these circumstances V_m was about 0.5–1 mV depolarized compared to the control potential before exposure to ouabain.

The experiments of MOORE and RABOVSKY (1979) and of FLATMAN and CLAUSEN (1979) were criticized (ZIERLER and ROGUS 1981 b) on the basis that ouabain was exhibited at too high a concentration and for too long. As explained in Sect. A.I, it is the disequilibrium of distribution of ions between the internal and external aqueous phases that is responsible for V_m, whether or not pumps are electrogenic. Ouabain invades this disequilibrium, and in the high concentrations used and over the sufficiently long time period ouabain alters the distribution of ions, increasing the entropy of the system, and depolarizes. Depolarization by ouabain under these circumstances is, then, not unambiguous evidence that hyperpolarization in the absence of ouabain is the result of stimulation of an *electrogenic* pump. That is, in Eq.(2), under the circumstances in which ouabain was used, not only was the V_P component of V_m diminished, but so also was the V_D component.

CLAUSEN and FLATMAN (1987) responded to this criticism by extending earlier observations (CLAUSEN and HANSEN 1977) on the time course of ouabain binding to muscle. In rat soleus muscle, CLAUSEN and HANSEN (1977) had found that maximum [^{3}H] ouabain binding, at $5 \mu M$ ouabain, required 1 h. The first measurement, after about 20 min, showed about half-maximal binding. (Incidentally, insulin did not alter the maximum, but did increase the rate at which maximum was approached.) In the more recent report, binding of $1 \mu M$ ouabain did not reach maximum in 1 h. Binding of $10 \mu M$ ouabain was much more rapid, but was also not maximum in 1 h. These authors argued also that the K concentration of their standard buffer solution was 5.9 mM and that this is high enough to compete with ouabain for access to the ATPase binding sites.

The difficulties with accepting these explanations are that V_m is measured only on surface fibers of the excised muscle, but the ouabain binding studies are measured in the whole muscle of some 30 or 40 fibers radius. A substantial part of the reported delay in ouabain binding is due to diffusion delay through the entire muscle. This does not apply to surface fibers which are exposed immediately to any change in the bathing solution. AKAIKE (1975), using rat soleus muscle, the same muscle used by FLATMAN and CLAUSEN (1979), found that within 15 min 3×10^{-5} M ouabain prevented about half the hyperpolarization due to stimulation of the Na$^+$,K$^+$-pump and 10^{-4} M prevented it completely. For an example of the difficulty in making interstitial fluid in a whole muscle identical to the bathing solution see a study of muscle V_m, as a function of fiber depth in the muscle, in response to serial alteration of bath [K] (ZIERLER et al. 1985). As for competition between ouabain and K$^+$, which does occur, 5.9 mM K is abnormally high for mammalian extracellular fluid. Our own studies are done routinely with 4.8 [K]$_o$.

MOORE (1983) cites the substantial number of published reports in which insulin has stimulated radiosodium efflux or radiopotassium influx, both of which are presumptive evidence that insulin stimulates an Na$^+$,K$^+$-exchange pump.

In addition to the studies on skeletal muscle, LaMANNA and FERRIER (1981) reported that IIH in dog and kitten hearts was prevented by small concentrations of acetylstrophanthidin. However, they did not add insulin "until effects characteristic of digitalis intoxication appears," which took at least 20 min. In the dog acetylstrophanthidin depolarized by 11 mV before insulin was added, so that there already were changes in the diffusive component of V_m.

Since MOORE's review (1983) MARUNAKA (1986, 1987) has added two interesting reports. He has been studying effects of various [K]$_o$ and [Na]$_i$ on the coupling ratio of the Na$^+$,K$^+$-pump and effects of insulin, 500 mU/ml, on bullfrog sartorius muscle. [Na]$_i$ was increased by soaking muscles for 15 h at 4° C in a normal Na, low K solution, and it was decreased by the same treatment in low Na, normal K solution. The coupling ratio, the ratio of Na efflux/K influx, was calculated from the observed ^{22}Na efflux, $[Na]_i$, membrane resistance, and changes in V_m produced by insulin, which MARUNAKA attributed entirely to the pump. The coupling ratio was less when muscle $[Na]_i$ was low and more when muscle $[Na]_i$ was high. The actual value depended on an assumption of a ratio in normal $[Na]_i$. If the ratio was 1.5 normally (if there is a normal after 15 h at 4° C), then the low was about 1.3 and the high about 1.6. Under no condition did the coupling ratio reverse. Insulin hyperpolarized low Na muscles by 1.3 mV, which the author describes as significant at $P < 0.05$, but which, from the unpaired variances tabulated, is not significant ($t < 2$ for ten degrees of freedom). Insulin hyperpolarized normal Na muscles by 3.6 mV and high Na muscles by 11 mV. Insulin had no effect on input resistance. In his other report, in which $[K]_o$ varied and $[Na]_i$ was constant, MARUNAKA (1986) calculated that the pump ratio increased by about 25% at half-normal $[K]_o$ and fell to about one-third normal, almost to unity at twice normal $[K]_o$. But this follows, with the assumptions made, from his observation that insulin hyperpolarized to a slightly greater extent in half-normal $[K]_o$ and not at all in high $[K]_o$. Again, there was no effect of insulin on input resistance. MARUNAKA's experiments provide interesting data about some effects of altered ionic environment on IIH, and that IIH was not associated with a change in input resistance. Although the data are consistent with his assumption that IIH is a consequence of insulin stimulation of the ouabain-inhibitable electrogenic pump, they are subject to alternative interpretation, which will be discussed later.

Does insulin increase the activity of the enzyme, Na^+,K^+-ATPase? ERLIJ and GRINSTEIN (1977) reported that in frog sartorius muscle insulin increased the maximum observed binding of [^{3}H]ouabain. They concluded that insulin unmasked inactive pumps. MANERY et al. (1977) made a similar observation. MOORE (1983) questioned the conclusion on the grounds that no maximum has been reached in the 2-h period of observation and that the author's estimates of binding parameters were based on observed binding after only 50 min, hence not at equilibrium throughout the muscle. Furthermore, CLAUSEN et al. (1977, 1987) found that insulin did not alter maximum ouabain binding in studies in which a plateau was reached. MANERY et al. (1977) found no effect of insulin on Na^+,K^+-ATPase from guinea pig or rat muscle. ROGUS et al. (1969) found no effect on the enzyme from rat skeletal muscle, and LETARTE and RENOLD (1969) found none in rat adipose tissue.

GAVRYCK et al. (1975) reasoned that when an enzyme assay is done under optimum conditions, as one customarily seeks to do it, one cannot expect an added agent to increase activity further. Accordingly, they assayed ATPase activity of a plasma membrane fraction from frog muscle under conditions in which concentrations of Na,K, and ATP, one at a time and in concert, were suboptimal. Insulin, 2.4 μM, increased this ATPase activity by as much as 100%. ATPase activity nearly vanished in the presence of 10^{-3} M ouabain. This latter is taken as evidence that the activity, although submaximum, really was Na^+,K^+-ATPase and not some other ATPases, otherwise one could be unconvinced that the activity measured was only that of the Na^+,K^+-ATPase, or even if it included that

ATPase. Unfortunately, ouabain dose-response relationships were not investigated. Rogus et al. (1969) found the Na^+,K^+-ATPase from membranes of rat skeletal muscle to be as exquisitely sensitive to ouabain as rat brain, although activity in the latter was much greater than in muscle; ouabain inhibition was observable at 10^{-7}, half-maximal at 10^{-6}, and complete at 10^{-5} M. Therefore, it remains uncertain as to whether the assay under suboptimum conditions is what it was intended to be. It should be added that Rogus et al. found no effect of ouabain, even at high concentration on Mg-ATPase of rat muscle. When the report by Gavryck et al. (1975) appeared we reproduced the assay conditions they described, but used rat skeletal muscle as the source of the enzyme. There was no effect of insulin on ATPase activity under these conditions (Rogus, unpublished). Rat skeletal muscle differs in this regard from frog muscle.

Brodal et al. (1974) reported that preincubation with insulin, 0.6 μM, ~ 100 mU/ml, increased a sarcolemma ATPase. It was assumed that this was an Na^+,K^+-ATPase because its activity nearly vanished when K was omitted, but it was not challenged with ouabain.

In summary, there are many results compatible with the hypothesis that insulin stimulates an Na^+,K^+-ATPase, and indeed insulin might in some tissues. However, evidence that it does so in skeletal muscle is subject to alternative interpretations, and the experiments are troubled by the fact that very large concentrations of insulin were used and large concentrations of ouabain were applied for too long a time to avoid the possibility that the results were secondary to redistribution of ions and not only to inhibition of a putative electrogenic pump.

3. Evidence that Insulin Does Not Stimulate a Ouabain-Inhibitable Electrogenic Na^+,K^+-Pump in Muscle

Otsuka and Ohtsuki (1965) reported that 1.7×10^{-5} M ouabain did not abolish IIH of rat diaphragm, and Lantz et al. (1980) found that ouabain had no effect on IIH in cultured chick heart cells. We (Zierler and Rogus 1981 b) tested the hypothesis that IIH is caused by stimulation of the ouabain-inhibitable Na^+,K^+-pump. We took advantage of the fact that β-adrenergic agonists do stimulate this electrogenic pump and, thereby, do hyperpolarize skeletal muscle. Under the same conditions, that is, over the same range of concentrations and over the same time period, in which ouabain blocked isoproterenol-induced hyperpolarization of rat skeletal muscle, there was no effect of ouabain on IIH, not even at a ouabain concentration ten times greater than that causing complete block of isoproterenol-induced hyperpolarization. We showed that within 15 min of exposure to ouabain one could detect alterations in electrolyte composition of the whole muscle. Measurements of intracellular Na and K concentration require the whole muscle in order to have enough material for analysis. It is reasonable to assume that altered electrolyte composition must occur earlier and to a greater extent in surface fibers, which are the one in which V_m is measured. It is for this reason that we suggest that the depolarization of muscle membrane produced by ouabain in the experiments cited in the previous section was not due to blocking the V_P component, but to depolarizing the V_D component of V_m in Eq. (2).

Unequivocal evidence that insulin hyperpolarizes by stimulating the Na^+,K^+-pump would be that the resulting V_m was more negative then E_K. No one has reported such an observation in normal tissue bathed in a solution with normal electrolyte composition. In all such cases, insulin has hyperpolarized toward but not beyond E_K.

However, when normal rat caudofemoralis muscles were exposed to insulin, 100 μU/ml ~ 640 pM, while the muscles were bathed in normal, 4.8 mM $[K]_o$, then transferred to a solution containing eight times that $[K]_o$, 38.4 mM, they were hyperpolarized to -38.5 mV compared to -33.8 mV for those muscles exposed to the same change in $[K]_o$, but not treated with insulin (WU and ZIERLER 1985). E_K was -35 mV; that is, pretreatment with insulin while muscles were in normal $[K]_o$ led to hyperpolarization beyond E_K; the Na^+,K^+-pump must have been activated. Final proof that it was the pump was that ouabain, 10^{-5} M, partially inhibited, and 10^{-4} M inhibited nearly completely within minutes. Also, 10^{-4} M ouabain had no effect on V_m in the absence of pretreatment with insulin. If, in addition to pretreatment with insulin, insulin was present also during incubation at high $[K]_o$, results were the same as with insulin pretreatment alone. If insulin was not present while muscles were held in normal $[K]_o$, but was added only while muscles were bathed in high $[K]_o$, there was no insulin effect on V_m.

We have no definitive explanation for this apparently strange result, but it is clear that the signal initiating events leading to increased Na^+,K^+-pump activity was given while muscles were bathed in normal $[K]_o$, although electrogenic pump activity did not begin until muscles were transferred to high $[K]_o$. One possibility is that hyperpolarization while muscles were in normal $[K]_o$ may have been the initiating signal. To investigate this, WU and ZIERLER (unpublished) incubated rat caudofemoralis in Na-free solutions, which we (WU et al. 1989) have shown hyperpolarize by removing the depolarizing Na component of the Goldman–Hodgkin–Katz equation (Eq. 1). When these muscles were then transferred to the high [K] solution the results were the same as with pretreatment with insulin, even though there was no insulin; V_m hyperpolarized beyond E_K.

BOLTE and LÜDERITZ (1968) made an interesting observation in their study of diaphragms from K-deficient rats. V_m of these diaphragms was -94.6 mV. The reader can calculate from the data given that E_K of these K-impoverished diaphragms was only -81.4 mV; that is, if the calculations are correct, there was substantial contribution to V_m from the Na^+,K^+-pump. When these diaphragms were treated with insulin, unlike the case for normal diaphragms, there was depolarization, instead of hyperpolarization; V_m depolarized to -83.4 mV and E_K became -86.5 mV owing to an increase in $[K]_i$ caused by insulin under these circumstances of K-depletion. Thus, when insulin was added to the solution bathing muscle in which the Na^+,K^+-pump had been active, the pump seems to have been turned off.

III. Changes in Electrolyte Concentration

IIH cannot be the result of redistribution of electrolytes produced by insulin. When insulin is administered to the whole animal there is a decrease in extracellular [K], evident within 5 min, owing to translocation of K^+ from extracellular to intracellular fluid. However, in experiments on excised muscle, intracellular muscle volume is small compared to volume of the bath. In effect, this volume ratio clamps bath electrolyte concentrations. Any changes that alter values of concentrations given in the Goldman–Hodgkin–Katz equation [Eq. (1)] must occur only with respect to intracellular values. When Cl^- terms are removed from Eq. (1) on the grounds that Cl^- movement across muscle cell membrane is passive, there remains only $[K^+]_i$ and $(P_{Na}/P_K)[Na^+]_i$. $[Na^+]_i$ is no more than 7% of $[K^+]_i$, and P_{Na}/P_K is about 0.02 for frog sartorius and rat caudofemoralis. Insulin is known to decrease $[Na]_i$. Therefore, the only concentration of importance in the context of our question is $[K]_i$ and the time course of its change. In rat skeletal muscle (ZIERLER 1959) the increase in $[K]_i$ was too small to be statistically significant 1 h after exposure to insulin, was significant after 2 h, but even after 3 h was far too small to account quantitatively for the observed hyperpolarization. Indeed, rather than the increase in $[K]_i$ being the cause of hyperpolarization, it is the other way around. The observed hyperpolarization produced by insulin is quantitatively thermodynamically adequate to account for the observed translocation of K^+. MOORE and RABOVSKY (1979) found no change in $[K]_i$ of frog sartorius muscle over the period of time during which insulin hyperpolarized.

Finally, insulin hyperpolarizes rat caudofemoralis muscle in less than 1 s (ZIERLER and ROGUS 1981a). No change in intracellular electrolyte concentration, adequate to account for the observed hyperpolarization, could take place in so short a time.

IV. Changes in Permeability Coefficients

1. Increased P_K

LANTZ et al. (1980) determined conductance from slopes of voltage–current curves plotted from results of current-clamp experiments on cultured chick embryo cardiocytes. They concluded that insulin increased conductance across the cardiocyte membrane, and that, because an increase in g_{Na} could only act to depolarize, the increase in total conductance must have meant that g_K increased. This conclusion may be correct, but it is not the only interpretation one can place on their data. From their published current–voltage curve, one could conclude, alternatively, that insulin translated the curve along the voltage axis, in the hyperpolarizing direction, and that any change in slope conductance between the two zero-current points on the curves was too small to be detected.

2. Decreased Ratio P_{Na}/P_K

^{42}K washout from excised rat skeletal muscle, following its intraperitoneal injection, was decreased, maximally by 45% when insulin was added to the washout solution; that is, insulin decreased K efflux (ZIERLER 1960). Because the driving force for K^+ transfer is from intracellular to extracellular, decreased K efflux means that the permeability coefficient for K^+ decreased. It was calculated (ZIERLER 1972) that P_K decreased from 1.46×10^{-7} cm s^{-1} without insulin to

0.87×10^{-7} with insulin, although the absolute values may be a little low due to underestimate of V_m in earlier experiments.

De Mello (1967) calculated electrical resistance of frog sartorius muscle. In the presence of insulin, 80 mU/ml, there was a small increase $(23\% \pm 8.4\%,\ 3$ muscles, about 20 fibers in each), of questionable significance, in normal physiological salt solution. When $[Cl]_o$ was replaced by an impermeant anion, insulin increased resistance by $83\% \pm 28.3\%$, also in 3 muscles. In frog muscle at least two-thirds of resting conductance is carried through Cl^- channels and, of the remaining ion pathways at rest, about 50 times as much through K^+ channels as through Na^+ channels. De Mello's conclusion was that insulin decreased K^+ conductance, because the effect was greater in the absence of Cl and because the observed effect in the absence of $[Cl]_o$ was too great to be accounted for by a decrease in Na^+ conductance, even if that were totally obliterated.

In an effort to understand the apparently anomalous response of muscle from K-deficient rats to insulin (depolarization and paralysis) in low $[K]_o$ solution, Kao and Gordon (1975) measured input resistance of diaphragms from K-deficient rats. It was increased by $38\% \pm 6.9\%$ in the presence of insulin, 5 mU/ml. Because $[Na]_o$ was replaced by the impermeant tetramethylammonium ion and $[Cl]_o$ by the impermeant methylsulfate, the result was attributed to decreased K^+ conductance.

Frol'kis (1980) reported that insulin significantly increased specific membrane resistance of rat muscle fibers from 10.7 to 15.1 kΩ cm^2; interpretation is clouded by the fact that insulin was administered to the whole animal. Imanaga (1978) states, without presenting quantitative data, that insulin, 300 mU/ml, decreasing input resistance of canine papillary muscle. However, the single illustration seems to have been taken from measurements made during the plateau phase of the myocardial action potential, which is due to the slow inward Ca^{2+} current and which, as we shall see later, is inhibited by insulin.

Inspection of the Goldman–Hodgkin–Katz equation [Eq. (1)] reminds us that a decrease in P_K must depolarize, not hyperpolarize, all else being constant. But insulin hyperpolarizes. Therefore, all else cannot be constant. There needs to be an appropriate change in at least one other parameter. A likely candidate is P_{Na}. If insulin decreases P_{Na} by a greater proportion than it decreases P_K, there will be hyperpolarization, as shown by the following.

For skeletal muscle, on the grounds that Cl^- distribution is entirely passive, the Goldman–Hodgkin–Katz equation can be modified to exclude Cl^- terms

$$V_m = -\frac{RT}{F} \ln \frac{[K^+]_i}{[K^+]_o + (P_{Na}/P_K)[Na^+]_o}, \tag{3}$$

where also the term in $[Na^+]_i$ has been neglected because it is only of the order of 0.1% of $[K^+]_i$. From Eq. (3), when V_m and the indicated ion concentrations are measured, the permeability ratio, P_{Na}/P_K, can be calculated.

On the basis of observations in rat caudofemoralis muscle, the half-maximal hyperpolarizing concentration of insulin, 100 µU/ml, decreased the permeability ratio from 0.019 at $V_m = -78$ mV to 0.010 at $V_m = -82$ mV (calculated from data in Zierler et al. 1985). At a maximum hyperpolarizing concentration of insulin, 1000 µU/ml, the permeability ratio fell to 0.0024. Because insulin decreased

P_K, calculated from ^{42}K flux studies, by a maximum of 45%, it must have decreased P_{Na} by an even greater percent.

3. Decreased P_{Na}

Inspection of Eq. (1) and (3) tells us that, if the interpretation of the data is correct, that insulin decreases both P_K and P_{Na}, and hyperpolarizes by decreasing the permeability ratio, P_{Na}/P_K, then if $[Na^+]_o$ is replaced by an even more poorly permeable cation, insulin should depolarize, provided that the replacement is not totally impermeant, in which case insulin would have no effect on V_m. This experiment has the advantage of distinguishing among the various candidate mechanisms. If IIH is due to increased P_K, then, in the experiment described insulin should still hyperpolarize. If IIH is due to stimulation of an electrogenic Na^+,K^+-exchange pump, then insulin should still hyperpolarize, unless removal of $[Na]_o$ causes reversal of the pump ratio; that is, unless it pumps in more K^+ than it pumps out Na^+, an unlikely possibility. MARUNAKA (1987) did not find reversal of the pump ratio in muscles soaked in solutions in which $[Na]_o$ was reduced to 5 mM.

Wu et al. (1989) carried out such an experiment, in which the very poorly permeant cation replacement was either TRIS or N-methylglucamine. Results were about the same with either cation. V_m was more polarized than in normal $[Na^+]_o$ because both ions were less permeant than Na^+, hence did even less than Na^+ to displace V_m from E_K. Yet, the membrane was somewhat permeable to them because V_m was still less polarized than E_K. With insulin, 1 mU/ml, there was depolarization by 3.5 mV.

From these experiments one can calculate the effect of insulin on P_{Na}. Refer to Eq. (3), the modified Goldman–Hodgkin–Katz equation. Substitute for $[Na^+]_o$ a very poorly permeant cation, its concentration in the bath designated $[B^+]$ and its permeability coefficient designated P_B. Now,

$$V_m = -\frac{RT}{F} \ln \frac{[K^+]_i}{[K^+]_o + (P_B/P_K)[B^+]_o}.$$

With all parameters known except the permeability ratio, we solve for P_B/P_K with and without insulin. We then make the reasonable assumption that insulin had no effect on P_B, which lets us calculate the ratio (P_K with insulin)/(P_K without insulin). The P_K ratio was 0.62, from which we conclude that insulin decreased P_K by 38%, in good agreement with our conclusions from earlier studies of ^{42}K efflux.

With this calculation of insulin effect on P_K, one can calculate its effect on P_{Na} from measurements made in ordinary Krebs–Ringer–HCO$_3$ solution with normal [Na]. As noted earlier, in rat caudofemoralis muscle, from Eq. (3) it was calculated that $P_{Na}/P_K = 0.019$ and for maximum response to insulin $P_{Na}'/P_K' = 0.0024$, where a prime, $'$, designates a value obtained in the presence of insulin. From the ratio of these two calculations, $P_{Na}'/P_{Na} = (0.0024/0.019)(P_K'/P_K) = 1/13$. That is, insulin reduces P_{Na} of resting caudofemoralis muscle in the rat by about 93%.

E. Is Hyperpolarization a Step in the Transduction Chain Leading to Stimulated Glucose Uptake?

I. Criteria

If IIH is a step in the insulin transduction chain leading to stimulated uptake of D-glucose, the following three criteria must be met:

1. IIH must precede insulin-induced glucose uptake.
2. One must be able to bypass the transduction steps preceding IIH, hyperpolarize by some other means without using insulin, and demonstrate increased stereospecific uptake of D-glucose.
3. A method must be found to prevent IIH, excepting the trivial case in which insulin binding to its receptor is blocked, and then it must be observed whether the usual insulin-stimulated D-glucose uptake occurs.

II. Evidence that IIH is a Transduction Step Leading to Glucose Uptake

1. IIH Precedes Glucose Uptake

By puffing insulin toward the surface of a muscle fiber impaled with a microelectrode for recording V_m, one can reduce delay due to diffusion. In such an experiment insulin hyperpolarized in less than 1 s (ZIERLER and ROGUS 1981 a), in some instances in less than 500 ms. Insulin stimulation of glucose uptake requires nearly 1 min. Criterion 1 of Sect. E. I is met.

2. Hyperpolarization, in the Absence of Insulin, Can Stimulate D-Glucose Uptake

Criterion 2 was tested in a triple sucrose gap apparatus in which one approximately 4-mm length of rat caudofemoralis muscle was hyperpolarized by a voltage applied to the bath while its neighbor, separated by a gap containing circulating isotonic sucrose, was depolarized (ZIERLER and ROGUS 1980). Radiolabeled 2-deoxy-D-glucose (2DG) uptake was increased by about 40% in the segment hyperpolarized by about 2 mV, averaged over the length of the segment. 2DG uptake by the depolarized segment was the same as that of a segment not exposed to applied voltage. The effect was specific for the D-isomer. L-glucose uptake was not affected.

More recently we have taken advantage of the fact that rat muscle is hyperpolarized slightly in Na-free solution (see Sect. D. IV. 3). 2DG uptake of rat muscle in Na-free solution was 47% ±9% greater than in normal Na solution (WU et al. 1989). In summary, there is evidence that hyperpolarization of rat skeletal muscle in the absence of insulin can stimulate glucose uptake.

3. Correlation Between Amount of IIH and Amount of Glucose Uptake

Criterion 3 has been the most difficult to test because the steps between insulin association with its receptor and the immediate mechanism of hyperpolarization remain uncertain. Several investigators had reported that IIH diminished or disappeared when muscles from K-deficient rats were bathed in high $[K]_o$ (OTSUKA and OHTSUKI 1965) or that insulin stimulation of glucose transport waned in high $[K]_o$ (BIHLER and SAWH 1971; KOHN and CLAUSEN 1972). We designed experiments to see if the two responses correlated in muscles from normal rats. Both IIH and insulin-stimulated 2DG uptake decreased as $[K]_o$ increased (ZIERLER et al. 1985). Depending upon the method by which $[K]_o$ was increased, IIH diminished but did not disappear (constant $[K]_o[Cl]_o$ product) or disappeared at 38 mM [K] (equimolar substitution of KCl for NaCl). Insulin-stimulated 2DG uptake decreased as $[K]_o$ increased, but was still detected at the highest $[K]_o$, 38 mM. Correlation between insulin-induced changes in V_m and in 2DG uptake was nonlinear with regression coefficient = 0.992.

III. Evidence That IIH is Not a Transduction Step Leading to Glucose Uptake

1. Elimination of IIH Does Not Completely Prevent All Insulin-Stimulated Glucose Uptake

We saw in the previous section that there is still some insulin-stimulated 2DG uptake when IIH is prevented, apparently completely, by bathing rat muscle in high $[K]_o$. CLAUSEN and FLATMAN (1987) exposed rat soleus muscles to 1 mM ouabain for 90 min, before adding insulin, a condition in which they had previously reported that insulin was unable to hyperpolarize. Insulin-stimulated transport of radiolabeled 3-O-methylglucose persisted quantitatively as though there were only insulin present. Ouabain alone had no effect. (There was an interesting asymmetry in the results. Insulin increased the rate of transport of 3-O-methylglucose out of muscle by a factor of 10, but it increased the amount of 3-O-methylglucose transported into muscle by a factor of only 2–2.5.) A possible difficulty in interpreting results of experiments on rat soleus muscle is that it has two populations of fibers. MOLGAARD et al. (1980) found that about 60% of soleus fibers hyperpolarized to a mean V_m of -98 mV when $[K]_o$ was reduced to 1 mM and in the remainder V_m was only -70 mV. The difference in V_m was about 10 mV at $[K]_o \sim 5$ mM. The bimodal distribution of V_m in rat soleus is about the same as the relative proportion of red and white fibers. One wonders whether there might not be differences in responsiveness of these fiber types to insulin with respect to V_m and to glucose transport.

KLIP et al. (1986) interpreted their data on the L6 line of transformed rat muscle as showing that insulin did not hyperpolarize, but that it did stimulate glucose uptake. In discussing their results (see Sect. A. II), it was pointed out that the illustrated fluorescence signals, by which V_m was assessed, were consistent with the small hyperpolarization insulin might be expected to produce in a cell that responded relatively poorly to insulin with respect to glucose uptake, as this cell does. More telling is the observation by KLIP et al. (1986) that the ionophores monensin and ionomysin, added to bathing solutions in which all $[K]_o$ was replaced by Na, and the ionophore gramicidin added to bathing solution in which all $[Na]_o$ was replaced by N-methylglucamine hyperpolarized substantially, by about 15 mV, the ionomysin only transiently, yet under none of these conditions was 2DG uptake increased. Responses of the L6 cell to insulin are not brisk. KLIP et al. (1986) reported that a very large concentration of insulin, 0.2 µM, about 30 mU/ml, increased 2DG uptake by only 45%, that there was no effect on 2DG uptake by incubating L6 cells in Na-free solution, whereas in rat caudofemoralis muscle we find that Na-free solutions hyperpolarize and increase 2DG uptake. Furthermore, KLIP et al. reported that when all $[Na]_o$ was replaced by K, there was actually a slightly greater basal 2DG uptake and that insulin then doubled 2DG uptake; the largest response to insulin occurred under conditions in which, in freshly excised rat muscle, others have found decreased basal glucose transport and diminished response to insulin. It is, therefore, difficult to know how to incorporate these data on L6 cells into the body of data on freshly excised rat skeletal muscle.

On the basis of their observation that insulin stimulation of 3-O-methylglucose transport is not altered by a large concentration of ouabain over a long period of time, CLAUSEN and FLATMAN (1987) conclude that IIH is not part of the transduction chain leading to glucose uptake, but lies along a separate parallel path.

2. All Agents that Hyperpolarize Muscle Don't Stimulate Glucose Uptake Equally

CLAUSEN and FLATMAN (1977) reported that β-adrenergic agonists, in large concentration, 10^{-5} M, hyperpolarized rat soleus muscle by 7 or 8 mV, a result confirmed by FLATMAN and CLAUSEN (1979) and later by ZIERLER and ROGUS (1981 b) in rat caudofemoralis muscle, using 10^{-6} M isoproterenol. CLAUSEN and FLATMAN (1987) reported that, although both insulin and β-adrenergic agonists hyperpolarize to about the same maximum, insulin increased 3-O-methylglucose transfer out of muscle by more than a factor of 10, but epinephrine increased it by only a factor of about 2.5.

CLAUSEN and FLATMAN (1987) conclude, on the basis of this observation and their observation that insulin-stimulated glucose uptake is unaffected by a depolarizing dose of ouabain that "there appears to be no coupling between resting membrane potential and glucose transport."

IV. What May Be the Roles of IIH?

To state that there is no coupling between membrane potential and glucose uptake is to ignore some published results, and if one were to state that hyperpolarization accounted completely and quantitatively for insulin-stimulated glucose uptake one would have to ignore some apparent exceptions. There is no doubt that hyperpolarization alone can stimulate D-glucose uptake; in the absence of insulin, electrically produced hyperpolarization of rat muscle by about 2 mV increased 2DG uptake by about 40% (ZIERLER and ROGUS 1980) and bathing muscle in Na-free solution hyperpolarized by about 3 mV and increased 2DG uptake by 47% (WU et al. 1989; WU et al., unpublished). However, the magnitude of the stimulation has been small, by only about 40%–50%, which may be explained entirely or only in part by the fact that the hyperpolarizations imposed on the muscles were substantially less than half the maximum hyperpolarizing response to insulin. There is at least some, and it may even be relatively large, residual stimulation of glucose transport that may not yet be accounted for by hyperpolarization.

It is difficult not to accept the model of CUSHMAN and WARDZALA (1980) and SUZUKI and KONO (1980) that insulin causes recruitment of glucose transporters from some intracellular organelle, probably Golgi, to the cell membrane, and that in the basal state the fraction of total glucose transporters in the cell plasma membrane is small compared to that in intracellular storage. What is not yet assured is that insulin has no further effect on glucose transport. Are glucose transporters freshly inserted into the plasma membrane fully active?

The electric field strength through a muscle cell membrane is about 100 000 V cm^{-1}. This is a powerful force tending to orient dipoles inside the membrane and so stress configuration of proteins in the membrane between the internal and external aqueous phases (ZIERLER 1985). If there are two stable states of glucose transporters, conferring different degrees of transport activity, then under the influence of a change in electrical field strength, such as hyperpolarization, whether insulin-induced or not, a susceptible portion of glucose transporters already incorporated in the membrane may be stressed to shift from a configuration in which they are less active to one in which they are more active.

In this view, hyperpolarization may or may not play a role in the recruiting process from Golgi to plasma membrane. Arguing that hyperpolarization plays no role in recruitment, but does in activation, is a qualitative explanation for all reported observations, including those that β-adrenergic agonists increase glucose transport only slightly compared to insulin, although they hyperpolarize to the same degree. Whether this conjecture is also a quantitative explanation remains to be tested.

F. Insulin Effects on Some Voltage-Sensitive Currents

I. Voltage-Sensitive vs Resting Currents

In Sects. D. IV. 2 and 3, evidence was presented that insulin decreases conductance through resting Na$^+$ and resting K$^+$ channels. Resting currents are small and not well characterized. They are carried by channels that are in the open state when the muscle is at rest at its normal steady state V_m. Larger currents appear in response to a change in conditions. They undergo transition to an open (conducting) state because a ligand has bound to a membrane receptor, or there has been a change in V_m of sufficient size in the proper direction, or the internal ionic environment has changed.

Encouraged by our observations that insulin decreased the two resting conductances, we turned to study possible effects of insulin on some voltage-sensitive channels. All the experiments from our laboratory described in the following sections were carried out on rat myoballs. These are spherical cells, 30–40 µm diameter, prepared by colchicine treatment of primary culture of rat embryo hindlimb muscles. Measurements were made through patch-clamp instrumentation in the whole-cell mode; that is, the contents of the cell intermingled freely with the far larger volume of solution in the patch-clamp pipette, and recorded currents are summations of all currents over the entire surface of the cell. Currents were measured with the cell voltage-clamped, when a step change in voltage was imposed. Action potentials were measured with the cell current-clamped.

II. Insulin Increases an Na^+ Current-Dependent K^+ Current

WU and ZIERLER (1986) discovered in rat myoballs an early outward transient current, which proved to be a K^+ current with an unusual feature. It undergoes transition to the open state when current passes through the TTX-inhibitable Na^+ channel. We have, therefore, designated it $I_K(Na)$. It is voltage-sensitive; it appears in response to depolarization steps greater than 30 mV from a holding potential of -70 mV.

$I_K(Na)$ is recognized in about 75% of myoballs. When insulin is added to the bathing solution $I_K(Na)$ is present in 100% of myoballs, and insulin increases the size of $I_K(Na)$ in those myoballs in which $I_K(Na)$ was present before insulin was added (ZIERLER and WU 1986). Insulin had no effect on the voltage-sensitive, TTX-inhibitable Na^+ current or on the delayed rectifier K^+ current.

Because $I_K(Na)$ is the earliest outward current, and because it is so intimately associated with the voltage-sensitive, TTX-inhibitable Na^+ current responsible for the depolarizing phase of the action potential, we suspected that its function might be to initiate and accelerate the repolarization phase of the action potential. It did. In the presence of insulin the action potential spike amplitude was greater, presumably because IIH activated more voltage-sensitive Na^+ channels, but the spike was sharper and the initial descent of repolarization was steeper.

III. Insulin Decreases Both Ca^{2+} Currents in Rat Myoballs

1. Two Ca^{2+} Currents in Rat Myoballs

Because the Ca^{2+} concentration ratio across the cell membrane is about 10^4, outside/inside, Ca^{2+} currents are inward under ordinary voltages. In rat myoballs there are two voltage-sensitive Ca^{2+} currents, a fast and a slow or delayed. The fast current is obscured by the larger inward voltage-sensitive Na^+ current overlapping it in time. The slow, delayed Ca^{2+} inward current is overwhelmed by the much larger outward delayed rectifier K^+ current. In order to display these two Ca^{2+} currents, voltage-sensitive Na^+ and K^+ currents must be suppressed. This is accomplished by holding the myoball at a potential too depolarized to activate the Na^+ channels, but not depolarized enough to open the Ca^{2+} channels and by substituting Cs^+ for all the K^+ in the patch-clamp pipette (WU and ZIERLER 1989).

Under these conditions the two Ca^{2+} currents were isolated from other currents and from each other. The fast transient Ca^{2+} current is abolished by TTX and is diminished by removing Na from the bathing solution. It is therefore designated $I_{Ca}(Na)$. The other current is designated $I_{Ca(slow)}$.

2. Insulin Decreases Both Ca^{2+} Currents

Insulin decreases both Ca^{2+} currents. It does not alter the depolarizing voltage at which the currents appear, or the voltage at which they peak, or the voltage at which current returns to zero (the reversal potential) (Wu and ZIERLER 1989). Both Ca^{2+} currents in response to insulin were less than controls at every applied voltage except at threshold and reversal potential. This means that insulin either decreased single-channel conductance or decreased the number of open channels, or both.

NAKIPOVA et al. (1987) earlier reported that insulin inhibited Ca^{2+} currents in frog myocardium (atrial trabeculae). They interpreted their data as showing that insulin decreased the reversal potential as well as decreasing conductance. However, because they were never able to reverse the Ca^{2+} currents from inward to outward, it is difficult to pinpoint a reversal potential and therefore difficult to know whether insulin affected it.

3. Insulin Shortens Action Potentials Prolonged by the Slow Ca^{2+} Current

The action potential of rat myoballs with normal electrolyte composition inside and out resembles that of normal adult skeletal muscle. It rises rapidly, overshoots 0 V, spikes, falls a little less rapidly than it rises. Its total duration is of the order of 10 ms.

When Cs^+ replaces K^+ in the pipette, the myoball action potential looks like a cardiac action potential. Shortly after the spike begins its return toward normal its descent is abruptly halted by a long near-plateau, lasting from 500 ms to 2 s. As in the normal cardiocyte action potential, this plateau is due to the slow Ca^{2+} current, which insulin blocks.

Insulin decreased the duration of the plateau and shortened the action potential of rat myoballs in a concentration-dependent fashion (ZIERLER and WU 1988). Maximum effect, an 80% reduction in action potential duration, occurred with 1 mU/ml, about 6.4 nM. Half-maximum occurred with about 30 μU/ml, about 200 pM, and the effect was detected (about 10%–15% reduction in action potential duration) at 10 μU/ml, about 64 pM. This is the most sensitive response to insulin described for skeletal muscle.

IV. Functions of Insulin Effects on Voltage-Sensitive Channels

The combined effect of insulin in $I_K(Na)$ and the two voltage-sensitive Ca^{2+} channels is to accelerate the repolarization phase of the action potential in rat myoballs. There is experimental verification of this for normal action potentials (Wu and ZIERLER 1989).

In skeletal muscle, activation of the slow Ca^{2+} current is too slow to involve it in the action potential, unlike the heart, so that there are likely to be other functions of these channels. An obvious function is to elevate the Ca^{2+} of the cell's interior, which, in turn may serve a variety of purposes. In cardiocytes, it is a source of Ca^{2+} needed for excitation–contraction coupling. In many cells, it is a source of Ca^{2+} for Ca^{2+}-activated K^+ channels, and for many other Ca^{2+}-activated processes.

Whatever the functions may prove to be, the fact that some of these channels are so exquisitely sensitive to insulin, even at a concentration circulating in the basal state, makes one suspect that they will be important in the performance and health of the cell.

G. Electrical Steps in the Insulin Transduction Scheme

Figure 1 summarizes my current concept of the possible roles of hyperpolarization and altered ion currents in the insulin transduction scheme in muscle. The center chain leads to stimulated glucose uptake via near-obliteration of resting P_{Na} and consequent hyperpolarization. Whether this series of events is also con-

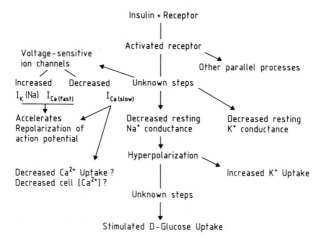

Fig. 1. A possible scheme for insulin transduction in muscle

cerned in recruitment of glucose transporters is unknown. Hyperpolarization alone accounts quantitatively for net K^+ uptake. In parallel with its decrease of resting P_{Na}, insulin decreases resting P_K to a lesser extent. The downstream effect of this is unknown. In another parallel chain, insulin acts on three voltage-sensitive ion channels. It increases $I_K(Na)$ and decreases current through two Ca^{2+} channels. The combined effect is to accelerate the repolarization phase of the action potential, a function important in both skeletal and cardiac muscle. In skeletal muscle, unlike the heart, the slow Ca^{2+} current is too late to participate in the action potential. Its function is at least to translocate extracellular Ca^{2+} into the cell. Insulin's block of this current, which is the most sensitive insulin effect demonstrated in skeletal muscle, is likely to cause downstream effects on the widespread intracellular functions sensitive to intracellular $[Ca^{2+}]$.

Acknowledgments. Original research from my laboratory cited here was supported by NIH grant DK17574, NSF grant DCB 8309232, and gifts from Samuel Shapiro & Company Inc.

References

Akaike N (1975) Contribution of an electrogenic sodium pump to membrane potential in mammalian skeletal muscle fibers. J Physiol (Lond) 245:499–520

Anderson GJ, Swartz J, Dennis SC, Reiser J (1983) The effect of insulin and potassium on infarcting canine tissue: an in vitro study. J Am Coll Cardiol 1:445–447

Beigelman PM, Hollander PB (1962) Effect of insulin upon resting electrical potential of adipose tissue. Proc Soc Exp Biol Med 110:590–595

Bihler I, Sawh PC (1971) Regulation of sugar transport in muscle: effect of increased external potassium in vitro. Biochem Biophys Acta 241:302–309

Bolte H-D, Lüderitz B (1968) Einfluß von Insulin auf das Membranpotential bei alimentärem Kaliummangel. Pflugers Arch 301:254–258

Brodal BP, Jebens E, Oy V, Iversen O-J (1974) Effect of insulin on (Na$^+$,K$^+$)-activated adenosine triphosphatase activity in rat muscle sarcolemma. Nature 249:41–43

Buéno L, Ruckebusch M (1976) Insulin and jejunal electrical activity in dogs and sheep. Am J Physiol 230:1538–1544

Cheng K, Groarke J, Ostimehin B, Haspel HC, Sonenberg M (1981) Effects of insulin, catecholamines, and cyclic nucleotides on rat adipocyte membrane potential. J Biol Chem 256:649–655

Clausen T, Flatman JA (1987) Effects of insulin and epinephrine on Na$^+$-K$^+$ and glucose transport in soleus muscle. Am J Physiol 252:E492–E499

Clausen T, Hansen O (1977) Active Na-K transport and the rate of ouabain binding. The effect of insulin and other stimuli on skeletal muscle and adipocytes. J Physiol (Lond) 270:415–430

Coker EN, Perry MC, Plummer DT (1981) Changes in the zeta potential of chick erythrocytes after the addition of insulin. Biochem Soc Trans 9:89–90

Cox M, Singer I (1977) Insulin-mediated Na$^+$ transport in the toad urinary bladder. Am J Physiol 232:F270–F277

Crabbé J (1969) Stimulation of an active sodium transport by insulin. In: Margoulies M (ed) Protein and polypeptide hormones. Excerpta Medica, Amsterdam, p 260

Crabbé J (1981) Stimulation by insulin of transepithelial sodium transport. Ann NY Acad Sci 372:220–234

Cushman SW, Wardzala LJ (1980) Potential mechanism of insulin action on glucose transport in the isolated adipose cell. J Biol Chem 255:4758–4762

Davis RJ, Brand MD, Martin BR (1981) The effect of insulin on plasma-membrane and mitochondrial-membrane potentials in isolated fat-cells. Biochem J 196:133–147

De Mello WC (1967) Effect of insulin on the membrane resistance of frog skeletal muscle. Life Sci 6:959–963

Dengler R, Hofmann WW, Rüdel R (1979) Effects of potassium depletion and insulin on resting and stimulated skeletal rat muscle. J Neurol Neurosurg Psychiatry 42:818–826

Enomoto K, Cossu MF, Edwards C, Oka T (1986) Induction of distinct types of spontaneous electrical activities in mammary epithelial cells by epidermal growth factor and insulin. Proc Natl Acad Sci USA 83:4754–4758

Erlij D, Grinstein S (1976) The number of sodium ion pumping sites in skeletal muscle and its modification by insulin. J Physiol (Lond) 258:13–31

Fidelman ML, May JM, Biber TUL (1982) Insulin stimulation of Na$^+$ transport and glucose metabolism in cultured kidney cells. Am J Physiol 242:C121–C123

Flatman JA, Clausen T (1979) Combined effects of adrenaline and insulin on active electrogenic Na$^+$-K$^+$ transport in rat soleus muscle. Nature 281:580–581

Friedmann N, Dambach G (1980) Antagonistic effect of insulin on glucagon-evoked hyperpolarization. A correlation between changes in membrane potential and gluconeogenesis. Biochim Biophys Acta 596:180–185

Frol'kis VV (1980) Hormonal regulation of electrical properties of cell membrane. Neurosci Behav Physiol 10:92–98

Frol'kis VV, Martynenko OA, Khilobok IY, Gol'dshtein NB (1977) Effect of insulin on membrane electrical properties and transcribing activity of cell chromatin of rats of different ages. Dokl Akad Nauk SSSR 232:1201–1203

Gavryck WA, Moore RD, Thompson RC (1975) Effect of insulin upon membrane-bound (Na^+K^+)-ATPase extracted from muscle. J Physiol (Lond) 252:43–58

Goldman DE (1943) Potential, impedance, and rectification in membranes. J Gen Physiol 27:37–60

Herrera FC (1965) Effect of insulin on short-circuit current and sodium transport across toad urinary bladder. Am J Physiol 209:819–824

Herrera FC, Whittembury G, Planchart A (1963) Effect of insulin on short-circuit current across isolated frog skin in the presence of calcium and magnesium. Biochim Biophys Acta 66:170–172

Hodgkin AL, Katz B (1949) The effect of sodium ions on the electrical activity of the giant axon of the squid. J Physiol (Lond) 108:37–77

Hofmann WW, Adornato BT, Reich H (1983) The relationship of insulin receptors to hypokalaemic periodic paralysis. Muscle Nerve 6:48–51

Imanaga I (1978) Effects of insulin on mammalian cardiac muscle. In: Kobayashi T, Sano T, Dhalla NS (eds) Heart function and metabolism. University Park Press, Baltimore, p 441 (Recent advances in studies on cardiac structure and metabolism, vol II)

Kanno T, Saito A (1976) The potentiating influence of insulin on pancreozymin-induced hyperpolarization and amylase release in the inner pancreatic acinar cell. J Physiol (Lond) 261:505–521

Kao I, Gordon AM (1975) Mechanism of insulin-induced paralysis of muscles from potassium-depleted rats. Science 188:740–741

Kernan RP (1961) Insulin and the membrane potential of frog sartorius muscle. Biochem J 80:23P

Kinoshita K, Yatani A (1982) After effects of perfusion with glucose-insulin-potassium solution on the membrane current and tension component of frog atrial muscle. Jpn Heart J 23:791–804

Klip A, Ramlal T, Walker D (1986) Insulin stimulation of glucose uptake and the transmembrane potential of muscle cells in culture. FEBS Lett 205:11–14

Kohn PG, Clausen T (1972) The relationship between the transport of glucose and cations across cell membranes in isolated tissues. Biochim Biophys Acta 255:798–814

Krishna G, Moskowitz J, Dempsey P, Brodie BB, McCallum Z (1970) The effect of norepinephrine and insulin on brown fat cell membrane potentials. Life Sci 9:1353–1361

LaManna VR, Ferrier GR (1981) Electrophysiological effects of insulin on normal and depressed cardiac tissues. Am J Physiol 240:H636–H644

Lantz RC, Elsas LJ, DeHaan RL (1980) Ouabain-resistant hyperpolarization induced by insulin in aggregates of embryonic heart cells. Proc Natl Acad Sci USA 77:3062–3066

Letarte J, Renold AE (1969) Ionic effects on glucose transport and metabolism by isolated mouse fat cells incubated with and without insulin. III. Effects of replacement of Na^+. Biochim Biophys Acta 183:366–374

Malinow MR (1958) The effect of insulin on the electrical activity of isolated ventricular muscle of the rat. Acta Physiol Latinoam 8:125–128

Manery JF, Dryden EE, Still JS (1977) Enhancement (by ATP, insulin, and lack of divalent cations) of ouabain inhibition of cation transport and ouabain binding in frog skeletal muscle; effect of insulin and ouabain on sarcolemmal (Na+K) MG ATPase. Can J Physiol Pharmacol 55:21–33

Marunaka Y (1986) Effects of external K concentrations on the electrogenicity of the insulin-stimulated Na,K-pump in frog skeletal muscle. J Membr Biol 91:165–172

Marunaka Y (1987) The effect of internal Na concentration on the electrogenicity of the insulin-stimulated Na,K-pump in frog skeletal muscles. Comp Biochem Physiol 86A:133–136

Miller JE, Constant RA (1960) The measurement of rabbit ciliary epithelial potential in vitro. Am J Ophthalmol 50:855–862

Molgaard H, Stürup-Johansen M, Flatman JA (1980) A dichotomy of the membrane potential response of rat soleus muscle fibres to low extracellular potassium concentrations. Pflugers Arch 383:181–184

Moore RD (1965) The ionic effects of insulin. Abstracts Biophys Soc Annual Meeting, 9th, p 111

Moore RD (1983) Effects of insulin upon ion transport. Biochim Biophys Acta 737:1–49

Moore RD, Rabovsky JL (1979) Mechanism of insulin action on resting membrane potential of frog skeletal muscle. Am J Physiol 236:C249–C254

Morrill GA, Weinstein SP, Kostellow AB, Gupta RK (1985) Studies of insulin action on the amphibian oocyte plasma membrane using NMR, electrophysiological and ion flux techniques. Biochim Biophys Acta 844:377–392

Nakipova OV, Kokoz YM, Freidin AA, Safronova VG, Lazarev AV (1987) Effect of insulin on calcium current in frog myocardium. Sechenov Physiol J USSR 73:492–498

Otsuka M, Ohtsuki I (1965) Mechanism of muscular paralysis by insulin with particular reference to familial periodic paralysis. Nature 207:300–301

Otsuka M, Ohtsuki I (1970) Mechanism of muscular paralysis by insulin with special reference to periodic paralysis. Am J Physiol 219:1178–1182

Petrozzo P, Zierler K (1976) Membrane potential-sensitive fluorescent changes in isolated rat adipocytes: effects of external K^+, epinephrine, and insulin. Fed Proc 35:602

Rehm WH, Schumann H, Heinz E (1961) Insulin on frog gastric mucosa. Fed Proc 20:193

Rogus E, Price T, Zierler K (1969) Sodium plus potassium-activated ouabain-inhibited adenosine triphosphatase from a fraction of rat skeletal muscle, and lack of insulin effect on it. J Gen Physiol 54:188–202

Schoen HF, Erlij D (1987) Insulin action on electrophysiological properties of apical and basolateral membranes of frog skin. Am J Physiol 252:C411–C417

Singh J (1985) Mechanism of action of insulin on acetylcholine-evoked amylase secretion in the mouse pancreas. J Physiol (Lond) 358:469–482

Sodi-Pollares D, Testelli MR, Fishleder BL, Bisteni A, Medrano GA, Friedland C, DeMicheli A (1962) Effects of an intravenous infusion of a potassium-glucose-insulin solution on the electrocardiographic signs of myocardial infarction. A preliminary report. Am J Cardiol 9:166–181

Stark RJ, O'Doherty J (1982) Intracellular Na^+ and K^+ activities during insulin stimulation of rat soleus muscle. Am J Physiol 242:E193–E1980

Stark RJ, Read PD, O'Doherty J (1980) Insulin does not act by causing a change in membrane potential or intracellular free sodium or potassium concentration of adipocytes. Diabetes 29:1040–1043

Suzuki K, Kono T (1980) Evidence that insulin causes translocation of glucose transport activity to the plasma membrane from an intracellular storage site. Proc Natl Acad Sci USA 77:2542–2545

Takamori M, Ide Y, Tsujihata M (1981) Pharmacological study on muscle treated with calcium ionophore A23187. J Neurol Sci 50:89–94

Van der Kloot W, van der Kloot TE (1986) Catecholamines, insulin and ACTH increase quantal size at the frog neuromuscular junctions. Brain Res 376:378–381

Williams JA, Bailey AC, Preissler M, Goldfine ID (1982) Insulin regulation of sugar transport in isolated pancreatic acini from diabetic mice. Diabetes 31:674–682

Wondergem R (1983) Insulin depolarization of rat hepatocytes in primary culture. Am J Physiol 244:C17–C23

Wu F-S, Zierler K (1985) Insulin stimulation of an electrogenic pump at high extracellular potassium concentration. Am J Physiol 249:E12–E16

Wu F-S, Zierler K (1986) A Na^+-conduction dependent K^+ channel in skeletal muscle. Fed Proc 45:1009

Wu F-S, Zierler K (1989) Calcium currents in rat myoballs and their inhibition by insulin. Endocrinol 125:(in press, November)

Wu F-S, Rogus E, Zierler K (1989) Insulin depolarization of skeletal muscle in the absence of external Na^+. Diabetes 38:333–337

Zemkova H, Teisinger J, Vyskocil F (1982) The comparison of vanadyl (IV) and insulin-induced hyperpolarization of the mammalian muscle cell. Biochim Biophys Acta 720:405–410

Zierler KL (1957) Increase in resting membrane potential of skeletal muscle produced by insulin. Science 126:107–1068

Zierler KL (1959) Effect of insulin on membrane potential and potassium content of rat muscle. Am J Physiol 197:515–523

Zierler KL (1960) Effect of insulin on potassium efflux from rat muscle in the presence and absence of glucose. Am J Physiol 190:1066–1070

Zierler KL (1972) Insulin, ions, and membrane potentials. In: Greep RO, Astwood EB (eds) Handbook of physiology. Endocrinology. I. Endocrine pancreas. American Physiological Society, Washington, p 347

Zierler K (1985) Membrane polarization and insulin action. In: Hollenberg MD (ed) Insulin. Its receptor and diabetes. Dekker, New York, p 141

Zierler K (1987) Insulin hyperpolarizes rat myotube primary culture without stimulating glucose uptake. Diabetes 36:1035–1040

Zierler K, Moore RD (1984) Insulin-induced hyperpolarization of skeletal muscle. Am J Physiol 242:E833–E835

Zierler K, Rogus EM (1980) Hyperpolarization as a mediator of insulin action: increased muscle glucose uptake induced electrically. Am J Physiol 239:E21–E29

Zierler K, Rogus EM (1981 a) Rapid hyperpolarization of rat skeletal muscle induced by insulin. Biochim Biophys Acta 640:687–692

Zierler K, Rogus EM (1981 b) Insulin does not hyperpolarize rat muscle by means of a ouabain-inhibitable process. Am J Physiol 241:C145–C149

Zierler K, Wu F-S (1986) Insulin promotes activation of a newly recognized K^+ channel in skeletal muscle. Clin Res 34:727A

Zierler K, Rogus EM, Scherer RW, Wu F-S (1985) Insulin action on membrane potential and glucose uptake: effect of high potassium. Am J Physiol 249:E17–E25

CHAPTER 20

Insulin Regulation of Metabolism Relevant to Gluconeogenesis

R. Bressler and J. J. Bahl

A. Introduction

In this chapter the role of insulin in the regulation of intermediary metabolism relevant to gluconeogenesis and the glucostat function of the liver will be reviewed. Changes in plasma blood glucose levels are moderated by the actions of the liver primarily under the control of insulin and glucagon. Minimizing hyperglycemic and preventing hypoglycemic states requires short- and long-term regulation of hepatic and peripheral metabolism. An appreciation continues to develop of how the relevant biochemical pathways are regulated acutely at the level of enzyme activity and chronically through gene expression. Understanding the role of insulin in regulating the relevant intermediary metabolism requires discussion of hepatic glucose catabolism and anabolism and knowledge of the antagonistic actions of insulin and glucagon which are inseparable from the physiological functioning of the liver to maintain normal levels of blood glucose.

B. Overview of Blood Glucose Regulation

Hormonal regulation of gluconeogenesis depends on the function of three hormones of the islets of Langerhans (UNGER and ORCI 1981; CHERRINGTON et al. 1987). The alpha cells produce glucagon, the beta cells insulin, and the delta cells somatostatin (UNGER and ORCI 1981). An internal regulation system exists in the pancreatic islets wherein glucagon stimulates secretion of insulin and somatostatin (SAMOLS et al. 1965; PATTON et al. 1977; UNGER and ORCI 1981), somatostatin inhibits the secretion of glucagon and insulin (ALBERTI et al. 1973; KOERKER et al. 1974; FELIG et al. 1976), and insulin inhibits glucagon secretion (SAMOLS et al. 1972; CLAUS and PILKIS 1981). The cells are anatomically arranged so as to function in a coordinated manner. Blood glucose is maintained through interactive secretions of insulin and glucagon, both of which are inhibited by somatostatin, whose secretion is stimulated by glucagon (PATTON et al. 1977; UNGER and ORCI 1981).

Following a meal glucose not extracted by the liver causes a rise in blood glucose and the release of insulin from the pancreas. Insulin acts to increase liver and muscle glycogen stores and is discussed in Chap. 16. Additionally insulin acts to blunt the rise in blood glucose by increasing utilization of glucose by stimulating pyruvate oxidation and lipogenesis as it decreases lypolysis and gluconeogenesis.

In the fasting state, plasma glucose concentrations are stable. They reflect the balance between hepatic glucose production by the liver, and glucose utilization by extrahepatic tissues. Hepatic glucose production derives from glycogenolysis and gluconeogenesis. Glycogen stores are depleted after 8–12 h of fasting and subsequent gluconeogenesis rather than glycogenolysis becoming the major source of plasma glucose. Both the liver and kidney contain the enzymes involved in gluconeogenesis, but the liver is quantitatively the more important site.

All cells possess the enzymes of glycolysis and are therefore capable of generating energy from the anaerobic conversion of glucose to pyruvate. In the fasted state glucose is the primary substrate of the brain and renal medulla and the sole energy source for erythrocytes and the retina. However, fatty acids and ketone bodies are utilized extensively as energy sources by other tissues. The most important role of gluconeogenesis is the production of blood glucose during periods of depleted glycogen stores and deficient food intake (EXTON 1972). It is also the means by which lactate, produced by erythrocyte glycolysis and exercising muscle, is reconverted to glucose (CHERRINGTON and LILJENQUIST 1981; LICKLEY et al. 1983; CHERRINGTON et al. 1987). The process can also be considered to conserve glycerol derived from triglyceride lipolysis in adipose tissue and amino acids derived from either proteolysis or absorbed from the gastrointestinal tract.

C. Biochemical Pathways of Glycolysis and Gluconeogenesis

The biochemical steps involved in glycolysis and gluconeogenesis are summarized in Fig. 1. Many of the detailed aspects of these important reactions are available in standard biochemistry texts (e.g., WHITE et al. 1978).

I. Enzymes of Glycolysis and Gluconeogenesis

Gluconeogenesis is not a simple reversal of glycolysis. Glycolysis and gluconeogenesis share common biochemical compounds and reactions. In Fig. 1 the extramitochondrial enzyme reactions indicated by double-headed arrows are shared by the two processes. They are not important points of regulation and are readily reversible. The two systems are however uniquely different. Glycolysis converts glucose to pyruvate/lactate and yields energy, whereas gluconeogenesis converts pyruvate/lactate to glucose and consumes energy. Direct reversal of glycolysis is impeded by several thermodynamic barriers (KREBS 1964; SCRUTTON and UTTER 1968). The enzymatic steps regulating glycolysis and differentiating it from gluconeogenesis are: (a) glucokinase conversion of glucose to glucose-6-phosphate; (b) fructose-6-kinase conversion of fructose-6-phosphate to fructose-1,6-diphosphate; (c) pyruvate kinase conversion of phosphoenolpyruvate to pyruvate. Gluconeogenesis is only possible because of four enzymatic steps that specifically regulate substrate flow and oppose the catabolic activity of the key enzymes of glycolysis. The four enzymes regulating the pyruvate–glucose pathway are: (a) pyruvate carboxylase, which carboxylates pyruvate to form oxaloacetate; (b) phosphoenolphosphate carboxykinase, which converts oxaloacetate to phosphoenolpyruvate; (c) fructose-1,6-diphosphatase, which

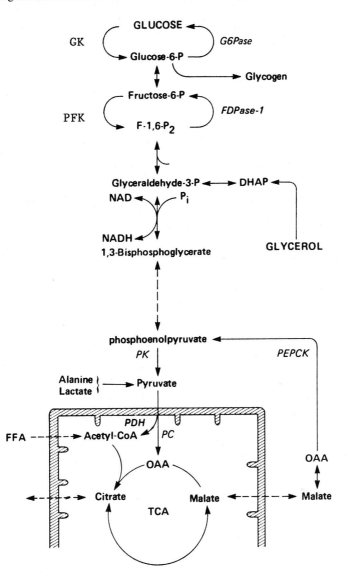

Fig. 1. Biochemical pathway of glycolysis and gluconeogenesis, a summary. Abbreviations: *GK* glucokinase; *G6Pase* glucose-6-phosphatase; *PFK* phosphofructokinase; *FDPase-1* fructose-1,6-diphosphatase-1; *F-1,6-P$_2$* fructose-1,6-diphosphate; *Glyceraldehyde-3-P* glyceraldehyde-3-phosphate; *DHAP* dihydroxyacetone phosphate; *PK* pyruvate kinase; *PEPCK* phosphoenolpyruvate carboxykinase; *PDH* pyruvate dehydrogenase; *FFA* free fatty acids; *PC* pyruvate carboxylase; *OAA* oxaloacetic acid; *TCA* tricarboxylic acid cycle

catalyzes the formation of fructose-6-phosphate from fructose-1,6-diphosphate; (d) glucose-6-phosphatase, which catalyzes the conversion of glucose-6-phosphate to glucose (STRUCK et al. 1966; WILLIAMSON 1967; SOLING et al. 1968; WILLIAMSON et al. 1968a).

II. Futile Cycles, Regulation by Enzyme Pairs

The three pairs of nonequilibrium reactions that control glycolysis and gluconeogenesis are shown in Fig. 2. A futile cycle results from simultaneous operation of paired anabolic and catabolic enzymes, consumes ATP, and generates heat without a net change in the carbon substrate. Substrate cycling in an apparent futile cycle may provide for enhanced in vivo regulation by allosteric effectors that change over a relatively small concentration range. Such amplification would not be possible if a single enzyme were responsible for both reactions (NEWSHOLME and STANLEY 1987). The enzymes involved in substrate cycling respond rapidly to allosteric and hormonal inputs, reflecting changing conditions and demands allowing for the integrated regulation of complex and competing metabolic processes.

III. Hepatic Gluconeogenesis

Hepatic gluconeogenesis in the fasting and/or insulin-deficient state is regulated by substrate availability and the actions of glucagon. Pyruvate primarily derived from peripheral lactate and alanine is carboxylated to oxaloacetate by the mitochondrial pyruvate carboxylase. This reaction appears to be rate limiting in the gluconeogenic process and is stimulated by administration of either free fatty acids or glucagon (STRUCK et al. 1966; WILLIAMSON 1967; WILLIAMSON et al. 1968a). The carboxylation of pyruvate takes precedence over the decarboxylation of pyruvate to acetyl coenzyme A and NADH by mitochondrial pyruvate dehydrogenase because the increased oxidation of free fatty acids inhibits the activity of pyruvate dehydrogenase and stimulates that of pyruvate carboxylase. Most of the oxaloacetic acid formed is converted to malate and transported to the cytosol where it is converted to phosphoenolpyruvate by phosphoenolpyruvate carboxykinase. Pyruvate kinase removal of phosphoenolpyruvate is inhibited by alanine and cAMP-dependent protein phosphorylation. Active fatty acid oxidation produces a higher NADH/NAD ratio favoring the production of glyceraldehyde-3-phosphate from 1,3-diphosphoglycerate catalyzed by phosphoglycerate kinase. Glycerol is incorporated into the gluconeogenesis pathway at this point by phosphorylation by glycerol kinase to glycerol phosphate and oxidation to dihydroxyacetone phosphate by glycerophosphate dehydrogenase. Subsequent steps include cleavage of fructose-1,6-diphosphate by fructose diphosphatase and the hydrolysis of glucose-6-phosphate to glucose by microsomal glucose-6-phosphatase.

D. Hormonal Regulation

Hormonal regulation of gluconeogenesis is complex and occurs at several levels. These include: (a) supply of glucose precursors and fatty acids (SCRUTTON and UTTER 1968; EXTON et al. 1970; SAMOLS et al. 1972; EXTON 1979, 1987; CLAUS and PILKIS 1981); (b) adaptive changes in enzyme activity due to regulation of protein anabolism and/or catabolism (WEBER 1972; TAUNTON et al. 1974); and (c) more rapid regulation of gluconeogenesis by glucagon, insulin, and catecholamines. These three hormones are thought to be most important in gluconeogenesis control (SAMOLS et al. 1972; PILKIS et al. 1978; CLAUS and PILKIS 1981). However, adrenal steroids (EXTON and PARK 1965), thyroid hormone (MENAHAN and WIELAND 1969; SINGH and SNYDER 1978), growth hormone (TOLMAN et al. 1973; JEFFERSON et al. 1973), and angiotensin II and vasopressin (HEMS and WHITTON 1973; WHITTON et al. 1978), have all been shown to regulate gluconeogenesis to some extent.

I. Glucagon

Glucagon is an important determinant of hepatic glucose output deriving from both glycogenolysis and gluconeogenesis, even in the presence of basal insulin concentrations (UNGER and ORCI 1981). Glucagon secretion varies little throughout a normal day, but is augmented by hypoglycemia, strenuous exercise, protein ingestion, amino acid infusions, and adrenal corticosteroids (SHERWIN and FELIG 1977; UNGER and ORCI 1981; UNGER 1985). The hyperglucagonemia of starvation results from decreased catabolism rather than increased secretion.

Glucagon enhancement of hepatic glucose output is a frequency-dependent pulsatile effect of the hormone (FRADKIN et al. 1980; KOMJATI et al. 1986; WEIGLE and GOODNER 1986; WEIGLE et al. 1987). Changes in hepatic glucose output are brought about by changes in glucagon concentrations rather than by the absolute hormone concentrations. The liver exhibits continuing responsiveness to pulsatile changes in plasma glucagon concentrations. These pulsatile cycles range 13–20 min in humans (WEIGLE and GOODNER 1986; WEIGLE et al. 1987), but the effects on hepatic glucose output persist for 30–60 min (SHERWIN and FELIG 1977; FRADKIN et al. 1980; KOMJATI et al. 1986). A fall in glucagon below basal levels results in a decrease in hepatic glucose output (CHERRINGTON et al. 1976). When an increment of glucagon concentration is accompanied by small elevations of plasma insulin concentration (0.5- to 1.5-fold) hepatic glucose output does not change.

Glucagon's effect on gluconeogenesis is at three sites: (a) hepatic gluconeogenic enzymes; (b) uptake of gluconeogenic amino acids by liver without augmentation of peripheral precursors release; and (c) increasing fatty acid oxidation (CHIASSON et al. 1974, 1975; LILJENQUIST et al. 1977; CHERRINGTON et al. 1978, 1982; CLAUS and PILKIS 1981; KILBERG 1982; DAVIS et al. 1985; KRAUS-FRIEDMAN 1986).

1. Effects on the Enzymes of Gluconeogenesis

Glucagon's actions in the liver are mediated by cAMP-dependent and cAMP-independent actions (RODBELL 1983, 1985; BERRIDGE 1984; WAKELAM et al.

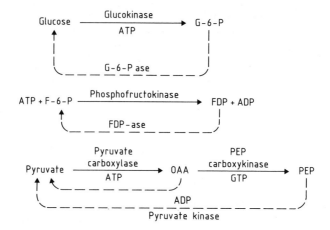

Fig. 2. Nonequilibrium reactions that control glycolysis and gluconeogenesis. Abbreviations: *G-6-P* glucose-6-phosphate; *G-6-Pase* glucose-6-phosphatase; *F-6-P* fructose-6-phosphate; *FDP* fructose-1,6-diphosphate; *FDP-ase* fructose-1,6-diphosphatase; *OAA* oxaloacetic acid; *PEP* phosphoenolpyruvate

1986). They are largely due to increasing cAMP and subsequently activating cAMP-dependent protein kinases. Thus, pyruvate kinase is inactivated (Fig. 2, bottom) and phosphofructo-6-kinase activity falls (Fig. 2, middle) as the level of the positive effector fructose-2,6-bisphosphate declines due to cAMP-dependent phosphorylation of the single enzyme 6-phosphofructo-2-kinase/fructose-2,6-bisphosphatase (Richards and Uyeda 1982; Pilkis et al. 1983). This effectively prevents two futile cycles (Fig. 2, middle and bottom) from occurring during gluconeogenesis. If gluconeogenesis has resulted in glycogen being an appreciable source of glucose the concentration of fructose-2,6-bisphosphate increases and substrate cycling then results.

2. Effects on Substrate Uptake

Glucagon increases the rate of alanine conversion to glucose before stimulation of alanine uptake can be observed (Chiasson et al. 1975). Stimulation of alanine uptake is via system A of amino acid transport and can be blocked by cycloheximide inhibition of protein synthesis (Kletzien et al. 1975; Felig 1975; Pariza et al. 1976; Davis et al. 1985). Augmentation of lactate uptake by glucagon is slower than the stimulatory effect on hepatic amino acid uptake and occurs after liver glycogen stores are depleted (Samols et al. 1972; Claus and Pilkis 1981; Cherrington et al. 1987).

3. Effects on Fatty Acid Oxidation

Glucagon also regulates ketogenesis in the liver, and the process by which it does so supports the increased intrahepatic gluconeogenic activity (Bressler 1970a; McGarry and Foster 1980). Glucagon lowers hepatic levels of malonyl coenzyme A and thereby decreases fatty acid synthesis and increases fatty acid oxidation (MᴄGᴀʀʀʏ and Fᴏsᴛᴇʀ 1980, 1981). The increase of fatty acid oxidation represents a rapid adaptation of gluconeogenic stimuli, whereas changes in gluconeogenic enzymes constitute a slower adapting mechanism (Wᴇʙᴇʀ et al. 1965, 1966; Sᴛʀᴜᴄᴋ et al. 1966; Wɪʟʟɪᴀᴍsᴏɴ et al. 1968 b; Sᴏʟɪɴɢ et al. 1968; Wɪʟʟᴍs et al. 1970). The acceleration of fatty acid oxidation is rapid and produces more acetyl coenzyme A, which allosterically activates pyruvate carboxylase, a rate-determining enzyme in gluconeogenesis (Uᴛᴛᴇʀ and Sᴄʀᴜᴛᴛᴏɴ 1969; Bʀᴇssʟᴇʀ 1970 b; Sᴛᴜᴄᴋɪ et al. 1976; ᴠᴏɴ Gʟᴜᴛᴢ and Wᴀʟᴛᴇʀ 1976; Bᴀʀ-ʀɪᴛᴛ et al. 1976).

II. Insulin Effects on Gluconeogenesis

Whereas the primary effects of glucagon on gluconeogenesis are localized to the liver, the effects of insulin are manifested at both hepatic and extrahepatic sites (Mᴏʀᴛɪᴍᴏʀᴇ and Mᴏɴᴅᴏɴ 1970; Cʜᴇʀʀɪɴɢᴛᴏɴ and Lɪʟᴊᴇɴǫᴜɪsᴛ 1981; Uɴɢᴇʀ and Oʀᴄɪ 1981; Cʜᴇʀʀɪɴɢᴛᴏɴ et al. 1987). Insulin's inhibitory effects on gluconeogenesis derive from a number of its actions. These include: (a) restraint of glucagon secretion by the pancreatic alpha cell and temporizing of glucagon's effects on the liver; (b) decreased availability of peripheral substrates such as pyruvate, lactate, alanine, and other glucogenic amino acids, glycerol, and long-chain fatty acids, resulting in a decreased hepatic uptake of glucose precursors; (c) enhanced synthesis of glycolytic and glycogen-synthesizing enzymes and in-hibition of synthesis of gluconeogenic enzymes. Actions (a) and (b) are acute aspects of insulin action, whereas (c) is generally a more chronic one.

1. Antiglucagon Activity

Insulin output decreases that of glucagon and thus decreases the hormonally driven stimulation of the gluconeogenic pathway. Hyperglycemia per se does not suppress glucagon unless sufficient insulin is elicited in response to the blood glucose rise (Bᴇʀɢᴍᴀɴ 1977; Cʜᴇʀʀɪɴɢᴛᴏɴ et al. 1978; Uɴɢᴇʀ and Oʀᴄɪ 1981; Isʜɪᴅᴀ et al. 1983).

The importance of insulin deficiency in glucagon output and the metabolic ef-fects deriving from it have been demonstrated in studies on insulin-dependent diabetic patients in whom normal patterns of insulin secretion were generated by means of an open-loop insulin delivery system (Isʜɪᴅᴀ et al. 1983). In these patients, both glucagon and blood glucose concentrations normalized, suggest-ing the essential role of physiological insulin levels in control of glucagon secre-tion and glucagon effects on hepatic glucose output.

Insulin regulates the supply of gluconeogenic precursors available for uptake by the liver through its anticatabolic and anabolic effects on adipose tissue and muscle. Basal insulin concentrations which maintain the glycolytic and lipogenic liver enzymes and repress the gluconeogenic enzymes may be the most important aspect of insulin's role in gluconeogenesis. Although part of insulin's inhibitory effect on gluconeogenesis has been ascribed in part to its decreasing supplies of amino acids and lactate to the liver, other studies take issue on this point (Fᴇʟɪɢ and Wahren 1971; Felig et al. 1975; Cherrington et al. 1987). The more important effect of insulin on hepatic enzymes is agreed upon.

Insulin blocks the production of cAMP induced by suboptimal doses of glucagon (shifts the dose-response curve of glucagon to the right). Insulin's short-term action as an antigluconeogenic agent is a result of restraint of glucagon secretion by the pancreatic alpha cell, thus preventing glucagon-induced cAMP stimulation of gluconeogenesis. Insulin in the absence of glucagon has little if any direct effect on the short-term activities on the enzymes of gluconeogenesis.

The opposing actions of glucagon and insulin on hepatic gluconeogenesis have been demonstrated experimentally in humans and in dogs (Felig et al. 1970a; Chiasson et al. 1974; Unger and Orci 1981; Claus and Pilkis 1981; Mohan and Bessman 1985; Cherrington et al. 1987). In studies where there was combined insulin and glucagon deficiency, gluconeogenesis was limited (Gerich et al. 1975; Unger and Orci 1981).

Insulin deficiency rapidly activates gluconeogenesis in animals and humans (Chiasson et al. 1976, 1979; Jennings et al. 1977; Cherrington et al. 1978). In the absence of insulin, there is an increased peripheral supply and hepatic uptake of amino acids and glycerol (Felig et al. 1970a; Shaw et al. 1976; Cherrington et al. 1987). Studies in diabetic animals and humans have shown that the untreated diabetic state is associated with increased concentrations of plasma branched-chain amino acids (Ivy et al. 1951; Carlsten et al. 1967; Felig et al. 1970a). Plasma alanine is however, depressed in untreated diabetics, and this could reflect an enhanced extraction by the liver (Ivy et al. 1951; Carlsten et al. 1966; Sestoft et al. 1977). Insulin deficiency accelerates the peripheral release of glycerol in rats and dogs (Winegrad 1965; Shaw et al. 1976), and splanchnic glycerol extraction in humans (Wahren et al. 1975, 1976). In the overnight-fasted dog both the fractional extraction of alanine by the liver and its conversion to glucose increase significantly (Cherrington et al. 1978, 1987).

Deficiency of basal fasting levels of insulin such as occurs in the untreated type I diabetic state rapidly results in a mobilization of peripheral gluconeogenic substrates and increases their extraction by the liver, and conversion to glucose. However, small increments in insulin above basal fasting levels do not suppress gluconeogenesis, in contrast to the marked stimulatory effect of small decrements of plasma insulin (Chiasson et al. 1980; Cherrington et al. 1987).

Insulin deficiency engenders an increased hepatic net extraction of glucogenic amino acids, lactate, and glycerol and their conversion to glucose. However, the major direct action of insulin on the liver is a slower modulation of enzymes of glycolysis and gluconeogenesis. Insulin deficiency stimulates both quantity and activity of the gluconeogenesis enzymes (Weber 1972; Taunton et al. 1974). These changes are discussed in Sect. D. III.

2. Anabolic and Catabolic Activity

Insulin plays a role in regulating both glycogenolysis and gluconeogenesis in liver (Chiasson et al. 1974; Claus and Pilkis 1981; Cherrington et al. 1987). These actions are aspects of insulin's anabolic and anticatabolic functions (Cahill 1971). Its anticatabolic effects on gluconeogenesis are in part a function of its restraint of lipolysis in adipose tissue, proteolysis in muscle, and glycogenolysis in liver decreasing hepatic glucose output. The absence or deficiency of insulin's effects results in a greater net influx of gluconeogenic precursors to the liver (BERGMAN 1977; CHERRINGTON et al. 1978, 1987).

a) Substrate Availability

A major effect of insulin on the regulation of gluconeogenesis is via its control of substrate availability (FELIG et al. 1970a; CHERRINGTON et al. 1978; MOHAN and BESSMAN 1985). The liver is the organ primarily responsible for maintaining blood glucose, but the amino acids which serve as glucose precursors are derived from muscle (FELIG et al. 1970b; FELIG 1973, 1975; MOHAN and BESSMAN 1985). Proteolysis in muscle provides amino acids for gluconeogenesis and for muscle energy during periods of insufficient quantities of glucose, as occurs in fasting or limited utilization of glucose in diabetes. Amino acids undergo transamination in muscle, which produces a large amount of amino nitrogen. The muscle, unlike liver, cannot synthesize urea and transaminates pyruvate to form alanine. Alanine is released to the circulation and is extracted by the liver (FELIG et al. 1970a; FELIG 1973; MOHAN and BESSMAN 1985; CHERRINGTON et al. 1987). In the liver, the amino group of alanine is removed via a transamination reaction which generates pyruvate. Pyruvate is converted to glucose in the liver. The glucose is secreted into the circulation, taken up by muscle, and catabolized to pyruvate, which again becomes available as an acceptor for amino groups. These reactions constitute a cycle that transports amino groups from muscle to liver (FELIG 1973, 1975).

   In the nonfasting euinsulinemic state or in the postprandial state, insulin stimulates the transport of most amino acids into muscle, stimulates protein synthesis, inhibits protein catabolism, and decreases amino acid oxidation (MORTIMORE and MONDON 1970; SHERWIN et al. 1976; FELIG et al. 1977; DAHLMAN et al. 1979).

b) Fatty Acid Oxidation

Gluconeogenesis requires the input of energy. It requires 6 mol adenosine triphosphate (ATP) to synthesize 1 mol glucose from lactate and 10 mol from alanine where additional ATP is required for the elimination of the amino nitrogen as urea. In the fasted or insulin-deficient state, fatty acid oxidation is an energy source for gluconeogenesis (RANDLE et al. 1966; DELISLE and FRITZ 1967; RUDERMAN et al. 1969; BRESSLER 1970b; TUTWILER and DELLEVIGNE 1979). A basal rate of fatty acid oxidation is necessary for the genesis of new glucose (BRESSLER 1970b; FROHLICH and WIELAND 1971; TUTWILER and DELLEVIGNE 1979; HUE 1987). Fatty acid oxidation provides ATP for energy, acetyl coenzyme A for activation of pyruvic carboxylase (BRESSLER 1970b; VON GLUTZ and WALTER 1976; STUCKI et al. 1976), and reduced pyridine nucleotide (NADH) for stimulation of glyceraldehyde-3-phosphate dehydrogenase activity (HUE 1987).

   Long-chain fatty acid oxidation in the liver is controlled by insulin in several ways. These include inhibition of lipolysis and fatty acid reesterification in adipose tissue, which regulate substrate access to the liver, and control of fatty acid oxidation in the liver by means of regulation of malonyl coenzyme A levels (MAHLER et al. 1964; COOK et al. 1977; MCGARRY and FOSTER 1980, 1981; CHERRINGTON et al. 1987).

The degradation of adipose tissue triglycerides is controlled by lipases which are at least in part acutely stimulated by a number of hormones. Adipose tissue lipolysis is stimulated by norepinephrine, epinephrine, adrenocorticosteroid-stimulating hormone, and to some extent glucagon (EXTON 1987; HUE 1987). Glucocorticoids, thyroxine, and growth hormone also stimulate lipolysis, but are slower acting and act through increased production of regulatory proteins rather than by activating existing adenylate cyclase as do the more acutely acting agents (HUE 1987).

Lowered blood glucose and insulin concentrations in the fasting state or insulin deficiency in the diabetic state, enhance fatty acid mobilization from adipose tissue by permitting levels of cAMP to elevate, thus keeping the adipose tissue triglyceride lipase in its active form (MAHLER et al. 1964; KHOO et al. 1973; COOK et al. 1977; BROWN and GINSBURG 1982; EXTON 1987).

The stimulatory effect of insulin deficiency on lipolysis in adipose tissue is due to an augmentation of lipolysis and a decrease in reesterification of fatty acids. Lipolysis in adipose tissue produces free fatty acids and glycerol and the reaction is irreversible (WINEGRAD 1965; TASKINEN and NIKKILA 1979). Glycerol cannot be reconverted to α-glycerol phosphate, which serves as an acceptor for long-chain fatty acyl coenzyme A moieties, to form triglycerides because adipose tissue does not possess a glycerol kinase. Glycerol is released into the circulation, taken up by the liver, and serves as substrate for gluconeogenesis (SHAW et al. 1976). Free fatty acids are capable of reesterification to triglycerides if α-glycerol phosphate is generated from insulin-stimulated glucose transport into adipose tissue (HAGEN 1963; MAHLER et al. 1964; ZIERLER and RABINOWITZ 1964; KHOO et al. 1973; COOK et al. 1977).

c) Fatty Acid Synthesis

The catabolic state associated with fasting and insulin deficiency is characterized by glycogenolysis, gluconeogenesis, and markedly impaired hepatic long-chain fatty acid synthesis (MCGARRY and FOSTER 1980; WAKIL et al. 1983; HUE 1987). The decreased fatty acid synthesis has been ascribed in part to depressed activity of the acetyl coenzyme A carboxylase, a rate-limiting enzyme in overall fatty acid synthesis (WAKIL et al. 1983; BREMER and OSMUNDSEN 1984).

Long-chain fatty acid synthesis in mammalian liver is greatest in association with high carbohydrate intakes and adequate insulin (WAKIL et al. 1983). Acetyl coenzyme A carboxylase is inactive in the phosphorylated state, but is dephosphorylated in the presence of glucose and insulin and thereby activated (SAGGERSON 1980; LEE et al. 1985). The diminished activity of this enzyme results in a lowered level of hepatic malonyl coenzyme A, a compound which restrains the long-chain carnitine acyltransferase of the outer portion of the inner mitochondrial membranes (MCGARRY and FOSTER 1980; BREMER and OSMUNDSEN 1984). This enzyme is the rate-limiting step in long-chain fatty acid oxidation and, therefore, its less restrained activity stimulates fatty acid oxidation (McGARRY and FOSTER 1980; BREMER and OSMUNDSEN 1984), and gluconeogenesis. The essential permissive role of fatty acid oxidation in gluconeogenesis has been demonstrated by means of pharmacologic inhibition of the long-chain carnitine acyltransferase by compounds which markedly inhibit gluconeogenesis and cause decreases in blood glucose in fasting normal and diabetic mice, rats, and dogs (TUTWILER et al. 1978; TUTWILER and DELLEVIGNE 1979; SAGGERSON 1980; EISTETLER and WOLF 1982; LEE et al. 1985).

III. Hepatic Gene Expression and Metabolic Zonation

Insulin and glucagon both exert long-term control of the glycolytic and gluconeogenic processes through the regulation of gene expression. The effect of insulin on mRNA production is reviewed in this volume in Chap. 17 and the action of insulin-like growth factors in Chap. 14.

Insulin integrates hepatic metabolism by inducing or suppressing enzyme levels. Key enzymes of glycolysis are induced: glucokinase, phosphofructokinase, and pyruvate kinase (WEBER et al. 1966) as enzymes of gluconeogenesis are suppressed; glucose-6-phosphatase, fructose-1,6-diphosphatase, phosphoenolpyruvate carboxykinase, and pyruvate carboxylase (WEBER 1972). Insulin control of phosphoenolpyruvate carboxykinase synthesis has been shown to be at the level of rapid and complete inhibition of RNA transcription and not translation, and occurs via stimulation of the insulin receptor (GRANNER et al. 1986). The stimulatory actions of insulin on the enzymes of gluconeogenesis may be due to a direct and specific increase in mRNA or may involve intervening protein synthesis (GRANNER et al. 1986).

1. Glycolytic Perivenous/Gluconeogenic Periportal Regions

The actions of insulin and glucagon on gene expression are readily demonstrated in vitro (AGIUS et al. 1986; CHRIST et al. 1986; HARTMANN et al. 1987). A single population of isolated hepatocytes maintained in tissue culture express predominantly glycolytic or gluconeogenic enzymes, depending upon whether insulin or glucagon is predominant (PROBST et al. 1982).

The liver in vivo displays zones of predominantly glycolytic or gluconeogenic activity corresponding to the perivenous and periportal regions, respectively. Functional heterogeneity of periportal and perivenous hepatocytes has been reviewed by JUNGERMANN (1986). While glycolysis and gluconeogenesis are more readily achieved in cells of one region, both metabolic processes do occur to some extent in both. Metabolic zonation predicts that glycogen would originate from glucose removed from the plasma by the perivenous hepatocytes and that periportal cells would produce glycogen from substrates containing three carbons. The ability to utilize pyruvate or glucose for glycogen synthesis, and the time course of glycogen accumulation and breakdown during periods of feeding and fasting correlates with the periportal–perivenous distribution that is predicted by metabolic zonation (BARTELS et al. 1987).

Hormonally induced metabolic zonation provides a mechanism for the segregation of activities that are interrelated. Oxygen tension is greater in the periportal region where fatty acid oxidation is a prerequisite for the ATP requirements of gluconeogenesis from three-carbon precursors. Glycolysis provides the necessary elements for lipogenesis and again the distribution of enzymes is accompanied by the zonation of metabolic function.

2. Induction and Maintenance of Metabolic Zonation

Metabolic zonation does not develop until shortly after birth. Contributing to the development and the maintenance of metabolic zonation is the ratio of the hormones insulin and glucagon across the liver and the temporal pattern of hormone levels. The liver clearance of glucagon is independently regulated from that of insulin and occurs at a different rate so the periportal and perivenous hepatocyte genome is exposed to signals that vary in amount rather than kind (Balks and Jungermann 1984). Other factors contribute to the maintenance of metabolic zonation. Neuronal influences, redox states, and oxygen tension also affect gene expression that would be expected to provide a mechanism for short-term responsiveness and longer-term adaptation.

The glucostat functioning of the liver may be derivative of this distribution of function (Jungermann 1986). Periportal cells extract pyruvate, lactate, and alanine and release glucose; perivenous cells take up glucose and release lactate, the activity of these reciprocal systems of glycolysis and gluconeogenesis being dependent in both the short- and long-term on substrate availability and both hormonally and neuronally directed activities.

References

Agius L, Chowdhury MH, Davis SN, Alberti KGMM (1986) Regulation of ketogenesis, gluconeogenesis, and glycogen synthesis by insulin and proinsulin in rat hepatocyte monolayer cultures. Diabetes 35:1286–1293

Alberti KGMM, Christensen NJ, Christensen SE, Prange-Hansen AA, Iversen J, Lundbaek KA, Sever-Hansen K, Orskov H (1973) Inhibition of insulin secretion by somatostatin. Lancet 2:1299–1301

Balks HJ, Jungermann K (1984) Regulation of the peripheral insulin/glucagon ratio by the liver. Eur J Biochem 141:645–650

Barritt GJ, Zander GL, Utter M (eds) (1976) Gluconeogenesis. Its regulation in mammalian species. Wiley, New York, pp 3–46

Bartels H, Vogt B, Jungermann K (1987) Glycogen synthesis from pyruvate in the periportal and from glucose in the perivenous zone in perfused livers from fasted rats. FEBS Lett 221:277–283

Bergman RN (1977) Integrated control of hepatic glucose metabolism. Fed Proc 36:256–261

Berridge MJ (1984) Inositol triphosphate and diacylglycerol as second messengers. Biochemistry 220:345–352

Bremer J, Osmundsen H (1984) Fatty acid oxidation and its regulation. In: Numa S (ed) Fatty acid metabolism and its regulation. Elsevier, Amsterdam, pp 113–154

Bressler R (1970a) Physiological-chemical aspects of fatty acid oxidation. In: Wakil SJ (ed) Lipid metabolism. Academic, New York, pp 44–77

Bressler R (1970b) Fatty acid oxidation. In: Florkin M, Stotz EH (eds) Lipid metabolism. Elsevier, Amsterdam, pp 331–359 (Comprehensive biochemistry, vol 18)

Brown WV, Ginsburg HH (1982) Diabetes and plasma lipoproteins. In: Brodoff BN, Bleicher SJ (eds) Diabetes mellitus and obesity. Williams and Wilkins, Baltimore, pp 192–199

Cahill GF Jr (1971) Physiology of insulin in man. Diabetes 20:785–799

Carlsten A, Hallgren B, Jagenburg R, Svanborg A, Werko L (1966) Amino acids and free fatty acids in plasma in diabetes. Acta Med Scand 179:361–369

Carlsten A, Hallgren B, Jagenburg R, Svanborg A, Werko L (1967) Arterio-hepatic venous differences of free fatty acids and amino acids. Acta Med Scand 181:199–205

Cherrington AD, Liljenquist JE (1981) Role of glucagon in regulating glucose production in vivo. In: Unger RH, Orci L (eds) Glucagon. Elsevier/North-Holland, New York, pp 221–253

Cherrington AD, Chiasson JL, Liljenquist JE, Jennings AS, Keller U (1976) The role of insulin and glucagon in the regulation of basal glucose production in the postabsorptive dog. J Clin Invest 58:1407–1418

Cherrington AD, Lacy WW, Chiasson JL (1978) Effect of glucagon on glucose production during insulin deficiency in the dog. J Clin Invest 62:664–677

Cherrington AD, Diamond MP, Green DR, Williams PE (1982) Evidence for an intrahepatic contribution in the waning of the effect of glucagon on glucose production in the conscious dog. Diabetes 31:917–922

Cherrington AD, Stevenson RW, Steiner KE, Davis MA, Myers SR, Adkins BA, Abumrad NH, Williams PE (1987) Insulin, glucagon and glucose as regulators of hepatic glucose uptake and production in vivo. Diabetes Metab Rev 3:307–332

Chiasson JL, Cook J, Liljenquist JE, Lacy WW (1974) Glucagon stimulation of gluconeogenesis from alanine in the intact dog. Am J Physiol 227:19–23

Chiasson JL, Liljenquist JE, Sinclair-Smith BC, Lacy WW (1975) Gluconeogenesis from alanine in normal postabsorptive man. Intrahepatic stimulatory effect of glucagon. Diabetes 24:574–584

Chiasson JL, Liljenquist JE, Finger FE, Lacy WW (1976) Differential sensitivity of glycogenolysis anbd gluconeogenesis to insulin infusions in dogs. Diabetes 25:283–291

Chiasson JL, Atkinson RL, Cherrington AD, Keller U, Sinclair-Smith BC, Lacy WW, Liljenquist JE (1979) Insulin regulation of gluconeogenesis from alanine in man. Diabetes 28:380

Chiasson JL, Atkinson RL, Cherrington AD, Keller U, Sinclair-Smith BC, Lacy WW, Liljenquist JE (1980) Effects of insulin at two dose levels on gluconeogenesis from alanine in fasting man. Metabolism 29:810–818

Christ B, Probst I, Jungermann K (1986) Antagonistic regulation of the glucose/glucose 6-phosphate cycle by insulin and glucagon in cultured hepatocytes. Biochem J 238:185–191

Claus TH, Pilkis SJ (1981) Hormonal control of hepatic gluconeogenesis. In: Litwack G (ed) Biochemical actions of hormones, vol 8. Academic, New York, pp 209–271

Cook GA, Nielsen RC, Hawkins RA, Mehlman MA, Laskshman MR, Veech RL (1977) Effect of glucagon on hepatic malonyl coenzyme A concentration and on lipid synthesis. J Biol Chem 252:4421–4424

Dahlman B, Schroeter C, Herbertz L, Reinauer H (1979) Myofibrillar protein, degradation and muscle proteinase in normal and diabetic rats. Biochem Med 21:33–39

Davis MA, Williams PE, Cherrington AD (1985) Effect of glucagon on hepatic lactate metabolism in the conscious dog. Am J Physiol 248:E463–470

Delisle G, Fritz IB (1967) Interrelations between hepatic fatty acid oxidation and gluconeogenesis: a possible regulatory role of carnitine palmityltransferase. Proc Natl Acad Sci USA 58:790–797

Eistetler K, Wolf PO (1982) Synthesis and hypoglycemic activity of phenylalkyloxiranecarboxylic acid derivatives. J Med Chem 25:109–113

Exton JH (1972) Gluconeogenesis. Metabolism 21:945–990

Exton JH (1979) Hormonal control of gluconeogenesis. In: Klachko DM, Anderson RR, Heimberg M (eds) Hormones and energy metabolism. Plenum, New York, pp 125–167

Exton JH (1987) Mechanism of hormonal regulation of hepatic glucose metabolism. Diabetes Metab Rev 3:163–184

Exton JH, Park CR (1965) Preliminary communication. J Biol Chem 240:PC955–PC957
Exton JH, Mallette LE, Jefferson LS, Wong EHA, Friedman N, Miller TB Jr, Park CR (1970) The hormonal control of hepatic gluconeogenesis. Recent Prog Horm Res 26:411–461
Felig P (1973) The glucose-alanine cycle. Metabolism 22:179–207
Felig P (1975) Amino acid metabolism in man. Annu Rev Biochem 44:933–955
Felig P, Wahren J (1971) Influence of endogenous insulin secretion on splanchnic glucose and amino acid metabolism in man. J Clin Invest 58:1702–1711
Felig P, Wahren J, Hendler R (1975) Influence of oral glucose ingestion on splanchnic glucose and gluconeogenic substrate metabolism in man. Diabetes 24:468–475
Felig PE, Posefsky T, Marliss E, Cahill GF Jr (1970a) Alanine: key role in gluconeogenesis. Science 167:1003–1004
Felig PE, Marliss, E, Ohman JL, Cahill GF (1970b) Plasma amino acid levels in diabetic ketoacidosis. Diabetes 19:727–736
Felig P, Wahren J, Sherwin RS, Hendler R (1976) Insulin, glucagon and somatostatin in normal physiology and diabetes mellitus. Diabetes 25:1091–1099
Felig P, Wahren J, Sherwin RS, Palaiologos G (1977) Protein and amino acid metabolism in diabetes mellitus. Arch Intern Med 137:507–512
Fradkin J, Shamoon H, Felig P, Sherwin RS (1980) Evidence for an important role of changes in rather than absolute concentrations of glucagon in the regulation of glucose production in man. J Clin Endocrinol Metab 50:698–703
Frohlich J, Wieland O (1971) Glucagon and the permissive action of fatty acids in hepatic gluconeogenesis. Eur J Biochem 19:557–562
Gerich J, Lorenzi M, Bier D, Schneider V, Tsalikia E, Karam JH, Forsham PH (1975) Prevention of human diabetic ketoacidosis by somatostatin: evidence for an essential role of glucagon. N Engl J Med 292:985–989
Granner DK, Sasaki K, Andreone T, Beale E (1986) Insulin regulates expression of the phosphoenolpyruvate carboxykinase gene. Recent Prog Horm Res 42:111–141
Hagen JH (1963) The effect of insulin on the concentration of plasma glycerol. J Lipid Res 4:46–51
Hartmann H, Probst I, Jungermann K, Creutzfeldt W (1987) Inhibition of glycogenolysis and glycogen phosphorylase by insulin and proinsulin in rat hepatocyte cultures. Diabetes 30:551–555
Hems DA, Whitton PD (1973) Stimulation vasopressing of glycogen breakdown and gluconeogenesis in the perfused rat liver. Biochem J 136:705–709
Hue L (1987) Gluconeogenesis and its regulation. Diabetes Metab Rev 3:111–126
Ishida T, Chap Z, Chou J, Lewis R, Hartlev C, Entman M, Field JB (1983) Differential effects of oral peripheral intravenous and intraportal glucose on hepatic glucose uptake and insulin and glucagon extraction in conscious dogs. J Clin Invest 72:590–601
Ivy JH, Svec M, Freeman S (1951) Free plasma levels and urinary excretion of eighteen amino acids in normal and diabetic dogs. Am J Physiol 167:182–188
Jefferson LS, Robertson JW, Tolman EL (1973) Effects of hypophysectomy on lactate metabolism in the perfused rat liver. J Biol Chem 248:4561–4567
Jennings AS, Cherrington AD, Liljenquist JE, Keller U, Lacy WW, Chiasson JL (1977) The roles of insulin and glucagon in the regulation of gluconeogenesis in the postabsorptive dog. Diabetes 26:849–856
Jungermann K (1986) Functional heterogeneity of periportal and perivenous hepatocytes. Enzyme 35:161–180
Khoo JC, Steinberg D, Thompson B (1973) Hormonal regulation of adipocyte enzymes. the effects of epinephrine and insulin on the control of lipase, phosphorylase kinase, phosphorylase and glycogen synthetase. J Biol Chem 248:3823–3830
Kilberg MS (1982) Amino acid transport in isolated rat hepatocytes. J Membr Biol 69:1–12
Kletzien RF, Pariza MW, Becker JE (1975) Hormonal regulation of amino acid transport and gluconeogenesis in primary cultures of adult rat liver parenchymal cells. J Cell Physiol 89:641–646

Koerker DJ, Ruch W, Chideckel E, Palmer J, Goodner CJ, Ensinck J, Gale CC (1974) Somatostatin: hypothalamic inhibitor of the endocrine pancreas. Science 184:482–484

Komjati M, Bratusch-Marrain P, Waldhausl W (1986) Superior efficacy of pulsatile versus continuous hormone exposure on hepatic glucose production in vitro. Endocrinology 118:312–319

Kraus-Friedmann N (ed) (1986) Hormonal control of gluconeogenesis, vol 1: Function and experimental approaches. CRC, Boca Raton

Krebs HA (1964) Gluconeogenesis. Proc R Soc Lond [Biol] 159:545–564

Lee SM, Bahl JJ, Bressler R (1985) Prevention of the metabolic effects of 2-tetradecylglycidate by octanoic acid in the genetically diabetic mouse (db/db). Biochem Med 33:104–109

Lickley L, Kemmer F, Wasserman DH, Vranic M (1983) Glucagon and its relationship to other glucoregulatory hormones in exercise and stress in normal and diabetic subjects. In: Lefebvre PJ (ed) Glucagon II. Springer, Berlin Heidelberg New York, pp 297–350

Liljenquist JE, Mueller GL, Cherrington AD, Keller U, Chiasson JL, Perry JM, Lacy WW, Rabinowitz D (1977) Evidence for an important role of glucagon in the regulation of hepatic glucose production in normal man. J Clin Invest 59:369–374

Mahler R, Stafford WS, Tarrant ME (1964) The effect of insulin on lipolysis. Diabetes 18:297–302

McGarry JD, Foster DW (1980) Regulation of hepatic fatty acid oxidation and ketone body production. Annu Rev Biochem 49:395–579

McGarry JD, Foster DW (1981) Ketogenesis. In: Unger RH, Orci L (eds) Glucagon. Elsevier/North-Holland, New York, pp 273–295

Menahan LA, Wieland O (1969) The role of thyroid function in the metabolism of perfused rat liver with particular reference to gluconeogenesis. Eur J Biochem 10:188–196

Mohan C, Bessman SP (1985) Anabolic regulation of gluconeogenesis by insulin in isolated rat hepatocytes. Arch Biochem Biophys 242:563–573

Mortimore GE, Mondon CE (1970) Inhibition by insulin of valine turnover in liver. J Biol Chem 245:2375–2383

Newsholme EA, Stanley JC (1987) Substrate cycles: their role in control of metabolism with specific references to the liver. Diabetes Metab Rev 3:295–305

Pariza MW, Butcher FR, Kletzien RF (1976) Induction and decay of glucagon induced amino acid transport in primary cultures of adult rat liver cells: paradoxical effects of cycloheximide and puromycin. Proc Natl Acad Sci USA 73:4511–4515

Patton GS, Ipp E, Dobbs RE, Orci L, Vale W, Unger RH (1977) Pancreatic immunoreactive somatostatin release. Proc Natl Acad Sci USA 74:2140–2143

Pilkis SJ, Park CR, Claus TH (1978) Hormonal control of hepatic gluconeogenesis. Vitam Horm 36:383–460

Pilkis SJ, Chrisman TD, El-Maghrabi MR, Colosia A, Fox E, Pilkis J, Claus TH (1983) The action of insulin on hepatic fructose 2,6-bisphosphate metabolism. J Biol Chem 258:1495–1503

Probst I, Schwartz P, Jungermann K (1982) Induction in primary culture of "gluconeogenic" and "glycolytic" hepatocytes resembling periportal and perivenous cells. Eur J Biochem 126:271–278

Randle PJ, Garland PB, Hales CN, Newsholme EA (1966) Interactions of metabolism and the physiological role of insulin. Recent Prog Horm Res 22:1–48

Richards CS, Uyeda K (1982) Hormonal regulation of fructose-6-P,2-kinase and fructose-2,6-P$_2$ by two mechanisms. J Biol Chem 257:8854–8861

Rodbell M (1983) The actions of glucagon at its receptor: regulation of adenylate cyclase. In: Lefebvre P (ed) Glucagon I. Springer, Berlin Heidelberg New York, pp 263–290 (Handbook of experimental pharmacology, vol 66/1)

Rodbell M (1985) Programmable messengers: a new theory of hormone action. Trends Biochem Sci 10:461–464

Ruderman NB, Toews CJ, Shafrir E (1969) Role of free fatty acids in glucose homeostasis. Arch Intern Med 123:299–313

Saggerson ED (1980) Regulation of lipid metabolism in adipose tissue and liver cells. In: Clemens MJ (ed) Biochemistry of cellular regulation. CRC, Boca Raton, pp 207–256

Samols E, Marri G, Marks V (1965) Promotion of insulin secretion by glucagon. Lancet 2:415–416

Samols E, Tyler JM, Marks V (1972) Glucagon-insulin interrelationships. In: Lefebvre PJ, Unger RH (eds) Glucagon: molecular physiology, clinical and therapeutic implications. Pergamon, Oxford, pp 151–173

Scrutton MC, Utter MF (1968) The regulation of glycolysis and gluconeogenesis in animal tissue. Annu Rev Biochem 37:249–302

Sestoft L, Trap-Jensen J, Lyngsoe J et al. (1977) Regulation of gluconeogenesis and ketogenesis during rest and exercise in diabetic subjects and normal men. Clin Sci Mol Med 53:411–418

Shaw WA, Issekutz TB, Issekutz B Jr (1976) Gluconeogenesis from glycerol at rest and during exercise in normal diabetic, and methylprednisolone treated dogs. Metabolism 25:329–339

Sherwin RS, Felig P (1977) Glucagon physiology in health and disease. In: McCann SM (ed) Endocrine physiology. Int Rev Physiol 16:151–171

Sherwin RS, Hendler RG, Felig P (1976) Effect of diabetes mellitus and insulin on the turnover and metabolic response to ketones in man. Diabetes 25:776–784

Singh SP, Snyder AK (1978) Effect of thyrotoxicosis on gluconeogenesis from alanine in the perfused rat liver. Endocrinology 102:182–187

Soling HD, Willms B, Friedrichs D, Kleineke J (1968) Regulation of gluconeogenesis by fatty acid oxidation in isolated perfused livers of non-starved rats. Eur J Biochem 4:364–372

Struck E, Ashmore J, Wieland O (1966) Effects of glucagon and long chain fatty acids on glucose production by isolated perfused rat liver. Adv Enzyme Regul 4:219–224

Stucki JW, Brawand F, Walter P (1976) Regulation of pyruvate metabolism in rat liver mitochondria by adenine nucleotides and fatty acids. Eur J Biochem 27:181–191

Taskinen MR, Nikkila EA (1979) Lipoprotein lipase activity of adipose tissue and skeletal muscle in insulin deficient human diabetes. Diabetologia 17:351–356

Taunton DO, Stifel FB, Greene HL, Herman RH (1974) Rapid reciprocal changes in rat hepatic glycolytic enzyme and fructose diphosphatase activities following insulin and glucagon injection. J Biol Chem 249:7228–7239

Tolman EL, Schworer CM, Jefferson LS (1973) Effects of hypophysectomy on amino acid metabolism and gluconeogenesis in the perfused rat liver. J Biol Chem 248:4552–4560

Tutwiler GF, Dellevigne P (1979) Action of the oral hypoglycemic agent 2-tetradecylglycidic acid ion hepatic fatty acid oxidation and gluconeogenesis. J Biol Chem 254:2935–2944

Tutwiler GF, Kirsch T, Morhbacher RJ, Ho W (1978) Pharmacologic profile of methyl 2-tetradecylglycide (McN-3716) an orally effective hypoglycemic agent. Metabolism 27:1539–1548

Unger RH (1985) Glucagon physiology and pathophysiology in the light of new advances. Diabetologia 28:574–578

Unger RH, Orci L (1981) Glucagon and the A cell. N Engl J Med 304:1518–1524, 1575–1580

Utter MF, Scrutton MC (1969) Pyruvate carboxylase. Curr Top Cell Regul 1:253–296

Von Glutz G, Walter P (1976) Regulation of pyruvate carboxylation by acetyl-CoA in rat liver mitochondria. FEBS Lett 72:299–303

Wahren J, Hagenfeldt L, Felig P (1975) Splanchnic and leg exchange of glucose, amino acids and free fatty acids during exercise in diabetes mellitus. J Clin Invest 55:1303–1314

Wahren J, Felig P, Hagenfeldt L (1976) Effect of protein ingestion on splanchnic and leg metabolism in normal man and in patients with diabetes mellitus. J Clin Invest 57:987–999

Wakelam MJ, Murphy GJ, Hruby VJ, Houslay MD (1986) Activation of two signal-transduction system in hepatocytes by glucagon. Nature 323:68–71

Wakil SJ, Stoops JK, Joshi VC (1983) Fatty acid synthesis and its regulation. Annu Rev Biochem 52:537–579

Weber G (1972) Integrative action of insulin at the molecular level. In: Shafrir E (ed) Impact of insulin on metabolic pathways. Academic, New York, pp 151–166

Weber G, Singhal RL, Srivastava SK (1965) Action of glucocorticoid as inducer and insulin as suppressor of biosynthesis of hepatic gluconeogenic enzymes. Adv Enzyme Regul 3:43–75

Weber G, Singhal RL, Stamm NB, Lea MA, Fisher EA (1966) Synchronous behavior pattern of key glycolytic enzymes: glucokinase, phosphofructokinase and pyruvate kinase. Adv Enzyme Regul 4:59–81

Weigle DS, Goodner CJ (1986) Evidence that the physiological pulse frequency of glucagon secretion optimizes glucose production by perifused rat hepatocytes. Endocrinology 118:1606–1613

Weigle DS, Sweet IR, Goodner CJ (1987) A kinetic analysis of hepatocyte responses to a glucagon pulse: mechanism and metabolic consequences of differences in response decay times. Endocrinology 121:732–737

White A, Handler P, Smith E, Hill R, Lehman IR (eds) (1978) Principles of biochemistry, 6th edn. McGraw-Hill, New York

Whitton PD, Rodrigues LM, Hems DA (1978) Stimulation by vasopressin, angiotensin and oxytocin of gluconeogenesis in hepatocyte suspensions. Biochem J 176:893–898

Williamson JR (1967) Effects of fatty acids, glucagon and anti-insulin serum on the control of gluconeogenesis and ketogenesis in rat liver. Adv Enzyme Regul 5:229–255

Williamson JR, Browning ET, Olson MS (1968a) Interrelations between fatty acid oxidation and the control of gluconeogenesis in perfused rat liver. Adv Enzyme Regulation 6:67–100

Williamson JR, Browning ET, Scholz R, Kreisberg RA, Fritz IB (1968b) Inhibition of fatty acid stimulation of gluconeogenesis by (+)-decanoylcarnitine in perfused rat liver. Diabetes 17:194–208

Willms B, Kleineke J, Soling HD (1970) The redox state of NAD/NADH systems in guinea pig liver during increased fatty acid oxidation. Biochim Biophys Acta 215:438–448

Winegrad AI (1965) Adipose tissue in diabetes. In: Renold AE, Cahill GF Jr (eds) Adipose tissue. Williams and Wilkins, Baltimore, pp 319–329 (Handbook of physiology, sect 5)

Zierler KL, Rabinowitz D (1964) Effect of very small concentrations of insulin on forearm metabolism. Persistence of its actions on potassium and free fatty acids without an effect on glucose. J Clin Invest 43:950–962

Subject Index

Handbook of Experimental Pharmacology
Eds.: G. V. R. Born, P. Cuatrecasas, H. Herken, A. Schwartz

Volume 91: **L. E. Bryan** (Ed.)
Microbial Resistance to Drugs
1989. XVII, 451 pp. 39 figs. Hardcover
DM 460,– ISBN 3-540-50318-8

Volume 90: **U. Trendelenburg, N. Weiner** (Eds.)
Catecholamines
Part I: 1988. XX, 571 pp. 43 figs. Hardcover
DM 450,– ISBN 3-540-18904-1
Part II: 1989. XXII, 488 pp. 34 figs. Hardcover
DM 450,– ISBN 3-540-19117-8

Volume 89: **E. M. Vaughan Williams** (Ed.)
Antiarrhythmic Drugs
Coeditor: T. J. Campbell
1989. XXVII, 650 pp. 113 figs. Hardcover
DM 480,– ISBN 3-540-19239-5

Volume 88: **D. B. Calne** (Ed.)
*Drugs for the Treatment
of Parkinson's Disease*
1989. XXIV, 599 pp. 62 figs. Hardcover
DM 480,– ISBN 3-540-50041-3

Volume 87:
Pharmacology of the Skin

Part I: M. W. Greaves, S. Shuster (Eds.)
*Pharmacology of Skin Systems.
Autocoids in Normal and Inflamed Skin*
1989. XXIX, 510 pp. 78 figs. Hardcover DM 480,–
ISBN 3-540-19403-7

Part II: M. W. Greaves, S. Shuster (Eds.)
*Methods, Absorption, Metabolism and
Toxicity Drugs and Diseases*
1989. XXXVII, 587 pp. 55 figs. Hardcover
DM 580,– ISBN 3-540-50277-7

Volume 86: **V. P. Whittaker** (Ed.)
The Cholinergic Synapse
1988. XXV, 762 pp. 98 figs. Hardcover DM 680,–
ISBN 3-540-18613-1

Volume 85: **M. A. Bray, J. Morley** (Eds.)
*The Pharmacology
of Lymphocytes*
1988. XXII, 626 pp. 69 figs. Hardcover DM 680,–
ISBN 3-540-18609-3

Volume 84: **K. Bartmann** (Ed.)
Antituberculosis Drugs
1988. XXIV, 566 pp. 68 figs. Hardcover DM 590,–
ISBN 3-540-18139-3

Volume 83: **P. F. Baker** (Ed.)
Calcium in Drug Actions
1988. XXVI, 567 pp. 123 figs. Hardcover
DM 590,– ISBN 3-540-17411-7

Volume 82: **C. Patrono, B. A. Peskar,** (Eds.)
*Radioimmunoassay in Basic
and Clinical Pharmacology*
1987. XXII, 610 pp. 129 figs. Hardcover
DM 580,– ISBN 3-540-17413-3

Volume 81: **G. R. Strichartz** (Ed.)
Local Anesthetics
1987. XII, 292 pp. 52 figs. Hardcover
DM 280,– ISBN 3-540-16361-1

Springer-Verlag Berlin
Heidelberg New York London
Paris Tokyo Hong Kong

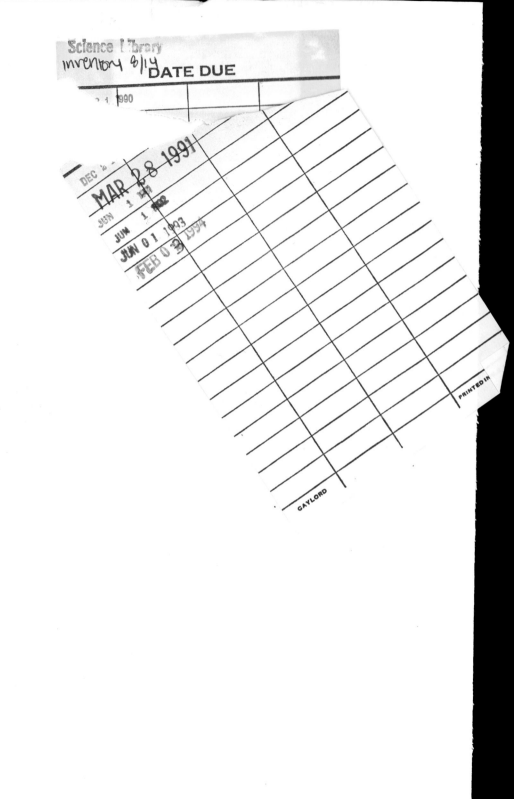